To Katy and Clara, Mum and Dad, and my friend Ivor Rosaire,
who taught me about elephants

You can't build a tortoise.

Gerald Durrell

COMPANION WEBSITE

This book has a companion website:

www.wiley.com/go/rees/zoo

with Figures and Tables from the book for downloading

An Introduction to Zoo Biology and Management

An Introduction to Zoo Biology and Management

Paul A. Rees

Senior Lecturer, School of Environment and Life Sciences,
University of Salford, UK

WILEY-BLACKWELL
A John Wiley & Sons, Ltd., Publication

Blackwell Publishing was acquired by John Wiley & Sons in February 2007. Blackwell's publishing program has been merged with Wiley's global Scientific, Technical and Medical business to form Wiley-Blackwell.

Registered Office
John Wiley & Sons Ltd., The Atrium, Southern Gate, Chichester, West Sussex, PO19 8SQ, UK

Editorial Offices
9600 Garsington Road, Oxford, OX4 2DQ, UK
The Atrium, Southern Gate, Chichester, West Sussex, PO19 8SQ, UK
111 River Street, Hoboken, NJ 07030-5774, USA

For details of our global editorial offices, for customer services and for information about how to apply for permission to reuse the copyright material in this book please see our website at www.wiley.com/wiley-blackwell.

Library of Congress Cataloging-in-Publication Data

Rees, Paul A.
An introduction to zoo biology and management / Paul A. Rees.
p. cm.
Includes index.
Summary: "This book is intended as an introductory text for students studying a wide range of courses concerned with animal management, zoo biology and wildlife conservation, and should also be useful to zookeepers and other zoo professionals. It is divided into three parts. Part 1 considers the function of zoos, their history, how zoos are managed, ethics, zoo legislation and wildlife conservation law. Part 2 discusses the design of zoos and zoo exhibits, animal nutrition, reproduction, animal behaviour (including enrichment and training), animal welfare, veterinary care, animal handling and transportation. Finally, Part 3 discusses captive breeding programmes, genetics, population biology, record keeping, and the educational role of zoos, including a consideration of visitor behaviour. It concludes with a discussion of the role of zoos in the conservation of species in the wild and in species reintroductions. This book takes an international perspective and include a wide range of examples of the operation of zoos and breeding programmes" – Provided by publisher.
Summary: "The aim of the book is to provide an introductory text on the biology of zoo animals and the management, regulation, organisation and conservation role of zoos"– Provided by publisher.
ISBN 978-1-4051-9349-8 (hardback) – ISBN 978-1-4051-9350-4 (paper)
1. Zoos–Management. 2. Zoo animals–Behavior. 3. Zoos–History. I. Title.
QL76.R44 2011
636.088′9–dc22

2010048253

A catalogue record for this book is available from the British Library.

This book is published in the following electronic formats: ePDF [9781444397826]; Wiley Online Library [9781444397840]; ePub [9781444397833]

Set in 9/11pt Photina by SPi Publisher Services, Pondicherry, India
Printed and bound in Malaysia by Vivar Printing Sdn Bhd

2 2012

Contents

COMPANION WEBSITE

This book has a companion website:

www.wiley.com/go/rees/zoo

with Figures and Tables from the book for downloading

Preface

Zoo biology is the scientific study of animals that live in zoos. This book is intended to provide students and zoo professionals with an introduction to zoo biology and an understanding of the way the zoo community functions. In particular it considers the role of zoos in conservation, including education, captive breeding and research, along with their obligation to assure high standards of animal welfare. It was designed to meet the needs of students following programmes of study in a wide range of animal-related subjects, including animal care, animal management, and zoo biology. I hope it will also be useful to zoo professionals, especially those who are in the early stages of their careers.

This is a small book and zoo biology has become a large subject. No textbook is ever perfect. I have attempted to include a wide range of topics of interest to those who either work in zoos or aspire to do so. This has meant that relatively little space has been devoted to some areas in order to make room for others. I hope I have succeeded in summarising the major principles of each area covered, but in any work of this size the content is inevitably a compromise between what I should like to have included and what can reasonably fit into a relatively small volume. Readers may detect in the text a small bias towards my own areas of interest. I make no apologies for this and hope that this book will encourage you to develop your own special interests.

I have drawn examples of good practice from zoos around the world, but have concentrated on zoos in the UK, Europe, the USA and Australasia. It goes without saying that this is not because these are the only countries with good zoos. I have also discussed some of the controversy surrounding zoos because it is important to remember that some people oppose the very existence of zoos in principle and some sophisticated organisations campaign vociferously against them.

I have made reference to a number of veterinary, enrichment and other practices, and the use of particular drugs and other products, techniques and procedures in zoos. This should not be taken as a recommendation or endorsement by the author or the publisher. I have also referred to the law in various countries. To the best of my knowledge and belief the statements made were accurate at the time of writing. However, it is in the nature of law that it changes from time to time and the reader should check authoritative sources when necessary.

Paul A. Rees *BSc (Hons) LLM PhD Cert Ed*
Senior Lecturer
Centre for Environmental Systems
and Wildlife Research
School of Environment and Life Sciences
University of Salford
United Kingdom
October 2010

Acknowledgements

I am indebted to a number of people for their encouragement and assistance during the production of this book. At the University of Salford, Professor David Storey and Denise Rennie, both former Heads of the School of Environment and Life Sciences, encouraged me to develop Wildlife programmes, including a BSc (Hons) in Wildlife Conservation with Zoo Biology. This was the first undergraduate course of its kind in the UK and its success was a major impetus to the writing of this book. Also at Salford, Cath Hide kindly assisted me in taking the photographs of parasites (Figs 11.1 and 11.2), Dr Glyn Heath provided the photograph of giant pandas (Fig. 12.26) and Andy Callen (Information and Learning Services) helped with locating information on wildlife documentaries. I am grateful to my students Louise Oates, Bethan Shaw, Rebecca Dasan and Simon Campion for providing photographs and access to their work, and to many other students who have unwittingly drawn my attention to some of the research referred to in the text.

My contacts with zoo professionals over many years have been invaluable. Louise Bell (Research Officer, Blackpool Zoo) provided Fig. 12.27 and allowed me access to various parts of Blackpool Zoo to take photographs. Many other zoo professionals have helped me to gain an insight into the day-to-day operation of zoos during my visits with students and to conduct research. These include Maggie Esson (Education Programmes Manager, Chester Zoo), Mick Jones (Elephant Specialist Keeper, Chester Zoo), Caroline Parsons (Education Officer, Dudley Zoo), Denise Chorley (Head of Education, Knowsley Safari Park), Claire Pipe (Conservation Officer, Twycross Zoo), Sophie Stevens (Education Officer, Twycross Zoo) and Nic Masters (Veterinarian, International Zoo Veterinary Group).

Many of the images of historical importance which are reproduced here were provided by Chetham's Library in Manchester, and I am most grateful to Jane Muskett for access to, and permission to copy, archive materials relating to Belle Vue Zoo. I am also grateful to Henry McGhie, Curator of Zoology at Manchester Museum, for providing information about a number of items of historical interest held by the museum. Extracts of the law of the United Kingdom are reproduced with the permission of the Controller of the Office of Public Sector Information.

At Wiley-Blackwell I should like to express my thanks to Ward Cooper (Senior Commissioning Editor, Ecology, Conservation and Evolution), Delia Sandford (Managing Editor), Camille Poire, and Kelvin Matthews (Project Editor). I am particularly grateful to Jessminder Kaur (Production Editor) who was responsible for the design of the book and Erica Schwarz (Schwarz Editorial) whose careful copyediting gave order and consistency to my words. Arunkumar Aranganathan (Project Manager at SPi) was responsible for overseeing the production of the book through typesetting and I am indebted to him for the care he has taken in the layout of the work.

Finally, my wife Katy and daughter Clara kindly helped to check and assemble the manuscript. Naturally, any errors that remain are mine.

Part 1

HISTORY, ORGANISATION AND REGULATION

Part 1 of this book provides background information about zoos by examining their historical development, their organisation at national, regional and international levels, the legal regulation of zoos, the ethics of keeping animals in zoos, and some general principles of wildlife conservation.

1 THE PURPOSE AND POPULARITY OF ZOOS

Walking in the Zoo, walking in the Zoo,
The O.K. thing on Sunday is the walking in the Zoo.

Lyrics of a song sung by Alfred Vance (Victorian music-hall artist, 1867)

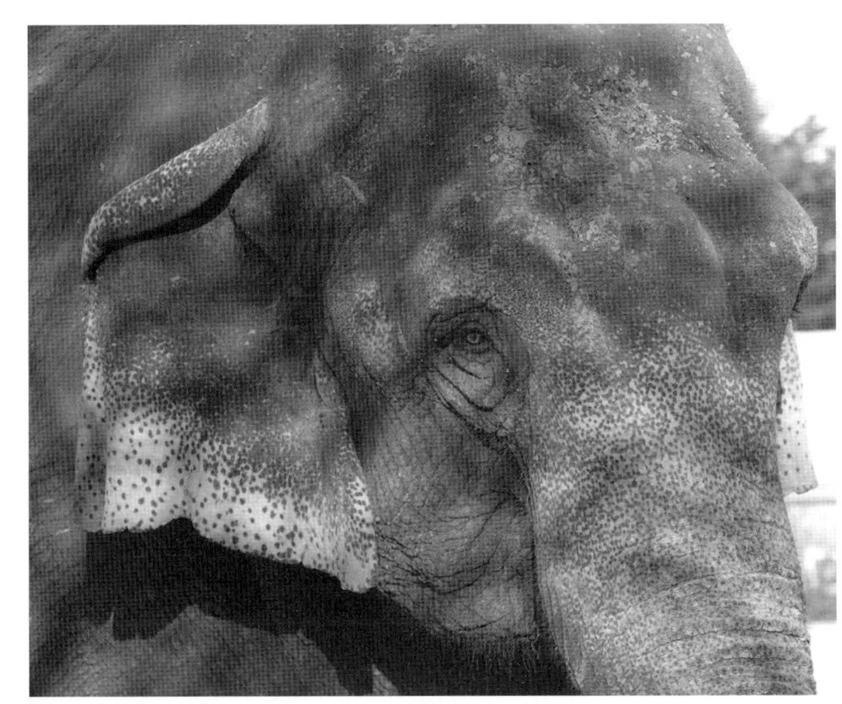

Conservation Status Profile

Asian elephant
Elephas maximus
IUCN status: Endangered A2c
CITES: Appendix I
Population trend: Decreasing

An Introduction to Zoo Biology and Management, First Edition. Paul A. Rees.

1.1 INTRODUCTION

Zoo biology is a relatively new discipline. Prior to 1982 there was no academic journal specifically dedicated to the scientific study of zoo animals. The study of animal welfare and the biology of zoo animals is an increasingly important component of college and university courses, and the work of zoo keepers is becoming increasingly scientific. The purpose of this book is to provide an introduction to the biology of animals in zoos. In addition it considers other aspects of zoos which are important in understanding how zoos have evolved, how they function and their role in conservation. Before we can begin to study zoos we need to define what a zoo is.

1.2 WHAT IS A ZOO?

The term 'zoo' is an abbreviation of zoological gardens and was probably first used as an abbreviation for the Clifton Zoo in Bristol, England, in 1847. In 1867 a music-hall song called *Walking in the Zoo*, sung by Alfred Vance, popularised the use of the term.

In this book 'zoo' is taken to encompass a very wide range of institutions, reflecting the legal definition used within the European Union:

> ..."zoos" means all permanent establishments where animals of wild species are kept for exhibition to the public for 7 or more days a year...
>
> Council Directive 1999/22/EC (Zoos Directive), Article 2

This definition encompasses traditional zoos, drive-through safari parks, aviaries, snake parks, insect collections, aquariums, birds of prey centres, and all manner of other animal collections which are open to the public, but excludes pet shops, circuses and laboratories that keep animals (Fig. 1.1). Other dictionary and legal definitions of a zoo are listed in Box 1.1.

1.2.1 What's in a name

In the minds of many people, the term 'zoo' has a strong association with animals being kept in poor conditions in old-fashioned iron cages. Some zoos have adopted new names in an attempt to disassociate themselves from these outdated notions and to emphasise their role in wildlife conservation and environmental education. Paignton Zoo has become *Paignton Zoo Environmental Park*, Jersey Zoo is now *Durrell* (named after its founder) and Marwell Zoo has been renamed *Marwell Wildlife*. However, many famous zoos show no sign of following this trend. Like traditional zoos, modern aquariums have begun to turn their attention to conservation. To reflect this, the New York Aquarium changed its name to the *Aquarium for Wildlife Conservation* in 1993.

1.3 THE INCREASING PUBLIC INTEREST IN WILDLIFE

There is considerable evidence of an increase in public interest in wildlife, zoos and conservation. In the last 40 years there has been a spectacular increase in the membership of organisations concerned with the protection of wildlife and wild places (Table 1.1).

People who spend the majority of their lives in urban areas seek out opportunities to observe animals in the wild. In Liverpool, the Royal Society for the Protection of Birds (RSPB) organises cruises around the Mersey Estuary for its members and other bird-watchers (Fig. 1.2). Every year, during a few weeks in autumn and winter, over 60,000 people visit the colony of grey seals (*Halichoerus grypus*) at the Donna Nook National Nature Reserve in Lincolnshire, England, when they come ashore to have their pups (Fig. 1.3). During Easter 2010, some 45,000 people visited Lake Hornborga in Sweden to witness the annual return of over 12,000 migrating Eurasian cranes (*Grus grus*).

Before the ownership of televisions was widespread films such as *Where No Vultures Fly* (1951) – about one man's struggle to establish a national park in East Africa – and *Born Free* (1966) – the true story of Elsa the lioness and George and Joy Adamson – drew the attention of the cinema-going public to the plight of African wildlife. The stars of *Born Free*, Bill Travers and Virginia McKenna, went on to found the pressure group *Zoo Check* (see Section 6.2.2.3).

Nowadays, wildlife documentaries are increasingly popular among television viewers and, in recent years, there has been an increase in the number of TV programmes about the work of vets and zoo keepers. In 2001, a total of 2284 TV programmes were broadcast which had 'wildlife' in the title or in the description of the programme. In 2008 this had increased to 6983. Only 196 programmes with 'zoo' in the title or description

(a)

(b)

(d)

(c)

Fig. 1.1 Examples of different types of zoos. (a) A traditional zoological gardens, Colombo Zoo, Sri Lanka. (b) A modern aquarium, *The Deep*, Hull, UK. (c) The Butterfly House, Lancaster, UK. (d) A safari park, West Midlands Safari Park, UK.

Box 1.1 What is a zoo? – some dictionary and legal definitions.

Some dictionary definitions:

The Concise Oxford Dictionary (1976)

Zoological garden – public garden or park with a collection of animals for exhibition or study.

Collins Dictionary and Thesaurus (2000)

Zoo – a place where live animals are kept, studied, bred, and exhibited to the public.

Some legal definitions:

European Union

Zoos Directive (Council Directive 1999/22/EC of 29 March 1999 relating to the keeping of wild animals in zoos)

Article 2: ... "zoos" means all permanent establishments where animals of wild species are kept for exhibition to the public for 7 or more days a year, with the exception of circuses, pet shops and establishments which Member States exempt from the requirements of this Directive on the grounds that they do not exhibit a significant number of animals or species to the public and that the exemption will not jeopardise the objectives of this Directive.

England

Zoo Licensing Act 1981, as amended by the Zoo Licensing Act 1981 (Amendment) (England and Wales) Regulations 2002 (SI 2002/3080)

Section 1(2): ... "zoo" means an establishment where wild animals ... are kept for exhibition to the public otherwise than for purposes of a circus ... and otherwise than in a pet shop.

(2A): This Act applies to any zoo to which members of the public have access, with or without charge for admission, on seven days or more in any period of twelve consecutive months.

United States of America

Animal Welfare Act 1966 (9 Code of Federal Regulations, Ch. 1)

Section 1.1: Zoo means any park, building, cage, enclosure, or other structure or premise in which a live animal or animals are kept for public exhibition or viewing, regardless of compensation.

Arizona State, USA (Arizona Revised Statutes, 17-101)

Section A. 23: "Zoo" means a commercial facility open to the public where the principal business is holding wildlife in captivity for exhibition purposes.

Australia

New South Wales, Zoological Parks Board Act 1973

Section 4(1): ... "zoological park" means a zoological garden, aquarium or similar institution in which animals are kept or displayed for conservation, scientific, educational, cultural or recreational purposes.

India

Wildlife (Protection) Act 1972

Section 2(39): "Zoo" means an establishment, whether stationary or mobile, where captive animals are kept for exhibition to the public but does not include a circus and an establishment of a licensed dealer in captive animals.

Table 1.1 Membership of selected conservation organisations.

Organisation	Membership 1971	Membership 2010
Royal Society for the Protection of Birds (RSPB)	98,000	> 1,000,000
WWF (UK)	12,000	168,417
WWF (Worldwide)	Unknown	c.5,000,000
Sierra Club	Unknown	1,300,000
Greenpeace	30,000 (1981)	2,900,000 (2007)
Wildlife Trusts (formerly the Royal Society for Nature Conservation)	64,000	791,000
Friends of the Earth	1,000	100,000

Sources: Church (1995), RSPB (2010), WWF (2010), Sierra Club (2010), Wildlife Trusts (2010) and FOE (2010).

Fig. 1.2 A bird-watching trip on the River Mersey, Liverpool, organised by the Royal Society for the Protection of Birds.

were shown in 2001, but by 2008 this had increased to 2119. Although many of these programmes were episodes of series and many will have been repeated on different channels – the number of which increases relentlessly – nevertheless there has been a spectacular increase in the amount of broadcast time devoted to this type of programme, and this must reflect a greater public interest (Table 1.2). Twelve million people watched the BBC's *Planet Earth* series, narrated by David Attenborough, and it received the highest audience appreciation score of any British programme on TV in 2006 (BBC, 2007).

Biography 1.1

Sir David Attenborough (1926–)

David Attenborough is a naturalist, author and broadcaster whose name is synonymous with the production of high quality wildlife documentaries. His highly acclaimed BBC series include *The Life of Mammals, The Life of Birds, Life in the Undergrowth, Planet Earth, The Blue Planet, Life in the Freezer, Life in Cold Blood* and *Planet Earth.* Although Attenborough does not generally film captive animals, some scenes, such as one of a polar bear inside its den, were recorded in a zoo because of the technical difficulties this would create in the wild. Attenborough's work has undoubtedly been extremely important in increasing awareness of conservation issues worldwide.

Fig. 1.3 The grey seal (*Halichoerus grypus*) colony at Donna Nook, Lincolnshire, UK. Inset: Adult grey seal.

Table 1.2 Examples of recent TV programmes about zoos.

Series/Title	Zoo/Subject	TV channel
Animal Arrivals	Various zoos	Animal Planet
Animal Cops Houston	Roadside zoo USA	Animal Planet
Animal Rescue Squad	Various zoo items	Fiver
Chaos at the Zoo	A small zoo near Tenby, Wales	SC4
Chimp Diaries	Mona Foundation Sanctuary, Spain	Adventure One
Crocodile Hunter	Australia Zoo	Animal Planet
E-Vets: The Interns	Alameda East Veterinary Hospital, USA	Animal Planet
Hero Animals	Gorillas, Jersey Zoo	Five
Kabul Zoo Rescue	Kabul Zoo, Afghanistan	Adventure One
Little Zoo That Could	Alabama Gulf Zoo	Animal Planet
Michaela's Zoo Babies	Various UK zoos	Fiver
Our Child the Gorilla	St Martin la Plaine Zoo, France	Animal Planet
Panda Is Born	America's National Zoo, Pandas	Animal Planet
Safari Park	West Midlands Safari Park	Animal Planet
The Keepers	San Diego Zoo, USA	Animal Planet
The Zoo	Auckland Zoo, New Zealand	Sky Real Lives
The Zoo UK	Bristol Zoo, UK	Sky Three
Trent's Wildcat Adventures	A zoo keeper travels to South Africa	Adventure One
Zoo Days	Colchester Zoo and Chester Zoo, UK	Five
Zoo Story	Paignton Zoo, UK	Sky One Mix
Zoo Tales	Taronga Park and Western Plains Zoo, Australia	Adventure One
Zoo Vet at Large	Zoo vet Matt Brash	Sky Real Lives

Table 1.3 Visits made in 2008 to selected visitor attractions in membership with the Association of Leading Visitor Attractions.

Attraction	Visitor numbers	Rank
British Museum	5,932,897	1
Natural History Museum	3,698,500	4
Chester Zoo	1,373,459	14
Kew Gardens	1,306,401	15
Eden Project	1,093,510	18
ZSL London Zoo	1,039,030	20
ZSL Whipsnade Zoo	468,669	42

Many UK zoos are not members of this organisation and will therefore not appear in the statistics.
Source: ALVA (2009).

In the UK, zoos are among the most popular visitor attractions (Table 1.3) and visitor numbers in UK zoos – following a decline in the 1970s – have increased in recent years (Fig. 1.4). The World Association of Zoos and Aquariums (WAZA) estimates that 600 million people visit zoos each year (Olney, 2005) and many institutions attract very large numbers. There is considerable variation in changes in visitor attendance at zoos with time. London Zoo suffered a dramatic decline in visitors from over two million in 1970 to around 866,000 in 1995 and almost had to close. By 2008 visitor numbers had risen to over 1.1 million. Chester Zoo experienced a less dramatic decline through the 1970s and now regularly receives over a million visitors each year (Fig. 1.5). In 2008, San Diego Zoo (California, USA) had over 3.3 million visitors, Cologne Zoo (Germany) had almost 1.5 million visitors, Colombo Zoo (Sri Lanka) had almost 1.4 million visitors and Chester Zoo (UK) was visited by over 1.3 million people (Fisken, 2010).

The American Association of Zoos and Aquariums estimates that its member institutions have over 175 million visitors each year (AZA, 2010a). Visitors are primarily women/mothers aged 25–35 years, and two out of three adults who visit a zoo or aquarium do so with a child. However, demographic changes in the developed countries of the world mean that zoos are faced with declining numbers of children, and therefore adults with young families, that are the mainstay

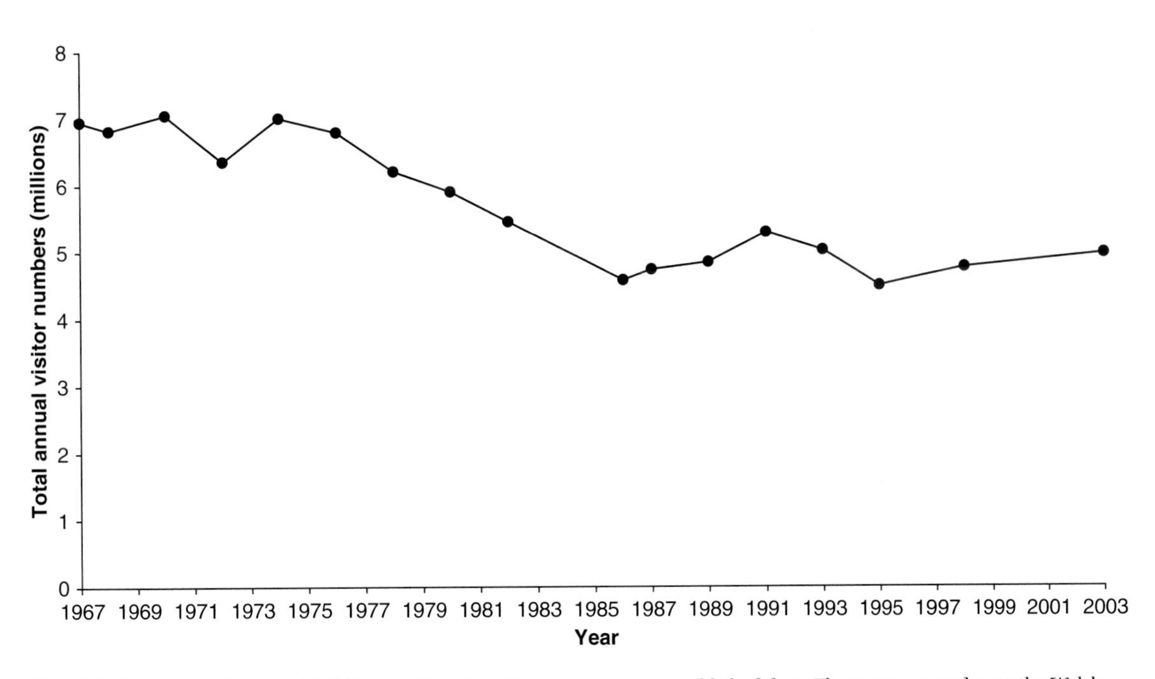

Fig. 1.4 Visitor attendance at 11 UK zoos (1967–2003). (*Source*: Rees, unpublished data. The zoos surveyed were the Welsh Mountain Zoo, Twycross, Paignton, Dudley, Jersey, Edinburgh, Bristol, London, Chester, Belfast and Whipsnade. Data were extracted from the *International Zoo Yearbook* volumes for the period.)

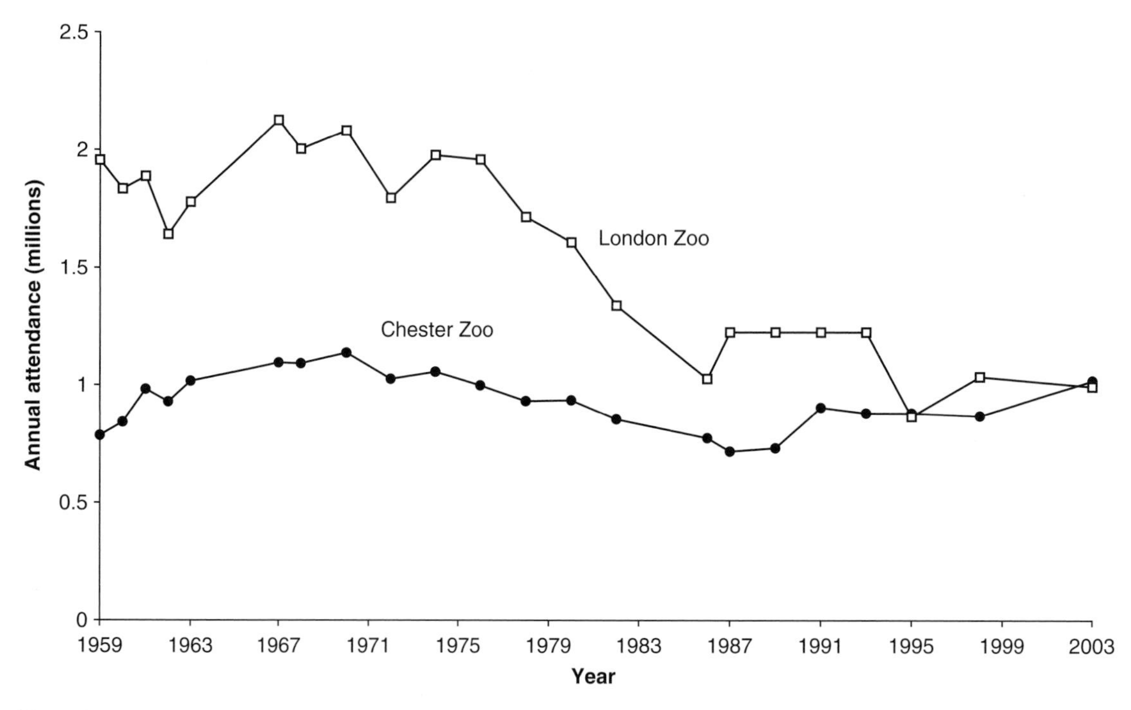

Fig. 1.5 Visitor attendance at London Zoo and Chester Zoo (1959–2003). (*Source*: Rees, unpublished data. Based on data extracted from the *International Zoo Yearbook* volumes for the period.)

of zoo visitor populations. In their place zoos have an opportunity to attract older visitors, and changing their focus away from child-friendly attractions and towards the interests of the older visitor will be a major challenge (Turley, 2001).

Although there is great public interest in wildlife, unfortunately this does not tend to influence voting decisions. In a survey of 1503 adults in the UK conducted immediately before the 2010 general election only 5% considered protecting the natural environment to be 'very important' in influencing their voting decision, way behind the economy (32%), health care (26%) and education (23%) (Ipsos MORI, 2010).

1.4 WHO OWNS ZOOS?

Some zoos are owned privately by individuals, while others are owned by multi-national entertainment companies. Some zoos are charities, some are run as businesses, primarily for profit, while others are run by national governments or by municipal authorities (Table 1.4).

Until 2007, the singer Michael Jackson had a private zoo at his Neverland Ranch, where he kept lions, tigers, giraffes, monkeys, orangutans, flamingos, snakes, a crocodile, an elephant and a chimpanzee called 'Bubbles'.

The drugs baron Pablo Escobar had a private zoo consisting of hundreds of exotic animals at his ranch, Hacienda Napoles, near Medellin in Colombia. When the zoo was seized by the state after he was killed by police, many of the animals died of hunger. Hippopotamuses escaped and were eventually shot.

Some privately owned zoos are open to the public, others are not. Knowsley Safari Park in England is open to the public but is owned by Lord Derby. Many well-established and famous zoos began as the privately owned ventures of individuals and have evolved into institutions of international importance, for example Gerald Durrell founded *Durrell* in Jersey, and George Mottershead founded Chester Zoo in England.

Table 1.4 The ownership of selected zoos.

Zoo	Location	Status/Ownership
Colombo Zoo	Sri Lanka	National government
St Louis Zoo	USA	Government with aid from non-profit trust
San Francisco Zoological Gardens	USA	Municipality with aid from zoological society
San Diego Zoo	USA	Non-profit corporation, administered by zoological society
Moscow Zoo	Russia	Municipality
Parc Zoologique to Paris	France	National government
Twycross Zoo	UK	Charitable trust
Burgers Zoo	Netherlands	Private
Banham Zoo	UK	Private limited company
Blackpool Zoo	UK	Spanish leisure company
Blue Planet Aquarium	UK	Private limited company

Zoos begin for all sorts of reasons. In his *Guide to British Zoos*, Schomberg (1970) describes Banham Zoo as follows:

> This is a small collection of monkeys and an aviary started by Mr H. Goymour and opened in February 1968. The purpose of the collection is to interest customers who come to buy farm produce at Grove Farm ... it is certainly a novel method of attracting business.

Some zoos in the UK were started by private collectors (Paignton Zoo), pet shop owners (Twycross), nobility (Harewood Bird Garden, Lions of Longleat, Woburn Park), businessmen (Chessington Zoo), market gardeners (Norton Petsenta), farmers (Ilfracombe Zoo Park), animal dealers (Weyhill Zoo Park) and circus owners (Southampton Zoo, Plymouth Zoo). Others were opened by zoological societies (London, Glasgow, Edinburgh), private companies (Marineland Oceanarium and Aquarium, Flamingo Park), or were originally municipal zoos (Belfast Zoo, Newquay Zoo). Some of these institutions no longer exist, while others have become important zoos, in some cases under new names.

Zoos have become big business in recent years. *Parques Reunidos* is the second largest operator of leisure parks in Europe. The company owns 68 parks, including 13 'zoo and nature parks':

- L'Oceanográfic, Valencia, Spain – the largest marine park in Europe
- Zoo Aquarium de Madrid, Spain – a zoo, aquarium, dolphinarium and aviary
- Faunia, Spain – a nature theme park which includes the biggest polar ecosystem in Europe
- Selwo Aventura, Spain – a nature theme park
- Selwo Marina, Spain – a marine park with a South American theme, including seals, sea lions, penguins and a dolphinarium
- Delfinario Costa Daurada, Spain – a dolphinarium and sea-lion exhibit
- Mar del Plata Aquarium, Argentina – a marine park containing fish, penguins, sea lions and dolphins
- Marineland – the largest marine park in France, containing sea lions, dolphins and killer whales
- Aquarium of the Lakes, England – an aquarium in the Lake District of England
- Blackpool Zoo, England – a traditional zoo with a dinosaur park
- Bournemouth Oceanarium, England – a coastal aquarium
- Silver Springs, Florida, USA – a nature theme park constructed around natural springs and exhibiting mostly native species
- Sea Life Park, Oahu, Hawai'i – a marine park containing sea lions, dolphins, sea turtles, penguins and sharks.

The Merlin Entertainments Group owns Chessington World of Adventures in England (which includes a zoo) and the chain of 26 Sea Life Centres in England, Belgium, Finland, France, Germany, the Netherlands, Ireland, Denmark, Scotland, Spain, Portugal, Italy and the USA. Some zoos and aquariums are charities. Others are operated as a business but have established charities to support their conservation work. South Lakes Wild Animal Park (UK) is a business but includes in its admission price a donation to its charity, the Wildlife Protection Foundation. This allows it to reclaim tax paid by visitors from the UK government under the Gift Aid scheme.

1.5 WHAT ARE ZOOS FOR?

Many western zoos have their origin in zoological societies that were established with an educational and scientific purpose in mind. The Bristol, Clifton and West of England Zoological Society was founded in 1835. At the meeting which established the Society, a local physician, Dr Henry Riley, proposed:

> ...that this meeting being persuaded that a Zoological Society in Bristol and Clifton will tend to promote the diffusion of knowledge by facilitating observations of the habits, form and structure of the animal kingdom as well as affording rational amusement and recreation to the visitors of the neighbourhood do determine that a Society be accordingly established.
>
> Warin and Warin (1985)

With the rise of the animal welfare movement and increasing concern for the environment in general, and the loss of biodiversity in particular, zoos have moved the focus of their activity towards conservation. The potential role of zoos in wildlife conservation has been recognised for at least 60 years (Delacour, 1947) and since the 1960s zoos have considered the conservation of endangered and threatened species as one of their most important functions (Mench and Kreger, 1996).

In 1986 the American Association of Zoological Parks and Aquariums (AAZPA) – now the American Zoo and Aquarium Association (AZA) – described the purpose of zoos as research, education, recreation and conservation, especially captive breeding (AAZPA, 1986). However, evidence for a significant education role for zoos is equivocal (Kellert, 1984; Jamieson, 1986; Marcellini and Jenssen, 1988; Spotte and Clark, 2004; Bickert and Meier, 2005) and many of their scientific contributions are largely unrelated to conservation *per se* (Rees, 2005a, 2005b). By the 1990s some commentators were questioning the value of breeding programmes (e.g. Varner and Monroe, 1991), arguing that money would be better spent on *in-situ* conservation efforts.

Reintroductions of zoo-bred animals have a very limited impact and many of the successful projects zoos quote were undertaken many years ago (Bertram, 2004). Stanley Price and Fa (2007) have concluded that zoos are not keeping the priority species for reintroduction, nor are species kept in adequate numbers as sources for reintroduction.

Wehnelt and Wilkinson (2005) have suggested that zoos have now moved the focus of their activity away from captive breeding and reintroduction and towards helping species *in-situ*, including the raising of funds to support these efforts (e.g. Smith and Hutchins, 2000; Christie, 2007). However, zoo mission statements rarely reflect this (Rees, 2005c), and a recent study of 136 North American zoos found that only 14% mentioned the support of global, national or local conservation programmes as one of their functions (Patrick *et al.*, 2007). The original World Zoo Conservation Strategy (IUDZG, 1993) made no mention of fund-raising as a function of zoos. However, its successor, the World Zoo and Aquarium Conservation Strategy, states that 'cooperative breeding programmes serve many purposes ... [including] ... providing fund-raising material...'(Olney, 2005).

Modern zoos have legal and moral obligations to care for the animals they keep and the species to which they belong. These obligations form the basis of much of the content of this book, and include the provision of:

- suitable housing
- an environment free from stress
- a high level of veterinary care
- the opportunity to express normal behaviour.

In addition zoos should engage in the *ex-situ* conservation of biodiversity – an obligation placed on Parties to the UN Convention on Biological Diversity (see Section 5.3.1.1). This may be defined as 'keeping components of biodiversity alive away from their original habitat or natural environment' (Heywood, 1995). A zoo's conservation activities include:

- captive breeding and reintroduction to the wild
- education
- record keeping
- information exchange
- research
- *in-situ* conservation.

Zoos need to balance their welfare obligations to the animals they keep while maintaining a conservation role (Fig. 1.6). These themes are explored in more detail below.

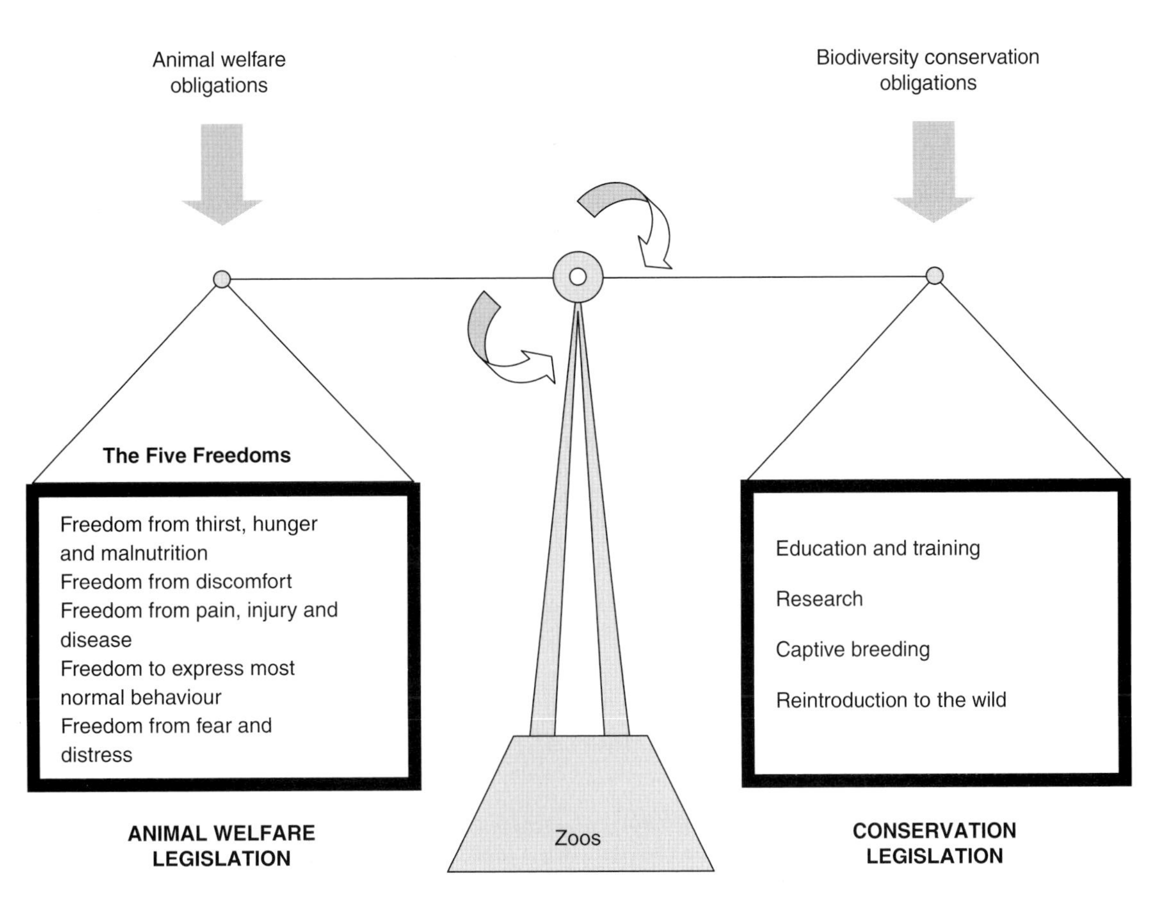

Fig. 1.6 Balancing the animal welfare and biodiversity conservation obligations of zoos.

1.6 ZOO MISSION STATEMENTS – WHAT DO ZOOS SAY THEY DO?

What do zoos say they do? Many zoos have mission statements and these tell us something about what they think they do, or at least what they aspire to do. A historical perspective on the changing role of zoos can be gained by examining how their mission statements have changed over time.

Until relatively recently many zoos referred to their role in captive breeding in their mission statements. In some zoos this was still the case long after their management was downplaying this particular role (Rees, 2005c). Bristol Zoo still retains a reference to breeding in its mission statement:

> Bristol Zoo Gardens maintains and defends biodiversity through breeding endangered species, conserving threatened species and habitats and promoting a wider understanding of the natural world. [2008]

However, Chester Zoo's mission has changed a number of times since 1995, and no longer specifically refers to breeding animals:

> ...to promote conservation by breeding of rare and endangered animals and by educational, recreational and scientific activities. [1995–96]

> ...to support and promote conservation by breeding rare and endangered animals, by excellent animal welfare, high quality public service, recreation, education and science. [1997]

> ...to support and promote conservation by breeding threatened species, by excellent animal

Box 1.2 Zoo mission statements.

Zoological Society of London, UK (2008)

To achieve and promote the worldwide conservation of animals and their habitats.

National Marine Aquarium, Plymouth, UK (2008)

To inspire everyone to enjoy, learn and care about our oceans through amazing, memorable experiences.

Dublin Zoo, Ireland (2008)

To work in partnership with zoos world-wide to make a significant contribution to the conservation of the endangered species on Earth.

Smithsonian National Zoological Park, USA (2008)

We are the Nation's Zoo, demonstrating leadership in animal care, science, education and sustainability. We provide the highest quality animal care. We advance research and scientific knowledge in conserving wildlife. We teach and inspire people to protect wildlife, natural resources and habitats. We practice conservation leadership in all we do.

Zoological Society of San Diego, USA (2008)

The Zoological Society of San Diego is a conservation, education, and recreation organisation dedicated to the reproduction, protection, and exhibition of animals, plants, and their habitats.

Minnesota Zoo, USA (2008)

Our mission is to create a conservation culture by incorporating conservation into the everyday fabric of our activities and programs.

National Zoological Gardens, Colombo, Sri Lanka (2008)

Resourceful conservation of animals by means of learning, achieved through the exhibition of species which were adopted with loving care.

Singapore Zoo (2008)

Our mission is to preserve biodiversity and to undertake public education, research and collaboration as well as maintain and update the exhibits in all our parks in the most humane, naturalistic and yet efficient manner.

Taronga Zoo, Sydney, Australia (2006)

We will demonstrate a meaningful and urgent commitment to wildlife, our natural environment and the pursuit of excellence in our conservation, recreation and scientific endeavours.

Zoos Victoria (Healesville Sanctuary, Melbourne Zoo and Werribee Open Range Zoo), Australia (2008)

Our zoos will be world leading centres for wildlife experience, education, conservation and research – on site, off site and online.

welfare, high quality public service, recreation, education and science. [1998–2003]

...to be a major force in conserving biodiversity worldwide. [Since 2004]

Early references to the role of zoos often mentioned the advancement of the study of zoology and the promotion of interest in animals. Modern mission statements usually refer to a role in education and conservation, but, as may be seen from the list in Box 1.2, this is not always the case. The original pur-

pose of the Royal Zoological Society of Scotland, as stated in its Charter, was:

> ...to advance the study of zoology and foster and develop amongst the people an interest in and knowledge of animal life. [1913]

Its current mission is:

> ...to inspire and excite our visitors with the wonder of living animals, and so to promote the conservation of threatened species and habitats. [2008]

Edinburgh Zoo has a separate *educational* mission statement:

> ...to inspire in our visitors an understanding of the value, the complexity and the fragility of the natural world by fully utilising the educational potential of our unique variety of living animals. [2008]

1.7 BIOPHILIA, HUMAN WELL-BEING AND ZOOS

Edward Wilson has defined 'biophilia' as 'the innately emotional affiliation of human beings to other living organisms' (Wilson, 1993). He believes that a desire to be close to other living things is an essential biological need; an element of human nature which is part of our genetic makeup. It has been suggested that zoos help people to fulfil this natural desire to feel more connected to nature (Davey, 2005). It is clearly important that zoos create places that people want to visit, because without an income from visitors most zoos could not exist.

Being close to nature may be important to human well-being. Studies have shown that interactions between animals and people can have positive health benefits. Physical contact with non-threatening animals has been shown to reduce blood pressure, respiration rate and heart rate (Cusack and Smith, 1984). Watching fish in an aquarium also appears to lower blood pressure (Kamberg, 1989).

1.8 ZOO ENTHUSIASTS

Some people collect stamps, others collect zoo memorabilia, including old zoo guides, tickets, postcards and souvenirs. Many people enjoy visiting zoos and may make such visits an integral part of a holiday. A number of books have been written to provide information about good zoos, including *The Good Zoo Guide* (Ironmonger, 1992), which describes UK zoos but is now out of date, and *America's Best Zoos* (Nyhuis and Wassner, 2008). *The Good Zoo Guide* now exists as a website: *The Good Zoo Guide Online* (www.goodzoos.com).

The Bartlett Society

The Bartlett Society is a group of enthusiasts who collect, study, preserve and record as much as possible of the history of wild animal keeping in zoos and elsewhere. It is named after Abraham Dee Bartlett, Superintendent of London Zoo from 1859 until his death in 1897.

Among other things, members of the Society publish a list of *First and Early Breeding Records for Wild Animals in the UK and Eire.*

The Independent Zoo Enthusiasts Society (IZES)

The Independent Zoo Enthusiasts Society was founded in 1995. It exists to foster interest in all good zoos and to counter misinformation about zoos. It has produced *The IZES Guide to British Zoos & Aquariums* (Brown, 2009) and has a number of other publications.

ZooChat

ZooChat is an online community of zoo and animal conservation enthusiasts. Its website hosts a number of special interest forums, a zoo photo gallery, information on zoo webcams, and maps and satellite images of many zoo locations.

Biography 1.2

Edward O. Wilson (1929–)

Professor Wilson is a world authority on the biology of ants, established the study of island biogeography and is an influential proponent of biodiversity conservation. He developed the controversial science of socio-biology: the study of the biological basis for all social behaviour. Wilson is Professor of Entomology at Harvard University and has won the US National Medal of Science and many awards for his writing, including the Pulitzer Prize twice, for his books *The Ants* and *On Human Nature.*

1.9 FURTHER READING AND RESOURCES

Many zoo websites contain information about their ownership and mission statements. The websites of major zoo associations such as the British and Irish Association of Zoos and Aquariums (BIAZA), the European Association of Zoos and Aquaria (EAZA) and WAZA also provide a great deal of general information about the purpose of zoos.

1.10 EXERCISES

1. Using zoo mission statements, discuss the changing role of zoos.
2. Is it more beneficial to conservation for a zoo to be operated as a business or as a charity?
3. Does visiting a zoo fulfil a basic biological need?
4. What evidence exists that people are increasingly becoming interested in wildlife and conservation?
5. Are wildlife documentaries a substitute for zoos?

2 CONSERVATION

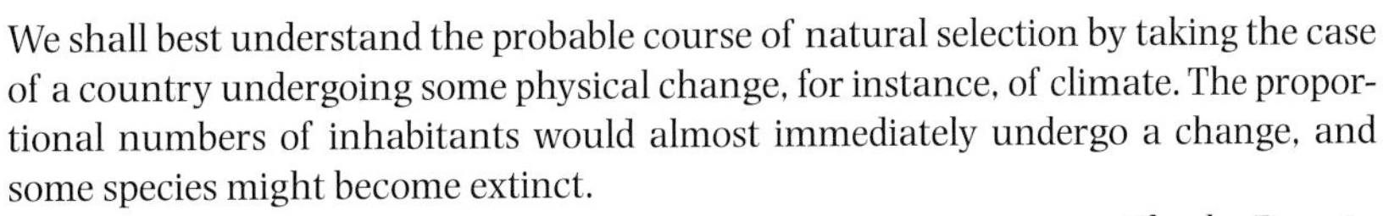

We shall best understand the probable course of natural selection by taking the case of a country undergoing some physical change, for instance, of climate. The proportional numbers of inhabitants would almost immediately undergo a change, and some species might become extinct.

Charles Darwin

Conservation Status Profile

Przewalski's horse *Equus ferus przewalskii*
IUCN status: Endangered D
CITES: Appendix I
Population trend: Increasing

An Introduction to Zoo Biology and Management, First Edition. Paul A. Rees.

2.1 INTRODUCTION

Modern zoos have adopted the conservation of biodiversity as their major function. In order to understand the role of zoos in the conservation of species and habitats we need first to examine some basic principles.

One definition of 'conservation' is the wise use of natural resources for the benefit of mankind. It is not the same as 'preservation', which implies that resources may not be used but merely protected. According to the World Conservation Strategy (IUCN, 1980) conservation is:

> The management of human use of the biosphere so that it may yield the greatest sustainable benefit to the present generation while maintaining its potential to meet the needs and aspirations of future generations.

Hambler (2004) has suggested a broader definition:

> Conservation is the protection of wildlife from irreversible harm.

Conservation is not just about protecting a small number of individuals of a particular species. It is essential to conserve genetic diversity in order that the species has a good chance of surviving into the foreseeable future.

2.2 WHY BOTHER WITH CONSERVATION?

Modern international laws often list the reasons why we should conserve biodiversity in their 'Preamble'.

> ...Conscious of the intrinsic value of biological diversity and of the ecological, genetic, social, economic, scientific, educational, cultural, recreational and aesthetic values of biological diversity and its components.
>
> Conscious also of the importance of biological diversity for evolution and for maintaining life sustaining systems of the biosphere.
>
> Affirming that the conservation of biological diversity is a common concern of humankind,...
>
> Preamble, UN Convention on Biological Diversity, 1992

> ...Recognising that wild flora and fauna constitute a natural heritage of aesthetic, scientific, cultural, recreational, economic and intrinsic value that needs to be preserved and handed on to future generations;
>
> Recognising the essential role played by wild flora and fauna in maintaining biological balances;...
>
> Preamble, Convention on the Conservation of European Wildlife and Natural Habitats, 1979

Fitter (1986) has discussed the usefulness of the world's genetic resources to man in terms of:

- food, drink and nutrition
- clothing and fabrics
- building and manufacturing
- fuel and energy
- human health
- ornamentation
- scientific knowledge and the safeguarding of knowledge
- biological control
- pets and gardens
- recreational hunting and field sports
- the enjoyment of wildlife
- wildlife as a social symbol.

This is a useful justification for conserving genetic resources but the approach is clearly anthropocentric and does not help us to decide which species we should seek to protect since just one or two of these categories could be used as a reason to preserve *all* forms of life.

Why should we conserve species and ecosystems? It is possible to justify conserving nature for a number of reasons:

a. Moral or ethical
As the dominant organisms on the planet we have a moral responsibility to ensure the survival of other species, and a duty to act as guardians.

b. Aesthetic
Animals and plants are important in many cultures. They give us great pleasure and feature widely in our art and our music.

c. Psychological
As human society becomes increasingly urbanised people appear to have a psychological need to find seclusion and peace in wild places. For many people this is difficult and zoos provide a means by which we can reconnect with nature in an urban setting.

d. Scientific
Ecological research has a value in its own right, but it also may be of value to people. An understanding of the workings of natural ecosystems helps us to understand the effects of human activity and to better predict future repercussions on the landscape and on biodiversity.

e. Pollution monitors
Some species are useful indicators of the state of the environment and may act as an early warning of when environmental stresses are likely to have an effect upon humans. For example, our attention was drawn to the toxicity of the insecticide DDT when birds of prey and other bird species began to decline. The chemical reduced egg shell thickness and led to a reduction in egg survival.

f. Balance of nature
This argument claims that any interference with nature is unwarranted because it might alter the 'balance of nature', and the repercussions might be undesirable for humans. The removal of predators may cause their prey to increase, and if their prey are large herbivores, this may eventually result in overgrazing. This assumes that all systems are in some kind of balance and that we could not tolerate any deviation from this. Although large scale disturbances to ecosystems and loss of species are undesirable, ecosystems are dynamic systems and constantly changing. We could undoubtedly lose many species without affecting human survival. Indeed, we already have. Of course, what we do not know is how many more we could lose without any significant effect on the sustainability of the planet.

g. Productive species
We ought to preserve some species that are particularly good energy converters or food sources because we may need them in the future. In the UK, rare breeds farms endeavour to prevent certain breeds of cattle, pigs, sheep and other farm animals from dying out. As climate change alters the agricultural landscape of the Earth it may be that currently unpopular breeds once more become useful. Many wild herbivores use poor pastures more efficiently than introduced cattle. Many species of antelope are already ranched in Africa.

h. Source of useful compounds
Many important compounds are obtained from plants. Hormones such as oestrogens, progesterone and cortisone can be produced from compounds extracted from soyabeans (*Glycine soya*) and the rhizomes of yams (*Dioscorea* spp.). Other economically important plants produce rubber, resins, oils, fibres, dyes, foodstuffs, alcoholic drinks and drugs.

i. Genetic conservation
When a species becomes extinct its genes may be lost forever and cannot be used in the breeding of new varieties of plants or animals. In the past, cultivated strains of wheat have benefited from interbreeding with wild varieties. The establishment of gene banks and the development of cryopreservation techniques for gametes can help to alleviate this problem.

j. Economic
Conservation is of great importance in some countries, for example Kenya, where a very high proportion of its foreign earning comes from tourism. Even in highly developed countries, wildlife can bring visitors to areas which are economically depressed such as the Highlands of Scotland. Some ecosystems have other types of economic value because they provide useful services. For example, wetlands can remove contaminants from water or, if they occur on the coast, act as buffers to the effects of tidal waves.

2.3 BIODIVERSITY

The word 'biodiversity' is a contraction of the term 'biological diversity'. The UN Convention on Biological Diversity defines biological diversity as:

> ...the variability among living organisms from all sources including ... terrestrial, marine and other aquatic ecosystems and the ecological complexes of which they are part; this includes diversity within species, between species and of ecosystems.

This definition emphasises the need to protect variation within species as this is the raw material of evolution. Ultimately, from this variation new species will arise.

Biodiversity should be thought of as a 'common good', that is it belongs to all of the people of the Earth, like the oceans or the atmosphere. However, recent analyses of the global status of animal populations have reported significant declines in many species. Global Biodiversity Outlook 3 (GBO-3, 2010), published by the Secretariat of the UN Convention on Biological Diversity, indicated that the populations of wild vertebrate species fell on average by 31% globally between 1976 and 2006, especially in the tropics (59%) and in freshwater ecosystems (41%).

The extinction risk has been reduced for some species by conservation interventions. However, many more species are moving closer to extinction.

2.4 BIODIVERSITY HOTSPOTS

The term 'biodiversity hotspot' is widely used to indicate a geographical area where there is a high diversity of endemic animal and plant life and a high level of threat (Myers *et al.*, 2000). However, some authors refer to 'hotspots' as areas of high species richness or high endemicity, without any reference to the degree of threat. Based on the definition used by Myers *et al.*, the hotspots that have been identified include areas of:

- Madagascar
- Indonesia
- New Zealand
- Sri Lanka
- the Western Ghats (India)
- Central America
- the Amazon
- Tanzania
- the north Mediterranean coast
- the Caribbean islands
- parts of the west coast of the United States

More than two-thirds of endangered mammals and 80% of the most endangered birds come from hotspots, which have collectively lost 88% of their original habitat (Hambler, 2004). Some modern multi-species zoo exhibits focus attention on the importance of hotspots, for example the Bronx Zoo's *Madagascar*, and the *Amazonia* exhibit at the National Zoo, Washington DC.

2.5 EXTINCTION

Extinction is an essential part of the process of evolution (Raup, 1986). Virtually all of the animal and plant species that have ever lived are now extinct. An analysis of almost 20,000 fossil genera by Raup and Boyajian (1988) found congruent rises and falls in extinction intensity, with taxonomic as well as ecological groupings exhibiting the same extinction profiles. This suggests that environmental stresses were the cause. For many mass extinctions there is evidence that climatic cooling occurred at the same time (Stanley, 1984).

There are five mass extinctions evident in the fossil record of the Earth, where there have been catastrophic losses of species from a wide range of phyla. The present rates of species loss appear to be comparable to those during these mass extinctions, leading some authorities to refer to the current loss of biodiversity as the sixth extinction (Leakey and Lewin, 1996).

2.5.1 Nice animals on the verge of extinction

The zoologist Dr Desmond Morris has claimed that the first rule of conservation is:

> the protection of nice animals on the verge of extinction.

He claims that conservationists have played on the emotions of the public by focusing on the plight of the most appealing failures. To qualify an animal should be:

> cuddly, attractively coloured, extremely rare, or better still all three.

Morris (1990) claims that this attitude is rooted in the ancient totem animals and in the good and evil animals of medieval times. He believes that wildlife will only be valued for what it is and honoured for its own sake if conservationists tackle the root cause of the problem: human population growth. He does not see this view as anti-human, but claims that carefully limited populations in the developing countries would increase the affluence of the people and leave room for wildlife. He is undoubtedly right.

Biography 2.1

Desmond Morris (1928–)

Dr Desmond Morris was formerly Curator of Mammals at London Zoo. He is the author of numerous books including *The Naked Ape, The Human Zoo* and *The Animal Contract.* Dr Morris was the first presenter of *Zoo Time*, the first television wildlife series in the world aimed at children. A total of 331 weekly episodes were transmitted between 1956 and 1968, broadcast from a specially built TV studio in London Zoo.

Table 2.1 Examples of mammal and bird species that have become globally extinct since 1600.

Mammals	
Gilbert's rat kangaroo	*Potorous gilberti*
Christmas Island shrew	*Crocidura trichura*
Giant aye-aye	*Daubentonia robusta*
Sardinian pika	*Prolagus sardus*
Assam rabbit	*Caprolagus hispidus*
Darwin's rice rat	*Oryzomys darwini*
Sea mink	*Mustela macrodon*
Stellar's sea cow	*Hydrodamalis stelleri*
Tarpan	*Equus gmelini*
Aurochs	*Bos primigenius*
Birds	
Great elephant bird	*Aepyornis maximus*
Rodriguez blue pigeon	*Alectroenas rodericana*
Cuban red macaw	*Ara tricolor*
Forest spotted owlet	*Athene blewitti*
Huia	*Heteralocha acutirostris*
Grand Cayman thrush	*Turdus ravidus*
Lesser koa finch	*Psittirostra flaviceps*
Townsend's bunting	*Spiza townsendi*
Lord Howe white-eye	*Zosterops strenua*
Black mamo	*Drepanis funerea*

Environmentalists often perpetuate the myth that before the arrival of modern man the Earth was teeming with wildlife and that primitive prehistoric societies lived in harmony with nature, only taking from the environment what they required for their immediate needs. This is far from the truth. At the end of the last Ice Age, over 200 genera of large animals became extinct: mostly large terrestrial herbivores, carnivores and scavengers. There is a considerable body of scientific evidence that links these extinctions to the geographical spread of prehistoric man and his development as a big-game hunter (Martin, 1971).

Conservationists are not generally concerned with the extinction of dinosaurs and woolly mammoths, but instead focus our attention on those species that have become extinct in relatively recent times, especially those that have disappeared as a result of human activity. The Falkland Island wolf (*Dusicyon australis*) was only discovered in 1690 but had been hunted to extinction by 1876 and the dodo (*Raphus cucullatus*) – a flightless bird – was hunted to extinction in Mauritius by around 1681. Between 1600 and 1959 we lost 40 mammal species and 95 birds to extinction (Table 2.1).

2.5.2 Case study: The extinction of the passenger pigeon

One of the most dramatic extinctions recorded in recent times is that of the passenger pigeon (*Ectopistes migratorius*) in North America. In the early 1800s the bird expert Alexander Wilson is reputed to have recorded a single flock of passenger pigeons that was a mile (1.6 km) wide and extending over 240 miles (386 km). He estimated that the flock contained over two billion birds. By 1914 the passenger pigeon was extinct. They were widely used for food, their feathers were used for pillows and their bones were ground up to make fertiliser. From 1858 passenger pigeon hunting became big business and hunters used traps, shotguns, artillery and even dynamite to kill them. In a period of just 40 days in 1869 almost 12 million pigeons were sent to market from Hartford, Michigan (Inskipp and Wells, 1979). In 1878 one professional trapper made $60,000 by killing three million birds. Commercial hunting of the species ceased in the early 1880s, when just a few thousand remained. The passenger pigeon laid just a single egg and was doomed to extinction. The last wild pigeon was shot on 24 March 1900 by a young boy in Ohio. The last passenger pigeon on Earth died in Cincinnati Zoo in 1994. Her stuffed body is at the National Museum of Natural History in Washington DC (Miller, 1994).

2.5.3 Keystone species

We have clearly lost many species from the planet without complete catastrophe and the survival of life on Earth does not depend upon saving every species. However, some species appear to be more important to the survival of particular communities than others. Ecologists call these 'keystone species'.

The concept of a 'keystone species' was first suggested by Paine (1969). He used the term to describe a species which has a disproportionate effect on the diversity of a biological community for its size and abundance. Paine found that when he removed the starfish *Pisaster* from a section of shore the animal community was reduced from a 15-species to an eight-species community. This was because the starfish controlled the numbers of a bivalve (*Mytilus*) that otherwise dominated the area.

In tropical forests ants, bees, bats and hummingbirds play keystone roles in pollination and seed dispersal. Loss of such species may lead to population crashes and extinctions in other dependent species. Ants are probably ecologically more important than most whale species and yet, unlike whales, they are not protected by international treaties. Wilson (1992) has emphasised the importance of insects to human survival, asserting that:

> So important are insects and other land-dwelling arthropods that if all were to disappear, humanity could not last more than a few months.

In African grassland ecosystems elephants are keystone species and as such have an enormous influence upon the vegetation and consequently the other animal species. It has been suggested that they are the causative agent in a cyclical change in the ecosystem caused by their wholesale destruction of trees which eventually causes a crash in their own population due to lack of food.

In Florida's Everglades the American alligator (*Alligator mississippiensis*) plays a keystone role. It digs deep depressions known as 'gator holes' which collect water and fish during dry spells. They also provide water and food for birds and mammals and as the water becomes fertilised by alligator dropping the hole becomes filled with lilies and other aquatic plants. Alligator nesting mounds provide nesting sites for egrets and herons and the alligators' movements between gator holes and nesting mounds help to maintain areas of open water free from invading vegetation. Alligators help to maintain populations of game fish like bream and bass by preying upon large numbers of predatory fish.

2.6 THREATS TO WILDLIFE

In the wild, animals are exposed to a wide range of threats (Fig. 2.1). Most of the current threats are the result of human activity and human population growth. At 11.00 GMT on 17 March 2010 the United States Census Bureau estimated the human population of the world to be 6,809,016,322, or approximately 6.8 billion. More people inevitably means less space and fewer resources for wildlife.

Some animal species are threatened because they have highly restricted distributions. The giant panda (*Ailuropoda melanoleuca*) occurs in six small mountainous areas in Sichuan, Shaanxi and Gansu provinces in southwest China. Its total range is approximately 5900 km^2 (2277 square miles). In Indonesia palm oil plantations are fragmenting habitats and threatening the survival of orangutans (*Pongo* spp.).

Traditional hunting by indigenous peoples generally does not threaten the survival of species because they take relatively few individual animals and practise hunting in a sustainable manner. Indeed, many international laws recognise the rights of such peoples to exploit wildlife resources (e.g. the UN Convention on Biological Diversity). However, commercial hunting must be regulated to prevent over-exploitation of species, and over-hunting and poaching have led to the extinction of a number of species. As loggers push deeper into the forests of Africa the local people gain access to forest animals that were previously inaccessible. Primates and other species are killed for bushmeat. This may be sold and eaten locally or it may even be exported to developed countries. In the oceans over-fishing is damaging fish stocks and the accidental bycatch of cetaceans is threatening their survival. Rare animals and animal products are traded as trophies, as part of the pet trade and as Chinese medicines.

Diseases threaten a wide range of taxa from amphibians attacked by the chytrid fungus to seal populations decimated by viruses. As the available space for wildlife contracts, more and more species – particularly large mammals such as elephants and large carnivores – come into conflict with people.

Anthropogenic (human-induced) climate change is now threatening a number of species such as polar bears (*Ursus maritimus*) and others that depend upon the ice sheets of the polar regions. In addition, the distributions of some animal diseases are changing because their vectors are extending their ranges into areas that they were previously unable to occupy. Bluetongue is a devastating disease of ruminants which is transmitted by midges (*Culicoides* spp.). In 2005, it was reported that since 1998, six strains of the bluetongue virus had spread across 12 countries in Europe and 800 km further north than previously reported (Purse *et al.*, 2005).

Biography 2.2 **Carolus Linnaeus (1707–78)**

Linnaeus was a Swedish botanist and physician. He devised the binomial system of nomenclature used for assigning scientific names to organisms. This was first published in his *Systema Naturae* in 1735. In 1741 he was appointed Professor of Practical Medicine at the University of Uppsala and then in 1742 Professor of Botany, Dietetics and Materia Medica. Linnaeus named many thousands of animals and plants using his system, and although some have since been reclassified, many still retain his original names.

Traditional medicines e.g. tiger body parts used in Chinese medicine	**Pollution** e.g. oil pollution, pesticides, fertilisers
Climate change e.g. loss of polar bear habitat	**Exotic pet trade** e.g. birds, reptiles
Disease e.g. chytrid fungus in amphibians	**Hunting for food** e.g. songbirds in Cyprus, bushmeat in Africa
Habitat loss and fragmentation e.g. logging of forests, draining of wetlands	**Bycatch** e.g. dolphins accidentally caught in fishing nets
Trophy hunting/poaching e.g. ivory, skins, horns, etc.	**Wildlife–human conflict** e.g. elephants destroying crops, wolves taking cattle
Jewellery e.g. ivory	**Over-exploitation** e.g. over-fishing, whaling
Cultural beliefs e.g. the aye-aye is considered a symbol of death	**Alien species** e.g. competition and predation by feral cats
Fur trade e.g. big cats, carnivores	**Collecting** e.g. birds, eggs, insects
Natural disasters e.g. earthquakes, tsunamis, El Niño	**Interbreeding with domestic animals** e.g. Scottish wildcat with domestic cats

Fig. 2.1 Major threats to the survival of wildlife.

2.7 NAMING ANIMALS

In order to create inventories of species present in particular ecosystems in the wild or animals kept in zoological collections it is essential that we have a standardised and universal naming system. Before records can be kept of individual animals in a zoo the species (and possibly subspecies) to which they belong must be established.

Each organism on Earth has been assigned a scientific or Latin name. In theory the scientific name of an organism is the same in every country in the world.

Table 2.2 Examples of animal species named after famous people.

Vernacular name	Scientific name	Person commemorated
Adélie penguin	*Pygoscelis adeliae*	Adélie Dumont d'Urville (1798–1842), wife of Admiral Dumont d'Urville, French explorer.
Agouti	*Dasyprocta azarae*	Félix Manuel de Azara (1746–1811), Spanish military officer, engineer, explorer and naturalist.
Audubon's oriole	*Icterus graduacauda*	John James Audubon (1785–1851), American ornithologist, collector, artist and writer.
Bewick's swan	*Cygnus bewickii*	Thomas Bewick (1753–1826), English ornithologist and engraver.
Burchell's zebra	*Equus burchellii*	William John Burchell (1781–1863), English explorer and naturalist.
Cook's petrel	*Pterodroma cookii*	Captain James Cook (1728–79), English explorer, captain of HMS *Endeavour*, discoverer of eastern Australia and New Zealand.
Cuvier's toucan	*Ramphastos tucanus cuvieri*	Georges Cuvier (1769–1832), French naturalist and palaeontologist, warden of the menagerie of the Jardin des Plantes, Paris.
Darwin's cactus ground finch	*Geospiza scandens*	Charles Darwin (1809–92), naturalist on HMS *Beagle* who proposed the theory of evolution.
Daubenton's bat	*Myotis daubentonii*	Dr Louis Jean-Marie d'Aubenton (1716–1800), French naturalist, first director of Museum of Natural History, Paris.
De Brazza's monkey	*Cercopithecus neglectus*	Count Pierre Savorgnan de Brazza (1852–1905), French naval officer and explorer, Governor of French Congo.
Goeldi's monkey	*Callimico goeldii*	Emil August Goeldi (1859–1917), Swiss zoologist who taught at the University of Bern.
Humboldt's penguin	*Spheniscus humboldti*	Baron Friedrich Wilhelm Heinrich Alexander von Humboldt (1769–1859), Prussian naturalist and explorer.
Lady Amherst's pheasant	*Chrysolophus amherstiae*	Sarah Countess Amherst (1762–1838), wife of the Governor General of Bengal.
Lear's macaw	*Anodorhynchus leari*	Edward Lear (1812–88), English poet, artist and traveller.
Livingstone's lourie	*Tauraco livingstonii*	Dr David Livingstone (1813–73), Scottish doctor, missionary and explorer.
Montagu's harrier	*Circus pygargus*	Colonel George Montagu, (1751–1815), British military officer and natural history writer.
Natterer's bat	*Myotis nattereri*	Dr Johann Natterer (1787–1843), Austrian naturalist and collector.
Okapi	*Okapia johnstoni*	Sir Harry Hamilton Johnston (1858–1927), English explorer, cartographer and colonial administrator.
Pallas' cat	*Otocolobus manul*	Peter Simon Pallas (1741–1811), German zoologist, geographer and explorer.
Père David's deer	*Elaphurus davidianus*	Fr. Jean Pierre Armand David (1826–1900), French Lazarist priest, zoologist and traveller.
Przewalski's horse	*Equus ferus przewalskii*	Colonel Nikolai Mikhailovitch Prjevalsky (1839–88), Russian Cossack naturalist, geographer and explorer.
Racey's pipistrelle bat	*Pipistrellus raceyi*	Prof. Paul Racey (1944–), Professor of Natural History, University of Aberdeen.
Raffles' malkoha	*Phaenicophaeus chlorophaea*	Sir Thomas Stamford Bingley Raffles (1781–1826), founder of Singapore and first president of the Zoological Society of London.
Rothschild's giraffe	*Giraffa camelopardalis rothschildi*	Lord Lionel Walter Rothschild (1868–1937), founder of the Natural History Museum at Tring.
Rüppell's griffon vulture	*Gyps rueppellii*	Wilhelm Peter Eduard Simon Rüppell (1794–1884), German collector and explorer.
Spix's macaw	*Cyanopsitta spixii*	Dr Johann Baptist von Spix (1781–1826), German naturalist, explorer and collector.
Verreaux's eagle	*Aquila verreauxii*	Jean Baptiste Edouard Verreaux (1810–68), French naturalist, collector and animal dealer.
Victoria crowned pigeon	*Goura victoria*	Queen Victoria (1819–1901), British queen and Empress of India.

However, the common or vernacular names of animals vary from country to country; for example, the animal known as a caribou in North America is usually called a reindeer in Europe. *Equus burchellii* is known as the common zebra, plains zebra, Burchell's zebra and bontkwagga, while *Nyctalus leisleri* is called the lesser noctule, hairy-armed bat or Leisler's bat. Many species have names which commemorate people (Table 2.2).

It is useful to note that, to avoid confusion, the law only recognises the scientific name of a species for the purposes of identification, and lists of protected species that appear in the Appendices to CITES and other laws contain these names.

The system for assigning scientific names to organisms was devised in the 18th century by Carolus Linnaeus and is known as the binomial system because it consists of two parts: the generic name and the specific name. The name is normally written in italics. The first name (genus) always begins with an upper case letter. The second name (species) always begins with a lower case letter, even if it is derived from a country or person's surname, for example *africana* or *johnstoni*. In older publications it is common to see 'L.' after the scientific name to indicate that the name was assigned by Linnaeus, or the surname of an alternative authority followed by the date, for example *Felis catus* L., *Loxodonta africana* Blumenbach 1797.

The scientific name of the tiger is:

Panthera	*tigris*
genus	species

Similar species belong to the same genus, for example:

- *Panthera leo*
- *Panthera tigris*
- *Panthera pardus*.

Scientists often abbreviate the scientific names of organisms by using only the initial letter of the genus once it has been written out in full, for example *P. leo*. Some authorities recognise subspecies and where these occur a third name is added, making it trinomial.

Six extant subspecies of tigers are recognised on the basis of distinctive molecular markers (Luo *et al.*, 2004):

Panthera tigris tigris	Bengal tiger
P. t. corbetti	Northern Indochinese tiger
P. t. amoyensis	South China tiger (possibly extinct)
P. t. altaica	Amur (Siberian) tiger
P. t. sumatrae	Sumatran tiger
P. t. jacksoni	Malayan tiger.

A further three subspecies were previously recognised on the basis of morphology but are now extinct:

P. t. balica	Bali tiger
P. t. sondaica	Javan tiger
P. t. virgata	Caspian tiger

The binomial system is hierarchical in nature, with similar species being collected together into larger and larger groups. Related species are grouped into genera; similar genera are grouped into families; similar families are grouped into orders and so on. The main divisions (or taxa) of this hierarchical system are:

- Kingdom
- Phylum (plural: phyla)
- Class
- Order
- Family
- Genus (plural: genera)
- Species (plural: species).

The complete system is much more complex than this and includes groupings such as suborders, superfamilies and infraclasses which fall between those listed above. A classification of the grey wolf (*Canis lupus*) is presented in Fig. 2.2.

It is sometimes convenient to use the name of higher taxa, instead of listing each species individually. The order Cetacea includes all whale, dolphin and porpoise species. The family Falconidae includes all species of falcons. The abbreviation 'spp.' is generally used to denote all species of a higher taxon. For example, *Panthera* spp. means all species of the genus *Panthera*, which includes many – but not all – of the big cats.

Although in theory scientific names are the same all over the world and remain the same forever, in practice animal and plant groups are occasionally reclassified. This is essential where new evidence has emerged about the relationships between taxa, for example when DNA analysis of a group of species indicates relationships which are not apparent from morphological features alone. The breadcrumb sponge (*Halichondria panacea*) is the marine organism that holds the record for the largest number of Latin synonyms. Since it was first described in 1766 it has been given 56 different scientific names. The naming of animal species is regulated by the International Commission on Zoological Nomenclature (see Box 2.1).

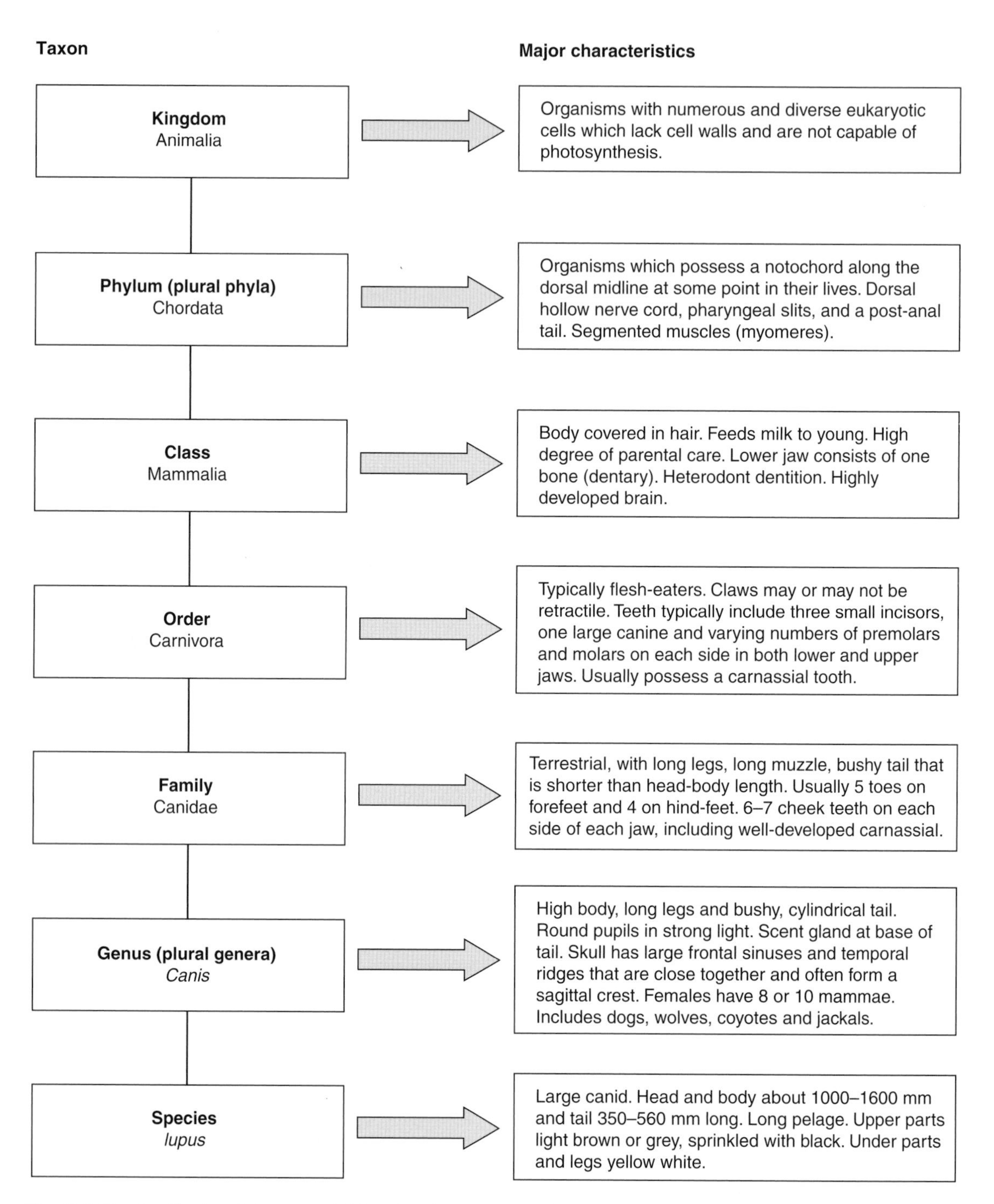

Taxon
Major characteristics
Kingdom
Animalia
Organisms with numerous and diverse eukaryotic cells which lack cell walls and are not capable of photosynthesis.
Phylum (plural phyla)
Chordata
Organisms which possess a notochord along the dorsal midline at some point in their lives. Dorsal hollow nerve cord, pharyngeal slits, and a post-anal tail. Segmented muscles (myomeres).
Class
Mammalia
Body covered in hair. Feeds milk to young. High degree of parental care. Lower jaw consists of one bone (dentary). Heterodont dentition. Highly developed brain.
Order
Carnivora
Typically flesh-eaters. Claws may or may not be retractile. Teeth typically include three small incisors, one large canine and varying numbers of premolars and molars on each side in both lower and upper jaws. Usually possess a carnassial tooth.
Family
Canidae
Terrestrial, with long legs, long muzzle, bushy tail that is shorter than head-body length. Usually 5 toes on forefeet and 4 on hind-feet. 6–7 cheek teeth on each side of each jaw, including well-developed carnassial.
Genus (plural genera)
Canis
High body, long legs and bushy, cylindrical tail. Round pupils in strong light. Scent gland at base of tail. Skull has large frontal sinuses and temporal ridges that are close together and often form a sagittal crest. Females have 8 or 10 mammae. Includes dogs, wolves, coyotes and jackals.
Species
lupus
Large canid. Head and body about 1000–1600 mm and tail 350–560 mm long. Long pelage. Upper parts light brown or grey, sprinkled with black. Under parts and legs yellow white.

Fig. 2.2 The classification of the grey wolf (*Canis lupus*). Descriptions of the genus *Canis* and the species *lupus* are based on Nowak (1999).

Box 2.1 The International Commission on Zoological Nomenclature.

The International Commission on Zoological Nomenclature (ICZN) was founded in 1895. Its purpose is to provide and regulate a uniform system of zoological nomenclature to ensure that every animal has a unique and universally accepted scientific name. This is an important task because more than 2000 new genus names and 15,000 new species names are added to the zoological literature every year.

The Commission publishes the *International Code of Zoological Nomenclature* which contains the universally accepted rules for allocating scientific names to animals. The Code regulates nomenclature only – the way names are created and published – not taxonomy (classification). The Commission publishes a journal – the *Bulletin of Zoological Nomenclature* – which contains papers about problems related to the naming of animals which are resolved by the Commission. The Commission's work directly affects studies of biodiversity and conservation as, clearly, it is essential that scientists can properly name and classify the animals with which they work.

ZooBank

ZooBank is the official registry of Zoological Nomenclature according to the ICZN. It contains information about the original descriptions of new scientific names for animals, publications containing these descriptions, their authors and information about the registration of type specimens (the specimen originally used to define the species).

International Code of Zoological Nomenclature

Article 5: Principle of Binomial Nomenclature

5.1: Name of species. The scientific name of a species, ..., is a combination of two names (a binomen), the first being the generic name and the second being the specific name. The generic name must begin with an upper-case letter and the specific name must begin with a lower-case letter.
5.2: Names of subspecies. The scientific name of a subspecies is a combination of three names (a trinomen, i.e. a binomen followed by a subspecific name). The subspecific name must begin with a lower-case letter.

Article 23: Principle of Priority

23.1: Statement of the Principle of Priority. The valid name of a taxon is the oldest available name applied to it, unless that name has been invalidated or another name is given precedence by any provision of the Code or by any ruling of the Commission.

Care must be taken in the interpretation of animal names used in laws. When laws list individual protected species, the scientific names are used to prevent confusion (e.g. in the Wildlife and Countryside Act in England and the Endangered Species Act in the United States). However, often a general term is used for convenience to describe a group of animals which differs from its zoological usage. For example, the term 'seafish' is defined under English law for the purposes of the Sea Fisheries Regulation Act 1966 as fish of any kind found in the sea, and includes shellfish (defined as crustaceans and molluscs of any kind) but excludes salmon and migratory trout. Zoologically this is clearly nonsense but having a single term to refer to a diverse range of taxa is sometimes convenient in a legal context. Where such a zoologically inaccurate term is used a definition would normally occur within the legal instrument concerned.

A basic classification of the Animal Kingdom and a more detailed taxonomy of mammals, birds, reptiles, amphibians and fishes, down to the level of order, are provided in Appendix 1.

2.8 HOW ARE NEW SPECIES DISCOVERED?

We continue to lose species from the Earth at an alarming rate. But new species are also being discovered (Table 2.3). New species may be discovered in a number of different ways:

Table 2.3 Examples of species discovered since 1865.

Species	Year of discovery	Discoverer	New (N)/Rediscovered (R)/ Reclassified (C)	Location
Black dwarf lowland tapir (*Tapirus pygmaeus*)	2008	Dr Marc van Roosmalen	N	Lowland Amazonia, Brazil
Kipunji (*Rungwecebus kipunji*) – a new monkey genus	2004	Wildlife Conservation Society scientists	N	Montane forest in SW Tanzania
Giant muntjac (*Megamuntiacus vuquangensis*)	1994	MacKinnon *et al.*	N	Northern Vietnam and Laos
Mbaiso tree kangaroo (*Dendrolagus mbaiso*)	1994	Dr Tim Flannery	N	Irian Jaya, Indonesia
Vu Quang ox (*Pseudoryx nghetinhensis*)	1992	Dung, Giao, Chinh, Tuoc, Arctander and MacKinnon	N	Vietnam/Laos
Hairy-eared lemur (*Allocebus trichotis*)	1989	B. Meier	R	N Madagascar
Fijian crested iguana (*Brachylophus vitiensis*)	1979	Dr John Gibbons	N	Yadua Taba, Fiji
Long-whiskered owlet (*Xenoglaux loweryi*)	1976	Drs J. O'Neill and Gary Graves	N	Peru
Greater bamboo lemur (*Hapalemur simus*)	1972	Dr A. Peyriéras	R	SE Madagascar
Iriomote cat (*Felis iriomotensis*)	1967	Yokio Togawa	N	Iriomote, Japan
Mountain pygmy possum (*Burramys parvus*)	1966	Dr K. Shortman	R	Victoria and NSW, Australia
Coelacanth (*Latimeria chalumnae*)	1938	Marjorie Courtenay-Latimer	N	South Africa
Kouprey (*Bos sauveli*)	1937	Prof. Achille Urbain	N	Cambodia
Congo peacock (*Afropavo congensis*)	1936	Dr James Chapin	N	Congo
Bonobo (*Pan paniscus*)	1929*	Dr Ernst Schwarz	C	Congo
Scaly-tailed possum (*Wyulda squamicaudata*)	1917	W. B. Alexander	N	Australia
Komodo dragon (*Varanus komodoensis*)	1912	Major P. A. Ouwens	N	Komodo, Indonesia
Okapi (*Okapia johnstoni*)	1901	Sir Harry Johnston	N	Ituri Forest, Zaire
Giant panda (*Ailuropoda melanoleuca*)	1869	Père Armand David	N	China
Père David's deer (*Elaphurus davidianus*)	1865	Père Armand David	N	China

* Originally classified as a subspecies, *Pan troglodytes paniscus*, in 1929 but elevated to a new species five years later.

• New species found in the wild and described by science for the first time, for example Vu Quang ox (*Pseudoryx nghetinhensis*) in 1992.
• Rediscovery of species that were thought to be extinct; for example in 1992 the whistling Anjouan scops owl (*Otus capnodes*) was rediscovered having not been reported since its original discovery in 1886.
• Reclassification of existing species due to new evidence, especially as a result of genetic studies; for example the bonobo (*Pan paniscus*) was originally considered to be a subspecies of the chimpanzee (*P. troglodytes*) and classified as *P. t. paniscus*.
• The discovery of preserved specimens in museum collections; for example in 2009 a new species of flying fox bat from Samoa was discovered in the collections of the Academy of Natural Sciences in Philadelphia. It had been preserved in alcohol since 1856.

Many new species have been discovered as a result of the exploration of the deep oceans and expeditions into tropical rainforests. Occasionally, new species are recorded by camera traps and film crews. In 2009, a BBC film unit discovered a new species of giant rat – the Bosavi woolly rat (*Mallomys* sp.) – in the crater of Mount Bosavi, an extinct volcano in Papua New Guinea (Anon., 2009a).

2.9 DATABASES OF ANIMAL SPECIES

The rapid development of, and increasing access to, the internet has resulted in a number of organisations producing extensive databases of animal species. These include:

- *Species 2000 and ITIS Catalogue of Life*. The goal of this project is to create a validated checklist of all the world's species by bringing together a collection of species databases.
- *The Global Biodiversity Information Facility (GBIF)*. This is an international organisation whose purpose is to make global biodiversity data accessible everywhere in the world via the internet. GBIF uses the 'Catalogue of Life' as the taxonomic basis of its web portal.
- *The World Register of Marine Species (WoRMS)*. A new World Register of Marine Species was inaugurated in June 2008 with the objective of consolidating world databases of marine organisms. At this time it contained about 122,500 validated marine species names. Some 56,400 aliases had been identified, representing 32% of all the names reviewed (WoRMS, 2008). The World Register will clarify for all time the valid name of all marine species (WoRMS, 2008).
- Wilson and Reeder's *Mammal Species of the World* is an online database of mammal species which provides information on the taxonomy and distribution of mammals and is based on Wilson and Reeder (2005).

2.10 FURTHER READING

Hambler, C. (2004). *Conservation*. Cambridge University Press, Cambridge, UK.

Shuker, K. (1993). *The Lost Ark: New and Rediscovered Animals of the 20th Century*. Harper Collins Publishers, London.

Wilson, E. O. (2001). *The Diversity of Life*. Penguin Books, London.

Zimmerman, A., Hatchwell, M., Dickie, L. and West, C. (eds.) (2007). *Zoos in the 21st Century: Catalysts for Conservation?* Cambridge University Press, Cambridge, UK.

2.11 EXERCISES

1. How can we justify spending money on the protection of biodiversity?
2. For a named species, discuss the threats to its future survival in the wild.
3. Why is it important to assign a scientific name to each animal species?
4. 'From an ecological point of view all species are not equally important.' Discuss.

3 A SHORT HISTORY OF ZOOS

Despite their obvious limitations, the early zoos did, in their time, play a major role in reminding the urban world of the wild wonders beyond the suburbs.

Desmond Morris

Conservation Status Profile

Visayan warty pig
Sus cebifrons
IUCN status: Critically Endangered A4cde
CITES: Not listed
Population trend: Decreasing

An Introduction to Zoo Biology and Management, First Edition. Paul A. Rees.

3.1 INTRODUCTION

The keeping of wild animals by humans has a very long history, probably extending back at least some 4500 years. The first captive animals were almost certainly domesticated species kept for food – sheep, cattle, goats – and those kept for companionship, such as dogs and cats. Others were kept for religious reasons, as symbols of wealth and because powerful animals conferred great status on their owners. Some species were believed to be incarnations of gods or of dead ancestors, and others were used as sacrifices. Some South American Indians kept some bird species as a currency.

Even within the last hundred years, some cities took great pride in the animals kept by their local zoo and sometimes new animals were purchased by public subscription. Exotic animals have long been considered fitting gifts for powerful people. Some species, such as lions and leopards, were considered so noble that only royalty could own them. Heads of state have given animals to each other as symbols of friendship for many thousands of years and this practice has occurred in recent times (see Section 12.1.1).

Modern zoos have evolved into sophisticated scientific organisations and many have a long history. The following account is intended as a very brief introduction to the history of zoos from their earliest beginnings in ancient times. More detailed accounts of the history of zoos can be found in Hoage and Deiss (1996) and Bostock (1993).

3.2 A BRIEF HISTORY OF ZOOS

Loisel (1912) divided the history of the development of menageries and zoos into five periods (Fig. 3.1). The earliest evidence of animal keeping appears to be the illustrations lining the tomb of the wealthy Egyptian nobleman Ti in Saggara (5th Dynasty, 2495–2345 BC). A neighbouring tomb contains wall sculptures that appear to be illustrations of a zoo where antelopes are tethered to their mangers and some are being fed by attendants or led by their horns. The earliest wild animal keeping may have occurred for religious reasons. The Egyptians regarded many animal species as sacred – including hippopotamuses, owls and crocodiles – and often kept them in temples. Tame lions were kept by pharaohs and some animals were mummified. There was an ancient zoo at Alexandria, founded by Ptolemy Philadelphus, but little is known about it.

There is earlier evidence suggesting a close association between people and wild animals in Egypt. A double bird-shaped slate palette dated from before 3100 BC is decorated with relief carvings of a man and three ostriches. The façade of the tomb of Atet, the wife of Nefermat, shows men with cranes, an ibex and a tethered ox, and dates from *c.*2600 BC. Both of these items are on display at the Manchester Museum.

King Shulgi of the 3rd Dynasty of Ur, in Mesopotamia (2094–2047 BC), probably owned the first zoo that contained large carnivores like lions. It was in Sumer – in what is now Iraq – and clay tablets found in several Sumerian cities record the receipt and distribution of livestock. Wild animals were received as gifts by many Babylonian and Assyrian kings.

There are ancient records of substantial numbers of wild animals being taken by man in the Middle East. In 879 BC, Assurnasirpal II, king of Assyria, collected entire herds of elephants and kept them in a 'zoo'. One of his most impressive hauls of animals is recorded on a stele (stone tablet) from the early 9th century BC and included 30 elephants, 450 tigers and 200 ostriches. Thutmose III, king of Egypt (1504–1450 BC), is said to have slain 120 elephants for their tusks. Large scale organised trade in ivory is not a recent phenomenon and is thought to have begun with the Phoenicians (Delort, 1992).

By the 4th century BC most of the city states in Greece probably had their own animal collections. Alexander the Great sent many specimens back to Greece from his military campaigns. In his *History of Animals* Aristotle – Alexander's tutor – described many species that were not native to Greece, suggesting that he must have visited many of these menageries.

The Romans slaughtered very large numbers of animals in the Colosseum and other similar stadia. In the 2nd century BC, as Rome's power increased in northern Africa, very large numbers of African animals were displayed and killed. In 55 BC a show organised by Pompey involved the slaughter of 20 elephants. Other events resulted in the deaths of many thousands of African animals, mostly lions and leopards. However, sometimes animals were simply displayed. Augustus exhibited a rhinoceros, a tiger and a large snake to the public.

In ancient China, Emperor Wen Wang created an 'Intelligence Park', but little is known of this apart from a description in a collection of ancient Chinese poetry. Although it appears to have contained animals it did not seem to be a collection of exotic animals. However, the park created by Emperor Chi-Hang-Ti – of the

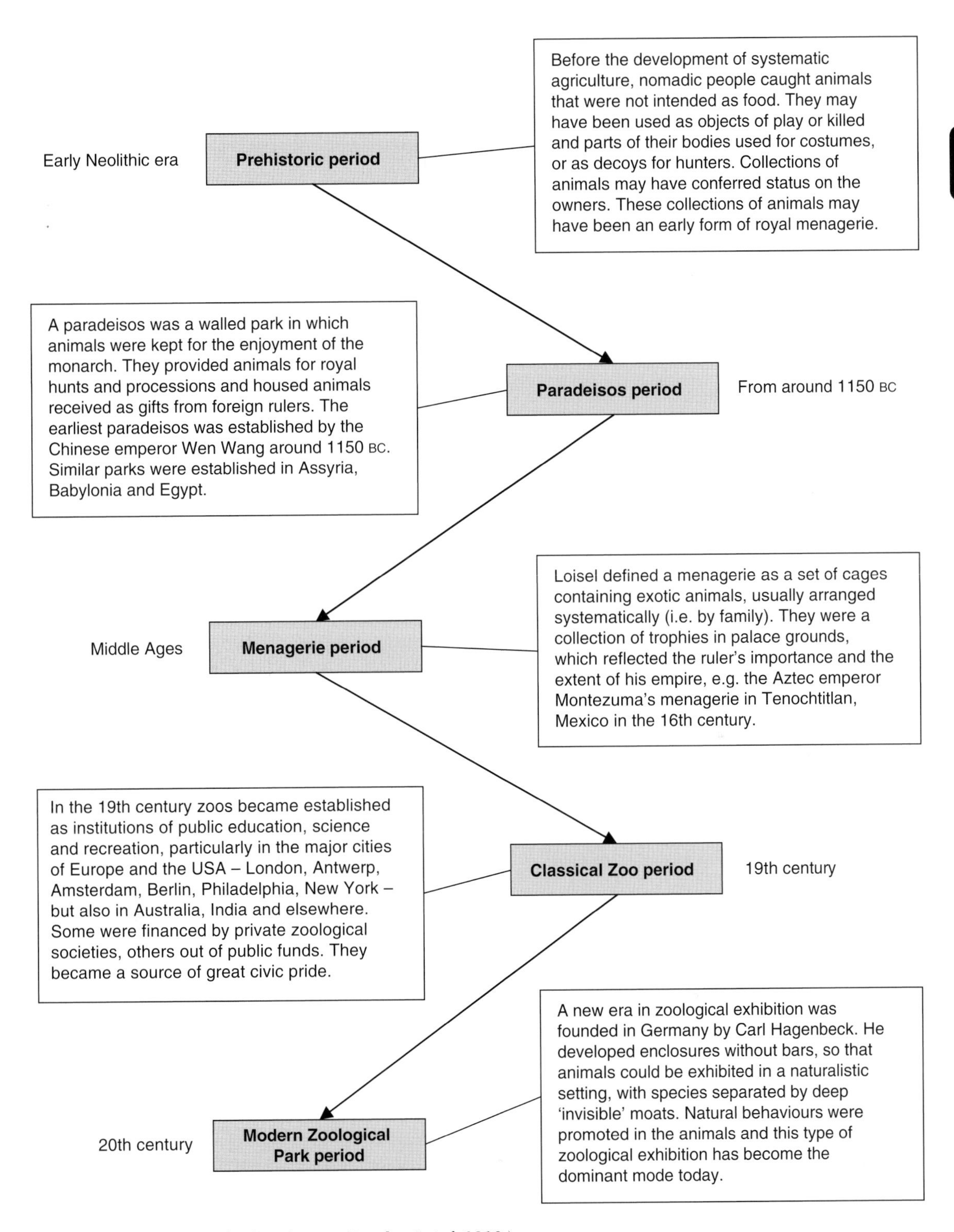

Early Neolithic era
Prehistoric period
Before the development of systematic agriculture, nomadic people caught animals that were not intended as food. They may have been used as objects of play or killed and parts of their bodies used for costumes, or as decoys for hunters. Collections of animals may have conferred status on the owners. These collections of animals may have been an early form of royal menagerie.
A paradeisos was a walled park in which animals were kept for the enjoyment of the monarch. They provided animals for royal hunts and processions and housed animals received as gifts from foreign rulers. The earliest paradeisos was established by the Chinese emperor Wen Wang around 1150 BC. Similar parks were established in Assyria, Babylonia and Egypt.
Paradeisos period
From around 1150 BC
Middle Ages
Menagerie period
Loisel defined a menagerie as a set of cages containing exotic animals, usually arranged systematically (i.e. by family). They were a collection of trophies in palace grounds, which reflected the ruler's importance and the extent of his empire, e.g. the Aztec emperor Montezuma's menagerie in Tenochtitlan, Mexico in the 16th century.
In the 19th century zoos became established as institutions of public education, science and recreation, particularly in the major cities of Europe and the USA – London, Antwerp, Amsterdam, Berlin, Philadelphia, New York – but also in Australia, India and elsewhere. Some were financed by private zoological societies, others out of public funds. They became a source of great civic pride.
Classical Zoo period
19th century
20th century
Modern Zoological Park period
A new era in zoological exhibition was founded in Germany by Carl Hagenbeck. He developed enclosures without bars, so that animals could be exhibited in a naturalistic setting, with species separated by deep 'invisible' moats. Natural behaviours were promoted in the animals and this type of zoological exhibition has become the dominant mode today.

Fig. 3.1 Loisel's five periods of zoo history. (Based on Loisel, 1912.)

Thsin Dynasty – was filled with animals and trees from all over his empire.

In Europe, from medieval times, large exotic animals tended to be the property of kings. They were often given as gifts from one monarch to another. The animals were kept in deer parks or menageries. King Charlemagne of France received an elephant from the caliph of Baghdad in 797. This practice of giving animals has continued in recent times, one head of state presenting another with animals as a token of friendship between their two countries. Monks also sometimes had menageries, such as the one at Saint-Gall in Switzerland, where bears, badgers, herons and silver pheasants were kept.

In England, about 1100, Henry I had a menagerie at Woodstock that contained lions, leopards, lynx, camels and a rare owl. This was later moved to the Tower of London, probably by Henry III. In 1251 Henry III received a polar bear that was allowed to fish in the River Thames. Three years later he received an elephant from the King of France, Louis IX. The elephant was the first in England and was put on display to the public.

In medieval Europe royalty and nobility often kept animals in deer parks. At this time 'deer' probably meant 'animal'. The Domesday Book records 35 around the year 1086. In England, King John (reigned 1199–1216) possessed almost 800 parks. Deer parks were originally used mainly for hunting and for food, but eventually became largely ornamental in function. Many of these parks still exist.

Wild and semi-wild animals were kept in medieval times in extensive areas for purposes other than hunting. White cattle were enclosed by a wall around the Chillingham Estate (now in Northumberland) erected in 1220. The cattle were probably brought from Italy in Roman times.

In the 13th century Marco Polo saw lions and tigers wandering freely through the rooms of a Chinese imperial palace. Around this time Kubilai Khan, the fifth Great Khan of the Mongol Empire, had animal parks that were used for hunting, and he also kept tame cheetahs, tigers and falcons.

In 1368 a traveller to China reported that he had seen 3000 monkeys in the park of a Buddhist pagoda. A little later, a closed garden was reported from near Peking, which contained a high mountain inhabited by monkeys and other animals.

Philip VI of France (reigned 1328–50) kept lions and leopards at the Louvre. French 15th century tapestries show monkeys and apes kept in the royal courts, and parrots were popular in the Vatican around this time. The Vatican menagerie expanded under Pope Leo X (1513–23) and included monkeys, civets, lions, leopards, an elephant and a snow leopard. As well as animals, one cardinal kept exotic foreign peoples including Moors, Indians, Turks and African negroes.

When Hernán Cortés conquered Mexico for Spain between 1517 and 1521 he discovered a magnificent zoo owned by the Aztec emperor Montezuma II, at his capital Tenochtitlan. It contained llamas, vicuna, antelopes, snakes, waterbirds, a large collection of birds of prey, an aquarium, large cats (including jaguars and pumas), and other carnivores. It also contained human albinos and deformed humans. The zoo appears to have had several hundred keepers and some of the birds appear to have been kept for their feathers. Similar menageries existed in other Aztec cities, including a large collection at Texcoco.

In 1552 Crown Prince Maximilian of Austria created a deer park and menagerie around the castle at Ebersdorf, near Vienna. He brought exotic animals to his park, including an Indian elephant. Maximilian established another park at Katterburg containing deer, birds and fish, and a menagerie at the castle of Neugebäude. After his death, Katterburg was absorbed into the imperial palace of Schönbrunn and, in 1752, Franz Stephan – the husband of Empress Maria Theresa – founded the first modern zoo here. It was essentially a private collection, although the public was admitted occasionally. Enclosures were arranged around a central rococo pavilion which afforded the best views of the animals, which were kept behind high walls. Later, Josef II established a Society for the Acquisition of Animals and he financed collecting expeditions to Africa and the Americas. In its day Schönbrunn was the best animal collection in Europe and the zoo still exists as Tiergarten Schönbrunn, or Zoo Vienna.

In 17th century France, Louis XIV built a menagerie in the grounds of his palace at Versailles (Fig. 3.2). The first animals were installed in 1665. As the collection grew the increasing number of visitors did so much damage that the king had to restrict admittance to members of his court. The menagerie fell into disrepair and closed in 1792. The remaining animals were offered to the former Jardin du Roi in Paris, which was renamed the Jardin des Plantes (Fig. 3.3). In 1793 the Jardin was incorporated into the new Muséum National d'Histoire Naturelle and in 1803 the zoologist Georges Cuvier assumed responsibility for the menagerie. The Jardin

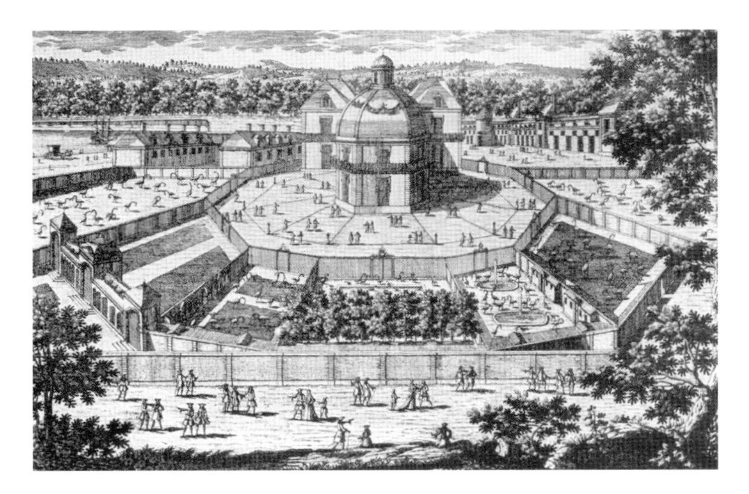

Fig. 3.2 Louis XIV's menagerie at Versailles. (From an engraving by Antoine Aveline.)

Biography 3.1

Georges Cuvier (1769–1832)

Cuvier was a French zoologist and naturalist. He was largely responsible for founding vertebrate palaeontology and comparative anatomy as scientific disciplines and demonstrated that past life forms had become extinct. In 1795 Cuvier moved to Paris and shortly thereafter he was appointed as Professor of Comparative Anatomy at the Muséum National d'Histoire Naturelle (National Museum of Natural History). In 1803 he became warden of the menagerie of the Jardin des Plantes, which at that time included a monkey and bird house, bear pits and a rotunda for herbivores including elephants and giraffes. Cuvier's most famous academic work was a systematic survey of zoology, *Le Règne Animal* (The Animal Kingdom), published in 1817.

Fig. 3.3 The Jardins des Plantes, Paris. (Courtesy of Library of Congress Prints and Photographs Division, Washington DC.)

des Plantes had broad walkways from which ordinary people could view the exhibits, unlike Schönbrunn where only the privileged occupants of the pavilion had a good view of the animals. Unfortunately, in 1870 the Prussian army laid siege to Paris and all of the edible animals were slaughtered for food.

The Zoological Society of London (ZSL) was founded in 1826 by Stamford Raffles. In 1826 London Zoo – in Regent's Park – opened to Fellows of the Society, and paying visitors were first admitted in 1847. London Zoo was the first scientific zoo in the world and its first superintendent was Abraham Dee Bartlett. The zoo opened the first reptile house (1849), the first insect house (1881) and the first Children's Zoo (1938). In 1931 the ZSL opened Whipsnade (Whipsnade Park Zoo), in Bedfordshire, in order to keep and study large animals in more natural surroundings. In 1960 the Society established the Institute of Zoology where scientists are employed to conduct zoological research.

Although London Zoo was extremely popular with the public it is important to remember that the only people who could visit it in its early days had to live relatively close by as they had no means of travelling long distances for a day out. When most people were unable to travel to zoos to see animals, travelling menageries were popular. Perhaps the best known of these was Bostock and Wombwell's Royal Menagerie which travelled widely in Britain and abroad from 1805 until 1932, when the animals were sold to London Zoo. The menagerie included a very wide range of animals including elephants, camels, lions and tigers (Fig. 3.4).

From around 1875 P.T. Barnum's Great Traveling Museum, Menagerie, Caravan, and Hippodrome exhibited animals across the United States. This eventually evolved into The Ringling Bros and Barnum & Bailey Circus ('The Greatest Show on Earth') which still exists today (Fig. 3.5). In 1882 Barnum and Bailey purchased the famous elephant 'Jumbo' from London Zoo (Box 3.1).

The founding of the Zoological Society of London was followed by the establishment of a large number of other major zoos around the world, particularly in Europe, the USA and Australia (Table 3.1), including the Royal Melbourne Zoological Gardens in 1872 and the National Zoological Park which opened in 1891 in Washington DC (Box 3.2). These zoos became an important focus of civic pride and in the United States a city was not considered to have a 'real' zoo unless it possessed an elephant.

Many new zoos opened in major cities in the UK, but most have now disappeared. In 1836 a zoo was opened at Belle Vue in Manchester (Fig. 3.7). It was also home to a large amusement park, an exhibition hall, a natural history museum, and later a football ground, speedway and greyhound racecourse. The zoo contained a number of large buildings including an impressive monkey house (Fig. 3.8).

At the end of the 19th century a zoo was established within Blackpool Tower (Fig. 3.9). The building also housed an aquarium. This was an unsuitable place for a zoo, but at the time it housed many species that performed in the Tower Circus. The zoo was moved to a new Blackpool Municipal Zoological Gardens in 1971.

In the second half of the 19th century many new zoos were founded in the United States, including Lincoln Park Zoo, New York Central Park Zoo, and the zoos in Cincinnati and Philadelphia (Table 3.1).

Without doubt the greatest influence on the future development of zoos was the creation of a new type of zoo in Germany at the beginning of the 20th century. In 1907, Carl Hagenbeck, an animal trainer and trader, founded Tierpark Hagenbeck in Hamburg, Germany

Biography 3.2

Sir Thomas Stamford Raffles (1781–1826)

Raffles was a British colonial administrator who was born in Jamaica and served in the British East India Company. He was at one time governor of Sumatra and founded Singapore in 1819. Raffles was a keen natural historian and became President of the Batavian Society which studied the natural history of Java and adjacent islands. Raffles returned to England in 1824. He founded the Zoological Society of London in 1826 and became its first president. Raffles published descriptions of some 34 bird species and 13 species of mammals, mostly from Sumatra. He named many new species including the sun bear (*Ursus malayanus*).

6 Catalogue of the Collection.

The Tasmanian Devil.

The Chimpanzee, or Missing Link.

Bonnet and Rhesus Monkeys.

280. Cotimondis, or Long Nosed Ant Eaters, from Brazil.

290. Paradoxure Cat, from South America.

300. Wallaby Kangaroo, from Western Australia.

310. Black Buck Antelope.
An animal well known to the sportsman of British India, it being a favourite object of pursuit.

320. Young Aswail, or Sloth Bear, from the Himalayas.

330. Civet, or Musk Cat, from Africa.

340. Ocelot, or Tiger Cat of India.
The most beautiful of the cat tribe.

350. Racoons, from Upper Canada.
These animals are remarkable for their cleanly habits, having been known to carry their food to wash it previous to eating it.

360. Young Leopards. Born in the Menagerie.

370. Caracal or Lynx, from South Africa.

380. Prairie Wolf, from North America.

390. Jackals, or Lions' Providers.
Found in Nubia, Caffraria, and many parts of Africa.

400. The Dingo, or Native Dog of Australia.
An Albino Specimen. Very Rare.

410. Capybara, from South America.
A very curious and rare rodent.

Catalogue of the Collection. 7

The Bengal Tiger.

The Cheetah, or Hunting Leopard.

The Spotted or Laughing Hyæna.

420. The Wombat.
The Wombats are short, clumsy and thick-legged Marsupials that are natives of Australia, and live entirely on vegetable matter, such as grass and roots, which they dig up with their strong claws. They are harmless creatures, and their flesh is said to be palatable.

430. Young Specimen of the Boomah.
The largest of the kangaroo species from Australia. By their peculiar formation kangaroos neither run nor walk but move along by a series of bounds. They are pouch-bearing animals and are much hunted in their native country.

440. Black-headed Sheep, from Port Said.

450. Zebu, or Sacred Brahmin Bull of Hindostan.
An animal still worshipped by some of the Hindoo Tribes, but principally used as a beast of burden.

460. Axis Deer, from Ceylon.
The handsomest of the deer family.

470. The Thar
Is a Himalayan Wild Goat that inhabits forests from Kashmir to Bhutan. It differs from typical Goats in its short, thick, and much compressed horns, and is of a ferocious disposition.

480. Hog Deer, from China.

490. Giant Male Hippopotamus.
About 8 years of age, added to the collection at enormous cost. It hails from the White Nile, and is positively the first living specimen ever seen in Scotland.

Fig. 3.4 Pages from the illustrated catalogue of Bostock and Wombwell's Royal Menagerie, 1917 edition. (Courtesy of Chetham's Library, Manchester.)

Fig. 3.5 Poster for the Jungle Menagerie of the Barnum & Bailey Greatest Show on Earth. (Courtesy of Library of Congress Prints and Photographs Division, Washington DC.)

Box 3.1 Jumbo – the most famous zoo animal of all time.

Fig. 3.6 Jumbo the elephant. (Courtesy of Library of Congress Prints and Photographs Division, Washington DC.)

Jumbo the African elephant is undoubtedly the most famous zoo animal of all time (Fig. 3.6). He was born in eastern Sudan, around Christmas 1860. In 1862 he was captured by Taher Sheriff and walked to Kassala, where he was delivered to Johan Schmidt, and then walked across the Sahara by Casanova. Jumbo was first sold to the Jardins des Plantes in Paris. Then, in 1865 he was traded with the Zoological Society of London for £450 (£30,000 today) and an Indian rhino, two dingoes, a black-backed jackal, a possum, a kangaroo and a pair of wedge-tailed eagles.

In 1882 Jumbo was sold by London Zoo to Barnum & Bailey Circus in America for £2000 (£138,000 today). The British public protested and pleaded with the zoo to keep Jumbo. In March of that year Matthew Berkley-Hill brought a lawsuit in the Court of Chancery against the Zoological Society, its Council and Barnum, claiming that the Society had no right to sell any of its animals without the consent of its Fellows. The suit failed.

Jumbo left for America after 16 years and 9 months at Regent's Park Zoo. On 15 September 1885 he was hit by a train and killed while crossing a railway line in St Thomas, Ontario, Canada. Barnum continued to make money out of the elephant by touring America with his stuffed body. On 4 April 1889 Jumbo's mounted skin was delivered to Barnum's museum at Tufts College where it remained until 14 April 1975 when it was destroyed by fire (Chambers, 2007).

Table 3.1 Dates of opening of selected zoos since 1752.

Date	Zoo	Country
1752	Schönbrunn Zoo, Vienna	Austria
1793	Jardin des Plantes, Paris	France
1828	London Zoological Gardens	England
1833	Dublin Zoological Gardens	Ireland
1836	Belle Vue Zoological Gardens, Manchester	England
1836	Bristol Zoological Gardens	England
1839	Royal Edinburgh Zoological Gardens	Scotland
1839	Amsterdam Royal Zoological Gardens	Netherlands
1843	Antwerp Zoological Gardens	Netherlands
1844	Berlin Zoological Gardens	Germany
1857	Rotterdam Zoological Gardens	Netherlands
1858	Frankfurt Zoological Gardens	Germany
1860	Jardin Zoologique d'Acclimatation, Paris	France
1860	Cologne Zoological Garden	Germany
1861	Dresden Zoological Gardens	Germany
1863	Hamburg Zoological Garden	Germany
1865	Breslau (now Wroclaw) Zoological Garden	Poland
1866	Budapest Zoological Garden	Hungary
1868	Lincoln Park Zoological Gardens	USA
1871	Stuttgart Zoological Garden	Germany
1872	Royal Melbourne Zoological Gardens	Australia
1873	New York Central Park Zoo	USA
1874	Basel Zoological Garden	Switzerland
1874	Philadelphia Zoological Garden	USA
1875	Cincinnati Zoo	USA
1876	Calcutta Zoological Gardens	India
1882	Ueno Zoological Gardens, Tokyo	Japan
1888	Cleveland Metroparks Zoological Park	USA
1888	Buenos Aires Zoo	Argentina
1888	Dallas Zoo	USA
1889	Atlanta Zoological Park	USA
1891	National Zoological Park, Washington DC	USA
1891	Giza Zoo, Cairo	Egypt
1892	St Petersburg Zoological Garden	Russia
1895	Baltimore Zoo	USA
1896	Düsseldorf Zoological Garden	Germany
1896	Königsberg Zoological Gardens	Kaliningrad (now in Russia)
1898	Pittsburgh Zoo	USA
1899	New York (Bronx) Zoological Park	USA
1899	Pretoria Zoo	South Africa
1899	Moscow Zoological Garden	Russia
1899	Toledo Zoological Gardens	USA
1907	Hagenbeck's Tierpark (Stellingen)	Germany
1913	Edinburgh Zoological Gardens	Scotland
1916	San Diego Zoo	USA
1923	Paignton Zoological Gardens	England
1931	Chester Zoological Gardens	England
1938	Dudley Zoological Gardens	England
1952	Arizona-Sonora Desert Museum	USA

(*continued*)

Table 3.1 *(Cont'd)*

Date	Zoo	Country
1959	Jersey Zoological Gardens	Jersey, Channel Islands
1963	Welsh Mountain Zoo, Colwyn Bay	Wales
1966	Longleat Safari Park, Wiltshire	England
1971	Knowsley Safari Park, Prescot	England
1994	South Lakes Wild Animal Park, Cumbria	England
1998	*Disney's Animal Kingdom*, Florida	USA
1998	*Blue Planet*, Ellesmere Port	England
2002	*The Deep*, Hull	England

Adapted from Hoage *et al.* (1996) and Kisling (1996) with additions by the author.

Box 3.2 A short history of the Smithsonian National Zoo.

The Smithsonian's National Zoo was created by an Act of Congress in 1889. It has two facilities: a 163 acre (66 ha) zoological park in northwest Washington, DC which is open to the public, and a non-public 3200 acre (1,295 ha) Conservation and Research Center in Fort Royal, Virginia.

The zoo became part of the Smithsonian in 1890. Plans for the zoo were drawn up by three men:

- William Temple Hornaday – a conservationist and head of the vertebrate division at the Smithsonian Institution
- Frederick Olmsted – the premier landscape architect of the day
- Samuel Langley – the third Secretary of the Smithsonian.

In its early years the zoo exhibited a small number of individuals of as many exotic species as possible. In the 1950s the zoo hired its first permanent vet, reflecting increasing concern for the welfare of the animals. The Friends of the National Zoo (FONZ) was founded in 1958 and persuaded Congress to fund the zoo through the Smithsonian, giving it a firmer financial base. It now operates as a not-for-profit organisation. FONZ has some 40,000 members and supports the activities of the zoo by assisting with guest services, education, outreach programmes, research and conservation activities.

In the early 1960s the zoo began to focus on the study and breeding of endangered species and a research division was created. The zoo's Conservation and Research Center (now the Smithsonian Conservation Biology Institute) was established in 1975.

Many of the zoo's aging facilities have been replaced by new exhibits including *Asia Trail* – which includes red pandas, sloth bears and clouded leopards – and a new facility for Asian elephants called *Elephant Trails*.

The zoo now runs educational programmes for teachers, students and the general public along with specialised training programmes for wildlife professionals. It has been at the forefront of the use of internet technology and is expanding its programmes so that it can become a conservation resource for a worldwide virtual audience.

(Box 3.3). Hagenbeck used hidden moats to create the illusion that animals were sharing the same landscape. As late as the beginning of the 20th century some zoos, including Hagenbeck's, thought nothing of exhibiting humans alongside animals (Boxes 3.3 and 3.4).

In 1931 George Mottershead opened Chester Zoo. He used many of Hagenbeck's ideas to create enclosures where animals were separated from the public by moats and ditches rather than bars. This has evolved into a world class zoo which contains many innovative exhibits. San Diego Zoo in California was founded in 1916 and is now one of the finest zoos in the world and home to a major conservation research institute (see Box 14.2).

There will undoubtedly be new zoos in the future, but the large scale developments are likely to be driven

Fig. 3.7 Cover of the 1904 guide to Belle Vue Zoological Gardens, Manchester, UK. (Courtesy of Chetham's Library, Manchester.)

Fig. 3.8 The Monkey House, Belle Vue Zoological Gardens, Manchester, UK. (Courtesy of Chetham's Library, Manchester.)

by the need to have a conservation focus. Bristol Zoo is planning a 55 ha *National Wildlife Conservation Park* which will be built separate from its current site. This will be the first conservation-led animal attraction in the UK. The intention is to link immersive exhibits of threatened species from around the world with *in-situ* conservation projects. The exhibits will include the *Congo Tropical Forest*, *Sumatra Rainforest*, *Indian Ocean Reef* and *Tanzania Savannah*. There will also be a *British Ancient Woodland* exhibit which will contain species that became extinct in Britain hundreds of years ago, including brown bears, lynx and wolves. Education facilities will be fully integrated into the project, with purpose-built classrooms located in each exhibit. Ranger stations will be manned by rangers who will act as tour guides. Sustainability principles will be built

Fig. 3.9 Blackpool Tower, UK. The building at the base of the tower once housed a zoo and an aquarium.

into every aspect of the development and operation of the park. It is rare for a large new zoo to be built from scratch, but developments like this clearly present an opportunity to construct a new type of modern conservation park for the future.

3.2.1 Safari parks

Academic works on the history of zoos tend to overlook the development of 'safari parks', which, although primarily developed for entertainment, nevertheless created a novel alternative to traditional zoos.

Africa USA (1953–61) was the first drive-through safari-style park in America. This 'cageless theme park' was created on a 121 ha (300 acre) plot of land in Florida purchased from the City of Boca Raton and Palm Beach County Commission in 1950. Visitors explored the park in a jeep safari train pulling open carriages and in boats. The landscape included artificial lakes and canals, a ('Zambezi') waterfall, a geyser and an African village. Some 55,000 plants were added to the site to give the appearance of a jungle. A large number of animals were imported from British East Africa (now Kenya) and others were purchased from zoos. They included giraffe, zebra, Asian elephants (rescued from circuses), wildebeest, cheetah, camels, ostrich, eland, chimpanzees, baboons, cranes, gazelles and many other species.

Jimmy Chipperfield (a circus owner) was largely responsible for the concept of safari parks in the UK, where visitors drive through enclosures in their own vehicles. In 1966 he established the first safari park in Britain at Longleat, in a partnership with Lord Bath. This was followed by Woburn Safari Park (1970), Blair Drummond Safari Park in Scotland (1970), Knowsley Safari Park (1971), West Midlands Safari and Leisure Park (1973) and Windsor Safari Park (1969, closed 1992). The Highland Wildlife Park is a safari park and zoo which specialises in keeping past and present Scottish wildlife. It was opened in 1972 and has been run by the Royal Zoological Society of Scotland since 1986. The safari park concept was popular in the decade between 1966 and 1975 and many other parks opened in Europe and North America, including the San Diego Wild Animal Park which opened in 1972.

3.2.2 Aquariums, oceanariums and marine parks

Aquariums are known from Roman times and a number of ancient civilisations kept fish in ponds as food. The 'Fish House' at London Zoo was the first aquarium – 'aquatic vivarium' – in the world. It was established in 1853 after two men approached the zoo for advice on keeping tropical fish in tanks. The Fish House contained over 300 types of fishes and marine invertebrates. Three years later, in 1856, the Division of Fishes was established at the Smithsonian's National Museum of Natural History in the United States. The New York Aquarium was opened in 1896 and the famous Shedd Aquarium in Chicago opened in 1930. By the early part of the 20th century most of the major cities in America and other countries in the developed world had their own aquarium. By the late 19th century there were major aquariums in Paris, Hamburg, Naples and many other major cities in Europe. In the UK the Victorian era saw the building of aquariums in Manchester, London and many seaside towns and cities such as Blackpool, Yarmouth, Southport and

Box 3.3 Carl Hagenbeck (1844–1913) and his Tierpark.

Carl Hagenbeck was a German animal trader and trainer who made a very significant contribution to the advancement of the design of enclosures and the exhibition of animals in zoos.

Originally, Hagenbeck owned a travelling exhibition in which he displayed animals alongside people from different parts of the world, including Lapps, Nubians and Eskimos (Inuit), often with their traditional homes and domestic animals. This was extremely popular and on 6 October 1878 around 62,000 people visited Berlin Zoo to see Hagenbeck's exhibition (Cherfas, 1984).

In 1907 Hagenbeck founded his own zoo at Stellingen, near Hamburg. The zoo contained cleverly designed exhibits which appeared to house carnivores and herbivores together, but in reality they were separated by hidden moats. Prior to building these moated enclosures Hagenbeck investigated the jumping abilities of the animals. As well as running his own zoo, he also supplied many well-known zoos with animals.

Hagenbeck's zoo was destroyed by bombing in 1943, but was rebuilt after the end of the Second World War. Edinburgh Zoo opened in 1913 and used many of Hagenbeck's innovations, as have many other zoos since.

Fig. 3.10 Carl Hagenbeck. (Courtesy of Library of Congress Prints and Photographs Division, Washington DC.)

Scarborough. As a consequence of its traditional links with the sea, Japan has a particularly impressive collection of aquariums.

A relatively new concept in the history of aquariums is the development of the 'oceanarium', 'marine park' and 'dolphinarium'. The world's first oceanarium was built at Marineland in Florida and opened in 1938. The tanks were seascaped and contained coral reefs and a shipwreck. This was the first attempt to capture large sea animals, particularly dolphins, and sustain them in captivity. Marineland was a popular tourist destination and was used as a set for a number of

Box 3.4 Zoos and racism.

The World's Fair 1904 was held at St Louis, Missouri and included exhibits from over 40 countries. Amongst the exhibits were some displaying the native inhabitants of different countries, including a group of pygmies one of whom was a young man called Ota Benga.

On 9 September 1906 *The New York Times* carried an article entitled 'Bushman shares a cage with Bronx Park apes' (Anon., 1906). It described a bizarre event whereby Ota Benga was exhibited alongside the primates at the Bronx Zoo. The director at the time was William Hornaday who helped to found the Smithsonian National Zoo in Washington.

In the early 20th century the German animal trader Carl Hagenbeck exhibited people alongside animals. He displayed Samoan and Sami people (Laplanders) alongside their tents, weapons and sleds with a group of reindeer. Between 1852 and 1969 Belle Vue Zoo in Manchester, England, held spectacular themed fireworks displays. The theme changed annually, and almost all were re-enactments of historical events such as 'The Burning of Moscow' and 'The Storming of Quebec' (Nicholls, 1992). However, in 1925 the display was entitled 'Cannibals' and featured black Africans depicted as savages (Fig. 3.11).

Augsburg Zoo, in Germany, held a four-day African Festival in 2005, featuring craft sellers, drummers, music groups, storytellers and traditional food from around Africa. Representatives of Germany's black community and some academics claimed that setting the festival in a zoo was racist and implied that non-whites were not really part of German society (Anon., 2005).

In July 2009 Dresden Zoo, in Germany, named a baby mandrill 'Obama' to commemorate the visit of the newly elected US president to the city. When challenged by a black advocacy group zoo staff claimed to be unaware of the racist tradition of using primates to caricature black people. The mandrill was renamed 'Okeke' (Anon., 2009b).

In August 2005, ZSL London Zoo exhibited a group of eight human volunteers in the *Bear Mountain* enclosure. They were treated like animals and kept amused with music, games and art. The purpose of the Human Zoo was to demonstrate the basic nature of man and to draw attention to man's impact on the rest of the Animal Kingdom.

Fig. 3.11 The 'Cannibals', Belle Vue, Manchester, 1925. (Courtesy of Chetham's Library, Manchester.)

films. The oceanarium still exists but much of the original structure and the original large tanks have been demolished.

There has been a resurgence of interest in aquariums in the last 25 years and a number of large new facilities have been opened including *Deep Sea World* in Fife, Scotland (1983), the *Blue Planet* at Ellesmere Port in Cheshire (1998), and *The Deep* in Hull (2002). The National Marine Aquarium in Plymouth, which is Britain's largest and Europe's deepest aquarium, was the first in the UK to be set up solely for the purpose of education, conservation and research. The first public aquarium in New Zealand opened in Napier as late as 1956, and in 2002 the country's National Aquarium opened here. *Underwater World* on Sentosa Island, Singapore, is a large oceanarium which was opened in 1991 and contains 2500 animals from 250 species. Visitors to *Underwater World* can dive with dolphins, sharks and dugongs (*Dugong dugon*).

3.2.3 Farm zoos

In some parts of Africa farmers have found that traditional farming is no longer viable and have established private nature reserves by purchasing big game species. Visitors pay to stay in tented camps or luxury lodges and view the animals from vehicles. Some of these areas are fenced and could be considered to be little more than large zoos containing native species.

In some developed countries a similar process has occurred. In the UK a number of farmers have established farm zoos in recent years. This has been at least partly a response to the damage done to farms after outbreaks of bovine spongiform encephalopathy (BSE) and foot and mouth disease, which resulted in large numbers of animals being destroyed.

Some farm zoos are still essentially working farms with a few exotic species. However, others have evolved into small zoos in their own right and keep a range of large exotics. Noah's Ark Farm Zoo, near Bristol, keeps over 85 species including tigers, rhinos, giraffes, zebra, tapirs, gibbons, camels and reptiles. The zoo opened in 1998, but for the 30 years prior to 1995 the site was operated as a dairy farm.

In the list of zoos in England published by the Department for Environment, Food and Rural Affairs (DEFRA) in 2007 there were 43 organisations listed as 'farm parks'. Some of these specialise in particular species that are of economic interest, such as wild boars and ostrich, both of which are used for meat.

1828

3.2.4 Specialist collections

Zoos traditionally contain a wide range of species. However, some small zoos specialise in particular taxa. Aviaries and falconry have a long history, but in more recent times, specialist birds of prey centres have been created where visitors can see displays of free-flying raptors. Some traditional zoos also include such displays in their summer visitor programmes. Some examples of other taxa which are the subject of specialist facilities are:

- otters (Chestnut Centre Wildlife Park, Derbyshire, UK)
- butterflies (*Butterfly World*, Hertfordshire, UK)
- insects (*Bugworld*, Liverpool, UK)
- Barbary macaques (e.g. *Monkey Forest*, UK).

3.2.5 Animal sanctuaries

Animal sanctuaries are places that keep animals that have been abandoned or rescued from zoos or unsuitable owners. Some contain a mixture of unwanted pets and native species that are recovering from injury. Others specialise in animals from particular taxa – for example chimps or bears – that have previously been kept by circuses or zoos, or have been confiscated by the authorities. Many of these animals live out their lives in the sanctuary, but sometimes sanctuaries are able to return animals to the wild. In June 2009, a group of bonobos cared for by staff at a sanctuary run by *Les Amis des Bonobos* were returned to the wild in a remote corner of the Democratic Republic of Congo.

Monkey World Ape Rescue Centre, Dorset, UK This sanctuary is home to over 250 primates from 15 species, including 60 chimpanzees. They have been rescued from a number of European and other countries where they were being used in laboratories, as pets, in circuses or as photographers' props. The chimps are managed in large social groups but the females receive contraception.

Elephant sanctuaries in the United States The Elephant Sanctuary® in Tennessee is America's 'natural habitat refuge' for African and Asian elephants. It occupies 1093 ha (2700 acres) of land in Hohenwald. Since 1995 the sanctuary has taken in 24 elephants that have retired from zoos and circuses. Construction costs to date (July 2009) have exceeded $10.7 million.

Riddle's Elephant and Wildlife Sanctuary is a smaller facility of around 134 ha (330 acres) in Arkansas which currently holds 12 elephants. It hosts an annual International School for Elephant Management. Elephant sanctuaries also exist in the range states of the species, for example in Thailand and South Africa.

Bear sanctuary, Brasov, Romania
This is a sanctuary for rescued bears set in 69 ha (170 acres) of Romanian woodland. The project is funded by the World Society for the Protection of Animals (WSPA). There are currently 34 bears at the sanctuary which have been rescued from roadside attractions and zoos. WSPA also operates the Kund Park Bear Sanctuary in Pakistan.

3.3 HOW MANY ZOOS?

It would be quite impossible to count all of the zoos in the world. Some countries do not license their zoos so official records may not exist. Even in those countries where a licensing system is operated, some small collections may not require a licence. Most of the world's leading zoos and aquariums are members of the World Association of Zoos and Aquariums (WAZA). The WAZA Network comprises some 1300 institutions, of which around 300 zoos and aquariums are institutional members. The membership of regional zoo organisations gives some indication of the distribution of the world's major zoos (Table 4.1).

In November 2007 there were 270 zoos in England listed by DEFRA (Fig. 3.12), an increase of 15 on the previous year. They included traditional zoological gardens, safari parks, birds of prey centres, farm parks, aviaries, aquariums, butterfly houses, wildlife rescue centres, museums, monkey, parrot and seal sanctuaries, rare breeds farms and other specialist collections. Two colleges have collections which are used for teaching purposes.

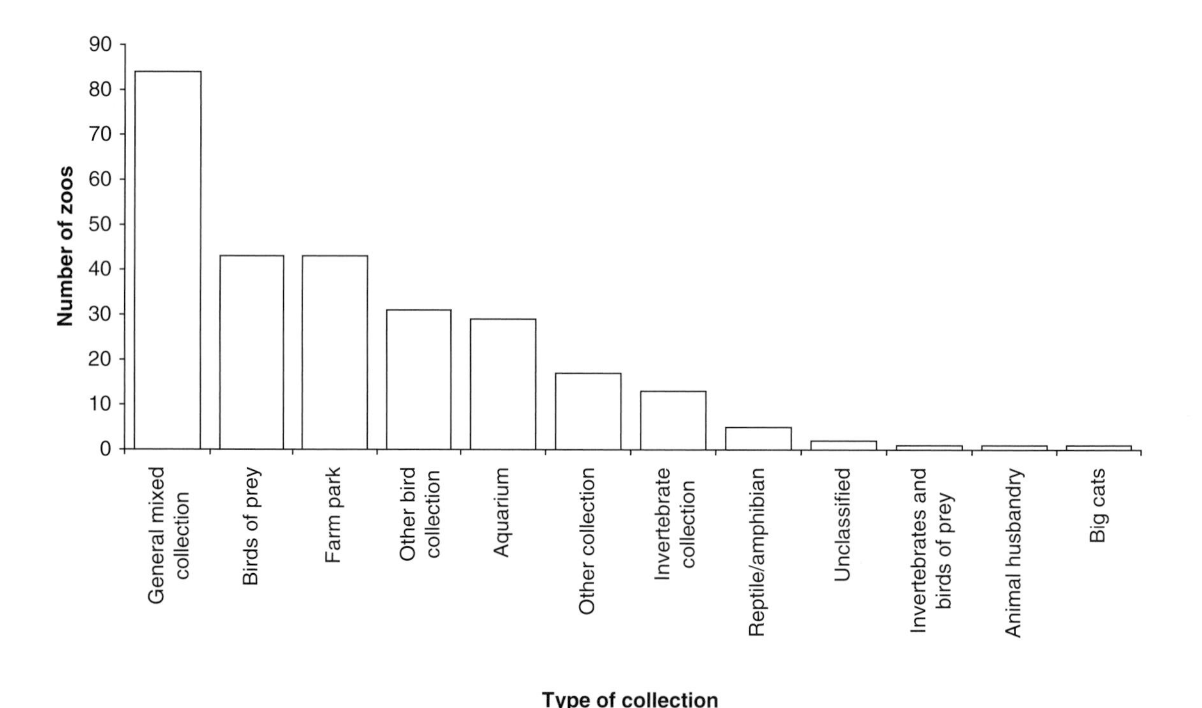

Fig. 3.12 Zoos in England (November 2007). (Based on data from DEFRA.)

3.4 WHERE ARE THE WORLD'S ZOOS?

Most of the major zoos in the world are located in the northern hemisphere in North America and Europe. Other important zoos exist in Australasia, Japan, China, South East Asia, South America and elsewhere.

Most large zoos are located within, or very near, urban areas. Indeed, many began as prestigious civic projects. A location near a large human population is essential if a zoo is to attract sufficient visitors to survive financially. For many zoos, this confines their activities to a relatively small area, perhaps within a walled compound (e.g. Bristol Zoo) or within a park (e.g. Dublin Zoo, ZSL London Zoo). However, zoos vary greatly in size from very small specialist collections – such as birds of prey centres or small aquariums – to extensive safari parks on large estates (Fig. 7.1). Obviously, the size and location of a zoo is likely to affect its potential as a conservation resource.

3.5 ZOO CLOSURES

Zoos are not like museums or art galleries. Their collections need actively to be maintained on a daily basis and their animals must be replaced when they die. Consequently, many zoos have disappeared when their owners, or their heirs, have lost interest, or when they have failed financially. Many of the major cities in the UK no longer have large zoos. The Liverpool Zoological Gardens Co. Ltd. went bankrupt in 1863, Belle Vue Zoological Gardens closed in 1979 (after more than 140 years), and in 1991 London Zoo almost closed because of a financial crisis.

3.6 FURTHER READING

Baratay, E. and Hardouin-Fugier, E. (2002). *Zoo: A History of Zoological Gardens in the West*. Reaktion Books Ltd., London.

Barrington-Johnson, J. (2005). *The Zoo: The Story of London Zoo*. Robert Hale Limited, London.

Bostock, S. St.C. (1993). *Zoos and Animal Rights: The Ethics of Keeping Animals*. Routledge, London and New York.

Cherfas, J. (1984). *Zoo 2000: A Look Beyond the Bars*. BBC, London.

Hoage, R. J. and Deiss, W. A. (eds.) (1996). *New Worlds, New Animals: From Menagerie to Zoological Park in the Nineteenth Century*. The Johns Hopkins University Press, Baltimore, MD and London.

3.7 EXERCISES

1. In what respects does a safari park differ from a traditional zoo?
2. Write a short account of the history of zoos.
3. With reference to a range of zoos from different periods in history, discuss the reasons why people build zoos.
4. Discuss the advantages and disadvantages of building a zoo and a theme park or fun fair within the same grounds.
5. Why do zoos disappear?

4 ZOO ORGANISATION AND MANAGEMENT

> Twenty six years after my coronation, I declared that the following animals were not to be killed: parrots, mynas, the aruna, ruddy geese, wild geese, the nandimukha, cranes, bats, queen ants, terrapins, boneless fish, rhinoceroses ... and all quadrupeds which are not useful or edible...
>
> Emperor Ashoka of India (in a decree issued in the 3rd century BC)

Conservation Status Profile

Crested (red-legged) seriema
Cariama cristata
IUCN status: Least concern
CITES: Not listed
Population trend: Unknown

An Introduction to Zoo Biology and Management, First Edition. Paul A. Rees.

4.1 INTRODUCTION

The casual zoo visitor may be forgiven for thinking that each zoo acts alone and is a completely independent entity. However, modern zoos would not be able to function without cooperating with each other. To this end, the zoo community is organised into a large number of associations at a global, regional and national level. At an institutional level, zoos have become increasingly complex as their role has evolved from being largely one of entertainment into diverse educational, research and conservation organisations. This has required zoos to adopt a management structure that reflects this new complexity.

4.2 THE INTERNATIONAL, REGIONAL AND NATIONAL ORGANISATION OF ZOOS

Zoos do not function in isolation but are organised into associations at national, subregional, regional and international levels (Table 4.1). This allows them to share best practice in animal husbandry, education, research and other aspects of their work and it provides a mechanism by which the interests of zoos can be represented to governments and international organisations. In addition, membership of an association provides a means by which minimum standards may be imposed in areas such as animal welfare.

4.2.1 World Association of Zoos and Aquariums (WAZA)

The World Association of Zoos and Aquariums is an umbrella organisation for the world zoo and aquarium community. Its members include leading zoos and aquariums, regional and national associations of zoos and aquariums and individual zoo professionals from around the world.

The organisation began as the International Union of Directors of Zoological Gardens (IUDZG) in 1946. In 1950 the IUDZG became an international organisation member of the International Union for the Conservation of Nature and Natural Resources (IUCN). In 2000 the IUDZG was renamed WAZA. Since 2001 its Executive Office has been located in the IUCN Conservation Centre in Gland, Switzerland.

WAZA's vision is that:

> The full conservation potential of world zoos and aquariums is realized.

Its mission is:

> WAZA is the voice of a worldwide community of zoos and aquariums and catalyst for their joint conservation action.

WAZA promotes cooperation between its member institutions with regard to conservation and the management and breeding of animals in captivity, and encourages the highest standards of animal welfare. It also represents zoos and aquariums in other international organisations, and promotes environmental education, wildlife conservation and research.

The international studbooks for rare and endangered species are kept under the auspices of WAZA. In addition WAZA has produced a world conservation strategy for zoos and aquariums.

4.2.1.1 World Zoo and Aquarium Conservation Strategy

The World Zoo Conservation Strategy was first published by the IUDZG in 1993 (IUDZG, 1993). This was superseded in 2005 by *Building a Future for Wildlife:*

Table 4.1 Membership of major zoo organisations.

Organisation	Number of institutional members	Number of countries/territories
World Association of Zoos and Aquariums (WAZA)	*c.*300	48 (worldwide)
European Association of Zoos and Aquaria (EAZA)	> 300 institutions	35 (Europe)
Association of Zoos and Aquariums (AZA)	221 zoos and aquariums + 17 'Certified Related Facilities'	USA, Canada, Mexico, Caribbean
Zoo and Aquarium Association (ZAA, formerly ARAZPA)	> 70 zoos and aquariums	Australia, New Zealand and South Pacific
African Association of Zoos and Aquaria (PAAZAB)	65 institutions	10 (Africa)
South East Asian Zoo Association (SEAZA)	30 institutions	—

The World Zoo and Aquarium Conservation Strategy (Olney, 2005). The purpose of the document is to provide a common set of goals for the zoo community and set out best practice in zoos in an attempt to allay the fears of those who are uncertain about the role of zoos and concerned about animal welfare.

The strategy addresses a wide range of issues:

- Integrating the work of zoos with conservation efforts in the wild
- The conservation of wild populations
- Science and research
- Population management
- Education and training
- Marketing and public relations
- Partnerships and politics
- Sustainability
- Ethics and animal welfare.

The document produces a very wide-ranging series of recommendations for zoos to help them improve their performance in these areas.

In 2009 WAZA published *Turning the Tide: A Global Aquarium Strategy for Conservation and Sustainability* (Penning *et al.*, 2009). The purpose of this document was to assist in the implementation of the World Zoo and Aquarium Conservation Strategy within the aquarium community.

4.2.2 European Association of Zoos and Aquaria (EAZA)

The European Association of Zoos and Aquaria has over 300 member institutions in 35 countries. It was founded in 1992 and its mission is to facilitate cooperation within the European zoo and aquarium community towards the goals of education, research and conservation. EAZA's Executive Office is located in Artis Zoo in Amsterdam. Each year since 2000 EAZA has organised a conservation campaign focusing on a particular issue or taxon. These campaigns increase cooperation between member institutions and also give smaller zoos an opportunity to take part in a project which they could not organise alone. Previous campaigns have focused on bushmeat, rainforest, tigers, rhinos, Madagascar, amphibians, and European carnivores.

EAZA coordinates European Endangered Species Programmes (EEPs) and European Studbooks (ESBs) which are run by its members. It also maintains the World Zoo and Aquarium Conservation Database – supported by Conservation Breeding Specialist Group Europe – to increase the number of conservation projects run and supported by EAZA members. The EAZA Academy runs training courses for its members in breeding programme management. The EAZA Nutrition Group promotes and supports zoos in improving nutrition in their institutions.

European Union of Aquarium Curators (EUAC) This organisation was formed in 1972 with the objective of organising regular symposia on topical issues, facilitating the exchange of specimens and improving communication between aquariums.

4.2.3 British and Irish Association of Zoos and Aquariums (BIAZA)

The British and Irish Association of Zoos and Aquariums (formerly the Federation of Zoological Gardens of Great Britain and Ireland) is an association for zoos and aquariums in the UK and Ireland. Its offices are in Regent's Park, London. BIAZA leads and supports its members in their conservation, education and research initiatives. It works closely with EAZA in its conservation campaigns.

BIAZA has a complex structure of groups and committees:

- Membership and Licensing Committee
- Conservation and Animal Management Committee
- Joint Working Groups Committee (representing groups concerned with Mammals, Birds, Reptiles and Amphibians, Terrestrial Invertebrates, Native Species, Plants and Aquariums)
- Veterinary Advisory Group
- Research Core Group
- Records Group
- Education and Training Committee
- Communications and Development Committee.

4.2.4 Association of Zoos and Aquariums (AZA) – USA

The Association of Zoos and Aquariums was founded in the United States in 1924 as a non-profit organisation dedicated to the advancement of zoos and aquariums in the areas of conservation, education, science and recreation. Some 20 years ago it established the Species Survival Plan Program to manage and conserve selected threatened or endangered species. The AZA is based in Silver Spring, Maryland.

Biography 4.1

William G. Conway (1929–)

Dr William Conway is a zoologist, ornithologist and conservationist who has played an important part in promoting the development of captive breeding programmes for endangered species. He was formerly Director of the New York Zoological Society and was responsible for modernising many of the exhibits at the Bronx Zoo. He later became President of the Wildlife Conservation Society. Conway led the development of the accreditation programme for the Association of Zoos and Aquariums and has written extensively on zoos. He is currently a Senior Conservationist with the Wildlife Conservation Society.

4.2.5 Zoo and Aquarium Association (ZAA) (formerly the Australasian Regional Association of Zoological Parks and Aquaria, ARAZPA)

The Association was established in 1990 as ARAZPA and is based in Mosman, New South Wales, Australia. The Association's species management arm is the Australasian Species Management Program (ASMP) and its 'Wildlife Conservation Fund' provides funding to *in-situ* conservation projects. The ZAA manages a network of over 100 environmental educators and more than 500 zoo and aquarium specialists, including vets and wildlife researchers.

4.2.6 African Association of Zoos and Aquaria (PAAZAB)

The African Association of Zoos and Aquaria was formed in 1989 at the National Zoological Gardens of South Africa, Pretoria, South Africa. It represents the interests of zoos and aquariums in African countries. In 1991 PAAZAB established a working group called the African Preservation Programme (APP) as a framework for cooperative breeding within the African region.

4.2.7 South East Asian Zoo Association (SEAZA)

The South East Asian Zoo Association is the regional zoo and aquarium association for institutions in South East Asia. It is primarily composed of members in Hong Kong, Indonesia, Kuwait, Malaysia, the Philippines, Vietnam, Taiwan and Thailand.

The South Asian Zoo Association for Regional Co-operation (SAZARC) was formed in 2000.

4.3 KEEPER ORGANISATIONS

A number of keeper organisations exist. Some are focused on particular taxa, for example elephants, while others are regional organisations for keepers regardless of their speciality.

4.3.1 International Congress of Zookeepers (ICZ)

The International Congress of Zookeepers is a global network of zoo keepers and other professionals in the field of wildlife care and conservation. It exists so that members may exchange their experience and knowledge and held its first conference in 2003. The ICZ assists with the development of individual keepers, particularly in developing countries, and helps to create new national keeper associations.

4.3.2 Association of British and Irish Wild Animal Keepers (ABWAK)

The Association of British and Irish Wild Animal Keepers was founded in 1974. Its members are interested in the keeping and conservation of wild animals and it seeks to achieve the highest possible standards in animal welfare. ABWAK is involved in the training, education and development of keepers and publishes a journal called *Ratel*. Its vision is:

> To unite the collective experience of wild animal keepers to continually advance the highest standards of animal care.
>
> ABWAK (2010)

4.3.3 American Association of Zoo Keepers (AAZK)

The American Association of Zoo Keepers is a non-profit volunteer organisation made up of professional zoo keepers and others dedicated to professional animal care and conservation. It was founded in 1967.

AAZK's mission is:

> To advance excellence in the animal keeping profession, foster effective communication beneficial to animal care, support deserving conservation projects, and promote the preservation of our natural resources and animal life.
>
> AAZK (2010)

The AAZK publishes *Animal Keepers' Forum* and other resources on CD-ROM.

4.3.4 Australian Society of Zoo Keeping (ASZK)

The Australian Society of Zoo Keeping was formed by eight keepers from Adelaide Zoo in 1976 as the Australasian Society of Animal Management. Its name was changed to the ASZK in 1980. The Society seeks to promote the exchange of information on all aspects of wild animal husbandry and in so doing contribute to the conservation of rare and endangered species. In pursuance of these objectives it produces journals and newsletters and organises workshops and conferences throughout the Australasian region.

4.3.5 Animal Keepers Association of Africa (AKAA)

The Animal Keepers Association of Africa describes itself as an African network for ethical and competent animal keepers.

The AKAA's core values are:

- Networking – To provide a forum to ensure productive and constructive interfacing, equity and communication between individuals in the field of animal care.
- Welfare – To provide the best care and most ethical animal management practices for animals in human care.
- Knowledge – To generate information sharing amongst the membership so that the industry as a whole remains professional and progressive.
- Service excellence – To exhibit animals in an ethical, educational and inspiring manner that serves to preserve their dignity and inspire the public to holistic conservation action and informed animal welfare.
- Conservation – To ethically and responsibly care for animals in our collections in a manner that serves and promotes conservation (AKAA, 2010).

4.3.6 European Elephant Keeper and Manager Association (EEKMA)

The aims of the European Elephant Keeper and Manager Association are:

- To create a platform for elephant keepers and represent their interests
- To improve the husbandry and breeding of elephants
- To train and educate elephant keepers
- To promote research on elephants.

The Association publishes a newsletter called *Elephant Journal*.

4.3.7 Elephant Managers Association (EMA)

The Elephant Managers Association is an international non-profit organisation of professional elephant handlers, administrators, veterinarians, researchers and elephant enthusiasts. It is dedicated to the welfare of elephants through improved conservation, husbandry, research, education and communication.

4.4 INTERNATIONAL UNION FOR THE CONSERVATION OF NATURE AND NATURAL RESOURCES (IUCN)

The International Union for the Conservation of Nature and Natural Resources was founded in 1948, originally as the International Union for the Protection of Nature (IUPN), but changed its name to the IUCN in 1956. From 1990 the organisation came to be known as the World Conservation Union, but this name is no longer commonly used. The IUCN is a partnership of states, government agencies and non-governmental organisations (NGOs) of over 1000 members and almost 11,000 volunteer scientists spread across more than 160 countries. The IUCN seeks to assist in the conservation of biological diversity and ensure the responsible and equitable use of the world's natural resources. It has Official Observer Status at the UN

Box 4.1 IUCN Red List categories (based on IUCN, 2001).

The International Union for the Conservation of Nature and Natural Resources (IUCN) produces a 'Red List' of species which places each into a category based on its conservation status, ranging from 'Extinct' to 'Least Concern', with some taxa listed as 'Data Deficient' because too little is known about their wild populations (Table 4.2). The status of a taxon may fluctuate with time and so it may move from one category to another (Table 4.3). Red List criteria can be applied to any taxonomic unit at or below the level of species.

Table 4.2 Summary of IUCN Red List categories and their definitions.

<table>
<tr><td rowspan="8">Evaluated</td><td rowspan="7">Adequate data</td><td colspan="2">Extinct (EX)
Exhaustive surveys throughout the historic range of the taxon have failed to record a single individual. No reasonable doubt that the last individual has died.</td></tr>
<tr><td colspan="2">Extinct in the Wild (EW)
Known only to survive in cultivation, captivity or as naturalised population(s). Exhaustive surveys throughout the historic range of the taxon have failed to record a single individual.</td></tr>
<tr><td rowspan="3"><u>Threatened</u></td><td>Critically Endangered (CR)
Best available evidence indicates that the taxon is facing an extremely high risk of extinction in the wild.</td></tr>
<tr><td>Endangered (EN)
Best available evidence indicates that the taxon is facing a very high risk of extinction in the wild.</td></tr>
<tr><td>Vulnerable (VU)
Best available evidence indicates that the taxon is facing a high risk of extinction in the wild.</td></tr>
<tr><td colspan="2">Near Threatened (NT)
The taxon does not qualify for Critically Endangered, Endangered or Vulnerable now, but is likely to qualify for a threatened category in the near future.</td></tr>
<tr><td colspan="2">Least Concern (LC)
The taxon does not qualify for Critically Endangered, Endangered, Vulnerable or Near Threatened. Widespread and abundant taxa are included in this category.</td></tr>
<tr><td colspan="3">Data Deficient (DD)
Inadequate data available to make a direct or indirect assessment of the risk of extinction of the taxon, based on its distribution and/or its abundance. A taxon in this category may be well studied and its biology well known. DD is not a category of threat. More information is needed. Future research may indicate that a threatened category is appropriate.</td></tr>
<tr><td colspan="4">Not Evaluated (NE)
The taxon has not yet been evaluated against the criteria.</td></tr>
</table>

There is a hierarchical alphanumeric numbering system of criteria and subcriteria under the categories CR, EN and VU which indicates the reason for the classification, such as declining numbers or a reduced geographical range. The system is extremely complex. For example, a species may be categorised as EN B1ab(v); D. This means that the species is Endangered (EN) due to:

- B – geographical range
- 1 – extent of occurrence estimated to be less than 5000 km^2
- a – severely fragmented or known to exist at no more than five locations
- b – continuing decline, observed, inferred or projected in (v) the number of mature individuals
- D – population size estimated to number fewer than 250 mature individuals.

Table 4.3 Changes in status of selected species between 2007 and 2008.

Scientific name	Vernacular name	Status 2007	Change	Status 2008
Mammals				
Loxodonta africana	African elephant	VU	▲	NT
Castor fiber	Eurasian beaver	NT	▲	LC
Prionailurus viverrinus	Fishing cat	VU	▼	EN
Pteropus niger	Mauritian flying fox	VU	▼	EN
Birds				
Apteryx owenii	Little spotted kiwi	VU	▲	NT
Sylvia undata	Dartford warbler	LC	▼	NT
Reptiles				
Crocodylus rhombifer	Cuban crocodile	EN	▼	CR
Amphibians				
Incilius holdridgei	N/A	CR	▼	EX
Invertebrates				
Hemiphlebia mirabilis	Ancient greenling	VU	▼	EN

▲, status improved; ▼, status deteriorated.

Further information about this system may be found in *IUCN Red List Categories and Criteria, Version 3.1* (IUCN, 2001). This is available from the IUCN website at www.iucn.org

General Assembly, and its headquarters are in Gland, near Geneva, Switzerland.

The IUCN's mission is:

> ...a just world that values and conserves nature.

The IUCN defines its role as:

- Knowledge – Developing and supporting conservation science, particularly on biodiversity and ecosystems and how they link to human well-being.
- Action – Running thousands of field projects around the world.
- Influence – Supporting governments, NGOs, international conventions, UN organisations, companies and communities to develop laws, policy and best practice.
- Empowerment – Helping to implement laws, policy and best practice by mobilising organisations, providing resources and training, and monitoring results.

4.4.1 Species Survival Commission (SSC)

The Species Survival Commission is the largest of IUCN's six volunteer Commissions, with a global membership of 7500 experts from almost every country in the world. The SSC advises IUCN and its members on the wide range of technical and scientific aspects of species conservation and is dedicated to securing a future for biodiversity.

Most members are deployed in more than 100 Specialist Groups and Task Forces which address conservation issues related to particular groups of animals or plants or topical issues such as reintroductions into the wild or wildlife health. The SSC produces a series of technical guidelines, for example:

- Guidelines for Reintroductions
- Management of *Ex-situ* Populations for Conservation
- Guidelines for the Prevention of Biodiversity Loss Caused by Alien Invasive Species
- IUCN Red List Categories and Criteria.

4.4.2 Red List categories

The IUCN produces a list of species – the Red List – based on their extinction risk. This is based on a series of eight categories: Extinct, Extinct in the Wild, Critically Endangered, Endangered, Vulnerable, Near Threatened, Least Concern and Data Deficient. A species is considered to be threatened if it is classified as Critically Endangered, Endangered or Vulnerable (see Box 4.1).

As of 2009 the IUCN had assessed 47,677 species (of animals and plants) and determined that 36% of these are threatened with extinction (GBO-3, 2010).

Many zoos indicate the IUCN Red List category of a species on signage on its enclosure. Of course, these categories are not fixed and may change from time to time as a species' position improves or deteriorates in the wild.

4.4.3 Conservation Breeding Specialist Group (CBSG)

The Conservation Breeding Specialist Group is a part of the Species Survival Commission of the IUCN and is supported by a non-profit organisation called the Global Conservation Network. The CBSG's mission is to save threatened species by increasing the effectiveness of conservation efforts worldwide through:

- innovative interdisciplinary methodologies
- culturally sensitive and respectful facilitation
- empowering global partnerships and collaborations.

The CBSG was founded in 1979, and has grown into a global volunteer network of 550 professionals. The CBSG began as a liaison between IUCN and the zoo community, and was instrumental in developing the tools and processes for the scientific management of captive animal populations. It has expanded its scope to small population management and the linking of scientific expertise in *in-situ* and *ex-situ* conservation. The CBSG has a small headquarters staff based in Minnesota, assisted by eight Regional and National Networks on five continents.

4.4.4 IUCN Technical Guidelines on the Management of *Ex-situ* Populations for Conservation

These guidelines are produced by the Species Survival Commission. The IUCN's vision in relation to *ex-situ* populations is:

> To maintain present biodiversity levels through all available and effective means including, where appropriate, *ex situ* propagation, translocation and other *ex situ* methodologies.

Its goal is that:

> Those responsible for managing *ex situ* plant and animal populations and facilities will use all resources and means at their disposal to maximise the conservation and utilitarian values of these populations, including:
>
> - increasing public and political awareness and understanding of important conservation issues and the significance of extinction
> - co-ordinated genetic and demographic population management of threatened taxa
> - re-introduction and support to wild populations
> - habitat restoration and management
> - long-term gene and biomaterial banking
> - institutional strengthening and professional capacity building
> - appropriate benefit sharing
> - research on biological and ecological questions relevant to *in situ* conservation, and
> - fundraising to support all of the above.
>
> IUCN (2002)

4.5 WILDLIFE NGOs

A very wide variety of NGOs are involved in the conservation of wildlife. Most of them are concerned primarily with protecting animals living wild and their habitats. Some are concerned with animal welfare and cruelty (see Section 6.2). Four examples are given below to give an indication of the work they are involved in.

4.5.1 World Wide Fund for Nature/World Wildlife Fund (WWF)

The World Wildlife Fund was founded in 1961 as an international fund-raising organisation that would work in collaboration with existing conservation groups to bring substantial financial support to worldwide conservation efforts, using the best scientific advice available from the IUCN and others. The WWF has since evolved into a worldwide network of 30 national organisations. Since 1985 the WWF network has invested over $1.165 billion in more than 11,000 projects in 130 countries.

Current WWF conservation projects are concentrating on 19 'priority places', that must be saved in the next 50 years. They include the Amazon, the Arctic, Borneo and Sumatra, the Congo Basin, Coastal East Africa, Galapagos, the Yangtze, Madagascar and the Mesoamerican reef in the Caribbean Sea.

In addition, the WWF concentrates its efforts on 'flagship species' that provide a focus for awareness-raising for broader conservation efforts in its priority places (e.g. polar bears), and 'footprint-impacted species' whose populations are declining because of unsustainable hunting, fishing or logging (e.g. many dolphin species).

4.5.2 Fauna and Flora International (FFI)

Fauna and Flora International is the oldest international conservation organisation in the world. It was founded as the Society for the Preservation of the Wild Fauna of the Empire in 1903. In 1950 it became the Fauna Preservation Society. In 1980 it changed its name to the Fauna and Flora Preservation Society, and in 1995 it became FFI. The organisation is perhaps best known for its role in helping to save the Arabian oryx (*Oryx leucoryx*) from extinction in the wild (see Section 15.6.6). Fauna and Flora International publishes the conservation journal *Oryx* and currently operates conservation programmes for a wide range of species including the jaguar (*Panthera onca*), the cebia tree (*Cebia pentandra*), the Iberian lynx (*Lynx pardinus*) and the pygmy hippo (*Choeropsis liberiensis*).

4.5.3 BirdLife International

BirdLife International is a global partnership of NGOs in over 100 countries and territories which works to conserve birds and their habitats. Each NGO partner represents a unique geographical area, including:

- Royal Society for the Protection of Birds (in the UK)
- Nature and Biodiversity Conservation Union (in Germany)
- Audubon (in the USA)
- Bird Studies Canada and Nature Canada
- BirdLife South Africa
- Wildlife Conservation Society of Tanzania
- Bombay Natural History Society (in India)
- Birds Australia
- Forest & Bird (in New Zealand).

BirdLife International is working on a number of international programmes including:

- Global Seabird Programme (protecting albatrosses and other seabirds)
- Flyways Programme (protecting migration routes)
- Forest of Hope Programme (protecting tropical forests).

4.5.4 Royal Society for the Protection of Birds (RSPB)

The Society for the Protection of Birds was founded in Manchester, England, in 1889 specifically to stop the slaughter of thousands of egrets, herons and birds of paradise each year for their plumes. Branches were set up overseas and the branch in India secured the first measure against the plumage trade: an order from the Indian government in 1902 that banned the export of bird skins and feathers. In 1904 the Society received a Royal Charter and became the Royal Society for the Protection of Birds. The Society raises funds for the conservation of birds and their habitats and manages 200 nature reserves in the UK covering 130,000 ha. It assists the police in the prosecution of people who commit crimes against birds and it is involved in programmes to reintroduce bird species in the UK, for example the Eurasian crane (*Grus grus*).

4.6 THE ORGANISATION AND MANAGEMENT OF A ZOO

4.6.1 Management structures

The management structure of a zoo will inevitably be determined largely by its size and complexity. A small zoo will have few staff and some may be required to undertake more than one role. For example, the education officer may also act as the conservation officer, and keepers may be expected to care for a wide range of species. In a large zoo staff are more likely to have specialised roles, with keepers working with a narrow range of taxa.

The staff in a large zoo may be allocated to departments or divisions within a complex organisational structure including those concerned with:

- Animals
- Conservation
- Research
- Education
- Horticulture
- Veterinary services
- Estates
- Administration
- Visitor services.

The organisational structure of the staff involved with conservation and education in a large zoo is

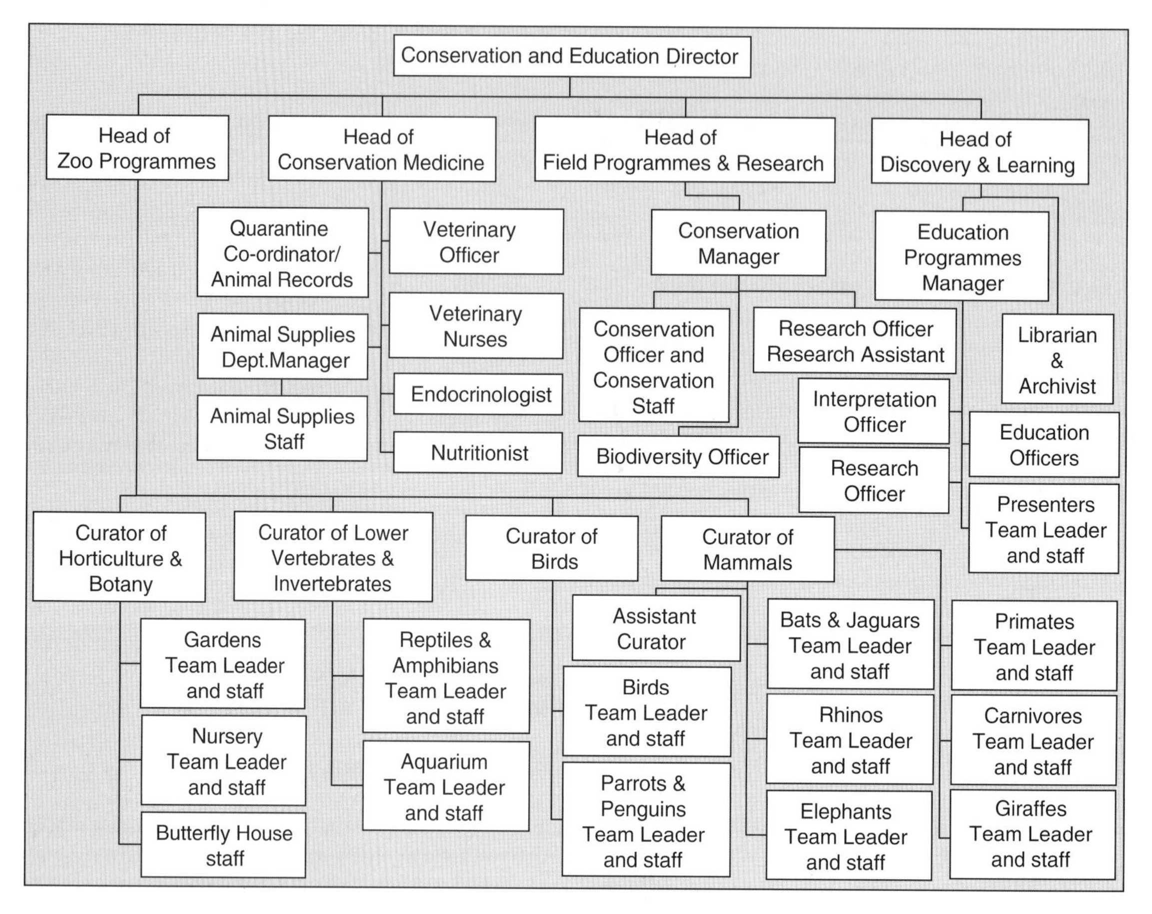

Fig. 4.1 The administrative structure of the Conservation and Education Directorate of Chester Zoo, UK. (Adapted from Chester Zoo Review 2009, Appendix 2.)

illustrated in Fig. 4.1. This structure is based on that developed by Chester Zoo. In addition to the Conservation and Education Directorate, this zoo also has a Corporate Services Directorate, concerned with finance, human resources and the zoo's estate, and a Commercial Services Directorate, concerned with visitor services, marketing and the zoo's retail and catering operations. These directorates operate under the guidance of the zoo's Director General who is also its Chief Executive. The zoo is owned by the North of England Zoological Society. The Society is a registered charity and its activities are controlled by its trustees.

In addition to their overall administrative structure, zoos may also operate a number of committees which oversee and direct the activities of the zoo. These might include committees concerned with:

- Research
- Enrichment
- Ethical review
- Environmental policy
- Conservation
- Education
- Planning
- Emergency response
- Zoo operations
- Animal health, welfare and husbandry
- Business operations
- Special events.

(a)

(b)

Fig. 4.2 A typical zoo keeper from (a) around 100 years ago and (b) 2010. A century ago most keepers were men and most wore a formal uniform. Nowadays, many keepers are young women and 'uniforms' are much less formal. (Courtesy of (a) Chetham's Library, Manchester and (b) Rebecca Dasan.)

4.6.2 Staffing

Zoos employ a wide range of staff from keepers (Fig. 4.2) to research scientists. In addition to their paid staff, many zoos also use volunteer staff, particularly in their educational activities during the summer.

Keepers in modern zoos are not only involved in basic animal husbandry, but also monitor the reproductive status of their animals, record their behaviour, design enrichment devices, take part in *in-situ* conservation projects and carry out a wide range of other tasks. *Durrell* describes its keepers as 'keeper-scientists', recognising this extended role. This is an important trend in zoo keeping, and zoos are increasingly employing graduate scientists to care for their animals.

4.6.3 Zoo economics

The economics of running a zoo will vary depending upon the type of zoo, its size, whether it is a charity or a business, and many other factors. Zoo income from visitors tends to be seasonal and very much influenced by the weather. A zoo with few indoor exhibits will see a drop in attendance figures during a wet winter, but an aquarium may see an increase during the same winter. In temperate latitudes the weather from Easter to the end of the school holidays in summer is critical. Other factors that might affect numbers are the birth of popular species such as an elephant or a polar bear, teachers' strikes – resulting in fewer school trips – or an outbreak of disease in the farming community, for example foot and mouth disease.

In the past, acquiring animals was a significant cost for zoos because they had to be bought from animal suppliers or other zoos. Captive breeding was not sufficiently well established for zoos to be able to replace their own animals or exchange their surplus animals with other zoos (see Section 12.1.1). However, in some cases zoos keep more individuals of a species than they need simply to attract visitors. A group of 20 Grevy's zebra (*Equus grevyi*) will not attract more visitors than a group of 10 but will cost more to feed and house. The popular species kept by zoos generate income and the surplus is used to help maintain the large groups of less popular species. This subsidisation process has been referred to as 'deficit financing' (Ironmonger, 1992).

Zoos generate income from a wide range of sources including:

- admission tickets
- annual membership fees
- trading (e.g. souvenirs, balloons, guide books, food, etc.)

- rides (e.g. monorails, boats, cable cars, miniature railways, etc.)
- special events (e.g. conferences, Christmas parties, etc.)
- donations, gifts and legacies
- investment income
- grants.

Zoo expenditure includes the cost of:

- staff (salaries, pensions, etc.)
- utilities (gas, water, electricity)
- investment management
- estates maintenance (buildings, roads, paths, signs, enclosures, etc.)
- gardens maintenance
- purchase of goods to be traded (food, drinks, souvenirs, etc.)
- animal food
- veterinary care
- research activities
- education services
- conservation programmes.

The Zoological Society of San Diego is a private non-profit corporation which operates San Diego Zoo, the San Diego Zoo's Wild Animal Park and the San Diego Zoo's Institute for Conservation Research (incorporating the Frozen Zoo®). The Society had an income of $200,264,000 in the fiscal year 2008 (ZSSD, 2008). The cost of its operations is shown in Table 4.4. Not surprisingly, 80% of the money spent by the Society was used to operate its two animal collections, with only a little more than 10% spent on research, conservation and education activities. This is not untypical of zoos, although the exact figures will obviously vary.

A summary of the financial statements of the Zoological Society of London (ZSL) for the year ended December 2008 is shown in Table 4.5. The Society is a charity but some of its commercial activities are carried out by wholly owned subsidiary companies.

Table 4.4 The cost of zoo operations, Zoological Society of San Diego.

Expenses	$000
Exhibition facility operations	149,567 (80.0%)
Research and conservation activities	17,046 (9.1%)
Education programmes	2,776 (1.5%)
Administration	17,497 (9.4%)
Total	186,886

Source: ZSSD (2008).

Table 4.5 Financial statements for the Zoological Society of London, 2008.

Income	£000
Voluntary income (gifts, donations, legacies)	2,727
Activities for generating income (subsidiaries' trading turnover, etc.)	9,789
Interest and investment income	842
Income resources from charitable activities	
Animal collections	20,053
Science and research	4,583
Conservation programmes	1,753
Other	539
Total income	40,286

Expenditure	£000
Cost of generating voluntary income	604
Fund-raising trading (cost of goods sold, etc.)	8,051
Investment management costs	28
Charitable activities	
Animal collections	22,220
Science and research	5,264
Conservation programmes	2,646
Governance costs	159
Other	615
Total expenditure	39,587

Income included over £5 million in grants. Expenditure included £18,154,000 in staff costs plus a further £250,000 for temporary staff.

The average full-time equivalent (FTE) numbers of staff working in each area were:

	FTEs	Percentage
Animal collections	371	60.1
Science and research	61	9.9
Conservation programmes	37	6.0
Cost of generating voluntary income	11	1.8
Cost of generating voluntary trading	80	13.0
Support costs	57	9.2
Total	617	100.0

Source for all these data: ZSL (2008).

The ZSL is comprised of ZSL London Zoo, ZSL Whipsnade Zoo and the Institute of Zoology. The Institute engages in research and *in-situ* conservation work, hence the large amount of expenditure in this area (see Box 14.1). Smaller zoos are unable to support these expensive activities so must make their contributions to conservation in other ways.

Utilities are a major cost for zoos especially those located in temperate areas that keep animals from tropical areas and have many heated houses. Dudley Zoo spent £150,000 on gas, water and electricity in 2008 which represented almost 6% of its income (DZG, 2009).

Some large zoos employ business development managers specifically to generate income for the zoo. Commercial sponsors may be persuaded to fund particular exhibits in high profile zoos. Jaguar cars paid £2 million for the construction of the *Spirit of the Jaguar* exhibit at Chester Zoo (see Section 14.5.2.2). The growing commercialisation of zoos in the United States has been discussed by Cain and Meritt (1998).

4.6.4 Manifesto for Zoos

The 'Manifesto for Zoos' is the result of a study by John Regan Associates Ltd. of the overall value of zoos to society in the UK (Regan, 2004). It was commissioned by a consortium of nine leading British zoos, facilitated by BIAZA, and aimed at persuading the government to work together with zoos on matters of mutual interest. The study examines the role of zoos in conservation, science and education. It also looks at the nature of zoo visitors, the economic environment in which zoos operate, their economic outputs and their potential role in regeneration policy. The report concludes that zoos have an enormous social, cultural, educational and economic impact on the British public and that they have the potential to do more given the right encouragement and working with appropriate partners.

4.6.5 Fund-raising

Zoos raise most of their revenue from admission fees and many have membership schemes that allow unlimited access to the zoo for an annual fee. Some zoos operate animal adoption or sponsorship schemes. The name of the adopter/sponsor is displayed on a board in the zoo and in return they contribute to the cost of keeping the animals. Some institutions offer unusual experiences to visitors, for a fee, such as being able to work as a 'keeper for a day', or being able to dive with sharks.

Zoos have developed many other ways of generating additional funding, including:

- conference facilities
- wedding receptions (e.g. at the Mappin Pavilion, ZSL London Zoo)
- night safaris (e.g. Singapore Zoo)
- behind the scenes tours
- wildlife festivals (e.g. the North West Bird Watching Festival at Martin Mere Wetland Centre).

At some zoos, the entrance fee income itself is supplemented by a voluntary donation sometimes added to the ticket price and only deducted at the visitor's request. In the UK the zoo may reclaim tax paid by tax-paying visitors through the government's Gift Aid scheme if the zoo is a registered charity.

Some zoos have installed machines to collect additional money from visitors. The *Gorilla Kingdom* exhibit at ZSL London Zoo contains a machine that will dispense a souvenir badge for a donation of one pound and asks the donor to decide how their money should be spent. An electronic scoreboard records the number of donors who have elected to support particular personnel, for example rangers or teachers. A similar machine at Chester Zoo asks visitors to donate money to the conservation of a particular species (Fig. 4.3).

4.6.6 Public relations

By their very nature, zoos are always in the public eye. They can use this to great advantage when newsworthy events occur. Animal births always attract attention and may be of great interest to the local, national and even international press, depending upon the event. An elephant birth is always something for a zoo to celebrate. When the birth of an Asian elephant (*Elephas maximus*) was imminent at Zurich Zoo in 2000, some 40,000 people joined an internet mailing list so that they could be notified of the live webcast of the birth (Rees, 2001a).

Some births, however, are controversial and may attract adverse attention. A polar bear (*Ursus maritimus*) called 'Knut', born at Berlin Zoo in 2006, attracted international media attention. The birth was controversial because many zoos believe that polar bears cannot be adequately cared for in captivity, and Knut was rejected by his mother. There was a public outcry after newspaper reports in *Der Spiegel* and *Bild* claimed that

Fig. 4.3 A vending machine selling badges to support wildlife conservation.

animal rights activists wanted Knut to be euthanised. However, the cub became an international celebrity and appeared on the front cover of *Vanity Fair*.

The zoo established a holding company to channel the various franchises established to sell T-shirts, toys, sweets, posters, books and other Knut memorabilia. Knut has made more than £7 million for the zoo (Boyes, 2007).

Animal deaths also attract media attention. Some may generate a sympathetic response when, for example, a much-loved zoo favourite dies of old age. However, deaths from accidents and deaths in species where there are particular welfare concerns may attract the attention of animal rights activists, giving new impetus to campaigns to close zoos or remove particular species from zoos.

Animal escapes usually attract negative attention for a zoo, particularly if the animals concerned are dangerous, and especially if they injure someone. Some escapes, however, may attract positive attention, for example if an animal appears to have out-witted keepers or found a particularly ingenious method of escape.

Keeper deaths in zoos are rare and tragic, but inevitably attract negative attention from the media, along with a response from local health and safety officials. Legal proceedings may follow, attracting yet more bad publicity, and possibly a substantial fine. These events are particularly damaging to the image of a zoo, especially if the keeper was obviously engaging in a dangerous practice at the time, for example entering the cage of a lion or tiger with the animal present.

The late John Aspinall established zoos at Howletts and Port Lympne in Kent. He encouraged his family, friends and keeping staff to enter cages with dangerous species such as wolves and tigers. A number of serious attacks on keepers and friends of the family brought a great deal of bad publicity to the zoos (Masters, 1989). Over a 20-month period five keepers were killed: three by tigers at Howletts and two by elephants at Port Lympne Wild Animal Park (Watson-Smyth, 2000).

4.6.7 Sustainability

Zoos should encourage their visitors to adopt a more sustainable lifestyle by setting a good example in their buildings and activities. Many new zoo exhibits contain a variety of environmentally friendly features (see Section 7.11). In addition, zoos have tried to improve their overall environmental credentials by making their operations more sustainable in a number of ways. Sustainability requires the adoption of the principles of reduce, reuse and recycle. Zoos should strive to reduce their use of materials and energy. Where additional materials have to be used – such as the construction of a new exhibit – they should be reclaimed materials where possible. Wherever possible, unwanted materials should be recycled or recyclable by others.

4.6.7.1 Energy and water

Energy is an important expense for zoos and anything that can be done to reduce its use will benefit the zoo and the environment. *Marwell Wildlife* uses small electric vehicles to move staff and goods around the zoo (Fig. 4.4). The Durrell Wildlife Conservation Trust (*Durrell*, formerly Jersey Zoo) uses a company that runs it vehicles on biofuel made from local recycled cooking oil to provide a free minibus service from various hotels to its zoo.

Fig. 4.4 An electric vehicle at Marwell Wildlife, UK. Such vehicles are quiet, pollution free, and send a strong environmental message to visitors.

Fig. 4.5 A rainwater butt collects rain from the roof of a building at ZSL London Zoo. The rain is used to water ornamental plants.

Fig. 4.6 Many zoos generate considerable income from the sale of helium balloons: here income generation takes precedence over environmental protection.

Many new zoo exhibits and other buildings incorporate energy saving or energy converting devices ranging from timers on lights in toilets to solar panels which generate electricity, rainwater collection systems (Fig. 4.5), water-saving taps and green roofs.

4.6.7.2 Recycling and organic food production

Some zoos provide separate waste bins for the collection of recyclable materials, especially from food outlets. *Durrell* has large recycling bins for plastics, cans and paper, and a collection point for old clothes and shoes in its car park. It also runs an organic farm which grows produce for many of its animals, using on-site manure and natural pest control. Some zoos sell elephant dung – 'Zoo Poo' – to visitors as manure.

4.6.7.3 Merchandise

Zoos can attempt to influence the purchasing habits of their visitors by offering for sale items that are environmentally friendly or which support indigenous communities and conservation projects in developing countries. These might include:

- Fair Trade confectionery
- Rainforest Alliance coffee
- traditional musical instruments
- traditional jewellery
- wooden carvings
- reusable shopping bags.

Most zoos target the majority of their merchandise at children. However, some sell specialist books and videos on conservation, wildlife and animal keeping.

Helium-filled balloons are, understandably, popular with children (Fig. 4.6). They are unpopular with many keepers because they can frighten animals and sometimes end up trapped in the roof spaces of animal houses. As they deflate they descend and may startle animals. These balloons are not environmentally (or animal) friendly. But they may generate a significant income for a large zoo.

4.7 FURTHER READING AND RESOURCES

The annual reports of individual zoos and zoo associations are a very useful source of detailed information about management structures and financial performance. Many institutions make these available on their websites.

The World Zoo and Aquarium Conservation Strategy is available from the WAZA website at www.waza.org/files/webcontent/documents/cug/docs/WAZA%20CS.pdf

The *Sustainability of Activities in Zoos and Aquariums* was the theme of volume 43 (2009) of the *International Zoo Yearbook*.

A list of zoo and keeper organisations, and wildlife NGOs, may be found in Appendix 2.

4.8 EXERCISES

1. Explain the role of regional zoo organisations.
2. What is the purpose of the World Zoo and Aquarium Conservation Strategy?
3. Describe a possible management structure for a large zoo.
4. How may a zoo encourage its visitors to live sustainably?
5. Discuss the role of keeper organisations.

5 ZOO LEGISLATION

...wild fauna and flora in their many beautiful and varied forms are an irreplaceable part of the natural systems of the earth which must be protected for this and the generations to come.

Preamble to CITES, 1973

Conservation Status Profile

Grevy's zebra
Equus grevyi
IUCN status: Endangered A2ac; C2a(i)
CITES: Appendix I
Population trend: Stable

An Introduction to Zoo Biology and Management, First Edition. Paul A. Rees.

5.1 INTRODUCTION

For the purpose of this book a zoo has a wide definition, encompassing traditional zoological gardens, aviaries, aquariums, safari parks and many other types of institutions. However, the legal definition of a zoo varies from jurisdiction to jurisdiction (Box 1.1). The activities of zoos are regulated by a number of laws at different levels:

- National or domestic laws – These are the laws of individual countries, for example the laws of the United States of America or the laws of the United Kingdom.
- European laws – These are the laws of the European Union and apply only to the Member States (Box 5.1).
- International laws – These are the laws which regulate the ways individual states deal with each other, for example in relation to animal movements between countries or illegal trade in endangered species.

In addition, some countries have laws which apply only within certain states within their territory, for example Texas state laws in the USA or the laws of the state of Queensland in Australia.

Apart from legislation specifically aimed at regulating their activity, zoos must also comply with a very wide range of other laws including employment law, health and safety law, disability discrimination law, consumer protection law, and environmental law.

5.2 WHO LICENSES ZOOS?

In most countries, zoos have to be licensed, either by a state authority or by local government. Zoos in the European Union must be licensed by the individual Member States.

Box 5.1 The Member States of the European Union and European law.

The European Union is an economic and political union of 27 Member States representing nearly half a billion people (Table 5.1).

European law consists largely of Directives and Regulations. These vary in the mechanism of their application.

Table 5.1 The Member States of the European Union (July 2010).

Austria	Germany	Netherlands
Belgium	Greece	Poland
Bulgaria	Hungary	Portugal
Cyprus	Ireland	Romania
Czech Republic	Italy	Slovakia
Denmark	Latvia	Slovenia
Estonia	Lithuania	Spain
Finland	Luxembourg	Sweden
France	Malta	United Kingdom

Directives

A Directive requires each Member State to comply with EU law by amending its national law. For example, the Zoos Directive was implemented in English law by changes made to the Zoo Licensing Act 1981 by the Zoo Licensing Act 1981 (Amendment) (England and Wales) Regulations 2002 (SI 2002/3080).

Regulations

A Regulation becomes law in each Member State without any change being necessary to national law. There are very few EU Regulations that are concerned with the environment. This mechanism is often used to implement commitments under international law, for example Council Regulation (EC) No. 338/97 on the protection of species of wild fauna and flora by regulating trade therein.

European Union Regulations should not be confused with Statutory Instruments in the UK, which are also referred to as 'regulations'.

European Union laws may be accessed via the EU website at http://europa.eu

Legal Box 5.1 The Zoos Directive 1999.

The detailed objectives of the Directive are stated in Article 1:

> *The objectives of this Directive are to protect wild fauna and to conserve biodiversity by providing for the adoption of measures by Member States for the licensing and inspection of zoos in the Community, thereby strengthening the role of zoos in the conservation of biodiversity.*

Article 2 defines a zoo (see Box 1.1).
Article 3 defines the conservation role of zoos and requires that:

> *Member States shall take measures ... to ensure all zoos implement the following conservation measures:*
>
> - *participating in research from which conservation benefits accrue to the species, and/or training in relevant conservation skills, and/or the exchange of information relating to species conservation and/or, where appropriate, captive breeding, repopulation or reintroduction of species into the wild,*
> - *promoting public education and awareness in relation to the conservation of biodiversity, particularly by providing information about the species exhibited and their natural habitats,*
> - *accommodating their animals under conditions which aim to satisfy the biological and conservation requirements of the individual species, inter alia, by providing species specific enrichment of the enclosures; and maintaining a high standard of animal husbandry with a developed programme of preventive and curative veterinary care and nutrition,*
> - *preventing the escape of animals in order to avoid possible ecological threats to indigenous species and preventing intrusion of outside pests and vermin,*
> - *keeping of up-to-date records of the zoo's collection appropriate to the species recorded.*

5.2.1 Europe – the Zoos Directive 1999

Council Directive 1999/22/EC of 29 March 1999 relating to the keeping of wild animals in zoos was intended to improve conditions for animals in European zoos while also requiring zoos to adopt a conservation role (Legal Box 5.1).

Note that while all zoos are required to promote public education, provide adequate housing, prevent escapes and keep records, a zoo is only required to engage in *one* of the following:

- Research
- Training in conservation skills
- Information exchange
- Captive breeding, repopulation or reintroduction to the wild (Art. 3) (Fig. 5.1).

In addition to the requirements listed in Article 3, Member States are required to establish a licensing and inspection system for zoos in order to ensure that the requirements in Article 3 are met. If a zoo fails to meet these requirements the Directive makes provision for the closure of the zoo by a 'competent authority', which, in the UK, would be the local authority which is responsible for issuing the licence.

In many respects, because it allows zoos to choose their own conservation roles, the Directive may be considered to be a lost opportunity to implement the *ex-situ* conservation obligations imposed by the Biodiversity Convention (Rees, 2005b).

5.2.2 England – Zoo Licensing Act 1981

In England, zoos are licensed by individual local authorities under the Zoo Licensing Act 1981. This process is overseen by the Zoo Licensing Branch of the Department for Environment, Food and Rural Affairs (DEFRA). The devolution of responsibility for zoos to the individual countries which make up the UK means that zoos in Wales, Scotland and Northern Ireland are regulated by the equivalent government departments under their own legislation. Amendments have been made to the Zoo Licensing Act to comply with the Zoos Directive: Zoo Licensing Act 1981 (Amendment) (England and Wales) Regulations 2002 (SI 2002/3080); Zoo Licensing Act 1981 (Amendment) (Scotland) Regulations 2003 (SSI 2003/174); Zoo Licensing Act 1981 (Amendment) (Wales) Regulations 2003 (WSI 2003/992 (W.141)). In Northern Ireland zoos are regulated by the Zoo Licensing Regulations (Northern Ireland) 2003 (SR 2003/115).

Under the Zoo Licensing Act a zoo must be licensed by the local authority within whose boundaries it is

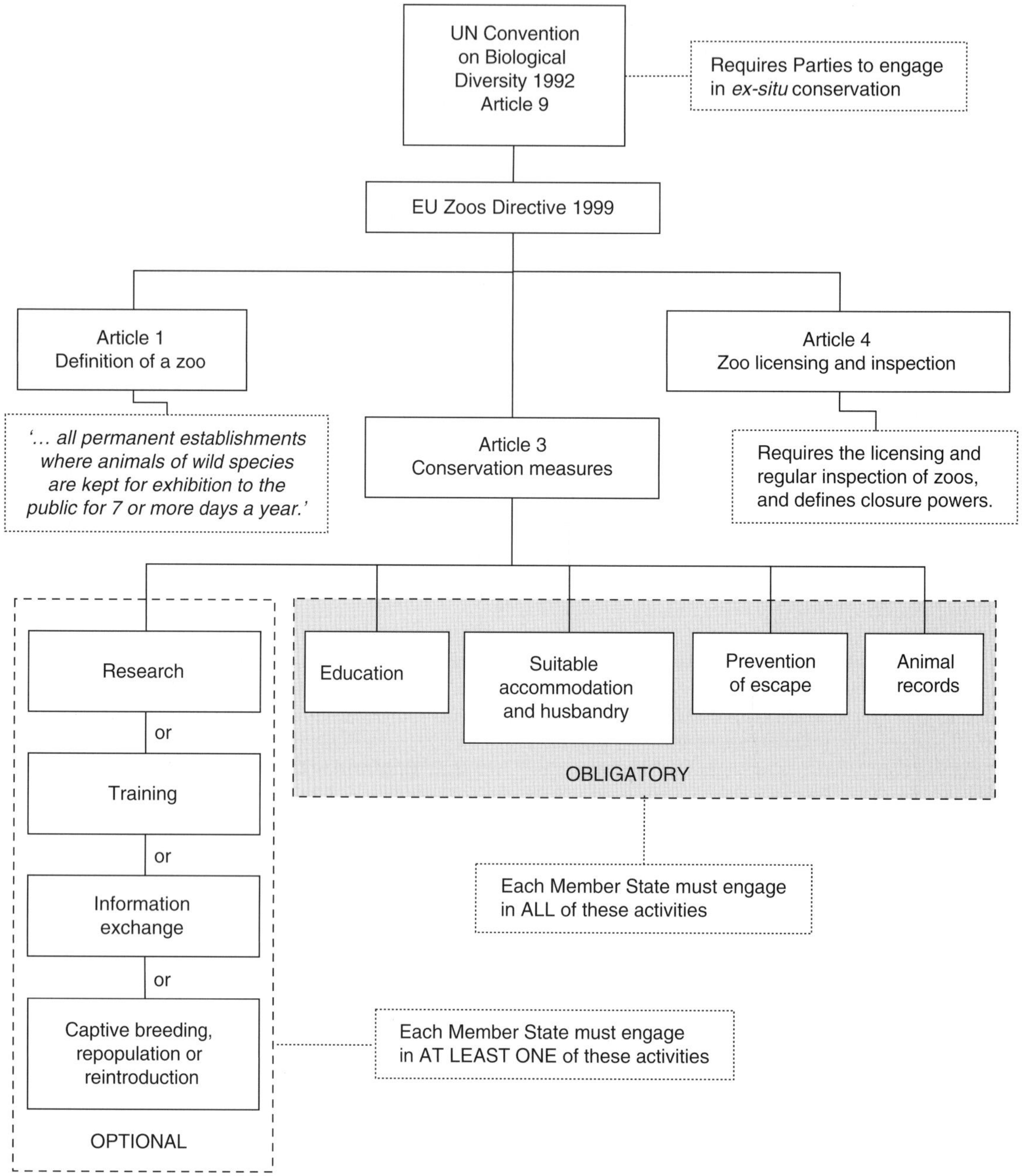

UN Convention on Biological Diversity 1992 Article 9

Requires Parties to engage in *ex-situ* conservation

EU Zoos Directive 1999

Article 1
Definition of a zoo

'... all permanent establishments where animals of wild species are kept for exhibition to the public for 7 or more days a year.'

Article 3
Conservation measures

Article 4
Zoo licensing and inspection

Requires the licensing and regular inspection of zoos, and defines closure powers.

Research
or
Training
or
Information exchange
or
Captive breeding, repopulation or reintroduction

OPTIONAL

Each Member State must engage in AT LEAST ONE of these activities

Education

Suitable accommodation and husbandry

Prevention of escape

Animal records

OBLIGATORY

Each Member State must engage in ALL of these activities

Fig. 5.1 The structure of the EU Zoos Directive.

Legal Box 5.2 The Zoo Licensing Act 1981.

Section 1: Licensing of zoos by local authorities
(1) Subject to this section it is unlawful to operate a zoo to which this Act applies except under the authority of a licence issued under this Act by the local authority for the area within which the whole or the major part of the zoo is situated.
(2) In this Act "zoo" means an establishment where wild animals (as defined by section 21) are kept for exhibition to the public otherwise than for purposes of a circus (as so defined) and otherwise than in a pet shop (as so defined).
(2A) This Act applies to any zoo to which members of the public have access, with or without charge for admission, on seven days or more in any period of twelve consecutive months.

Section 9: Secretary of State's standards
After consulting such persons on the list and such other persons as he thinks fit, the Secretary of State may from time to time specify standards of modern zoo practice, that is, standards with respect to the management of zoos and the animals in them.

Section 10: Periodical inspections
(1) The local authority shall carry out periodical inspections in accordance with this section of any zoo for which a licence granted by that authority is in force.

Section 21: Interpretation
(1) In this Act—
"animals" means animals of the classes Mammalia, Aves, Reptilia, Amphibia, Pisces and Insecta and any other multi cellular organism that is not a plant or a fungus and "wild animals" means animals not normally domesticated in Great Britain;

situated (Legal Box 5.2). However, the Secretary of State may make an exemption, for example, under s.14(1)(a) of the Act if a local authority makes an application for a dispensation. Such exemptions are normally available for:

- traditional deer parks
- collections of llamas and alpacas not exceeding five
- collections of small, non-hazardous and non-conservation sensitive wild species not normally exceeding 120 specimens (DEFRA, 2003).

A zoo operator may apply directly to the Secretary of State for a dispensation for a collection where the hazardous and/or conservation sensitive species component of the collection does not normally exceed 50 specimens (s.14(2)). Other situations may qualify for exemptions and each case is considered on its merits.

The Zoo Licensing Act makes provision for:

- the licensing of zoos by local authorities (s.1)
- conservation measures in zoos (s.1A)
- inspection by zoo inspectors (s.8)
- standards of modern zoo practice (s.9)
- zoo inspections (ss.10 and 11)
- enforcement of licence conditions (s.16A)
- zoo closures (ss.16B and 16C)
- the disposal of animals (ss.16D, 16E and 16F).

Of the 270 zoos listed by DEFRA in November 2007 (Fig. 3.12), only 47 (17.4%) required a full zoo licence; the majority qualified for dispensations under ss.14(1)(a), 14(1)(b) or 14(2) of the Act. An interesting development in Scotland has given rise to some challenging questions regarding the definition and licensing of zoos (Box 5.2).

5.2.2.1 The Zoo Inspectorate

The Zoo Inspectorate is part of DEFRA and has a small headquarters team based in Bristol. It employs and manages inspectors who regularly visit zoos to ensure compliance with the Zoo Licensing Act. Section 8 requires the Secretary of State to compile a list of zoo inspectors in two parts:

- The first part contains names of veterinary surgeons and practitioners who have experience of zoo animals and can advise on the conservation measures required by s.1A of the Act.
- The second part contains the names of persons competent to inspect animals in zoos, advise on their welfare, husbandry, the implementation of conservation measures (in s.1A of the Act) and the general management of zoos.

Box 5.2 When is a zoo not a zoo? – the Alladale Wilderness Reserve.

The Alladale Estate is owned by Paul Lister, a millionaire businessman, and is located in Sutherland, Scotland, some 30 miles north of Inverness. The owner has ambitious plans to recreate an ancient ecosystem within a fenced area of 11,000 ha which he hopes will include wolf (*Canis lupus)*, brown bear (*Ursus arctos*), lynx (*Lynx lynx*), beaver (*Castor fiber*), bison (*Bison bonasus*), elk (*Alces alces*) and wild boar (*Sus scrofa*). This would restore lost food chains and the wolves would help to control the red deer (*Cervus elephus*) population which currently has to be culled by shooting.

These plans raise a number of interesting legal problems.

- If the animals are to remain on the estate, as what amounts to a private zoo, the owner would need to apply for a licence under the Dangerous Wild Animals Act 1976.
- There are public rights of way running through the estate so the owner cannot prevent the public from entry and therefore possible exposure to dangerous animals. The owner would be legally liable for any harm caused by the animals.
- If animals were simply to be released onto the estate, with no fences to keep the animals in, a licence to release non-native animals into the wild would be required under s.14 of the Wildlife and Countryside Act 1981. This is unlikely to be granted for large carnivores as it has taken many years for the Scottish government to issue a licence for the release of a very small number of beavers elsewhere.
- If the public are allowed entry to the estate to view the animals – from vehicles –on more than seven days a year the owner would almost certainly need a zoo licence as dispensations are unlikely to apply.
- If a zoo licence was granted the owner could not allow some vertebrates to feed on others (e.g. wolves to feed on red deer); he would be required to provide adequate housing and veterinary care for the animals; he would need to keep records of the animals; and all of the other requirements of the zoo licensing legislation would need to be fulfilled.

Some opponents of this project have suggested that if people want to see animals that were once native to Scotland they can visit the Highland Wildlife Park which is operated by the Royal Zoological Society of Scotland. This is true. But it would be a great pity if we were to see keeping animals in zoos as an alternative to restoring lost ecosystems and returning once native species back to the wild.

The Inspectorate's mission statement is:

to inspect zoos and provide consistent advice to local authorities, with the aim of monitoring and promoting:

- high standards of animal care and husbandry in zoos
- high standards of health and safety for zoo visitors
- participation of zoos in proactive measures to conserve biodiversity, and
- participation of zoos in promoting public education and awareness in relation to conservation of biodiversity.

5.2.2.2 *Secretary of State's Standards of Modern Zoo Practice*

Section 9 of the Zoo Licensing Act 1981 gives a power to the Secretary of State to specify 'standards of modern zoo practice' after consulting with persons on the list of zoo inspectors and with others.

The *Secretary of State's Standards of Modern Zoo Practice* (SSSMZP, 2004) advises zoos on their responsibilities and appropriate standards in relation to:

- the provision of food, water, and a suitable environment
- the provision of animal health care and the opportunity to express most normal behaviour
- the provision of protection from fear and distress
- the transportation and movement of animals
- conservation and education
- public safety, insurance, and escapes
- stock records
- staff and training
- public facilities, first aid, toilets, and parking
- display of the zoo licence.

Appendices to the document include additional information about:

- the EC Zoos Directive
- the ethical review process

- conservation and education
- animal transaction, disposal and euthanasia
- veterinary facilities
- animal contact areas, walk-through exhibits, diving experience exhibits, touch-pools and drive-through enclosures
- animal training and animal demonstrations
- specialist exhibits (invertebrates, reptiles and amphibians, venomous species, pinnipeds and marine birds, aquariums, waterfowl and birds of prey)
- staff and staff training
- pre-inspection audit and the inspection report form
- hazardous animal categorisation.

The lists above are a summary only. The *Secretary of State's Standards of Modern Zoo Practice* can be obtained via the DEFRA website at www.defra.gov.uk/wildlife-pets/zoos/standards.htm. In Wales the equivalent document is the *National Assembly for Wales Standards of Modern Zoo Practice*, available from the Llywodraeth Cynulliad Cymru website (http://cymru.gov.uk).

5.2.2.3 Zoos Forum/committee of experts

The Zoos Forum was a non-departmental body that advised the UK government on zoo licensing matters. Its members included veterinary surgeons, scientists, educators and others with expertise in animal conservation and welfare. The Forum encouraged the role of zoos in conservation, education and scientific research and kept the operation and implementation of the zoo licensing system in the UK under review. In addition, it advised ministers of any changes that it considered necessary to legislation.

The Zoos Forum Handbook provided advice to zoos on:

- the ethical review process
- conservation, education and research
- sustainability initiatives in UK zoos
- animal welfare and its assessment in zoos
- diving in zoos and aquariums
- veterinary services.

The handbook is available from the DEFRA website at www.defra.gov.uk/wildlife-pets/zoos/zf-handbook.htm

In 2010 the UK government decided to abolish the Zoos Forum and reconstitute it as a committee of experts.

5.2.3 USA

In the USA, at the federal level, animals in zoos are protected by the Animal Welfare Act (AWA) (7 USC §2131 et seq.). The Act requires animal dealers and exhibitors to be licensed, to keep records and to mark the animals under their control. It also lays down requirements in relation to the care, handling and transport of animals. The Act is administered by the Department of Agriculture through the Animal and Plant Health Inspection Service (APHIS) whose inspectors inspect facilities and investigate complaints. In addition to the AWA, each state has its own anti-cruelty laws. The AZA regulates the activity of its accredited zoos via a system of voluntary standards.

5.2.4 India and South East Asia

In India the functioning of zoos is regulated by the Central Zoo Authority, which is an autonomous statutory body constituted under the Wildlife (Protection) Act 1972.

Standards for housing, husbandry and animal management are laid down in the Recognition of Zoo Rules, 2009. Since its inception in 1992 the Authority has evaluated 347 zoos. Of these, only 164 have been granted recognition. The Authority acts as a facilitator, providing technical and financial assistance to recognised zoos. It also regulates the exchange of endangered species listed under Schedules I and II of the Wildlife (Protection) Act between zoos and approves exports of animals to foreign zoos. The Authority coordinates and implements capacity building programmes for zoo personnel, breeding programmes and *ex-situ* research. It has established a Laboratory for the Conservation of Endangered Species at Hyderabad, which conducts biotechnology research.

The website of the Zoo Outreach Organisation (www.zooreach.org) contains a great deal of information on zoo legislation in South East Asia and Australasia. It also contains information on IUCN specialist group networks for South East Asia and publishes *Zoos' Print*.

5.2.5 Australia

In Australia, zoos are licensed at state level. In New South Wales zoo licences are issued under the Exhibited Animals Protection Act 1986 (No. 123). Zoo standards are contained in the Exhibited Animals Protection Regulation 1995. In Queensland, zoos are regulated under the Nature Conservation Regulation 1994.

5.2.6 New Zealand

In New Zealand, zoos are regulated under the Animal Welfare Act 1999. The Animal Welfare (Zoos) Code of

Welfare 2004 (Code of Welfare No. 5) is a code issued under this Act. The Code applies to all persons responsible for the welfare of animals in zoos, animal parks and aquariums, and animals held for rehabilitation. It contains minimum standards and recommended best practice. Only the minimum standards have legal effect. The Code covers a range of aspects of zoo management including animal acquisition and disposal, hygiene, management of reproduction, provision of housing, food and water, health and disease, behaviour and stress, transport and staffing.

In New Zealand, Standard 154.03.04 specifies the requirements for the containment and keeping of a range of species in zoos (Anon., 2007). It replaced the Zoological Gardens Regulations 1977 in 2003. This Standard was approved by the Environmental Risk Management Authority in accordance with the Hazardous Substances and New Organisms Act 1996, and pursuant to ss.39 and 40 of the Biosecurity Act 1993.

5.3 OTHER WILDLIFE AND ZOO-RELATED LEGISLATION

Apart from the legislation concerned with the licensing of zoos, zoo professionals also need to be aware of a wide range of international, European and national laws which may affect their activities. Some *in-situ* conservation projects may involve protected areas that have been designated under international law.

5.3.1 International law

International law is concerned with agreements made between states. Once a state signs up to a particular treaty or convention it is referred to as a 'Party' or 'Contracting Party' to the treaty and is expected to abide by its provisions. However, if a Party decides to ignore a provision of a treaty there is very little that other Parties can do to make it comply. International law is not directly applicable to individuals and cannot be enforced in national courts unless it has been implemented in national laws. For example, in the UK, CITES is enforced through the Control of Trade in Endangered Species (Enforcement) Regulations 1997 (SI 1997/1372), as amended, often referred to as COTES.

There is no international court that deals with crimes against wildlife. When individual states fail to meet their treaty obligations all that other Parties can do is to apply political pressure. In 1994, the United States sent an important signal to the international community when President Clinton imposed trade sanctions on Taiwan in order to prompt better control of the illegal trade in endangered species on the island, particularly in relation to rhinos and tigers. The prohibition applied to the importation to the USA from Taiwan of 'fish and wildlife products', which include any wild animal, whether alive or dead, whether or not bred in captivity, and any derivative or product of such animals. In June 1995 the USA lifted its sanctions on Taiwan in recognition of the progress it had made in combating illegal wildlife trade (Anon., 1995a).

There is no international law specifically aimed at regulating the establishment or operation of zoos. However, the UN Convention on Biological Diversity makes specific reference to the role of *ex-situ* conservation efforts in the overall protection of global biodiversity. The Convention on International Trade in Endangered Species of Wild Fauna and Flora (CITES) restricts international movements of protected species including movements involving zoo animals. Other treaties refer to the desirability of reintroducing species back into the wild and, for some species, zoos may play a part in this (see Legal Box 15.1). It should be noted that the obligations imposed by the treaties described below relate to the states that have ratified the treaties and not to zoos within those states. However, in many cases much of the work of *ex-situ* conservation undoubtedly falls to zoos.

5.3.1.1 UN Convention on Biological Diversity

The UN Convention on Biological Diversity (Biodiversity Convention) was signed by 150 government leaders in 1992 at the Rio Earth Summit in Brazil. Its purpose is to promote sustainable development while protecting the Earth's biological diversity. By March 2010 there were 193 Parties (192 countries and the European Union). The UK has been a Party since 1994 but the United States has never signed the treaty. The EU has been a Party since 1993.

The Convention requires the Contracting Parties to adopt measures for the *ex-situ* conservation of components of biological diversity (Art. 9(a)) and to establish and maintain facilities for the *ex-situ* conservation of and research on animals (Art. 9(b)) (Legal Box 5.3).

Legal Box 5.3 The UN Convention on Biological Diversity, 1992.

Article 9 of the Convention states that:

Each Contracting Party shall, as far as possible and as appropriate, and predominantly for the purpose of complementing in-situ measures:
(a) Adopt measures for the ex-situ conservation of components of biological diversity, preferably in the country of origin of such components;
(b) Establish and maintain facilities for ex-situ conservation of and research on plants, animals and micro-organisms, preferably in the country of origin of genetic resources;
(c) Adopt measures for the recovery and rehabilitation of threatened species and for their reintroduction into their natural habitats under appropriate conditions;
(d) Regulate and manage collection of biological resources from natural habitats for ex-situ conservation purposes so as not to threaten ecosystems and in-situ populations of species, except where special temporary ex-situ measures are required under subparagraph (c) above; and
(e) Cooperate in providing financial and other support for ex-situ conservation outlined in subparagraphs (a) to (d) above and in the establishment and maintenance of ex-situ conservation facilities in developing countries.

Under Article 2 of the Convention:

"Ex-situ conservation" means the conservation of components of biological diversity outside their natural habitats.

and

"Biological diversity" means the variability among living organisms from all sources including, inter alia, terrestrial, marine and other aquatic ecosystems and the ecological complexes of which they are part; this includes diversity within species, between species and of ecosystems.

The Convention makes a specific reference to an obligation to restore both ecosystems and species in the wild in Article 8 which is concerned with *in-situ* conservation. Article 8(f) requires that each Contracting Party shall, as far as possible and as appropriate:

Rehabilitate and restore degraded ecosystems and promote the recovery of threatened species, inter alia, through the development and implementation of plans or other management strategies;

The Convention requires Contracting Parties to promote technical and scientific cooperation between Parties in relation to conservation, in particular with respect to developing countries. Zoos have an important role to play in providing expertise for *in-situ* conservation programmes.
Article 18 states that:

1. The Contracting Parties shall promote international technical and scientific cooperation in the field of conservation and sustainable use of biological diversity, where necessary, through the appropriate international and national institutions.
2. Each Contracting Party shall promote technical and scientific cooperation with other Contracting Parties, in particular developing countries, in implementing this Convention, inter alia, through the development and implementation of national policies. In promoting such cooperation, special attention should be given to the development and strengthening of national capabilities, by means of human resources development and institution building.

Furthermore, this Convention commits signatories to the recovery of endangered species by creating an international legal obligation to reintroduce them into their former habitats (Art. 9(c)) (Rees, 2001b). Individual states are unlikely to provide new special facilities for the captive breeding of endangered species, so this task will generally fall to existing zoological gardens.

5.3.1.2 Berne Convention

The Convention on the Conservation of European Wildlife and Natural Habitats 1979 (Berne Convention) aims to ensure the conservation of wild animal and plant species and their natural habitats. Protected species are listed in three Appendices:

- I – Strictly protected flora species
- II – Strictly protected fauna species
- III – Protected fauna species (migratory species).

Appendix IV lists prohibited means and methods of killing and capture, and other forms of exploitation. The Convention does not specifically refer to zoos but Article 11(2) imposes an obligation upon states to encourage reintroductions of native species, and zoos may play an import part in this:

> Each Contracting Party undertakes:
>
> **a.** to encourage the reintroduction of native species of wild flora and fauna when this would contribute to the conservation of an endangered species, provided that a study is first made in the light of the experiences of other Contracting Parties to establish that such reintroduction would be effective and acceptable.

Species listed in Appendix II include the grey wolf (*Canis lupus*), the otter (*Lutra lutra*), the European wildcat (*Felis silvestris*), all species of the Falconiformes, the spur-thighed tortoise (*Testudo graeca*) and the Moor frog (*Rana arvalis*).

5.3.1.3 Convention on International Trade in Endangered Species of Wild Fauna and Flora (CITES)

International trade in wildlife is controlled by the Convention on International Trade in Endangered Species of Wild Fauna and Flora (CITES), which was signed in Washington DC in March 1973 and entered into force in July 1975. CITES prohibits international commercial trade in the rarest species and requires licences from the country of origin for exports of some other rare species (Legal Box 5.4). The convention regulates trade in whole animals and plants, living or dead, and recognisable parts and derivatives. The protected species are listed in three Appendices (Table 5.2):

- Appendix I – Includes all species threatened with extinction which are or may be affected by trade.
- Appendix II – Includes all species which may become threatened with extinction if trade is not strictly regulated (and other species which must be subject to strict regulation in order to achieve this objective).
- Appendix III – Includes other species which any Party strictly protects within its own jurisdiction and which requires the cooperation of other Parties in the control of trade.

Trade in endangered species is regulated by the requirement for import and export licences. The strictest restrictions apply to Appendix I species (Art. III).

Parties to CITES meet biennially and may agree to add or remove species from the Appendices or move them from one Appendix to another as their status improves or deteriorates.

Legal Box 5.4
Convention on International Trade in Endangered Species of Wild Fauna and Flora, 1973.

Article III states that:

> *3. The import of any specimen of a species included in Appendix I shall require the prior grant and presentation of an import permit and either an export permit or re-export certificate. An import permit shall only be granted when the following conditions have been met:*
> *a. a Scientific Authority of the State of import has advised that the import will be for purposes which are not detrimental to the survival of the species involved;*
> *b. a Scientific Authority of the State of import is satisfied that the proposed recipient of a living specimen is suitably equipped to house and care for it; and*
> *c. a Management Authority of the State of import is satisfied that the specimen is not to be used for primarily commercial purposes.*

The term 'specimen' is defined in Article I, and means an animal or plant, whether alive or dead, and includes recognisable parts and derivatives.

Table 5.2 Selected examples of species listed on the Appendices to the Convention on International Trade in Endangered Species of Wild Fauna and Flora (CITES), 1973 (The Washington Convention). Note that in some cases only certain geographical populations are listed.

Appendix I–Species threatened with extinction which are or may be affected by trade.	
Loxodonta africana (except the populations of Botswana, Namibia, South Africa and Zimbabwe, which are included in Appendix II).	African elephant
Elephas maximus	Asian elephant
Indriidae spp.	Indris
Pan spp.	Chimpanzees
Hylobatidae spp.	Gibbons
Falco punctatus	Mauritius kestrel
Grus americana	Whooping crane
Alligator sinensis	Chinese alligator
Varanus komodoensis	Komodo dragon
Bufo periglenes	Orange toad
Acipenser brevirostrum	Short nose sturgeon
Papilio chikae	Luzon peacock swallowtail
Lampsilis virescens	Alabama lamp pearly mussel

Appendix II–Species which may become threatened unless trade is regulated and species which are similar in appearance to them.	
Primate spp.	All primate species not included in Appendix I.
Equus kiang	Kiang
Fossa fossana	Malagasy civet
Anas formosa	Baikal teal
Tauraco spp.	Turacos
Iguana spp.	Iguanas
Allobates zaparo	Sanguine poison frog
Anguilla anguilla	European eel
Strombus gigas	A mesogastropod

Appendix III–Species which any Party to CITES identifies as subject to regulation within its jurisdiction to prevent or restrict exploitation and needing the cooperation of other Parties to control trade.	
Antilope cervicapra (Nepal)	Blackbuck
Canis aureus (India)	Golden jackal
Nasau nasau solitaria (Uruguay)	South American coati
Crax alberti (Colombia)	Blue-billed curassow
Hoplodactylus spp. (New Zealand)	Geckos
Macrochelys temminckii (USA)	Alligator snapping turtle
Colophon spp. (South Africa)	Cape stag beetles
Corallium konjoi (China)	A coral

The owner of Southport Zoo, in the UK, was prosecuted for displaying species listed on Appendix I of CITES without a licence in July 2001. They included cotton-top tamarins (*Saguinus oedipus*), ocelots (*Leopardus pardalis*), scarlet macaws (*Ara macao*), owls and tortoises. He was fined £5000 and 37 specimens were confiscated (Rees, 2002a). The zoo is now closed.

TRAFFIC – Trade Records Analysis of Flora and Fauna in Commerce TRAFFIC was established in 1976 and is the joint monitoring programme of the World Wide Fund for Nature (WWF) and the IUCN which works in cooperation with the CITES Secretariat to monitor trade in endangered species. It publishes the *TRAFFIC Bulletin* – which contains articles on many aspects of the global wildlife trade – and a number of identification guides on various taxa and wildlife products including ivory, bear gall bladders, crocodilians, turtles, tortoises, butterflies and seahorses. TRAFFIC International is based in Cambridge, UK, and the organisation has regional offices in many parts of the world.

Many zoos exhibit animal products that have been seized at ports of entry (usually airports) in order to educate the public about the species protected by CITES (Fig. 5.2; Box 5.3).

5.3.1.4 RAMSAR Convention

The Convention on Wetlands of International Importance Especially as Waterfowl Habitat (the Ramsar Convention) was signed in the Iranian town of Ramsar in 1971. Its aim is to protect wetlands, their flora and fauna, and to promote their wise use. Article 2.1 of the Convention requires the Contracting Parties to designate suitable wetlands for inclusion in a 'List of Wetlands of International

Fig. 5.2 A display of illegally traded cat skins and body parts.

Importance' which is maintained by the IUCN (Art. 8). Ramsar sites include the Danube Delta (Romania), Lake Naivasha (Kenya), Okavango Delta System (Botswana), Morecambe Bay (UK), Kakadu National Park (Australia) and the Everglades National Park (USA).

5.3.1.5 Agreement on the Conservation of African-Eurasian Migratory Waterbirds (AEWA)

The Agreement on the Conservation of African-Eurasian Migratory Waterbirds 1995 is concerned with the conservation of over 250 bird species that are ecologically dependent upon wetlands for their survival during at least part of their annual cycle. These species include ducks, gulls, spoonbills, pelicans, penguins and a wide range of other taxa. The Agreement requires Contracting Parties to take a variety of measures to protect migratory waterbirds including:

- protection of species
- conservation of habitats
- supporting research and monitoring
- cooperating in emergency situations threatening waterbirds
- preventing introduction of non-native waterbirds
- engaging in information exchange, training and education.

Box 5.3 Wildlife trade – some facts.

- The global illegal trade in wildlife is worth £5 billion annually.
- The Metropolitan Police have seized over 30,000 endangered species items under *Operation Charm*, mostly traditional Chinese medicines.
- The world's largest seizure of Shahtoosh shawls occurred in 1998 in London, and represented approximately 1000 critically endangered Tibetan antelope.
- The world's largest seizure of rhino horn occurred in London in 1996 and was valued at several million pounds.
- The biggest single threat to the survival of the tiger is the trade in illegal Chinese medicines.
- The Metropolitan Police continue to seize new ivory and ivory products, some of which have been manufactured in the UK.
- Weight for weight, musk, bear bile, rhino horn and many other endangered species products cost more than gold.
- In 2000, a man in London was sentenced to six months imprisonment for selling taxidermy specimens. His shop contained over 80 new specimens, including tigers, gorillas, chimpanzees, leopards, endangered birds and sea turtles.
- Many zoos take in endangered species that have been confiscated by police and customs officials.

Source: *Operation Charm*, Metropolitan Police, London.

Examples of wildlife trade prosecutions in the UK

On 14 April 2000 a 61-year-old man from Northallerton, North Yorkshire, was jailed for two-and-a-half years on four counts of smuggling endangered birds into the UK. Customs officers seized nine birds from his premises in 1998, including three Lear's macaws. The rare macaws were held at a secret location until after the trial and were then returned to the Brazilian government for release back into the wild.

In June 2001 two London shopkeepers were convicted of smuggling bushmeat into Britain and sentenced to four months imprisonment. They were arrested after selling a monkey to a journalist who was posing as a customer.

An animal dealer was jailed for six-and-a-half years by Isleworth Crown Court on 18 January 2002 after 23 endangered birds of prey were discovered in suitcases at Heathrow Airport. The birds had been on a flight from Thailand and included eagles, owls and kites, with a black-market value of £35,000. When RSPCA inspectors raided the defendant's home they recovered birds, mammals and reptiles from 14 endangered species. In total more than 60 animals, dead and alive, were seized, from 29 species including several primates. The trial followed *Operation Retort*, a joint operation against wildlife crime mounted between the police and HM Customs.

The above cases have been described in Rees (2002a).

In October 2010 a man was sentenced to 12 months in prison for attempting to smuggle rhino horn to China through Manchester Airport. The horn was hidden inside a bronze sculpture and had been taken from a white rhino (*Ceratotherium simum*) called 'Simba' that died at Colchester Zoo in 2009. The origin of the horn was determined by comparing its DNA with that from a blood sample from Simba (Gordon, 2010).

5.3.1.6 World Heritage Sites

World Heritage Sites are of international importance and are designated under the Convention for the Protection of the World Cultural and Natural Heritage (World Heritage Convention). The Convention was adopted by the United Nations Educational, Scientific and Cultural Organisation (UNESCO) in 1972. World Heritage Sites include Ngorongoro Crater in Tanzania, the Royal Botanic Gardens, Kew (UK), Yellowstone National Park (USA) and the Sichuan Giant Panda Sanctuaries in China, home to 30% of the world's pandas and the most important site for their captive breeding.

5.3.1.7 Biosphere reserves

Biosphere reserves are protected areas of terrestrial and coastal ecosystems which represent important examples of the world's biomes. The sites form a World Network, and are particularly useful in the monitoring of long-term ecological changes. Biosphere reserves were devised by UNESCO as part of their Man and the Biosphere (MAB) Programme (Project No. 8). Sites are nominated by national governments and must meet a minimal set of criteria and adhere to a minimal set of conditions before being admitted to the World Network. Biosphere reserves include Big Bend Biosphere Reserve and National Park (USA), North Norfolk Coast (UK), Niagara Escarpment (Canada), Uluru (Australia) and Mount Olympus (Greece).

5.3.1.8 Council of Europe Biogenetic Reserves

In 1976, the European Network of Biogenetic Reserves was established by the Council of Europe to conserve representative examples of European flora, fauna and natural areas. This resulted from a recommendation from the European Ministerial Conference on the Environment in 1973. The selection of biogenetic reserves is based on their nature conservation value – they must contain specimens of flora or fauna that are typical, unique, rare or endangered – and the effectiveness of their protected status (Council of Europe Resolution (76)17). Biogenetic reserves include the Camargue (France), Hartland Moor (UK), Reuss Valley (Switzerland) and Clara Bog (Ireland).

5.3.2 European law

European law is the law of the 27 Member States of the European Union and it is interpreted by the European Court of Justice. It takes the form of Directives and Regulations. Directives state the intention of the law and leave the method of implementation to each individual Member State. In order to comply with the Directive each state must amend its own laws. Regulations become law in the Member States as soon as they are issued and require no changes to national laws.

5.3.2.1 Wild Birds Directive

Council Directive 2009/147/EC of 30 November 2009 on the conservation of wild birds (the Wild Birds Directive) requires Member States to take various conservation measures to protect birds. It applies to the conservation of all species of birds which occur naturally in the wild within the European territory of the Member States. The Birds Directive requires Special Protection Areas (SPAs) to be established to protect birds.

5.3.2.2 Habitats Directive

The main aim of Council Directive 92/43/EEC of 21 May 1992 on the conservation of natural habitats and of wild fauna and flora (the Habitats Directive) is to promote the maintenance of biodiversity (Art. 2), while taking into account economic, social, cultural and regional requirements. It identifies certain priority species and habitats in need of special protection, and recognises the transboundary nature of many of the threats to our natural heritage. The reintroduction of species listed in Annex IV (Animal and plant species of community interest in need of strict protection) is encouraged by Article 22(a), where this may contribute to the re-establishment of these species at a favourable conservation status. Annex IV species include species that are being captive bred in zoos such as the dormouse (*Muscardinus avellanarius*) and the sand lizard (*Lacerta agilis*). The Habitats Directive has led to the establishment of a European network of protected areas (Special Areas of Conservation (SACs)) collectively known as Natura 2000. These include the SPAs established under the Wild Birds Directive.

5.3.2.3 CITES and the European Union

Trade in endangered species within the European Union is regulated by the Regulation on the protection of species of wild fauna and flora by regulating trade therein (338/97/EC). This Regulation replaces CITES Appendices with Annexes as follows:

- Annex A – This includes all CITES Appendix I species plus certain other species that have a similar appearance and therefore need a similar level of protection, or whose protection is necessary for the effective protection of rare taxa within the same genus.
- Annex B – This includes all of the remaining species listed in CITES Appendix II plus certain other species:
  - with a similar appearance; or
  - which experience a level of trade which may not be compatible with the survival of the species or the survival of local populations; or
  - which pose an ecological threat to indigenous species.

• Annex C – This includes all of the remaining species listed in CITES Appendix III.
• Annex D – This includes species not listed in Annexes A to C which are imported into the EU in such numbers as to warrant monitoring.

5.3.2.4 The Balai Directive

Although the EU has created a wide range of animal health rules relating to farm animals and fish, there are many other animal groups to which this legislation does not apply. The Balai Directive (Council Directive 92/65/EEC) was designed to fill this gap and applies to the movement of non-domestic animals, and their semen, ova and embryos, from other EU Member States and countries outside the EU, which are not covered by other EU animal health legislation. The Directive does not apply to pet animals, or domestic cattle, swine, sheep, goats, equids, poultry (including eggs), fish and fishery products, bivalves or aquaculture animals.

The Directive sets special conditions for:

- apes
- lagomorphs
- wild ungulates and non-domesticated varieties of sheep, goats, camels, pigs, cows and deer
- foxes, ferrets and mink
- cats and dogs
- captive birds and their hatching eggs
- all animals susceptible to rabies.

The Directive lays down the animal health requirements governing trade in and imports into the EU of animals, semen, ova and embryos which are not subject to other EU legislation. Animals must come from registered or approved premises (e.g. a breeder registered with their local authority) and such premises must meet stringent biosecurity requirements. The main requirements are that the site must:

- be clearly separated from the surrounding area, to avoid a health risk
- be capable of catching, confining and isolating animals
- have adequate quarantine and testing facilities
- be disease free
- keep records of the animals held, including records of species, age, sex, blood tests and diseases
- possess facilities for post-mortems
- be capable of the safe disposal of carcasses
- employ a vet approved by the competent veterinary authority.

5.3.3 UK law

The national law of the United Kingdom includes a very large number of laws concerning animals. The following is a brief account of some of the laws which may affect those that keep or work with captive wild animals.

5.3.3.1 Dangerous Wild Animals Act 1976

In Great Britain it is unlawful to keep any animal listed under the Dangerous Wild Animals Act 1976 without a licence (Legal Box 5.5).

A licence may be issued by the local authority to a suitable person only if it is satisfied that it would not be contrary to the public interest on the grounds of safety or nuisance, and the animal's accommodation is secure and safe. The Act does not cover animals held in zoos, circuses, pet shops or registered scientific establishments, all of which are covered by their own legislation.

The Act has recently been updated by the Dangerous Wild Animals Act 1976 (Modification) (No. 2) Order 2007 (No. 2465). This Order removed a number of species that were previously considered dangerous, giving rise to fears that more people could buy exotic pets without knowing how to look after them properly (RSPCA, 2008).

Animals currently listed by the Act include:

- most felids (e.g. lions, tigers, cheetahs, lynx, puma)
- most canids (e.g. wolves, wild dogs, jackals)
- all bears and pandas
- all Old World monkeys
- most New World monkeys
- elephants
- tapirs.

The list also includes a small number of species of dangerous spiders and scorpions.

Some farmers in England rear bison (*Bison bison*) for their meat, and they need a licence under this Act to keep these animals.

5.3.3.2 Animal Welfare Act 2006

The purpose of the Animal Welfare Act 2006 was to extend animal cruelty laws in England and Wales so that persons responsible for animals – in most cases vertebrates – are required to ensure that their welfare needs are met (Legal Box 5.6).

Legal Box 5.5
Dangerous Wild Animals Act 1976.

Section 1: (1) Subject to section 5 of this Act, no person shall keep any dangerous wild animal except under the authority of a licence granted in accordance with the provisions of this Act by a local authority.
(2) A local authority shall not grant a licence under this Act unless an application for it—
(a) specifies the species (whether one or more) of animal, and the number of animals of each species, proposed to be kept under the authority of the licence;
(b) specifies the premises where any animal concerned will normally be held;
(c) is made to the local authority in whose area those premises are situated;
(d) is made by a person who is neither under the age of 18 nor disqualified under this Act from keeping any dangerous wild animal; and
(e) is accompanied by such fee as the authority may stipulate…

Legal Box 5.6 Animal Welfare Act 2006.

Section 1: Animals to which the Act applies
(1) In this Act, …, "animal" means a vertebrate other than man.

Section 2: "Protected animal"
An animal is a "protected animal" for the purposes of this Act if—
(a) it is of a kind which is commonly domesticated in the British Islands,
(b) it is under the control of man whether on a permanent or temporary basis, or
(c) it is not living in a wild state.

Section 4: Unnecessary suffering
(1) A person commits an offence if—
(a) an act of his, or a failure of his to act, causes an animal to suffer,
(b) he knew, or ought reasonably to have known, that the act, or failure to act, would have that effect or be likely to do so,
(c) the animal is a protected animal, and
(d) the suffering is unnecessary.

Section 9: Duty of person responsible for animal to ensure welfare
(1) A person commits an offence if he does not take such steps as are reasonable in all the circumstances to ensure that the needs of an animal for which he is responsible are met to the extent required by good practice.
(2) For the purposes of this Act, an animal's needs shall be taken to include—
(a) its need for a suitable environment,
(b) its need for a suitable diet,
(c) its need to be able to exhibit normal behaviour patterns,
(d) any need it has to be housed with, or apart from, other animals, and
(e) its need to be protected from pain, suffering, injury and disease.

The Act gives new powers to local authority inspectors, but anyone, including a member of the public, may bring a prosecution under the Act. Although the Act does not specifically refer to zoos its provisions could be used to prosecute a zoo if welfare requirements were not being met.

5.3.3.3 Animals (Scientific Procedures) Act 1986

If a zoo wishes to conduct an experiment on an animal as part of a research programme it may need to comply with the Animals (Scientific Procedures) Act 1986. This

Legal Box 5.7 Animal (Scientific Procedures) Act 1986.

> *Section 1: (1) Subject to the provisions of this section, "a protected animal" for the purposes of this Act means any living vertebrate other than man and any invertebrate of the species Octopus vulgaris from the stage of its development when it becomes capable of independent feeding.*

The Act controls certain procedures which are not conducted for veterinary or husbandry reasons. 'Regulated procedures' are defined in s.2(1) of the Act:

> *Section 2: (1) Subject to the provisions of this Section, "a regulated procedure" for the purposes of this Act means any experimental or other scientific procedure applied to a protected animal which may have the effect of causing that animal pain, suffering, distress or lasting harm.*

The Home Office may issue a licence where a regulated procedure is to be performed for a *bona fide* scientific purpose. The performance of a regulated procedure without such a licence would constitute an offence under s.3 of the Act:

> *Section 3: No person shall apply a regulated procedure to an animal unless—*
> *(a) he holds a personal licence qualifying him to apply a regulated procedure of that description to an animal of that description;...*

Legal Box 5.8 Wildlife and Countryside Act 1981.

Under the Wildlife and Countryside Act 1981, s.14:

> *(1) ...if any person releases or allows to escape into the wild any animal which:*
> *(a) is of a kind not ordinarily resident in and is not a regular visitor to Great Britain in a wild state; or*
> *(b) is included in Part I of Schedule 9,*
> *he shall be guilty of an offence.*
> *(2) ...if any person plants or otherwise causes to grow in the wild any plant which is included in Part II of Schedule 9, he shall be guilty of an offence.*

In this context the term 'animal' applies to adult animals and includes any egg, larva, pupa, or other immature stage (s.27(3)).

Act transposes into UK law Council Directive 86/609/EEC, which makes provision for the protection of animals used for experimental or other scientific purposes. This Act applies to animals kept in zoos where they fulfil the criteria laid down in s.1(1) (Legal Box 5.7).

5.3.3.4 Wildlife and Countryside Act 1981

The Wildlife and Countryside Act 1981 protects a wide range of animals and plants in Great Britain. The degree of protection depends upon the species and the Act may be relevant to a zoo in a number of contexts, for example:

- if it takes native wild animal species into captivity
- when a new zoo development threatens the sheltering place of certain species, for example bats, rare birds or great crested newts (*Triturus cristatus*)
- if it controls the numbers of any native species
- if it reintroduces animals back into the wild in Britain.

The introduction of exotic animal species into the wild is prohibited under the Act (Legal Box 5.8). In addition, certain listed animal and plant species may not be released.

Animals which are established in the wild and may not be released except under licence are listed in Part I of Schedule 9 to the Act. They include:

- coypu (*Myocaster coypus*)
- red-necked wallaby (*Macropus rufogriseus*)
- American mink (*Mustela vison*)
- black rat (*Rattus rattus*)
- grey squirrel (*Sciurus carolinensis*)
- golden pheasant (*Chrysolophus pictus*)
- African clawed toad (*Xenopus laevis*)

- Canada goose (*Branta canadensis*)
- ruddy duck (*Oxyura jamaicensis*)
- white-tailed eagle (*Haliaetus albicilla*).

Section 14(3) of the Act provides a defence where an introduction has been caused unintentionally, provided that all reasonable steps were taken to prevent the escape. The law intended to prevent unwanted releases to the wild also hinders projects intended to reintroduce species to their former habitats. After a long absence, beavers (*Castor fiber*) have recently been introduced into the wild in Scotland under licence from the Scottish government.

5.3.3.5 Countryside and Rights of Way Act 2000 and Biodiversity Action Plans

Some of the international obligations imposed by the Biodiversity Convention have been given effect in England and Wales by the Countryside and Rights of Way Act 2000. Section 74(1) imposes a duty on government ministers, government departments and the National Assembly for Wales to have regard, in the carrying out of their functions, to the purpose of conserving biological diversity in accordance with the Convention. The Secretary of State (in England) and the National Assembly for Wales (the 'listing authorities') are required to publish lists of important species and habitats (s.74(2)) after consultation with Natural England or the Countryside Council for Wales, respectively.

Section 74(7) defines conservation as including 'the restoration or enhancement of a population or habitat'. This effectively imposes a legal duty on the government to restore threatened species and habitats. In practical terms this duty applies to those species and habitats included in the UK Biodiversity Action Plan (BAP) (Anon., 1995b) which was drawn up to comply with Articles 6 and 8 of the Biodiversity Convention. Some zoos in the UK are contributing to Biodiversity Action Plans. Dudley Zoo has been breeding red squirrels (*Sciurus vulgaris*), and Chester Zoo and *Durrell* have been breeding sand lizards (*Lacerta agilis*) as part of the BAPs for these species.

Fighting wildlife crime in the UK In the UK the following are responsible for combating wildlife crime:

- National Wildlife Crime Unit
- Partnership for Action Against Wildlife Crime (PAW)
- CITES Enforcement Teams at Heathrow Airport and the Port of Dover
- Customs CITES liaison officers.

5.3.4 United States law

The federal and state laws of the USA include a very wide range of laws concerned with the protection of animals. A good introduction has been written by Freyfogle and Goble (2009). Listed below are some of the laws concerned with the protection and conservation of wildlife, including some that are concerned with the establishment of funds to support *in-situ* projects outside the United States. Only laws which are concerned with exotic species and their importation, exportation and transportation have been included.

5.3.4.1 Endangered Species Act (16 USC §1531–1543)

This Act prohibits the importation, exportation, taking and commercialisation in interstate or foreign commerce of wildlife, fish and plant species that are listed as threatened or endangered. It also implements the provisions of CITES. Section 4(f) of the Act directs the Secretary of Interior or the Secretary of Commerce to develop and implement recovery plans for animal and plant species listed as endangered or threatened. American zoos play an important part in recovery plans for a number of native animal species.

The recovery programme for the California condor (*Gymnogyps californianus*) involves captive breeding facilities at San Diego Wild Animal Park, Los Angeles Zoo, and The Peregrine Fund's World Center for Birds of Prey. The project is overseen by the US Fish and Wildlife Service and the California Condor Recovery Team.

The US Fish and Wildlife Service has established a breeding colony of black-footed ferrets (*Mustela nigripes*) at its National Black-footed Ferret Conservation Center near Laramie, Wyoming, as part of the recovery programme for this species. In addition populations have been established at Phoenix Zoo, Louisville Zoological Garden, Henry Doorly Zoo, Smithsonian National Zoological Park's Conservation Research Center, and the Cheyenne Mountain Zoological Park and Toronto Zoo, Canada.

Both of these projects are supported by the International Species Information System (ISIS) (see Section 13.3.1). At 9 May 2008 ISIS listed 138 (58.78.2) black-footed ferrets in 15 North American zoos, and 79 (34.39.6) California condors held at Los Angeles Zoo, Portland Zoo and San Diego Wild Animal Park. At the end of May 2008 there were just 298 California condors left in the world.

5.3.4.2 Other US wildlife laws

Lacey Act (18 USC 42; 16 USC §3371–3378) This Act is concerned with the humane treatment of wildlife shipped to the United States. In addition the Act prohibits the importation, exportation, transportation, sale or purchase of wildlife or fish taken or possessed in violation of state, federal, tribal or foreign laws. The Act is an important tool in deterring the illegal trade in and smuggling of wildlife, and it allows for the provision of federal assistance to the states and foreign governments in the enforcement of their wildlife laws.

Marine Mammal Protection Act (16 USC §1361–1407) A moratorium on the taking and importation of marine mammals was established under this Act, including their parts and products. It defines federal responsibilities for marine mammal conservation and assigns management authority for the walrus, sea otter, polar bear, dugong and manatee to the Department of the Interior.

Rhinoceros and Tiger Conservation Act (16 USC §5301–5306) This Act prohibits the import, export or sale of any product, item or substance containing any substance derived from tiger or rhinoceros species. These prohibitions also apply to anything labelled or advertised as containing any substance from these species.

Wild Bird Conservation Act (16 USC §4901) This Act promotes the conservation of exotic birds by encouraging wild bird conservation and management programmes in countries of origin. In addition, it requires that all US trade in exotic birds is sustainable and of benefit to the species. Where necessary, the Act allows for the restriction or prohibition of imports of exotic birds.

Antarctic Conservation Act (16 USC §2401) This Act provides for the conservation and protection of Antarctic flora and fauna. It makes it unlawful for anyone in the USA to possess, sell, offer for sale, deliver, receive, carry, transport, import or export from the USA any native mammal or bird taken in Antarctica.

Migratory Bird Treaty Act (16 USC §703–712) Migratory species of birds are protected by this Act. It restricts the circumstances in which it is lawful to hunt, kill, capture, possess, buy or sell any migratory species, including its feathers, other parts, eggs, nests or products.

Other Acts A number of Acts provide financial support for *in-situ* conservations projects for particular taxa:

- African Elephant Conservation Act (16 USC §4201–4245)
- Asian Elephant Conservation Act (16 USC §4261–4266)
- Marine Turtle Conservation Act (16 USC §6601–6607)
- Great Apes Conservation Act (16 USC §6301–6305).

5.3.4.3 United States Fish and Wildlife Service (USFWS)

The Acts listed above are some of those enforced by the United States Fish and Wildlife Service (USFWS). The USFWS is a federal agency of the Department of the Interior. It operates a wildlife forensics laboratory in Oregon. The laboratory has the capacity to identify many protected species by DNA analysis, microscopic methods and other means. It produces identification keys for some taxa to assist its agents and others in the identification of protected species. In the USA, wildlife may generally only be exported or imported through one of 18 designated ports of entry, including New York, Miami, Los Angeles and Seattle.

In 2007 the USFWS investigated 12,177 cases involving breaches of wildlife laws, including 67 relating to African elephants and 70 cases concerned with rhino and tiger products. Overall, its investigations resulted in the imposition of fines of over $14 million (USFWS, 2008).

5.3.5 Wildlife and animal law in countries outside the UK, Europe and the USA

All countries have laws which are intended to protect their wildlife, promote nature conservation and restrict hunting. Many also have specific laws relating to the operation of pet shops, the use of animals in circuses, cruelty to domestic and wild animals, and the use of animals in experiments. These laws occur at the national, regional and international level, and no single book can cover all of them. Many countries now publish legislation on their government's website. This may often be found in the pages relating to the department responsible for the environment, forestry or agriculture. The departmental responsibility for animal wildlife laws varies from country to country. Some useful texts and websites are listed at the end of this chapter.

5.4 HEALTH AND SAFETY LAW

The law generally imposes a duty of care on the owner or occupier of premises to protect the health and safety of his staff and visitors. Under English law, where children visit premises, the standard of care which must be taken must consider the special risks to children (s.2(3)(a), Occupiers' Liability Act 1957). This is important because a zoo may not delegate the responsibility for a child's safety to a parent or guardian. Where visitor barriers are installed to keep people away from enclosure fences and other barriers it is essential that they are designed to exclude children. In some jurisdictions a zoo owner may even owe a limited duty of care to a trespasser who breaks into a zoo and is harmed as a result (e.g. in England and Wales, Occupiers' Liability Act 1984).

Local health and safety legislation is likely to require an employer to protect zoo staff and visitors from exposure to disease, risk of injury from equipment and vehicles and other sources of harm.

5.5 ENVIRONMENTAL PROTECTION

Zoos consume energy and materials, and they produce pollution. Most countries have a considerable amount of legislation relating to the release of pollutants into the environment and the control of waste. Zoos should design their facilities and manage their operations in compliance with this legislation. For example, waste from animal houses and enclosures should not be allowed to pollute watercourses. This material, general refuse and clinical waste should be disposed of according to local regulations.

5.6 CODES OF PRACTICE AND GOVERNMENT POLICIES

A code of practice is a set of rules written by an organisation, usually with a view to helping its employees, and others, understand a particular area of the law. It is not law itself, but is usually at least partly based on the law. In legal proceedings a court is likely to give a considerable amount of weight to any infringements of a code of practice, where its provisions cover matters that are the subject of those proceedings.

In the UK, the *Secretary of State's Standards of Modern Zoo Practice* (SSSMZP, 2004) is a code of practice for zoos in England issued under a power conferred by s.9 of the Zoo Licensing Act 1981.

Some governments may publish policies in relation to the management of zoo animals. In New South Wales, Australia, the government has produced a *Policy on the Management of Solitary Elephants in New South Wales* pursuant to Clause 8(1) of the Exhibited Animals Protection Regulation, 2005 (Anon., 2009c).

5.7 FURTHER READING AND RESOURCES

A global overview of zoo legislation can be found in Cooper (2003). Some of the information contained in this article is now out of date but, nevertheless, it provides a useful summary of the variation in approaches taken by different countries.

UK Acts of Parliament and Statutory Instruments are available at the website of the Office of Public Sector Information (www.opsi.gov.uk). The *Secretary of State's Standards of Modern Zoo Practice* and the Zoos Forum Handbook are available on the DEFRA website at www.defra.gov.uk/wildlife-pets/zoos/index.htm

European law is available from the website of the European Union (at http://europa.eu) and more information about US federal law may be obtained from the US Fish and Wildlife Service, Digest of Federal Law Sources (www.fws.gov/laws/lawsdigest/resourcelaws.htm), and the Animal Legal & Historical Center of the Michigan State University College of Law (www.animallaw.info). Information about zoo legislation in South East Asia can be found at the website of the Zoo Outreach Organisation (www.zooreach.org).

It should be remembered that it is important to consult up-to-date sources of law. Textbooks are not generally considered to be sources of law, and it is always more useful to consult the original legislation as published by the relevant government, and the associated subsequent amendments.

A more detailed treatment of many aspects of UK, European and international wildlife law can be found in the following texts:

Bowman, M., Davies, P. and Redgwell, C. (2010). *Lyster's International Law*. Cambridge University Press, Cambridge, UK.

Broom, S. and Legge, D. (1997). *Law Relating to Animals*. Cavendish Publishing Limited, London and Sydney.

Freyfogle, E. T. and Goble, D. D. (2009). *Wildlife Law: A Primer*. Island Press, Washington DC.

IUCN (1987). *African Wildlife Laws (Environmental Policy and Law)*. IUCN, Gland, Switzerland.

Le Prestre, P. G. (2002). *Governing Global Biodiversity: The Evolution and Implementation of the Convention on Biological Diversity*. Ashgate, Burlington, VT.

Rees, P. A. (2002). *Urban Environments and Wildlife Law: A Manual for Sustainable Development*. Blackwell Science, Oxford.

Reeve, R. (2002). *Policing International Trade in Endangered Species: The CITES Treaty and Compliance*. Earthscan Publications Ltd., London.

5.8 EXERCISES

1. To what extent does EU law require zoos to have a conservation function?
2. How does international law restrict the movement of wildlife from one country to another?
3. Describe how the law regulates the operation of zoos in a country of your choice.
4. To what extent does international law envisage a conservation role for zoos?

6 ETHICS AND ZOOS

If a being suffers there can be no moral justification for refusing to take that suffering into consideration.

Peter Singer

Conservation Status Profile

Cook Strait tuatara
Sphenodon punctatus
IUCN status: Lower risk/ least concern
CITES: Appendix I
Population trend: Unknown

An Introduction to Zoo Biology and Management, First Edition. Paul A. Rees.

Some people believe that zoos perform a useful function in a world where many species of animals (and plants) are threatened with extinction; others believe they have no conservation function and do harm to animals. Some of these people are badly informed and hold misconceptions about the nature of zoos, but others are very well informed and have serious and legitimate concerns about animal welfare.

6.1 ETHICS

6.1.1 Introduction

Ethics is the branch of philosophy which is concerned with the morality of an individual's actions. Whether or not such actions are considered to be morally acceptable depends upon whether we attach more importance to their motivation or to their consequences.

The idea that we have a moral responsibility to ensure the survival of other species for the benefit and enjoyment of future generations is enshrined in the preambles of a number of international treaties concerned with the conservation of biodiversity:

> Recognising that wild flora and fauna constitute a natural heritage of aesthetic, scientific, cultural, recreational, economic and intrinsic value that needs to be preserved and handed on to future generations;...
>
> Convention on the Conservation of European Wildlife and Natural Habitats, 1979

> Determined to conserve and sustainably use biological diversity for the benefit of present and future generations,...
>
> Convention on Biological Diversity, 1992

This is a curious notion. It is difficult to see how future generations can have rights to anything because the individuals who would have these rights have not yet been born. Nevertheless, the fact that so many states have signed treaties that refer to obligations to future generations is an important recognition of our collective responsibility for the biological resources of the planet.

6.1.2 Ideas of morality

It is possible for one's actions to be morally justified even if the consequences of these actions do harm. On the other hand, someone may behave in a manner which others may consider immoral in order to achieve a morally desirable objective.

A zoo may have the good intention of saving a species from extinction but in trying to achieve this it may cause harm to individuals of that species by keeping them in captivity.

Utilitarianism is the proposition that the moral worth of an action is solely determined by its contribution to overall utility. It requires that one should act in such a way as to do the greatest amount of good for the largest number of individuals. In other words, the end justifies the means. This school of thought is generally credited to Jeremy Bentham (1823) who, in considering which animals we should protect from injury, considered that:

> The question is not, can they reason? Nor, can they talk? But, can they suffer?

An advocate of utilitarianism might argue that the harm done to a small number of individuals is justified if a larger number of individuals of the same species benefits. For example, we could justify keeping a small number of gorillas in zoos if these zoos raised money to protect wild gorillas. This argument is often advanced by conservationists. However, others claim that the harm suffered by individual animals in zoos cannot be justified by any benefits to other individuals or to the species as a whole.

In the late 18th century Immanuel Kant argued that the act itself, to be moral, must have a pure intention behind it, regardless of the final consequences (Gregor, 1998). His 'counter-utilitarian' idea would give greater consideration to the rights of individual gorillas kept in a zoo than to any overall benefits to wild gorillas in general. This is generally the position taken by organisations whose primary interest is animal welfare rather than conservation.

Neither of these viewpoints is right or wrong; they are alternative ways of thinking about the same situation. Inevitably, conservationists will generally take the view that our primary concern should be the survival of the species and that it is acceptable for a relatively small number of individual animals to be kept in captivity if this helps to achieve this aim. On the other hand, those primarily concerned with animal welfare consider that the interests of the individual animal must be paramount and many therefore argue that it is unacceptable to keep animals in zoos even if this approach threatens the survival of the species in the wild.

Whether or not we believe that animals in zoos are worse off than wild individuals of the same species depends upon how they are treated (Box 6.1) and could be determined by applying the 'basic needs test' and the 'comparable life test'.

Box 6.1 What activities are acceptable in a modern zoo?

Consider the following activities that might be seen in zoos. Which of these do you find morally acceptable?

- Dressing a chimpanzee in human clothing (Fig. 6.1).
- Keeping common species that are of very low conservation value.
- Training animals to perform in shows for the entertainment of visitors (Fig. 6.2).
- Removing a young animal from its mother and sending it to another zoo.
- Selling helium balloons which may escape and contribute to a litter problem.
- Keeping tropical animals in a cold climate.
- Taking animals from the wild.
- Confining animals in spaces that are much smaller than the area of their home range in the wild.
- Performing procedures on a gorilla in order to artificially inseminate her.
- Housing an old bear alone.
- Culling surplus animals.
- Using birds of prey in free-flight displays.
- Using animals in 'animal encounter' experiences for the public.
- Reintroducing animals into the wild knowing that some individuals are likely to perish.
- Allowing animals to reproduce as an enrichment and culling the offspring after they have weaned.

Fig. 6.1 'Consul' the chimpanzee was popular with visitors to Belle Vue Zoo, Manchester. (Courtesy of Chetham's Library, Manchester.)

(*continued*)

Box 6.1 *(cont'd)*

Fig. 6.2 Are animal shows acceptable in a modern zoo?

The answer to many of these questions may depend upon the context. For example, it may be considered acceptable to take animals from the wild for breeding purposes if they are the last surviving members of a species (e.g. the California condor) but not if they are from a common species. It may be acceptable to keep an old bear alone if he does not get along with members of his own species.

6.1.3 The basic needs test and the comparable life test

A philosopher would apply two tests when considering whether or not an animal should be kept in a zoo:

- First, can a zoo provide for the basic physiological and psychological needs of animals (the basic needs test)?
- Second, can a zoo provide a life at least as good as the life the animal could expect in the wild (the comparable life test)?

De Grazia (2002) has argued that 'taking animals from the wild involves so many harms that it should rarely, if ever, occur'. Nowadays, of course, zoos rarely do this. He argues that if an animal is to be kept in a zoo there is an obligation to provide it with a life that is at least as good as it would have been likely to have in the wild. Making a zoo animal worse off would constitute unnecessary harm.

Some species live longer in zoos than in the wild. This must be a benefit as death is clearly a harm. However, some would argue, for example, that elephants in zoos often die young and many of those that survive live a miserable life.

Keeping animals in a zoo is considered by some people to be disrespectful. However, animals are unlikely to have a substantial understanding of their own best interests so the argument that captivity is disrespectful to them is invalid. De Grazia has concluded that keeping animals in zoos is neither intrinsically harmful nor necessarily disrespectful. He asserts that zoos have a responsibility to cultivate an attitude of respect for animals as beings with moral status; as beings of importance in their own right. This view would preclude the use of elephants in entertainment, but some zoos train their elephants to perform, claiming that this is a form of behavioural enrichment. In some cases it may also be important physical exercise for the animals. In other cases, training may be important for veterinary treatment, for example training an elephant to lift its foot so that the sole may be examined for infections or to open its mouth for a dental examination or procedure. However, teaching zoo elephants to play musical instruments or to perform other 'entertaining' activities is more difficult to justify.

De Grazia singles out the great apes and dolphins as taxa for which there should be a strong presumption against keeping them in zoos. Remarkably, he makes no specific reference to elephants.

Biography 6.1

Peter Singer (1946–)

Peter Singer is an Australian philosopher and currently Professor of Bioethics at Princeton University. He is the author of *Animal Liberation* which is perhaps the most influential book on animal rights. Singer is a supporter of the *Great Apes Project*, an international organisation of primatologists, psychologists, philosophers and others who advocate a UN Declaration of the Rights of Great Apes as a precursor to demanding the release of apes from captivity in research laboratories and elsewhere.

6.1.4 Animal rights

Some people believe that animals should have 'rights' based on their capacity to have 'interests'. Others do not. There is an extensive literature covering this issue and it is not my intention to reiterate all of the arguments here. I shall begin with the contention that at least some animals may have 'rights' and that in order to qualify for these rights a species must fulfil a number of conditions.

Singer (1995) has argued that:

> To avoid speciesism we must allow that beings who are similar in all relevant respects have a similar right to life...

'Speciesism' is a term used by Singer to describe a prejudice in favour of the interests of members of one's own species and against those of members of other species. The term was first coined by Richard Ryder in 1970 (Ryder, 1975).

It is now generally accepted that mammals (and members of other sophisticated animal taxa) feel pain and experience suffering in much the same way as humans, so inflicting physical suffering on such animals is morally difficult to justify. The morality of taking life is more complex; although it may not involve suffering, the loss of life is clearly a harm. Singer maintains that the life of an animal that is self-aware, capable of abstract thought, of planning for the future and of complex acts of communication is more valuable than the life of an animal without these capacities. He argues for a basic moral principle of 'equal consideration of interests' to apply to members of other species as well as to our own.

6.1.4.1 Autonoetic consciousness

The term 'autonoetic consciousness' is used by some psychologists as synonymous with what philosophers call 'phenomenal consciousness' of the past, present and future. The autonoetic consciousness paradigm supports a moral hierarchy of persons, near-persons and merely sentient animals. Members of each level within the hierarchy deserve some form of special respect vis-à-vis members of the lower levels.

Persons deserve the highest level of respect. They have a full-blown biographical sense of their own lives. This category probably only contains humans. Some animals, such as chimpanzees, are undoubtedly near-persons in the sense that they have a less expansive form of autonoetic consciousness involving at least their non-immediate past and non-immediate future. At the other end of the hierarchy, some animals are merely sentient: they are conscious of pain and pleasure but live in the present with no meaningful sense of their own past or future.

6.1.5 The mirror test (red-spot test or mark test)

The Latin name *Homo sapiens* translates to 'wise man', or perhaps 'thinking man'. The ability to solve complex problems has been uniquely associated with humans until relatively recently. Associated with this has been the capacity to recognise one's self; to be self-aware.

We now know that a number of species have capabilities that were once thought to be unique to humans and which, in the past, were used to justify a superior position for humans within the Animal Kingdom. This superior position has been used to justify our treatment of 'lower' forms in our agricultural systems, animal experimentation and in other circumstances where people interact with animals, including the keeping of animals in zoos.

When chimpanzees (*Pan troglodytes*), gorillas (*Gorilla gorilla*) or orangutans (*Pongo* sp.) see themselves in a mirror they appear to be able to recognise themselves (Fig. 6.3). However, when monkeys see themselves in a mirror they react to their own image as if it were another monkey.

A number of experiments have used the 'red-spot test' which involves a spot of red dye being placed on the animal's face under anaesthetic or while asleep, and then observing the animal's response to the spot when it looks in a mirror. Apes appeared to show self-awareness by touching the spot more often than a control area. For example, they touched an ear marked with a spot more often than the other (unmarked) ear (Rogers, 1997). Chimps also demonstrated self-awareness by using a mirror to examine parts of the body that could not be seen directly (Povinelli and Preuss, 1995).

Fig. 6.3 When chimpanzees see their own reflections they know they are looking at themselves because they are capable of self-recognition.

The experimental evidence for self-recognition in apes is inconsistent – because some of the apes tested in red-spot experiments did not respond to the spots – and is therefore inconclusive.

Gallup (1970) suggested that the capacity for self-recognition may 'not extend below' humans and the great apes. However, self-examination behaviour has been recorded in dolphins exposed to mirrors and video images. They will also examine marked areas of their bodies in a mirror (Marten and Psarakos, 1995). It has been suggested that the tendency of dolphins and killer whales to adorn themselves with seaweed around their fins or flukes, or by carrying dead fish on their snouts, might be evidence of self-awareness (de Waal, 1996).

What happens when elephants are given the opportunity to see themselves in mirrors? When Povinelli tested two Asian elephants (*Elephas maximus*) at the National Zoo in Washington DC with a mirror he found they paid little attention to their images and concluded they did not show self-recognition. However, an elephant's eyes are on the side of its head and the experiment has been criticised for using mirrors that were too small to allow the animals to see the entire side of their bodies at one time (Povinelli, 1989). A similar experiment conducted with three Asian elephants appeared to establish self-recognition in these animals when an X-shaped mark was placed on the head (Plotnik *et al.*, 2006). However, when a large feather was attached to the forehead of elephants they showed no evidence of self-recognition (Nissani and Hoefler-Nissani, 2007).

The significance of the results of mirror tests has been controversial. Experiments using mirrors have found that pigeons respond to coloured dots in a similar way to Gallup's chimps. At the time the experimenters concluded that this could not possibly be interpreted as complex behaviour or self-awareness in pigeons and offered a simpler explanation in terms of 'environmental events' (Epstein *et al.*, 1981). We now know that pigeons are capable of very complex behaviour. Recently, Prior *et al.* (2008) obtained evidence of self-recognition in the European magpie (*Pica pica*) using a mark test, suggesting that the ability has evolved separately in different vertebrate taxa.

6.1.6 Is extinction a harm?

If a species becomes extinct there is no meaningful sense in which that species is harmed because the concept of a species is a human creation. A species is merely a group of genetically similar individuals. Only individuals can have interests and only individuals can suffer harm. It is, therefore, important to consider whether species preservation is desirable if it conflicts with the welfare needs of individual animals.

The keeping of a rare species in captivity as part of a breeding programme may serve the interests of the humans who do not wish to see it disappear. But if this involves suffering for the individual animals concerned, such programmes clearly do not serve the interests of the animals themselves, unless they would have faced a certain death or considerably more suffering in the wild.

Animals – even intelligent animals like chimpanzees, dolphins and elephants – do not care if their species becomes extinct. They may care about the fate of individual friends and relatives, and be psychologically affected by their loss, but this is clearly not the same thing. Ultimately, species conservation serves human interests not the interests of the animals we seek to conserve. However undesirable the extinction of elephants might be from a human perspective, it is difficult to find a moral justification for keeping elephants in captive breeding programmes if they suffer more harm from the manner in which they are kept than they would if left in the wild. An individual elephant has interests and may suffer harm; elephant species have no interests and thus may not suffer harm.

The concept of a 'biographical life' could help us to determine which kinds of animals should have rights, but the animal rights approach does not appear to be able to accommodate the concept of species rights. Regan claims that the rights view is about the moral rights of individuals and does not recognise the moral rights of species to anything, including survival. He claims that rarity has no effect on rights and that if people are encouraged to believe that the harm done to animals matters only when they belong to endangered species, then this encourages the view that harm done to other (more common) species is morally acceptable (Regan, 1988).

Russow (1994) takes a different view. She claims that our obligations to vanishing species are not inconsistent with a general condemnation of speciesism. She proposes that we protect animals because of their aesthetic value, their rarity, their adaptations and for many other reasons, not because they belong to a particular species. We have moral obligations to protect individual animals and to ensure that there will continue to be such animals in the future. Russow would argue, for example, that it is not the species *Elephas maximus* that we admire, but individual elephants. We value encounters with rare animals because they are less frequent than encounters with common animals. Russow says we should preserve these animals because we value possible future encounters with other individuals of the same species.

Can we base rights for wildlife on a human need for the species to survive? International nature conservation law increasingly recognises the rights of future generations (of people) to exist in a world with a high degree of biodiversity (e.g. the Biodiversity Convention). However, this is difficult to argue in law. The traditional view of rights is that individuals who have not yet been born cannot have rights because

Box 6.2 Arguments against the existence of zoos (adapted from Bertram, 2004).

1. Zoos take animals from the wild and damage wild populations.	Most animals in modern western zoos have been born in zoos. CITES strictly controls imports and exports. Occasionally small numbers of animals are taken from the wild to supplement captive breeding programmes. This may be essential to increase genetic diversity and to prevent inbreeding.
2. Zoos distract attention from *in-situ* conservation efforts.	Modern zoos support *in-situ* conservation by providing funds, personnel and expertise. However, only a very small proportion of the income generated by zoos is directed towards *in-situ* projects. The best zoos link their exhibits to *in-situ* projects. For example, the jaguar exhibit at Chester Zoo is linked to the Cockscomb Basin Jaguar Project in Belize.
3. Money spent on zoos should be spent on *in-situ* conservation.	Zoos either earn or are given money for the work they do. This is not money which has been redirected from elsewhere. Zoos help to raise money for conservation. However, zoos are expensive to build and run and if the money raised to build new zoo exhibits could be diverted to *in-situ* conservation projects it would have a much greater overall impact.
4. Species should be allowed to die out with dignity.	Death in the wild is not dignified. In any case, dignity is a human concept.
5. It is wrong to restrict an animal's freedom.	Wild animals are constrained by habitat loss. We constrain farm animals. There is no fundamental difference between keeping animals in zoos and keeping them as domestic pets or as farm animals. However, some zoos clearly keep animals in very small enclosures.
6. Confining wild animals makes them suffer.	This is true if conditions are unsuitable. Food quality and veterinary care are good in zoos. Stereotypic behaviour may be an indicator of poor welfare but its origins are not yet clear. Wild animals may suffer from starvation, predation, diseases of old age, etc.
7. It is degrading for animals to be 'gawped at' in zoos.	Most species make little attempt to hide from public view. Most spend very little time trying to escape. Why are zoo animals different from birds visiting bird tables?
8. Animals should not be kept for entertainment.	Good zoos do not use their animals in any overt way to provide entertainment. However, most visitors go to zoos for a pleasant day out, so in a general sense, they must have an entertainment value. It is difficult to argue that it is acceptable to keep animals for food but not entertainment. Performing animals have been removed from most good zoos.
9. We should be conserving habitats, not single species.	We should do both. Protecting habitats alone will not save some species because they are killed by poachers. Species conservation provides a safety net.
10. It is more cost effective to conserve animals in the wild than in zoos.	This is true in many cases – it depends how you calculate the costs. The two are not alternatives. Conserving in the wild may be less safe. In some countries game guards have colluded with poachers.
11. Zoos can only save a small number of species.	This is true, but *in-situ* conservation programmes can also only save a few species.
12. Reintroduction back into the wild does not work.	Reintroductions do work but often require repeated attempts. They do not all work first time. Examples of successes are the Arabian oryx, golden lion tamarin and Hawaiian goose. The role of zoos in reintroductions so far has probably been exaggerated. Most zoo animals will live out their lives and eventually die in a zoo. Zoos often quote small species, for example Partula snails, as captive breeding success stories. Clearly, a large zoo facility is not required to breed snails.

13. Zoos give out the wrong educational messages.	This is because they display animals out of context. Most people gain something of educational value from a zoo visit. Hundreds of thousands of children visit zoos on school visits annually. We could read books or watch TV to get the same information. Most people who visit zoos now would not do this instead.
14. Films and TV programmes make zoos unnecessary.	These are not alternatives. These programmes and films do not breed animals.
15. People can see animals in the wild.	Most cannot. Most Indians have never seen a wild Asian elephant.
16. No useful research is possible in zoos.	Zoos conduct research on veterinary problems, husbandry, behaviour, etc. Most of this has little to do with conservation. Much of it is about welfare.
17. Some zoos are clearly bad.	This is true but not in the UK and most developed countries. The EU Zoos Directive aims to improve zoos in EU Member States.
18. Zoo staff do not care about animal welfare.	Most zoo keepers are very committed to the care of their animals. There have been occasional prosecutions of animal keepers for cruelty.
19. Zoos cull animals.	This does occasionally happen. Excess or sick animals may be killed. But this happens to pets and farm animals too.
20. Some species should never be kept in zoos.	This is true. Cetaceans cannot be adequately housed in zoos. Polar bears and elephants are also problematic.

they do not yet exist. Nevertheless, many international treaties clearly recognise the importance of protecting wildlife for the 'good of mankind' and for the 'benefit of future generations'.

Wheatley (1995) has examined the international obligations imposed upon the State to conserve species. He argues that the State must take into account the rights of other States, 'peoples' within the State and the rights of future generations, and concludes that species have a right to freedom from extinction. Although Wheatley takes a 'rights' approach these rights are held by States or by people rather than by the species themselves.

6.1.7 Misconceptions about zoos

A number of general objections have been raised to the existence of zoos. Some people believe that zoos are fundamentally wrong and have no legitimate purpose. They claim that they have no significant conservation or education function, and that animal welfare in zoos is poor. Those who support zoos have attempted to answer these objections and the discussion in Box 6.2 is based on Bertram (2004).

6.2 ANIMAL WELFARE ORGANISATIONS AND ANTI-ZOO GROUPS

6.2.1 Introduction

A large number of organisations have an interest in animal welfare, and some actively lobby to remove certain types of animals from zoos or even to close zoos down. Some are large, international, professionally run organisations, while others are lobby groups established to achieve one particular end, for example to secure the removal of a particular animal from a particular zoo (see Box 6.3). Large animal welfare organisations employ animal welfare scientists and lawyers, and their concerns about welfare should be taken seriously. They petition governments and give evidence to government committees and enquiries. Some of these organisations

Box 6.3 Recent campaigns to remove elephants from zoos.

A number of individual campaigns have been organised in the United States and elsewhere against zoos keeping elephants alone or in what some perceive as poor conditions (Table 6.1). On 20 June 2009 *In Defense of Animals* organised the *International Day of Action for Elephants in Zoos*. Protests were held at 23 zoos in the United States, two in Canada, one in Spain and one in Thailand.

Table 6.1 Recent campaigns to remove elephants from zoos.

Elephant/Zoo	Campaign/Organisation
ZSL London Zoo, England	Elephant-Free London/Born Free Foundation
Jenny, Dallas Zoo, USA	Concerned Citizens for Jenny
Henry Vilas Zoo, USA	Citizens for Human Treatment of Vilas Elephants
Woodland Park Zoo, Seattle, USA	Friends of Woodland Park Zoo Elephants
Billy, Los Angeles Zoo, USA; Lucky, San Antonio Zoo, USA	Save Billy Campaign/Free Lucky Campaign/Voice for the Animals Foundation
Los Angeles Zoo, USA	Elephant Sanctuaries Not Captivity/Last Chance for Animals
Philadelphia Zoo, USA	Friends of Philly Zoo Elephants
Reid Park Zoo, Arizona, USA	Save Tucson Elephants
Valley Zoo, Edmonton, Canada	Valley Zoo Elephant Campaign
Susi, Barcelona Zoo, Spain	Free Susi Campaign/Libera (Spain)
Taronga Zoo, NSW, Australia	Campaign to Stop Importation of Thai Elephants to New South Wales/World League for Protection of Animals/ Animal Liberation NSW

Some zoos in the United States have moved their elephants to sanctuaries because they no longer believe that they can provide suitable accommodation for them. These include Detroit Zoo, Frank Buck Zoo (Texas), Mesker Park Zoo (Indiana), Henry Vilas Zoo (Wisconsin), Chehaw Wild Animal Park (Georgia), Alaska Zoo and the Louisiana Purchase Gardens and Zoo. In the UK, zoos in Bristol, Dudley, Edinburgh, London, Cricket St Thomas and the Welsh Mountain Zoo no longer keep elephants. In November 2009 the government of India issued a circular banning the keeping of elephants in zoos (Central Zoo Authority Circular File No. 7-5/2006-CZA (Vol. II)).

have a large public membership and therefore represent the views of substantial numbers of people.

6.2.2 Examples of animal welfare and anti-zoo groups

6.2.2.1 Royal Society for the Prevention of Cruelty to Animals (RSPCA)

The Royal Society for the Prevention of Cruelty to Animals began life as the Society for the Prevention of Cruelty to Animals and was founded in England in 1824 to promote the humane treatment of work animals, such as horses and cattle, and of household pets. The RSPCA is not primarily an anti-zoo organisation but it has run campaigns aimed at zoos and funded research into the welfare of animals living in zoos. In 2002 it published *Live Hard, Die Young – How Elephants Suffer in Zoos*, which called for the phasing out of elephants in European zoos, based on controversial research conducted by Clubb and Mason (2002).

6.2.2.2 Universities Federation for Animal Welfare (UFAW)

The Universities Federation for Animal Welfare is a charity that works to develop and promote improvements in the welfare of all animals through scientific and educational activity worldwide. UFAW organises conferences on welfare issues, publishes books, produces videos and technical reports, and provides advice to the government on animal welfare including legislation. It publishes the journal *Animal Welfare*, which contains, among other things, papers on zoo animal welfare. UFAW funds research into animal welfare and makes an annual *Wild Animal Welfare Award*. In 2007 Bristol Zoo won the award for its development and testing of a new type of animal-friendly flipper band for identifying penguins (Fig. 13.5b). UFAW does not campaign against zoos.

6.2.2.3 Born Free Foundation and Zoo Check

The Born Free Foundation is an international wildlife charity that works throughout the world to stop individual animal suffering and to protect threatened animal species in the wild. The Born Free Foundation does not believe that the conservation benefits claimed by zoos justify the keeping of wild animals in captivity. *Zoo Check* is a programme operated by the Foundation which works to prevent captive animal suffering and phase out zoos. This organisation has done some good work in drawing attention to the plight of animals in bad zoos but is fundamentally opposed to the existence of all zoos as a matter of principle.

6.2.2.4 In Defense of Animals (IDA)

In Defense of Animals is an international non-profit animal protection organisation dedicated to ending the institutionalised exploitation and abuse of animals by defending their rights, welfare and habitats. It campaigns to stop the captive breeding and exhibition of dolphins and whales and in 2009 it organised a day of action for elephants in zoos (Box 6.3).

6.2.2.5 Animal Defenders International (ADI)

Animal Defenders International campaigns against the use of animals in fur farms, experiments and entertainment. Although it does not specifically target zoos in general, ADI has campaigned to help badly treated circus and zoo animals and spoken out about the fate of particular animals such as elephants, chimpanzees and dolphins in some zoos.

6.2.2.6 Captive Animals' Protection Society (CAPS)

The Captive Animals' Protection Society is based in the UK. It was founded in 1957 and campaigns on behalf of animals in circuses, zoos and the entertainments industry. One of its stated aims is to end the captivity of animals in zoos and the Society actively discourages the public from visiting zoos, claiming that they have no conservation or educational value.

6.2.2.7 Humane Society of the United States

The Humane Society of the United States (HSUS) was established in 1954 and is the largest animal protection organisation in the United States. It works to prevent cruelty, exploitation and neglect of animals. Although much of its work concerns domestic and farm animals, it has also taken action against roadside zoos in America.

6.3 FURTHER READING

There is a significant amount of published material on ethics, animal rights and the mental lives of animals. Some useful texts are listed below. Animal welfare and animal rights organisations produce a great deal of literature to support their arguments. This is usually available from their websites (see Appendix 2).

Bostock, S. St.C. (1993). *Zoos and Animal Rights: The Ethics of Keeping Animals*. Routledge, London and New York.

De Grazia, D. (2002). *Animal Rights: A Very Short Introduction*. Oxford University Press, Oxford.

de Waal, F. (1996). *Good Natured: The Origins of Right and Wrong in Humans and Other Animals*. Harvard University Press, Cambridge, MA and London.

Dolins, F. L. (ed.) (1999). *Attitudes to Animals: Views in Animal Welfare*. Cambridge University Press, Cambridge, UK.

Hills, A. (2005). *Do Animals have Rights?* Icon Books Ltd., Cambridge, UK.

Mellor, D., Patterson-Kan, E. and Stafford, K. J. (2009). *The Sciences of Animal Welfare*. Wiley-Blackwell, Oxford.

Regan, T. (2004). *The Case for Animal Rights*. University of California Press, Berkeley, Los Angeles, CA.

Singer, P. (1995) *Animal Liberation*, 2nd edn. Pimlico, London.

Sunstein, C. R. and Nussbaum, M. C. (eds.) (2004). *Animal Rights: Current Debates and New Directions*. Oxford University Press, Oxford.

Wemmer, C. and Christen, C. A. (eds.) (2008). *Elephants and Ethics: Toward a Morality of Coexistence*. The Johns Hopkins University Press, Baltimore, MD.

Wynne, C. D. L. (2001). *Animal Cognition: The Mental Lives of Animals*. Palgrave Macmillan, Basingstoke, Hampshire and New York.

6.4 EXERCISES

1. What is the moral justification for the existence of zoos?
2. Discuss the arguments for and against the existence of zoos.
3. Should zoos keep sentient species such as chimpanzees, dolphins and elephants?
4. Are zoos morally justified in keeping some animals in order to benefit the conservation of others?
5. To what extent do zoo animals suffer a 'harm' from being kept in zoos?
6. Should zoos have performing animals?

Part 2

ENCLOSURES, HUSBANDRY AND BEHAVIOUR

Part 2 of this book is concerned with the challenges of keeping exotic animals in captivity. It considers the design of zoos and animal enclosures, zoo animal husbandry – particularly nutrition and reproduction – veterinary care, and the behaviour of animals living in zoos.

Modern animal welfare practices are expected to provide animals with the 'Five Freedoms' identified by the *Report of the Technical Committee to Enquire into the Welfare of Animals kept under Intensive Livestock Husbandry Systems* (The Brambell Report) (HMSO, 1965). These freedoms are:

- Freedom from hunger and thirst
- Freedom from discomfort
- Freedom from pain, injury or disease
- Freedom to express normal behaviour
- Freedom from fear and distress.

In the UK, these principles are embodied in the Animal Welfare Act 2006, s.9 (Legal Box 5.6).

Husbandry guidelines exist for many species. Some of these have been written by individual experts or compiled from the combined efforts and experience of groups of specialists. In some cases regional zoo organisations publish a series of husbandry guidelines that are available to their members, usually via their websites. Others are made freely available on the internet. These guidelines are variable in content but should contain information about minimum enclosure sizes, nutrition, reproduction and other aspects of the biology of the species concerned.

Husbandry manuals and guidelines and other information regarding minimum standards for keeping animals are produced by a wide range of organisations including:

- British and Irish Association of Zoos and Aquariums (BIAZA) (www.biaza.org.uk)
- European Association of Zoos and Aquaria (EAZA) (www.eaza.net)
- Australasian Zoo Keeping (AZK) (www.australasianzookeeping.org)
- Association of Zoos and Aquariums (AZA) (www.aza.org/husbandry-manuals/)
- United States Department of Agriculture (USDA) guidelines (www.usda.gov)
- International Wildlife Rehabilitation Council (www.iwrc-online.org/pub/publications.html)
- Australian Society of Zookeeping Inc. (amphibians) (www.aszk.org.au/husbandry.amphibian.ews)
- Bear Information Exchange for Rehabilitators, Zoos & Sanctuaries (BIERZS) (www.bearkeepers.net).

There is no recognised standard format for husbandry guidelines. A typical list of chapter headings for such a document is shown in Table 7.1.

7 ZOO AND EXHIBIT DESIGN

[A management goal of zoos should be to] provide an environment in which the animal, if it *could* know, *would not* know that it was in captivity.

Michael Brambell

Conservation Status Profile

Cheetah
Acinonyx jubatus
IUCN status:
Vulnerable A2acd; C1
CITES: Appendix I
Population trend:
Decreasing

An Introduction to Zoo Biology and Management, First Edition. Paul A. Rees.

7.1 INTRODUCTION

The quality of the overall experience of zoo visitors is largely determined by the design of the zoo. Unfortunately, most zoos have evolved over a long period of time. Consequently, the design has developed piecemeal over many years and under the leadership of many directors with different visions and influenced by changing trends. In some cases a zoo may even contain protected buildings (e.g. ZSL London Zoo, Dudley Zoo) which cannot be demolished or significantly changed, or in some cases, must be painted particular colours.

Old-fashioned menageries often consisted of rows of similar cages where animals were grouped together based on taxonomy: big cats, monkeys, apes, bears. Many zoos still retain this approach with, for example, a traditional parrot house or reptile house. There are good practical reasons why similar species should be housed close together if they are managed by a dedicated keeper or team of keepers.

A good exhibit design should give visitors the illusion that they have encountered a group of animals in the wild. This can be achieved by careful positioning of trees, other vegetation and structures such as walls. Barriers may be designed so that they contain observation points that may include windows or hides built into the barrier itself. Ideally it should not be possible for visitors to walk all of the way around an exhibit or to see people all around the perimeter as they look into the enclosure. In practice, however, this last requirement is very difficult to achieve, especially when visitor numbers are high.

Visitors should be at eye level with the animals or looking up at them, not looking down on them. Exhibits should replicate natural environments and the ecological niche of the species as closely as possible. Ideally, they should contain a mixture of species that normally occur together. Part of the value of naturalistic exhibits is that they remind the visitor of the need to protect habitats as well as species.

Table 7.1 Typical contents for captive animal husbandry guidelines.

Platypus (*Ornithorhynchus anatinus*) Captive Husbandry Guidelines Jackson, S. (ed.) (2001). Healesville Sanctuary (Zoos Victoria), Melbourne, Australia. CONTENTS	
1.	Introduction
2.	Taxonomy
3.	Natural history
4.	Housing requirements
5.	Handling and trapping
6.	Health requirements
7.	Behaviour
8.	Captive dietary requirements
9.	General husbandry
10.	Breeding
11.	Artificial rearing
12.	Acknowledgements
13.	References
14.	Bibliography
15.	Appendices

7.2 ZOO DESIGN

Up until the end of the 19th century most zoos were little more than animal prisons. Then, in the period up to the middle of the 20th century zoos became living art galleries (the Modernist Movement). Thereafter, zoos took on a conservation and education ethos. Zoo design moved from having a 'classical' structure,

Biography 7.1

Berthold Lubetkin (1901–90)

Lubetkin was a Russian-born architect who formed a group called *Tecton* with six other architects. Tecton designed many iconic zoo enclosures and buildings which were characterised by their sweeping curves and constructed from reinforced concrete. Many of these buildings are now protected, including the Gorilla House and Penguin Pool at ZSL London Zoo.

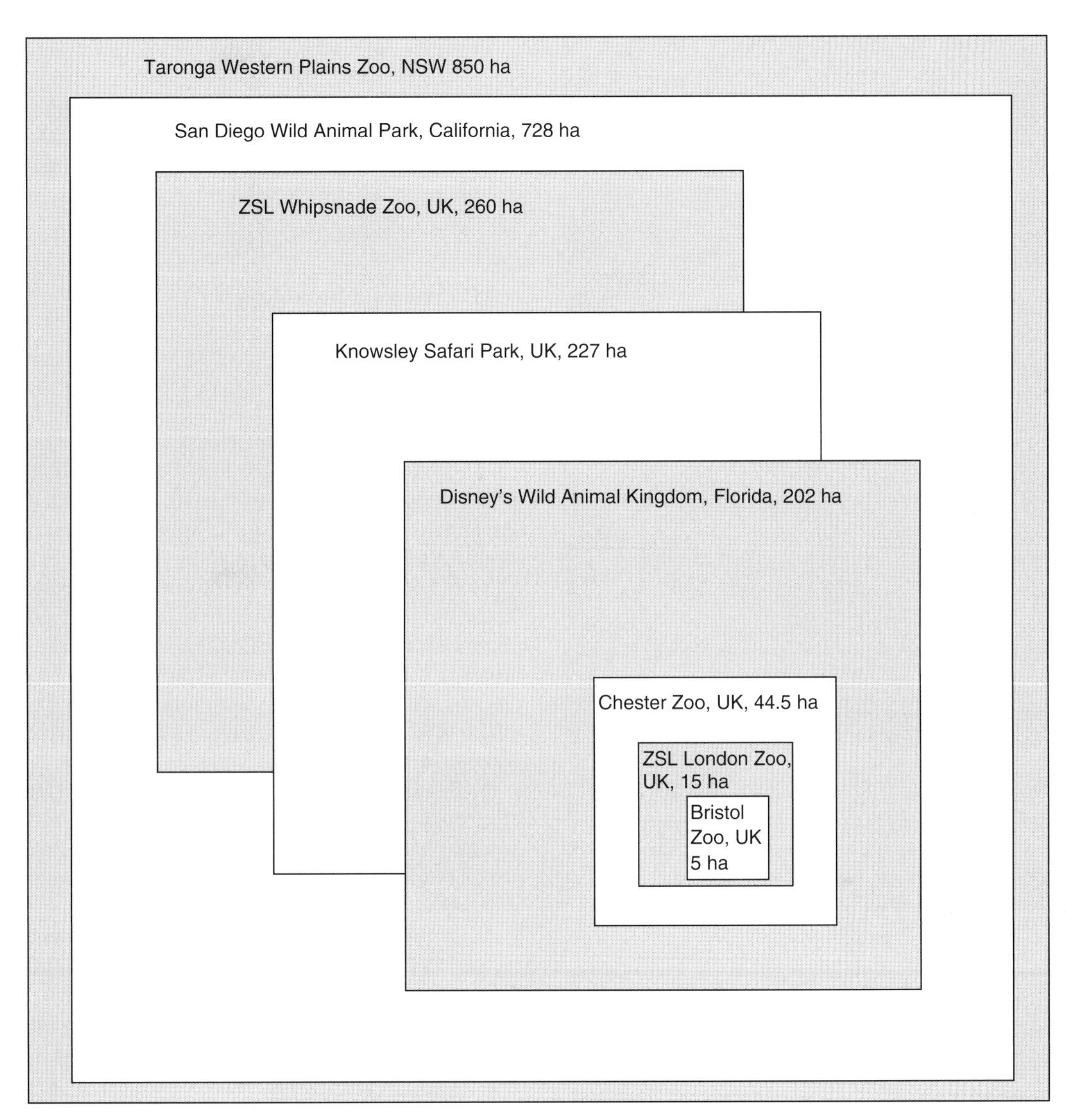

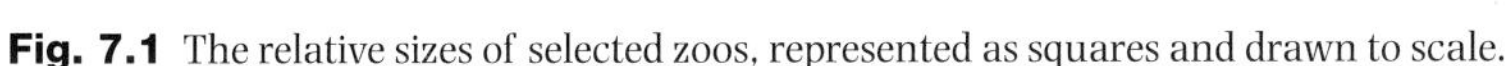
Fig. 7.1 The relative sizes of selected zoos, represented as squares and drawn to scale.

with exhibits arranged systematically, to 'experience oriented', with themed worlds based on different biomes or continents.

Most of the world's zoos are not the result of grand plans. They have developed from small collections and have grown in a piecemeal manner so that they contain a mixture of old, outdated exhibits and new state-of-the-art facilities. The size of zoos varies considerably from just a few hectares, or a small facility within a museum (e.g. the vivarium at the Manchester Museum (University of Manchester)) to huge facilities such as the 850 ha Taronga Western Plains Zoo in New South Wales, Australia (Fig. 7.1).

The piecemeal development of most zoos has precluded the possibility of a clear overall design. Some zoos still present their animals in taxonomic groups within dedicated buildings such as monkey houses. However, other institutions have attempted to group

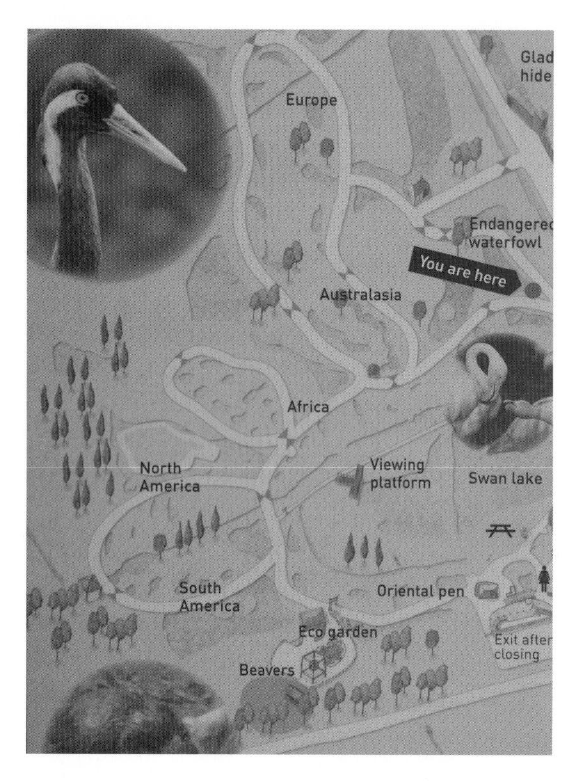

Fig. 7.2 Visitor map of the wildfowl collection at Martin Mere (Wildfowl and Wetlands Trust). This collection is divided into areas containing species from particular geographical regions, for example Australasia, Europe, North America and Africa.

their exhibits so that they represent biogeographical regions or ecosystems. This is easier to do when many of the species in the collection have similar enclosure requirements. For example, it is relatively easy to group waterfowl or fishes into exhibits representing geographical regions such as 'Europe' and 'North America', but much more difficult to regroup large mammals that may need specialist housing and barriers (Fig. 7.2). It is common to find a mixture of themed, sometimes multi-species, exhibits (e.g. *African Plains, Amazon Jungle, Islands in Danger*) and more traditional exhibits (e.g. *Giraffe House, Reptile House, Sea-lion Pool*) within the same zoo (Fig. 7.3).

Although, like most zoos, it originally developed in a piecemeal fashion, Chester Zoo is planning to create a 'SuperZoo', tripling its current size, and creating zones representing African savannah, grassland, forest and island and wetland habitats.

The arrangement of paths in a zoo fundamentally affects the visitor experience. If the exhibits on either side of a path contain species from different parts of the world or from different ecosystems, it is impossible to achieve a sense of immersion. The behaviour of visitors in relation to paths is discussed in Section 14.5.

7.2.1 Themed buildings

Many zoos are creating new themed exhibits. Some of these are based on a single species while others are multi-species exhibits based on a particular habitat. This is not a new idea. The Buffalo House at the Smithsonian National Zoological Park was built in 1891 to look like a log cabin, invoking memories of the American frontier (Fig. 7.4). It housed the American bison (*Bison bison*) in what was the first house built at the zoo. The house built for European bison (*B. bonasus*) at the Berlin Zoo in 1905 resembled a Russian wooden manor house. An Ostrich House opened in Berlin Zoo in 1901 was built and painted to look like an Egyptian temple. An Indian temple was built in Berlin for the Asian elephants and an Asian temple was constructed to house monkeys at Bristol Zoo.

At Edinburgh Zoo the approach to the new Amur tiger enclosure has bamboo walls and is intended to look like a border post at a Tiger Reserve on the Russia–China border. The bamboo hides the sides of the enclosure as the visitors approach the exhibit.

Many new themed exhibits have been given evocative names. San Diego Zoo, California, has exhibits entitled *Polar Bear Plunge, Elephant Odyssey, Tiger River, Ituri Forest, Gorilla Tropics, Sun Bear Forest* and *Wings of Australia*. In Sydney, Taronga Zoo has *Wild Asia, Great Southern Oceans, African Waterhole, Australian Nightlife* and *Wild Australia*. Some of these new exhibits have required considerable financial investment (Table 7.2; Fig. 7.5).

7.2.2 Reuse of old buildings and exhibits

Some exhibits are not designed from scratch but involve the reuse of old facilities designed for different species. Inevitably, the new design is a compromise between the requirements of the original and the current occupants. This is exacerbated in some old zoos by the existence of listed buildings that are no longer suitable for the animals for which they were intended

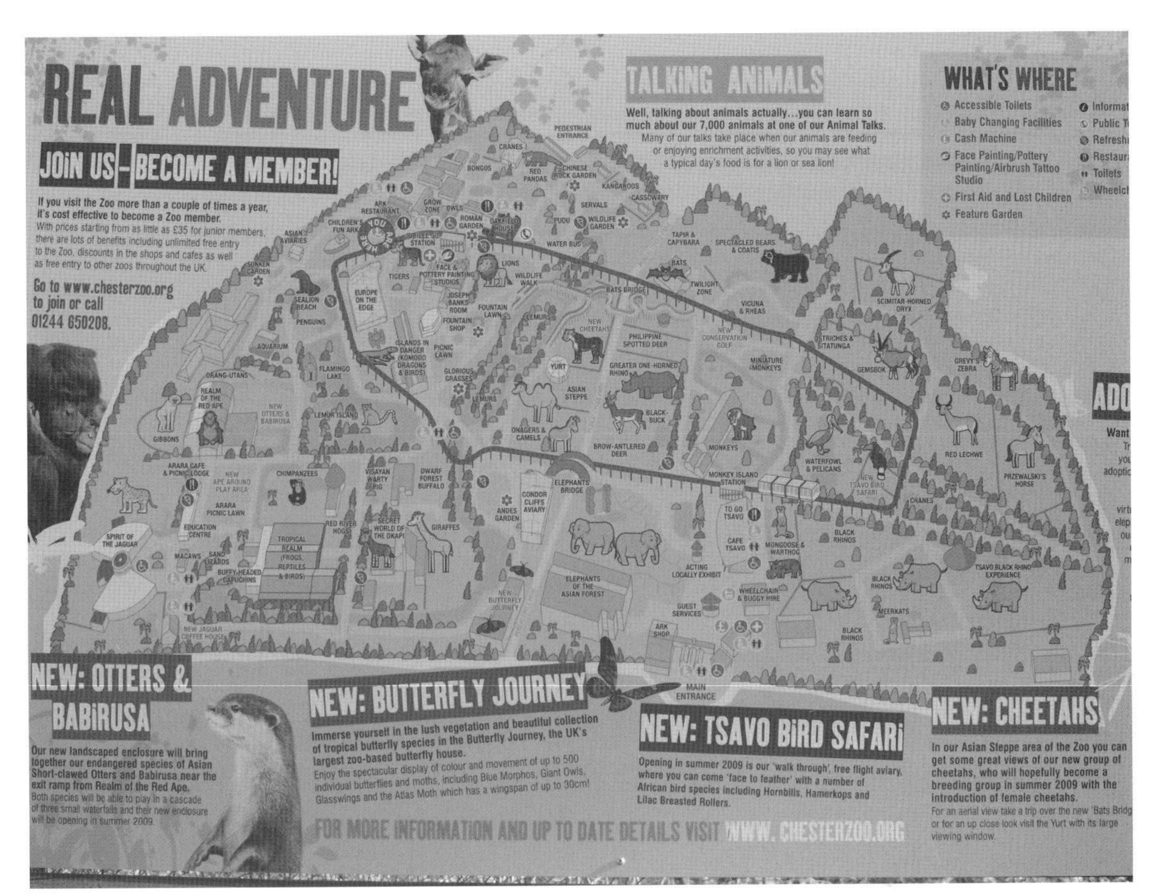

Fig. 7.3 Visitor map of Chester Zoo. This zoo contains a mixture of new themed exhibits and older exhibits of individual species. Some parts of the zoo have a biogeographical structure (e.g. South America), but most do not. This type of structure is common in long-established zoos.

Fig. 7.4 The Buffalo House at the Smithsonian National Zoological Park, Washington DC. (Courtesy of Library of Congress Prints and Photographs Division, Washington DC.)

(Table 7.3). The Giraffe House at Bristol Zoo is a Grade II listed building. It has previously housed elephants but now houses gorillas and okapi. The Mappin Terraces at ZSL London Zoo are an artificial mountain landscape constructed from reinforced concrete in 1913–14. At different times in the past they have been home to a wide variety of species including polar bears, ibex, sloth bears and Hanuman langurs. The terraces currently house kangaroos and emus in an Outback exhibit. The main structure cannot be substantially modified as it is a Grade II listed building. The original Gorilla House at ZSL London Zoo (Fig. 7.6) no longer holds gorillas and a new *Gorilla Kingdom* has been constructed (Fig. 7.5). The Raven's Cage at ZSL London Zoo is a Grade II listed building and is protected because of its

Table 7.2 The approximate cost of some major new zoo exhibits.

Exhibit	Zoo	Species	Cost (US$ millions)
Elephant Odyssey	San Diego	Asian elephant	45.0
Congo Gorilla Forest	Bronx	Western lowland gorilla, okapi, black and white colobus monkeys, red river hogs	43.0
Masoala Rainforest	Zurich	Lemurs	42.0
Gorilla Kingdom	ZSL London	Western lowland gorilla, black and white colobus monkeys	10.4
Realm of the Red Ape	Chester	Orangutan, lar gibbons	7.1
Spirit of the Jaguar	Chester	Jaguar	4.0

Source: Various, including Francis *et al.* (2007).

Fig. 7.5 The new *Gorilla Kingdom* at ZSL London Zoo opened in 2007.

Table 7.3 Some listed buildings in English zoos. (Listed buildings in England are recorded in the National Monuments Record by English Heritage.)

Building	Zoo	Designer	Date of construction	Listed status
Raven's Cage	ZSL London	Decimus Burton	*c.*1827	Grade II
Gorilla House	ZSL London	Lubetkin and Tecton	1932–33	Grade I
Elephant and Rhinoceros Pavilion	ZSL London	Sir Hugh Casson, Neville Condor and Partners	1962–65	Grade II*
Snowdon Aviary	ZSL London	Lord Snowdon and Frederick Price	1962–65	Grade II*
Giraffe House and Hippopotamus House	ZSL London	Decimus Burton	1836	Grade II
Mappin Terraces and Mappin Café	ZSL London	Belcher and Joas	1914 (Café 1920/27)	Grade II
Bear Pit	Dudley	Lubetkin and Tecton	1937	Grade II*
Elephant House	Dudley	Lubetkin and Tecton	1937	Grade II
Sea-lion Pools	Dudley	Lubetkin and Tecton	1937	Grade II
Tropical Bird House	Dudley	Lubetkin and Tecton	1937	Grade II
Elephant House	ZSL Whipsnade	Lubetkin and Tecton	1935	Grade II*
Oakfield House	Chester	Ould/Beswick	*c.*1885	Grade II

Fig. 7.6 The old Gorilla House at ZSL London Zoo designed by Lubetkin 1932–33.

historical interest even though it is no longer used to house birds (Fig. 7.7).

7.2.3 Horticulture

Horticulture is an important element of modern zoo design. Established zoos often have large trees within their boundaries, with signs indicating their species, for example Bristol Zoo (UK) and Colombo Zoo (Sri Lanka). In some cases they are non-native and this helps to create an exotic atmosphere. Some zoos have well-established formal gardens, often enhanced by statues, and sculptures. Chester Zoo has a formal Roman Garden and an area devoted to grasses from different parts of the world.

New exhibits are commonly planted with species that reflect the environment from which the animals

Fig. 7.7 The Raven's Cage at ZSL London Zoo no longer houses birds but is protected as a Grade II listed building.

originate; for example, ZSL Whipsnade Zoo has planted the approach to the area where its Asian elephants perform with bamboo. This helps to create an immersive experience for the visitor. Jackson (1996) has discussed some of the challenges that horticulturalists face in the design and construction of a new exhibit. These include:

- finding suitable plants that will grow in a zoo in a temperate climate but will realistically simulate a tropical forest
- accurately portraying forestry and agricultural practices in exhibits
- reducing the negative impacts that animals have on plants
- creating naturalistic ecological successions
- incorporating dead trees into exhibits.

Some animals can damage plants used for decorative purposes around the edge of an exhibit. Strategically placed electric fences (hot wires) can prevent this and should be positioned out of sight of visitors where possible. Care must also be taken to ensure that animals are not exposed to poisonous plants (see Table 11.4).

Fig. 7.8 Gerald Durrell, founder of Jersey Zoo (now *Durrell*).

7.2.4 Zoo art

Art can convey powerful messages in zoos, and many zoos contain excellent pieces of art within their grounds. Some are sculptures of people and commemorate their role in conservation (e.g. Gerald Durrell (Fig. 7.8) or Sir Peter Scott (Fig. 7.9)) while others are provocative pieces that draw attention to a particular environmental issue (e.g. a polar bear and her cub made out of old plastic milk bottles (Fig. 7.10) or a great white shark made of recycled car parts (Fig. 7.11)).

Some zoos have busts or even whole body sculptures of well-known animals (e.g. 'Guy' the gorilla in ZSL London Zoo). Some sculptures show animal behaviour that might otherwise be difficult to see in a zoo environment (e.g. cheetahs running at high speed), or give visitors the opportunity to get close to a representation of an animal that would otherwise be inaccessible (e.g. a Komodo dragon).

Fig. 7.9 Sir Peter Scott, founder of the Wildfowl Trust (now the Wildfowl and Wetlands Trust).

Fig. 7.10 Polar bear with cub made of plastic from recycled milk bottles.

Fig. 7.11 A shark made of old car parts.

The design of signage is extremely important in zoos as it must both attract attention and convey information (see Section 14.3). Yew (1991) has compiled an interesting collection of photographs of art from zoos around the world – from posters to signage systems – and has shown how this enhances the visitor experience.

7.3 ENCLOSURE AND EXHIBIT DESIGN

7.3.1 Introduction

In many countries the design of zoo enclosures is the subject of guidelines or codes of practice (e.g. SSSMZP (2004) in the UK), while in others it is regulated by legislation (e.g. *Standard 154.03.04: Containment Facilities for Zoo Animals* (Anon., 2007), in New Zealand). Before considering the detailed design of enclosures and

Fig. 7.12 Bars surrounding a bear enclosure at ZSL Whipsnade Zoo are typical of the type used to contain large carnivores at the beginning of the 20th century.

barriers it is useful to consider briefly the evolution of enclosure design.

7.3.2 The evolution of zoo enclosures

Victorian zoo enclosures tended to be simple structures and many animals were kept behind bars (Fig. 7.12) or in wire cages (Fig. 7.13). Often the design and strength of the cages far exceeded what was necessary merely to confine the animals.

The early 20th century saw the development of the first naturalistic panoramas surrounded by moats at Carl Hagenbeck's Tierpark. Then, in the 1920s and 1930s there followed a period known as the 'Disinfectant Era' during which enclosures and cages were designed for ease of cleaning – with concrete floors and tiled walls – rather than to meet the needs of the animals. Some zoos still have enclosures of this type to this day. Thereafter, in the 1970s the concept of 'landscape immersion' developed. The first exhibit to have adopted a landscape immersion design is considered to have been the gorilla exhibit opened in 1978 at the Woodland Park Zoo in Seattle, Washington, designed by Grant Jones and Jon Coe (Hyson, 2000). Since then zoos have developed as conservation centres and now promote the concept of sustainability in their enclosure designs, an excellent example of which – the *Rhinos of Nepal* at ZSL Whipsnade Zoo – is described at the end of this chapter.

Within the last 50 years zoo exhibits have been transformed from simple cages designed to retain and display animals, with little consideration given to their biological needs, to sophisticated enclosures which provide places where the animals may remain hidden from the public but which also offer the visitors an immersive experience. Large dangerous animals that were kept behind unnecessarily strong iron bars – often with visitors looking down on them – are now exhibited behind chain-link fencing and glass windows (Fig. 7.14). Good design can result in a rewarding experience for visitors with minimal disturbance to animals. Poor design can result in disappointed visitors staring into apparently empty enclosures or unnecessarily exposed animals that are continuously disturbed. Developments in the design and construction of enclosure barriers have been instrumental in the evolution of modern enclosures.

Biography 7.2

Jon Coe

Jon Coe is an influential zoo exhibit designer. He is a landscape architect and formerly held an academic post at the University of Pennsylvania. Coe worked for a number of major companies involved in zoo design, including Jones & Jones of Seattle, before founding his own company, Jon Coe Design Pty Ltd., in Victoria, Australia. He has worked on over 150 design projects for more than 60 zoos, aquariums, museums and similar organisations. These have included the *Gorillas of Cameroon* exhibit at Zoo Atlanta, the Asian Forest and Elephant Exhibit at Taronga Zoo, and the *African Plains* Master Plan at Metro Toronto Zoo. Coe has published a wide range of academic papers on zoo exhibit design.

7.3.3 The structure of an exhibit

Simple exhibits may be divided into a foreground, a middle ground and a rear ground. The foreground is the area in front of the exhibit, on the visitors' side of the front barrier. The middle ground is the central area

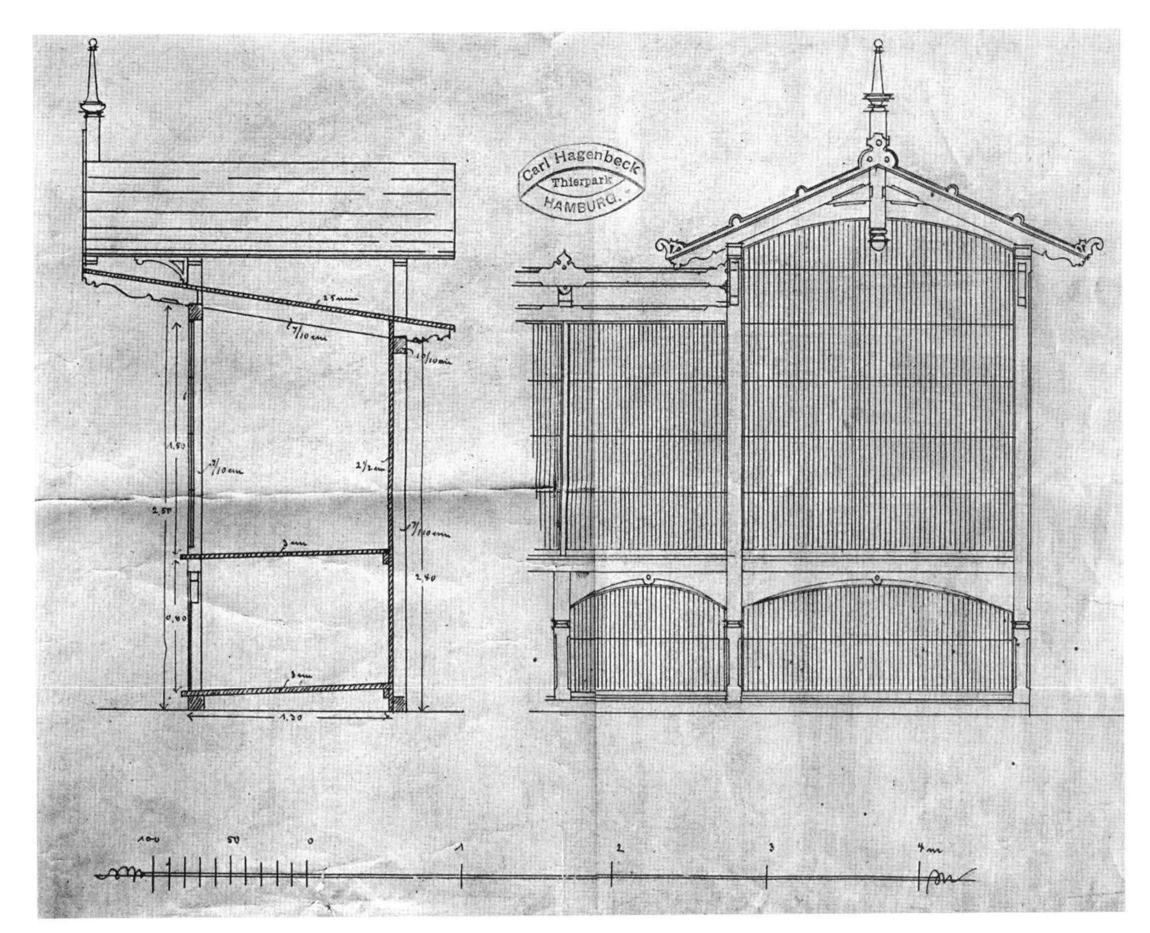

Fig. 7.13 Plans for a cage – possibly an aviary – stamped Carl Hagenbeck, found amongst documents from Belle Vue Zoo at Chetham's Library, Manchester (date unknown). (Courtesy of Chetham's Library, Manchester.)

of the enclosure that contains the animals themselves. The rear ground is the back of the exhibit.

Ideally, all of the barriers should be invisible from the viewpoint of the visitor, although in practice this is rarely achieved. The perception of 'landscape immersion' may be achieved if the vegetation and other features, such as rocks and fallen trees, are identical in the foreground and on the animals' side of the front barrier.

The middle ground is where the animals will be displayed and contains rock features, pools, trees and other components necessary to create a suitable simulated ecological niche. The design needs to be a compromise that allows the animals to feel secure and provides them with some privacy and yet ensures that they are not constantly able to avoid being on view to visitors. Side and rear barriers may be hidden from view by using carefully concealed moats or by heavy planting behind and through fences. A well-designed rear barrier can give an illusion of depth to the exhibit.

Many zoos have out-of-date exhibits that consist of fencing stretched between metal posts. Some are now 'softening' the appearance of these dated barriers by cladding the posts with wood, helping them to blend in with trees and other vegetation.

Every exhibit must take into account the needs of the animal, keepers, and visitors. Modern exhibits should also address sustainability considerations. Some basic design principles for the presentation of animals are listed in Box 7.1. The competing needs of the users of animal enclosures are summarised in Table 7.4.

A cage made of steel bars, typical of those used to confine lions in zoos and menageries well beyond the middle of the 20th century in developed countries, and still used in some zoos in some developing countries. Such cages emphasised the dominion of humans over ferocious wild beasts and were typically small and barren. Lions have also been exhibited in pits similar to those used for bears.
A chain-link fence enclosure allows visitors to see, hear and smell lions at relatively close quarters. Modern enclosures contain wooden platforms, pools, trees and other features. If high viewpoints are provided, visitors may be able to obtain uninterrupted views of the animals over the top of the fence.
Glass viewing windows are often added to fenced enclosures. Windows allow visitors to see lions at very close quarters, but they quickly become soiled and this may detract from the overall experience. They also act as a focus for visitors and may become very crowded and noisy at busy times. The overall visitor experience may seem less exciting and 'safer' than viewing lions behind fencing.
The most modern enclosures contain emersion elements such as this Land Rover viewpoint. Visitors may sit inside the vehicle and observe the lions through the windows as if on safari. The bonnet of the vehicle is heated to encourage lions to sit on it.

Fig. 7.14 The evolution of the lion cage.

Box 7.1 Exhibit design – definitions and principles.

Design principles for the presentation of animals (adapted from Jones, 1982)

1 Animals should be at or above the eye level of visitors.
2 Animals should not be surrounded by visitors; exhibits should include a number of smaller overlooks without overlapping lines of sight.
3 Allow the animal to remove itself from stressful situations; allow the animal to choose between hot and cool, high and low, wet and dry, and off/on show.
4 Display social animals in social groups.
5 Do not display deformed or disfigured animals.
6 Do not display animals using human artefacts; provide things for animals to do using features of their natural habitat.
7 Recreate as far as possible a landscape typical of the animal's natural habitat.
8 Make it impossible for the visitor to determine how the animal is retained within the exhibit; hide or disguise the barrier.
9 Immerse the visitor in the replicated landscape even before seeing the animal; make overlooks and adjacent circulation areas appear as extensions of the animal's habitat; do not build perceptual barriers by placing visitors in a man-made setting and the animals in a naturalistic setting.
10 Do not display animals from different habitats together in a natural habitat setting; combine compatible animals from the same habitat.
11 Relate adjacent exhibits into habitat complexes, forming transitional or ecotonal areas between exhibits of adjacent habitat zones.
12 Plan all of the elements of the exhibit concurrently as interrelated parts; do not design the buildings first.

Definitions for 'natural' zoo exhibit habitats (adapted from Polakowski, 1987)

Exhibit type	Definition
Realistic natural habitat	Reproduces the real habitat in appearance, land formation, vegetation and animal activity.
Modified natural habitat	Uses elements of the real habitat but uses substitute trees and other plants, uses existing or modified landforms, and integrates the habitat into the existing surroundings.
Naturalistic habitat	Makes little or no attempt to duplicate elements of the real habitat. Stylistic use of natural materials.

7.3.3.1 Viewpoints

The design of cages and enclosures in early European menageries and zoos reflected man's dominion over the animals. Dangerous animals, like lions and tigers, were kept behind unnecessarily thick iron bars, and bears were held in deep pits. High viewpoints meant that visitors looked down on the animals and the animals in turn looked up at the people.

Even today's modern zoos have not entirely rid themselves of these high viewpoints. Enclosures bounded by a ha-ha inevitably result in visitors looking down on rhinos, elephants, gorillas and many other species when they approach the barrier. Many modern exhibits incorporate high-level platforms for viewing big cats, wolves and other dangerous species, and also to allow visitors to see giraffes at their eye level.

Although it could be argued that some of these designs help to perpetuate old-fashioned attitudes towards animals, viewpoints that give visitors uninterrupted views of animals, especially dangerous animals, are clearly popular with visitors, particularly photographers. At South Lakes Wild Animal Park it is possible to take photographs of tigers climbing poles for food, over the top of the fences around their enclosure, from a wooden walkway (Fig. 10.16).

Many new exhibits have multiple viewpoints. They may be classified by their location or by the nature of their construction and design.

Viewpoint levels:

- Ground level
- High level
- Subterranean – naked mole rats, dens and setts
- Underwater – pools and ponds.

Table 7.4 Zoo exhibit design considerations

Animal needs	Keeper needs	Visitor needs	Sustainability needs
• How big are the animals? • How much space do they need? • Are they active or sedentary? • Are they social or solitary? • Where will they feed? • Where will they sleep? • Where will they breed? • What vegetation and substrate are suitable? • What resting places do they need? • Do they need structures to climb, water, rocks, open space? • Do individuals need to escape from conspecifics and other species? • Where can they hide from visitors? • What are their temperature, humidity, light, water quality and other environmental needs? • Are any toxic plants or other toxic materials present? • Are there any dangerous structures? • Is there any shade? • Are there any risks to the animals' safety? • Are there any risks to the animals' health? • Has suitable enrichment been provided?	• How will the keeper enter the enclosure? • How can the keeper escape from the enclosure in an emergency? • Is the enclosure secure? • Where will food be prepared? • How and where will food be provided? • Are there isolation areas for veterinary access? • Are there off-show areas for breeding and sick animals, etc.? • Can animals be easily viewed for health checks? • How will waste be removed and how will the enclosure be cleaned? • Can pathogens and parasites be controlled? • Is there access to electricity and water? • Are enrichment devices and other equipment easy to maintain? • Can contact with dangerous animals be controlled or avoided?	• Can the visitors see the animals clearly? • Are viewpoints accessible to young, elderly and disabled visitors? • Are hand-washing facilities available if the visitor enters the enclosure? • Is the exhibit attractive? • Does it reflect the animals' natural habitat? • Does the visitor feel immersed in the exhibit? • Are the barriers intrusive? • Are visitors safe? • Is there adequate information available about the animals? • Is the signage up-to-date? • Is the exhibit educational? • Where should the interpretation signs be located? • Are safety notices adequate? • Can the visitor view a video or CCTV link if the animals are off-show or hiding?	• Are the building materials from sustainable sources? • Can existing facilities be reused? • Can recycled construction materials be used? • Is the animal house properly insulated? • Can water in pools and other water features be recycled? • Can rainwater from roofs be collected? • Can solar heating be used? • Can solar panels supplement the electricity supply? • Can waste be recycled? • Can natural light be used to illuminate the animal house? • Can energy-efficient electrical devices be used? • Does the animal house incorporate a green roof? • Can the design of the enclosure be used to encourage visitors to think about sustainability?

Types of viewing points:

- Windows:
  - panoramic – e.g. aquarium, chimpanzees, lions
  - letterbox – e.g. orangutans, maned wolves, bongos (Fig. 7.15a)
  - porthole – e.g. hunting dogs (Fig. 7.15b)
  - ground-level observation – e.g. prairie dogs, lynx, tigers
  - underwater windows (animals viewed from the side) – e.g. hippos, polar bears, penguins, sea lions
  - underwater galleries (animals viewed from underneath) – e.g. polar bears
  - underwater tunnels – e.g. sharks
  - hides – e.g. wildfowl.
- Boardwalks:
  - ground level – e.g. lemur
  - high level – e.g. wolves, giraffes, cheetah (Fig. 7.15c)
  - low level, over water – e.g. alligators, waterbirds.
- Remote:
  - periscope (high or low level) – e.g. giraffes (Fig. 7.15d)
  - CCTV – for animal dens.

The safari park concept introduced a new type of viewpoint by allowing visitors to drive through enclosures ('reserves') in their own vehicles. This presented a number of safety issues because visitors would sometimes leave their vehicles to touch or photograph animals, or if their vehicle broke down, and occasionally animals are hit by cars as they cross the roads. Some animals were found to be unsuitable for drive-through enclosures. Tigers (*Panthera tigris*) at Knowsley Safari

(a)

(b)

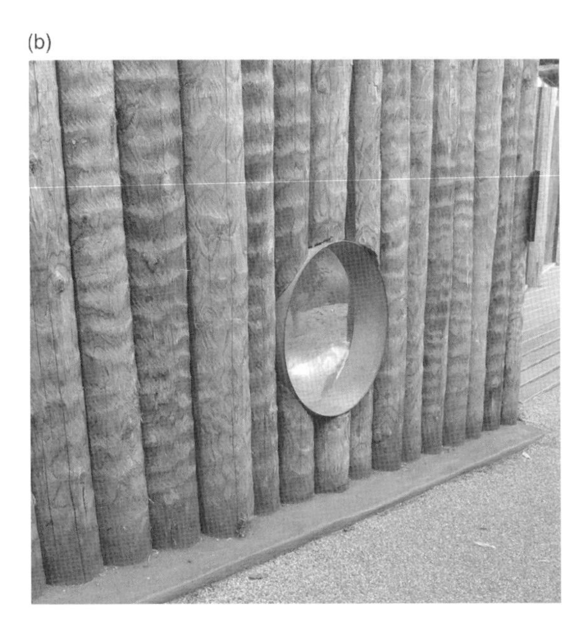

(c)

(d)

Fig. 7.15 Zoos create a variety of viewpoints for visitors depending upon the requirements of the species and access to the exhibit. (a) A bongo (*Tragelaphus eurycerus*) enclosure bounded by a wooden wall with 'letterbox' viewpoints, preventing disturbance to the animals. (b) A 'porthole' window allowing low-level viewing of hunting dogs (*Lycaon pictus*). (c) A high-level walkway overlooking a grey wolf (*Canis lupus*) enclosure at Kristiansand Zoo, Norway. (d) A periscope for low-level viewing of warthogs (*Phacochoerus africanus*).

Park attacked vehicle tyres when they were introduced in the 1970s and are now exhibited in a separate fenced enclosure. In the 1970s, African elephants (*Loxodonta africana*) at this park were exhibited in drive-through enclosures. They could move around freely, but a keeper patrolled the enclosure in a 4×4 vehicle. Later, safety considerations led to these animals being kept behind an electric fence in a drive-through enclosure. They have now been moved to a large enclosure to which visitors' vehicles have no access.

Traditional aquariums only offer visitors one type of viewpoint: an eye-level view through glass on one side of a tank. Modern aquariums offer many more viewpoints:

Fig. 7.16 A moving floorway conveys visitors through a tunnel under an artificial coral reef.

Fig. 7.17 An overhead monorail at Chester Zoo allows visitors to have a bird's eye view of some of the exhibits and to travel from one part of the zoo to another.

Fig. 7.18 Many zoos have miniature railways. This one is passing Asian great one-horned rhinoceroses (*Rhinoceros unicornis*) at ZSL Whipsnade Zoo.

Fig. 7.19 Walled enclosures need to be designed so that small visitors can see the animals.

- circular tanks accessible all around the perimeter
- touch pools where animals can be observed through the transparent side of the tank or from above; in some cases visitors may be allowed to touch and feed the fish
- large panoramic windows (e.g. at the *Blue Planet*)
- underwater tunnels which allow visitors to experience a special type of immersion (perhaps it should be called 'submersion'; Fig. 7.16)
- vertical access: a stairway may allow the visitor to travel from the surface of the exhibit to the bottom via a series of stairs (e.g. at *The Deep*).

Underwater World in Singapore has an 83-metre long travelator that moves visitors through a 6 mm thick underwater acrylic-windowed observation tunnel from which they may watch sharks, moray eels, and stingrays and other species that inhabit an artificial coral reef ecosystem.

Some zoos offer novel viewpoints by providing interesting ways of travelling around the zoo. These include:

- high-level monorail systems, e.g. Chester Zoo (Fig. 7.17)
- narrow gauge railways, e.g. ZSL Whipsnade Zoo, *Marwell Wildlife*, South Lakes Wild Animal Park (Fig. 7.18)
- boats, e.g. Chester Zoo
- land trains, e.g. *Marwell Wildlife*, Silver Springs, Florida
- cable cars, e.g. *Sky Safari*, San Diego Zoo
- moving 'underwater' floorways, e.g. *Blue Planet* aquarium.

Some zoos, for example Singapore Zoo, offer night safaris to their visitors, providing a new and exciting perspective for visitors.

Access for children and disabled

Viewpoints must accommodate the needs of children and the disabled (Fig. 7.19). Ramps should be used to give access to high-level viewpoints. Barriers should incorporate low-level windows for children and visitors

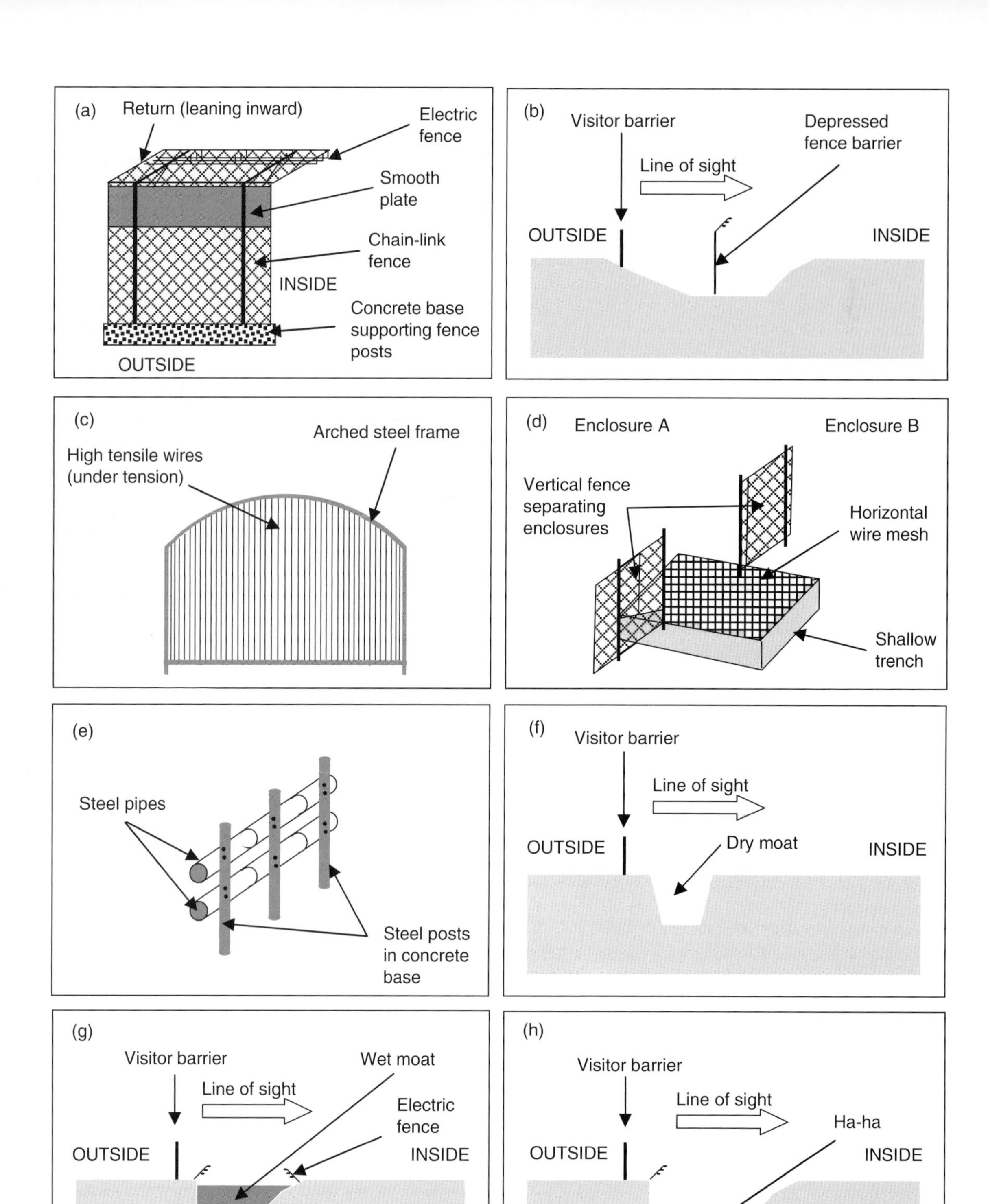

Fig. 7.20 Some common barriers found in zoos: (a) chain-link fence with return; (b) depressed fence; (c) high tensile vertical wire fence; (d) horizontal wire fence between adjacent enclosures; (e) reinforced pipe barrier; (f) dry moat; (g) wet moat; (h) ha-ha.

in wheelchairs. In some countries there is a legal obligation to provide suitable access for the disabled (e.g. the Equality Act 2010 in England and Wales).

7.3.3.2 The use of murals in exhibit design

Many zoos contain old exhibits that could benefit from renovation. One relatively inexpensive way of achieving this is to renovate them by using murals. Barker (2002) suggests that a good mural needs rocks, trees and interpretive elements to work. Murals can be used to cover up unsightly features or to enhance the three-dimensional aspects of an exhibit. Barker discusses some standard rules for mural design:

- Think of the mural as a 'stage' for the 'actors' (animals).
- The horizon should be at eye level.
- Avoid the use of animals as they detract from the real animals.
- Avoid any objects in motion, e.g. running water, flying birds.
- Replicate the habitat of the animals in the exhibit.
- Pay attention to detail to create accurate replicas, even if space is limited.
- Avoid or camouflage unsightly features, e.g. door handles, corners, soffits.
- Use lines in the composition, e.g. branches, to lead the eye towards something you want visitors to see.
- Arrange lighting so as to avoid unwanted shadows.
- Protect the mural from damage by the public by using rails, planters, etc.

Murals are often used for small terrariums housing reptiles or other small animals and can give an impression of depth to the exhibit.

7.4 BARRIER DESIGN

Zoos utilise a very wide range of barriers to contain animals (Fig. 7.20). Barriers are necessary to keep animals inside their enclosures and people out. They need to be safe for animals and people, and they should be as unobtrusive as possible. Barrier design needs to take into account the climbing and jumping abilities of the animals that are to be confined, and also their strength. The choice of barrier may be influenced by the amount of space available. Fences take up much less space than ha-has or moats. The use of a simple fence may maximise the area available to the animals while a naturalistic exhibit, of the same size, with a carefully designed wet moat may restrict the animals to a very small area of land.

Fig. 7.21 Rotating solar-powered bird scarer mounted on the top of a large aviary.

Some enclosures may need to include devices to deter predators and other native species. These range from electric fences and closely arranged thin wires stretched across the top of enclosures to prevent birds from entering, to solar-powered rotating bird scarers (Fig. 7.21).

7.4.1 Barrier standards

Design standards and 'suggestions' can be found in a number of documents. For example, in New Zealand *Standard 154.03.04: Containment Facilities for Zoo Animals* (Anon., 2007) specifies the enclosure standard for Canidae, Hyaenidae and cheetah as follows:

The perimeter of the enclosure shall be:

1. Vertical, 2.5 × mean species body length in height, OR the height of known jumping height plus 15%, whichever is the greater, AND one physical, OR one psychological failsafe.

OR

2. Horizontal, 2.5 × mean species body length wide, OR width of known jumping distance plus 20%, AND depth equal to 1.5 × mean species standing body height AND one physical, OR one psychological failsafe.

OR

3. Both vertical and horizontal – the vertical and horizontal dimensions shall be no less than 150% of the relevant formulas above.

In India, *Barrier Designs for Zoos* (Gupta, 2008), published by the Central Zoo Authority, contains recommendations and suggestions for barriers, rather than specifications, and provides examples of dimensions used by various zoos in India and elsewhere. For the striped hyena (*Hyaena hyaena*) its barrier suggestions are:

a. It is suggested that a dry moat of 3.5 m width and 2.5 m depth can be provided....

b. The rear barrier can be [a] wall of 2.5 m height or of 3.0 m chain link mesh of 5 cm × 5 cm × 10 g.

c. In case of space constraint, the viewer's side can have 3.0 m chain link mesh fence of the above specifications.

7.4.2 Fences

Fences in zoos come in a great variety of shapes and sizes and may be used to contain a very wide range of species from wildfowl to big cats. Most zoos have perimeter fences around their boundaries but some older zoos are contained within a perimeter wall (e.g. Bristol). The perimeter fence of a zoo should be designed to keep out predators and they are often electrified at the top and bottom.

Many enclosures are constructed of 'zoo mesh' made of flexible stainless steel cable. The cable is woven into a mesh and is available in a wide range of specifications with respect to cable diameter, mesh opening size and nominal breaking strength.

Vertical chain-link (wire-netting) fences

This is a type of woven fence made from steel wire which may be coated with polyethylene. Adjacent vertical wires are bent into a zig-zag pattern forming the characteristic diamond pattern associated with this type of fence. The fencing may be supported by steel, timber or concrete posts driven into the ground or set in concrete. The footing depth should be one third of the height of the fence and fencing may need to be buried underground to prevent animals from digging out. The fence colour should be matte black, dark green or other colour to reduce its visibility.

This type of fencing is suitable for a wide range of exhibits from large carnivores and hoofstock to large ratites and wildfowl, and for perimeter fences. Where possible it is best used as a side or rear fence for an exhibit. It is cheap and easy to construct and may be low, to retain small species, or high to retain larger species. However, it can be easily damaged by some animals, high winds and fallen trees. Ideally the fence line should be constructed away from large trees.

Obviously any large trees inside the enclosure must not allow escape over the top of the fence, so it may be necessary to cover the lower sections of the trunk with a smooth material or add a hot wire to prevent climbing.

It is essential to ensure that the fence remains secured to the posts, particularly at the bottom. In the 1970s a white rhino (*Ceratotherium simum*) from Knowsley Safari Park was found grazing with Lord Derby's cattle after it apparently lifted the fence with its horn, passed under the chain-link fence, and walked into adjacent farmland. The fence fell back into place so the means of escape was not immediately obvious. If species that dig, for example hunting dogs (*Lycaon pictus*), are kept behind mesh fences they should be embedded in a concrete foundation.

Chain-link fencing is typically used in safari parks to separate one enclosure from another. In traditional zoos, where dangerous animals are kept behind these fences, a secondary barrier is necessary to prevent visitors from touching the animals. This may be a low wall, hedge or railing.

This type of fence does not provide pleasing views of animals for visitors. It is difficult for visitors to take good photographs through fencing, but if large lens apertures are used it is possible to throw the fencing out of focus if the animal is some distance away. Some exhibits incorporate a viewing window or provide a high-level viewpoint. At South Lakes Wild Animal Park in Cumbria high-level wooden walkways allow visitors to look over the top of the fencing around the lion, tiger and jaguar enclosures. As animals must climb tall wooden posts to reach their food, visitors are able to see and photograph them with an uninterrupted view. At Kristiansand Zoo in Norway visitors are able to use a similar but more extensive elevated walkway to view wolves (*Canis lupus*), moose (*Alces alces*) and lynx (*Lynx lynx*) (Fig. 7.15c). Chester Zoo has an elevated monorail system that passes over a number of fenced enclosures and near others, offering clear views of

Fig. 7.22 A vertical fence with return.

Asian lions, wildfowl, antelopes and many other species (Fig. 7.17).

Vertical fence with return

If dangerous animals are enclosed by chain-link fencing it typically has a 'return' at the top: a section of fencing that leans inwards towards the animals (Figs. 7.20a and 7.22). This overhang prevents animals from climbing out and may be made of a series of parallel wires supported by metal bars, often incorporating an electric fence. Where climbing species are housed smooth panels made of sheet steel, fibreglass or some other suitable smooth material may be fixed vertically flush against the fencing. In some cases the return may also be covered with a smooth material. For some climbing species, for example snow leopard (*Uncia uncia*), it may be necessary to cover the top of the enclosure with netting or wire mesh.

Depressed vertical fence

A depressed fence barrier is a vertical fence whose base is lower than the general ground level so that its perceived height is reduced. Used at the front of an exhibit it may allow visitors to see into the enclosure with an uninterrupted view (Fig. 7.20b). A depressed vertical fence is a useful rear barrier as it may be located so far below the general ground level that it cannot be seen by visitors from the foreground of the exhibit. This type of fence allows the vegetation and scenery located beyond the rear boundary to become part of the display.

Vertical wire fence

Vertical wire fences are suitable for some large hoofstock species. The giraffe enclosure at Twycross Zoo is surrounded by a fence made of horizontal wires supported by metal posts which lean inwards. This prevents visitors from touching the lower parts of the animals, and being kicked – because the giraffes cannot stand close to the base of the fence – and makes it difficult for them to reach visitors over the top.

Fences made of strong steel posts and steel cable may be used to contain elephants (Fig. 7.23). However, there is a risk that they may attempt to climb such fences and this could result in injury.

High tensile wires

Barriers made from high tensile steel wires (piano wire) are useful for relatively small animals. They have the advantage that when the animal is positioned at some distance behind the wires the human eye focuses on the animal and the wires almost disappear from view. This phenomenon makes it possible to take good photographs through fences of this construction. These barriers are expensive to construct as the wires are typically held under tension between a pair of parallel horizontal heavy steel bars. An arched design allows a less obtrusive top bar to be used (Fig. 7.20c). These barriers are unsuitable for any species that is strong enough to force the wires apart.

Horizontal fence

This consists of a horizontal welded mesh attached to a metal frame and mounted over a shallow trench moat, similar in principle to a cattle grid (Fig. 7.20d). Animals such as tapirs and hyenas cannot cross this type of barrier but it is ineffective against felids.

Fine mesh netting

A fine nylon mesh netting may be used as a barrier for small primates, some birds and other small species.

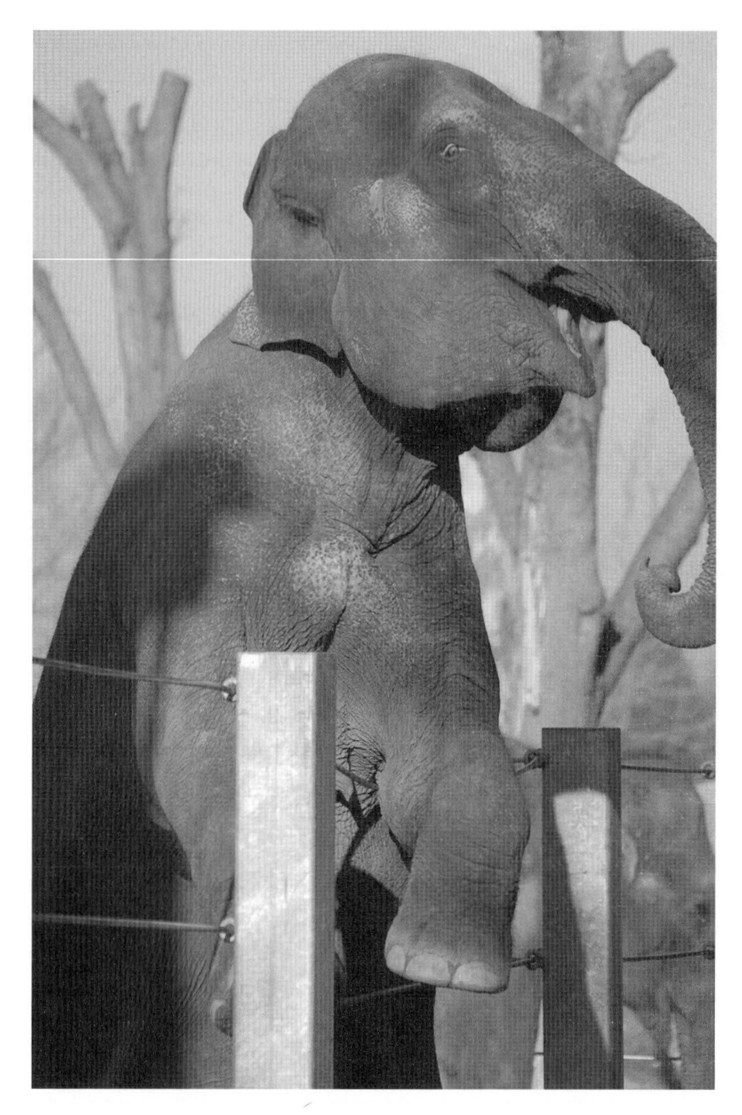

Fig. 7.23 A vertical fence made of steel posts and cable. Is this really suitable for retaining elephants?

Like high tensile wire fencing the visitors' eyes (and cameras) may focus beyond the mesh if animals stand far enough back, so the barrier almost disappears from view. This type of mesh is often used to cover the top of aviaries and may be suspended from poles in such a way as to make it possible to lower it for maintenance purposes.

Fine metal mesh may be used for other small species where nylon is unsuitable. It is very obtrusive and generally provides a very poor view for the visitor.

7.4.3 Bars, pipe barriers and posts

Bars

The use of steel bars as a barrier is uncommon in modern zoos, although some enclosures housing hoofstock may have them. Bars were commonly used in cages for big cats, bears, great apes and other large primates. Such cages still exist in zoos in some developing countries but are generally considerate inappropriate in a modern zoo. Bears and big cats can be safely housed behind chain-link fencing and large primates are now typically kept behind wet moats with a secondary hot wire barrier.

Reinforced pipe barrier
This barrier consists of horizontal steel pipes attached to steel posts embedded in a concrete foundation (Fig. 7.20e). It is suitable for large, powerful mammals such as rhinos and buffalo.

Vertical posts
Some large mammals may be safely retained behind a series of vertical posts. Part of the Asian elephant enclosure at Chester Zoo uses a row of artificial vertical tree trunks as a barrier. A secondary fence barrier prevents visitors from standing within reach of the animals.

7.4.4 Wall barriers

Low walls
Low wall barriers are suitable for enclosures containing small, relatively harmless animals such as otters, bush dogs (*Speothus venaticus*), meerkats (*Suricata suricatta*) and tortoises. They are cheap to construct but have the disadvantage that they provide poor viewing opportunities if the animals are close to the base of the wall. This difficulty may be overcome by incorporating windows into the design. It may be necessary to add a hot wire to the inside of the wall to prevent climbing and wires over the top of the exhibit to keep predators out; for example, widely spaced thin wires suspended in the shape of a cone over a prairie dog exhibit – where the main barrier is a low wall – may prevent birds of prey and other large birds from flying into the enclosure.

Many zoos have old low-walled enclosures that have been reused to house species for which they were not originally intended. An enclosure that originally housed beavers (*Castor fiber*) at Chester Zoo was later modified for coatimundis (*Nasau* sp.).

Tortoises may be retained behind low wooden barriers made of log rolls or old railway sleepers, because of their poor climbing ability.

High walls
Enclosures made with high concrete walls containing windows are used to house apes in a number of zoos. These facilities are unattractive and often the only way the animals can see out of the enclosure is to climb high up onto a pole or climbing frame. The relatively narrow barrier allows maximum use of the enclosure space by the animals. It is impossible for apes to escape across the smooth, high concrete walls, but these exhibits are generally poorly designed and it seems inconceivable that any modern zoo would choose to exhibit apes like this today.

7.4.5 Pits

Pits were traditionally used to exhibit bears, lions and some other large dangerous species in old zoos. They are now considered to be inappropriate. Bear pits were usually barren concrete structures in which the animals were forced to look up at the public, and, when feeding was allowed, the bears would stand on their hind legs begging for food. Although many of these bear pits still exist, in better zoos they have been converted to new uses as, for example, aviaries, by stretching netting across the top. An old bear pit at Chester Zoo is now home to a band of yellow mongooses (*Cynictis penicillata*).

7.4.6 Moats

Dry moat
This is a channel which separates the visitors from the animals (Fig. 7.20f). An inner vertical wall deters the animals from entering the moat, but in some cases this is little more than a raised 'lip'. The depth of the moat varies with the species. Moats around elephant enclosures are typically about 2 m deep and 2 m wide (Fig. 7.24). As elephants are unable to jump such an arrangement acts as an efficient but dangerous barrier. Elephants enjoy balancing on the edge of moats and occasionally they fall – or are pushed – in. Where moats exist it is essential that a ramp is constructed between the bottom of the moat and normal ground level so that animals that fall in can be recovered. Dry moats for elephants cannot be constructed merely by digging a ditch in the soil between the enclosure and the public. Such a ditch, combined with a hot wire, was originally used to retain African elephants at Knowsley Safari Park. The animals stepped on the post supporting the hot wire, walked down the sloping side of the ditch and up the other side.

Wet moat
Wet moats are useful for containing animals that do not like water (Fig. 7.20g). They often have an electric

Fig. 7.24 A dry moat once used to contain elephants at ZSL London Zoo.

fence on both sides, e.g. the gorilla enclosure at Bristol Zoo (Fig. 7.25). They should be shallow on the animals' side to prevent the possibility of an animal falling or being pushed into deep water. Wet moats can be designed to look like meandering rivers, but unfortunately many old exhibits have linear moats which detract from any sense of immersion.

7.4.7 Ha-has

A ha-ha is a concealed barrier which has been widely used by landscape architects and was an essential component of the designs of the 18th century English landscape architect Lancelot 'Capability' Brown. It consists of a section of ground that slopes down from the general ground level towards a sunken wall as it approaches the visitor barrier of an enclosure (Fig. 7.20h). This means that animals in the centre of the enclosure may be at the eye level of the visitors, but as they move closer to the perimeter they descend below this level. This type of barrier may, from certain positions, give the impression that there is no barrier present. For example, if there is a ha-ha at the rear of an exhibit it may be impossible to see it from the front. This effect is similar to that created by a depressed fence barrier. The correct choice between the two depends partly on the jumping or climbing ability of the animal.

A ha-ha is particularly useful for enclosures containing elephants, rhinos and other large ungulates. However, designers face a dilemma. Visitors want to get close to the animals. This is only possible if the slope of the ha-ha is steep, bringing the animals up to the visitors' eye level within a very short distance. However, this may create a steep-sided channel into which an animal may fall. This may be a particular hazard for very young or very old animals, especially if the channel is poorly drained and prone to flooding. If the slope of the ha-ha is too gentle, animals will inevitably be at a great distance from visitors before they reach eye level. Furthermore, regardless of the angle of the slope, when animals reach the edge of the enclosure they may disappear from view altogether from some viewpoints or be directly beneath visitors from others.

A ha-ha barrier requires a secondary barrier, such as a hedge, a low wooden fence or wall, to prevent visitors from falling into the enclosure. A window should be provided to allow viewing from a wheelchair or pushchair, and by small children.

Fig. 7.25 Western lowland gorillas (*Gorilla gorilla gorilla*) retained by a wet moat and an electric fence at Bristol Zoo.

7.4.8 Electric fences

Electric fences ('hot wires') consist of metal wires or plastic interwoven with wires (polywire and polytape) supported by insulated posts. Permanent fences are normally made of galvanised wire. Electric fences carry short duration high-voltage electrical pulses (e.g. 8500 volts at one or two pulses per second). This gives an animal the opportunity to move away from the wire once it has been shocked. The fence must be able to deliver sufficient voltage to penetrate the resistance of the animal's skin. Electric fences are psychological rather than physical barriers because animals learn not to touch them. However, they do not always act as a deterrent. Hancocks (1996) describes an orangutan (*Pongo* sp.) at Woodland Park Zoo who used to sit with his hand on an electrified window pane allowing his arm muscles to be jolted by electric shocks and chimpanzees (*Pan troglodytes*) at another facility that used to swing and climb on a live high-voltage electric fence.

When wires are located on the ground it is important to position them at the correct height for the species so that an animal will touch two wires simultaneously, and at such a height that an inexperienced animal cannot pass under the wires after it has been shocked. The location of electric fences should be clearly indicated by appropriate signage (Fig. 7.26).

Electric fences would not normally be used as a primary barrier. They are often used as a secondary barrier on the visitor side – or both sides – of a wet moat, or to prevent animals from climbing trees or destroying ornamental vegetation. They are also commonly used on the top of high fences to keep animals

Fig. 7.26 An electric fence ('hot wire').

from climbing out and to exclude predators such as domestic cats and foxes.

Rows of parallel horizontal wires may be installed just above the ground surface as a barrier for elephant enclosures. Elephants quickly learn not to touch these wires and cannot step over them if they cover a sufficiently wide area. They are used in drive-through enclosures to keep elephants away from vehicles and in outdoor enclosures to keep elephants away from visitor barriers.

Further information about the construction and use of electric fences may be found in McKillop (No date).

7.4.9 Windows

Glass and acrylic windows may give visitors of all ages good views of zoo animals. However, it is increasingly being recognised that this is not always in the interests of the animals. Visitors tap on windows and at busy times this may result in more or less continuous disturbance. Gorillas may charge at the glass when teased by visitors. Some zoos have covered large windows, creating smaller viewing slots for species that are particularly prone to disturbance, such as orangutans. Glass may be totally unsuitable for some species that may fail to perceive its presence, for example some bird species. One-way glass has been used to construct a large window on one side of the indoor accommodation for Chilean flamingos (*Phoenicopterus chilensis*) at Chester Zoo, allowing visitors to see them at close quarters without causing disturbance (Fig. 7.27).

Glass is frequently used as a barrier for great apes, in indoor and outdoor enclosures, because it allows viewing at close quarters and it prevents disease transmission between apes and human visitors. The outdoor gorilla enclosure at ZSL London Zoo may be viewed from inside the gorilla house through a large window which extends to the floor. The floor of the viewing area is at the same level as the floor of the enclosure, giving the illusion of continuity (Fig. 7.5). The same technique has been used in the lion exhibit at ZSL Whipsnade Zoo (Fig. 7.14).

A major disadvantage of glass is that it needs to be cleaned regularly on both sides. Reflection and the generally poor optical quality of the material used creates a barrier between the animals and the visitors and makes it difficult to take good photographs. Glass also makes it difficult for visitors to smell or hear the animals. Windows may be a hazard to wild native birds as they may fly into them and be killed. Some exhibits have silhouettes of birds of prey fixed to windows to deter them.

Windows can be used to give visitors experiences they would not otherwise have, for example, viewing penguins, seals, otters, hippos and other species swimming underwater. Transparent acrylic tunnels allow visitors to have an artificial underwater experience in aquariums and other water-based exhibits (Fig. 7.16). Unfortunately the curved shape of these tunnels causes an optical distortion of the appearance of the animals.

Some aquariums contain very large marine tanks in which they exhibit coral reef communities including sharks, rays, conger eels and a wide variety of smaller fish. The *Blue Planet* in Ellesmere Port, UK, has such a tank. Visitors can view the occupants via a large window, which is used for educational talks and feeding displays by divers, or by travelling along an underwater tunnel equipped with a moving floorway.

Fig. 7.27 Chilean flamingos (*Phoenicopterus chilensis*) viewed through a window made of one-way glass.

7.4.10 Free-range behavioural barriers and perceptual barriers

Some species have traditionally been allowed to wander around zoos such as peafowl, ducks and geese. At the South Lakes Wild Animal Park in Cumbria free-roaming species include kangaroos, wallabies, lemurs and emus. Other species may be allowed to roam freely in a zoo because they can easily be trained to return to their indoor accommodation if this is where they are routinely fed, for example lemurs, marmosets and macaws.

Where birds which are capable of flight are allowed to roam free they may be discouraged from leaving the zoo by providing them with a number of nesting, roosting and feeding places. Other large species may be 'pinioned'. The process of pinioning involves the cutting of one wing at the carpel joint, thereby removing the basis from which the primary feathers grow. This makes the bird permanently incapable of flight because it is lopsided. An alternative method is to clip the primary feathers of one wing ('clipping'). This makes the bird temporarily incapable of flight until the feathers are replaced at the next moult.

Some barriers act as a perceptual boundary to keep visitors away from animals, for example a very low wooden rail placed at the edge of a lawn where kangaroos and wallabies graze, or where lemurs or monkeys roam (Fig. 7.28).

7.4.11 Unacceptable barriers

Some barriers are unacceptable in a modern zoo simply because they are unnecessary and outdated. These include bear pits and cages of thick steel bars. Other barriers are unacceptable because they are dangerous. Dry moats are totally unsuitable for retaining elephants and other large mammals that may fall in. Some such structures have been modified by placing rows of large rocks at the edge to keep the animals away. Rows of short metal spikes embedded in a concrete floor have been used as a barrier for rhinos, but this is obviously extremely dangerous (Hancocks, 1996).

7.5 TYPES OF EXHIBITS

Most zoos contain many cages and enclosures in which individual species are kept alone and to which the public has no direct access. However, some exhibits allow the public closer contact with animals and some contain a mixture of species living together.

7.5.1 Walk-through exhibits

Walk-through exhibits such as aviaries, butterfly houses and bat caves require barriers that minimise the opportunity for flying species to escape through

Fig. 7.28 A perceptual barrier for free-ranging Barbary macaques (*Macaca sylvanus*). Inset: Female macaque with young.

the entrance (Fig. 7.29). Often these take the form of strips of thick plastic sheeting hung vertically as a screen. When visitors push through the displaced sections quickly fall back into place, sealing the entrance. Often two such screens will be placed in sequence, forming an 'air lock' arrangement. Barriers may also be formed using lengths of chain made of metal or plastic to cover a doorway. Traditionally doors may also be used in pairs. The entrance to a lemur exhibit at Bristol Zoo is constructed from two steel and mesh doorways in an 'air lock' arrangement. Each is operated with an electric switch. When visitors enter through the outer door the lock for the inner door will not operate until the outer door is closed. This makes it impossible for both doors to be open at the same time. The golden lion tamarin (*Leontopithecus rosalia*) exhibit at *Marwell Wildlife* uses such a system with red and green lights to indicate when visitors may enter.

Monkey Forest is a zoo which consists of a single free-range exhibit. It keeps approximately 140 Barbary macaques (*Macaca sylvanus*) in a 24 ha (60 acre) broad-leaved woodland in the UK (Fig. 7.28). The woodland is enclosed by a high fence and visitors are allowed to walk among the animals along paths bounded by a low wooden rail. The perimeter fence cannot be seen from many parts of the enclosure, giving the perception that the monkeys are completely free. The macaques are adapted to a temperate climate so do not need shelters or supplementary heating in winter. Keepers ensure that visitors get a good view of the animals by providing regular scatter feeds of fruit and other foods at specific locations, prior to giving a talk about their behaviour and conservation.

Fig. 7.29 Entrance to a free-flight aviary.

Keepers often need to be present continuously in free-range exhibits to prevent escapes and to protect the people from the animals and the animals from the people. Animals, such as lemurs, may be attracted to children's prams and buggies, and small children may stray too close to the animals.

Walk-through bat houses are a relatively new development in zoos. As bats are largely nocturnal, day and night must be reversed using timer switches to operate the lighting if visitors are to observe active animals. Bats will often fly very close to visitors, especially if parts of the ceiling of the exhibit have been lowered to create the illusion of a cave. As with other walk-through exhibits, it is essential that staff are present to prevent visitors from interfering with the animals.

7.5.2 Drive-through enclosures

Drive-through enclosures are a fundamental feature of safari parks. Barriers are typically high chain-linked fences with a return, linked by gates which may be manually or electrically operated. Cattle grids are used to prevent escapes and the mixing of animals in adjacent enclosures. Grids are effective for bovids and antelopes but are unsuitable for felids and other more agile species.

Double electric gates may be used to prevent the movements of dangerous animals such as big cats. As vehicles pass from one enclosure to the next they pass between the fences by entering a small area of fenced road. They drive up to the closed gate to the second enclosure and this is not opened until the gate to the enclosure they are leaving is closed behind them. This system works well but restricts vehicle flow and requires an operator who is usually located in a hut or tower overlooking the gates. Some parks have abandoned this system in favour of manually operated gates which are left open until the animals approach. Where enclosures contain lions a low fence may be erected on the inside of the enclosure on the approach to the gate. This will channel animal movements along the road and prevent an individual from charging at the entrance. In lion enclosures it would be normal practice to have keepers patrolling the gate area and an armed keeper in a vehicle with the pride at all times, moving animals away from the gates if they came too close.

Roads connecting drive-through enclosures need to be arranged so that visitors may avoid entering areas where their vehicles are at risk of being damaged. For example, some visitors may not wish to enter a baboon enclosure as these animals may damage vulnerable components such as windscreen wipers, mirrors and other exposed structures.

7.5.3 Petting zoos

Many zoos contain a petting zoo (Children's Zoo or Children's Farm) where visitors may touch and feed farm animals and some common exotic species, with food provided by the zoo. Sometimes the farm species are exhibited in simulated farm buildings including pig sties, chicken coups and cattle stalls. Barriers often resemble the types of fencing and gates associated with farms and in many zoos visitors are able to enter some of the enclosures and touch and feed the animals. These areas must be provided with hand-washing facilities.

Fig. 7.30 A multi-species exhibit at Dublin Zoo – *African Plains*.

7.5.4 Multi-species exhibits

Multi-species (or mixed-species) exhibits are increasingly popular in zoos. They have a number of advantages and disadvantages.

Advantages include:

- greater interest and educational value for the public by representing natural associations between species
- enrichment for the animals as a result of more complex interactions.

Some disadvantages are:

- possible competition for food between species
- negative interactions between species, e.g. aggression
- possibility of unnatural behaviours and interactions between species
- risk of disease and parasite transmission between species
- risk of hybridisation between closely related species
- possibility of poor educational value if species from different habitats or biogeographical regions are mixed.

Most of the potential problems listed above can probably be overcome with some thought and careful monitoring of health and behaviour by keepers and other staff. Species with similar food requirements may compete for food, but so may conspecifics. It may be necessary to construct places within multi-species exhibits where food can be hidden for particular species, and where one species can escape from another.

Thomas and Maruska (1996) have listed a number of combinations of species that have successfully been exhibited together. These include:

- gorilla (*Gorilla gorilla*) with colobus monkey (*Colobus* spp.)
- black-and-white lemur (*Varecia variegata*) with ring-tailed lemur (*Lemur catta*)
- chevrotain (*Tragulus napu*) with Bali mynah (*Leucopsar rothschildi*)
- koala (*Phascolarctos cinereus*) with echidna (*Tachyglossus aculeatus*).

At Dublin Zoo the *African Plains* exhibit contains plains zebra (*Equus burchellii*), giraffe (*Giraffa camelopardalis*) and ostrich (*Struthio camelus*) in the same enclosure, offering the visitor the opportunity to see a mammal community similar to that which naturally occurs in parts of East and Southern Africa (Fig. 7.30). However, the zoo has also added the rare scimitar-horned oryx (*Oryx dammah*) which is extinct in the wild. In contrast, Knowsley Safari Park exhibits wallabies, from Australia, in the same enclosure as blackbuck (*Antilope cervicapra*), from India. Although the animals may be biologically compatible, such a combination has little educational value.

Birds are often kept in mixed-species exhibits. However, some zoos keep species from the same continent together even though they may not have overlapping natural ranges. This again gives the impression of some ecological integrity to the exhibit while offering visitors sights of species mixtures which could never occur in nature. Whether or not this matters depends upon your point of view. On the whole, it is probably better for the animals to be kept in a series of very large mixed-species exhibits, where they can interact – even if they lack ecological accuracy – than in small single-species exhibits.

Some combinations of species may lead to unusual interspecific interactions. In 1994 Safari Beekse Bergen in the Netherlands began housing five female African elephants (*Loxodonta africana*) with a group of over 30 hamadryas baboons (*Papio hamadryas hamadryas*) in a 1.3 ha outdoor enclosure during the daytime (Deleu *et al.*, 2003). This provided considerable enrichment for both species. There were occasional agonistic encounters between the two species but baboons are able to escape to rock formations within the enclosure. A single incident of cross-contamination with salmonella occurred but close monitoring of the health of both species appeared to assure that this did not become a problem. The interspecific interactions observed included baboons searching for undigested food items in elephant dung and in holes dug by elephants. Remarkably, baboons (particularly juveniles and subadults) enjoyed riding on the backs of the elephants. The elephants invited the baboons to do this by stretching their trunks out to individuals sitting on rocks. The elephants may gain some benefit from this association because the baboons have been observed removing seeds and insects from their skin. Stereotypic behaviour was reported to be almost absent in both species but no comparison was made with a control situation.

Multi-species exhibits involving elephants are not new. In the 1970s Knowsley Safari Park, in the UK, kept seven African elephants (one bull and six cows) in paddocks through which visitors would drive their vehicles to view the animals. On some days they were kept with a large herd of buffalo (*Syncerus caffer*) and on others they associated with wildebeest (*Connochaetes taurinus*), zebra (*Equus burchellii*), giraffe (*Giraffa camelopardalis*) and a pair of hippopotamuses (*Hippopotamus amphibius*). Agonistic interactions were rare because a keeper in a 4×4 vehicle was always present to keep the species separate. However, sometimes agonistic encounters occurred between the elephants and a large bull buffalo, and on one occasion the bull elephant attempted to mount one of the hippos.

7.6 HEALTH AND SAFETY

Zoo staff and visitors may be potentially exposed to risk from zoo animals. These include the risk of injury and the possibility of contracting disease. Zoos should take basic precautions to protect visitors, for example by avoiding the use of vehicles or other machinery in the zoo when visitors are present. They must also ensure that enclosures and visitor barriers are designed to prevent contact with animals, with a few exceptions.

7.6.1 Visitor barriers

Visitors need to be kept away from the primary barriers around the perimeter of an enclosure for their own safety and for the safety of the animals. Visitor barriers include:

- concrete walls
- rope fences
- wooden guard rails
- low hedges
- chains
- chain-link fence or mesh.

The choice of barrier will depend upon the relative costs, attractiveness, maintenance requirements and safety considerations. Some zoos may use a combination of more than one of these barriers; for example, a low hedge may be grown against a chain-link fence and a guard rail may be mounted in front of the hedge.

Visitor barrier construction should take into account the need to keep very small children away from enclosure barriers and the need to deter adults from intentionally climbing into enclosures. Occasionally, visitors fall or climb into the enclosures of dangerous animals. In the summer of 1986 a small boy fell into the gorilla enclosure at Jersey Zoo (now *Durrell*). Although he was knocked unconscious by his fall, he was not harmed by the gorillas, and a large silverback named 'Jambo' appeared to stand guard over him until he was rescued by

Table 7.5 Hazardous animal categories – examples of category 1 and category 2 species.

Category 1 (Greater risk)	Category 2 (Less risk)
Red kangaroo (*Macropus rufus*)	Large opossums (*Didelphis* spp.)
Grey wolf (*Canis lupus*)	Fruit bats (Pteropodidae)
Giant panda (*Ailuropoda melanoleuca*)	Lemurs (*Lemur* spp.)
Seals (*Phoca* spp.)	Spider monkeys (*Ateles* spp.)
Elephants (Elephantidae)	Bat-eared fox (*Otocyon megalotis*)
Apes (Pongidae)	Aardvark (*Orycteropus afer*)
Rhinoceroses (Rhinocerotidae)	Tapirs (*Tapirus* spp.)
Giraffe (*Giraffa camelopardalis*)	Llama (*Lama glama*)
Lion (*Panthera leo*)	Okapi (*Okapia johnstoni*)
Ostrich (*Struthio camelus*)	Gazelles (*Gazella* spp.)
California condor (*Gymnogyps californianus*)	Cassowaries (*Casuarius* spp.)
Eagle owls (*Bubo* spp.)	Great white pelican (*Pelecanus onocrotalus*)
Komodo dragon (*Varanus komodoensis*)	Buzzards (*Buteo* spp.)
Alligators (*Alligator* spp.)	Macaws (*Ara* spp.)
Mambas (*Dendroaspis* spp.)	Australian snapping turtles (*Elseya* spp.)
Poison arrow frogs (*Phyllobates* spp.)	Aldabra giant tortoise (*Testudo gigantia*)
Scorpion fishes (Scorpaenidae)	Swift snakes (*Psammophis* spp.)
Grey and tiger sharks (Carcharhinidae)	Giant salamanders (Cryptobranchidae)
Blue-ringed octopus (*Hapalochlaena maculosa*)	Conger eels (Congridae)
Black widow or redback spiders (*Latrodectus* spp.)	Bird-eating spiders or tarantulas (Theraphosidae)

Source: SSSMZP (2004).

zoo staff. This incident made the news all over the world and helped to change the public's perception of gorillas.

Sometimes visitors enter zoo enclosures intentionally. In 2007 a man jumped into a tiger cage at Patna Zoo in India, and was badly injured. A video of the incident was posted on the internet. In April 2008 a mentally ill man, who apparently loved tigers, entered a tiger cage in a zoo in northeast China and was eaten. By the time he was found only his legs and skull were left (Reuters, 2008). In July 2010 a drunken Australian tourist climbed into an enclosure at Broome Crocodile Park, Western Australia, and was seriously injured after attempting to ride a saltwater crocodile (*Crocodylus porosus*).

In safari parks the visitors may travel through enclosures in their own vehicles or in an open 'land train' operated by the park, depending upon the species kept. In this type of facility the vehicle is the barrier between the animals and the visitor. Some parks have caged some of their animals within their drive-through enclosures either because they would be difficult to see or because they are dangerous. In the late 1970s Knowsley Safari Park kept cheetahs (*Acinonyx jubatus*) in a relatively small wire mesh pen within an enclosure because they would otherwise not have been seen in the long grass. The park currently keeps African wild dogs (*Lycaon pictus*) and tigers (*Panthera tigris*) in a similar fashion. It originally kept tigers in open enclosures but they attacked visitors' cars and bit tyres. Elephants used to be allowed to roam relatively freely in drive-through enclosures in British safari parks, but this is now considered too dangerous. At the West Midlands Safari Park they are kept away from the road by an electric fence, but at Knowsley they are now kept within their own facility away from vehicles and separate from the drive-through enclosures.

7.6.2 Hazardous animal categorisation

In the UK, zoo animals are grouped into three categories according to their risk to people (Table 7.5) (SSSMZP, 2004).

Category 1 (Greater risk)

Contact with the public is likely to cause a serious injury or be a serious threat to life because of the risk of injury, toxin or disease. Category 1 animals must usually be separated from the public by a physical barrier that prevents contact. Individual animals in this category can only be taken into the same areas as the public if it can be shown that they pose no risk.

Category 2 (Less risk)
Contact between category 2 animals and the public may result in injury or illness caused by injury, toxin or disease, but is not likely to be life threatening. These animals should normally be separated from the public by a barrier but it need not prevent all physical contact. Some category 2 species may be kept in free-ranging, free-flying or walk-through exhibits. Any animal that has caused injury or behaved in a way which could have caused injury or transmitted disease must be treated as a category 1 animal.

Category 3 (Least risk)
All species not listed in category 1 or category 2 are automatically category 3. This does not mean that they pose no risk to the public. Many taxa in this category are not well known and zoo operators should determine the appropriate barrier by undertaking a risk assessment. If any animal in this category has caused injury or behaved in a way which could have caused injury or transmitted disease it must be treated as a category 1 animal.

Further information on the definitions of the three hazardous animal categories and a complete list of animals in categories 1 and 2 are given in Appendix 12 of the SSSMZP (2004).

7.6.3 Hygiene and visitors

Walk-through exhibits are increasingly popular with zoos and there is a real risk of contamination with pathogens as visitors touch hand rails, fences, signs and, in some cases, the animals themselves. Hand-washing facilities, including bactericidal soap and driers, should be provided wherever people come into contact with animals or surfaces and objects touched by animals (Fig. 7.31).

In September 2009 a petting farm in Surrey, UK, was temporarily closed after a number of children were hospitalised because of an outbreak of a potentially fatal strain of *E. coli*, known as 0157:H7.

7.6.4 Animal escapes and visitor safety

Animals that are both strong and intelligent – such as primates and elephants – will often spend a great

Fig. 7.31 An anti-bacterial soap dispenser.

deal of time gradually damaging enclosure furniture, barriers and gates. Eventually this damage may allow their escape from the enclosure or from one area of the enclosure to another from which they are sometimes, or always, excluded. Orangutans will gradually unscrew nuts and bolts on climbing frames and elephants may shake gate mechanisms until the locks fail. It is essential that keepers regularly check barriers, gates and other structures with which animals may interfere so that they can be sure that the animals are safely contained at all times.

Zoo animals have a long history of escaping from their cages and enclosures. Sometimes they only escape into the zoo grounds, but occasionally they manage to escape beyond the zoo's perimeter fence or wall.

Warin and Warin (1985) describe a number of escapes from Bristol Zoo dating back to a leopard escape in 1892. Other escapes included a kangaroo, wallaby, California sea lion (*Zalophus californianus*), pelican, crowned crane, clawed otter and beavers (*Castor* sp.). In 1930 a pair of chimpanzees (*Pan troglodytes*) escaped into the zoo grounds and the male ended up in the zoo bar. On another occasion 36 monkeys escaped from the 'monkey temple' after a ladder was left against the inside wall. Some found their way into the nearby Clifton College Music School. It was three weeks before they were all recaptured.

Improved nutrition and fitness in some species of zoo animals may mean that enclosure specifications that were previously considered adequate are no longer so. As zoo animals become fitter and stronger it may be that zoos need to raise barrier heights accordingly. There have been a number of recent escapes in US zoos from exhibits that have not previously been breached. In 2003 a gorilla (*Gorilla gorilla*) at Franklin Park Zoo, Boston scaled an electrified wall and crossed a 3 metre wide, 3 metre deep moat before escaping from the zoo (MacQuarrie and Belkin, 2003). In 2004 a gorilla in Dallas Zoo escaped from an enclosure surrounded by a 4 metre high concave wall (Jalil, 2004). In 2008 an orangutan (*Pongo* sp.) escaped from his enclosure at Los Angeles Zoo by making a hole in the fence (Anon., 2008a).

7.6.4.1 Ecological effects of escapes

Most jurisdictions have legislation that prohibits the release of non-native species into the wild. In Great Britain the relevant legislation is contained within the Wildlife and Countryside Act 1981 (s.14) (see Legal Box 5.8).

Baker (1990) has documented 293 specimens of 39 exotic mammal species that were not normally kept as pets recorded out of captivity in the UK between 1970 and 1989. Between 1975 and 2001 DEFRA received 27 reports of non-native cats – including one specimen of *Lynx lynx* – that had escaped into the wild. Of these 12 were shot, eight were recaptured and the remainder were either found dead or their fate was never established (DEFRA, 2007). The introduction of the Dangerous Wild Animals Act in 1976 is believed to have been responsible for the release of many exotic species into the wild by their owners, possibly including large felids (Taylor, 2005).

Escapes from animal collections may have considerable ecological effects. The ruddy duck (*Oxyura jamaicensis*) is a North American species that escaped from Slimbridge (WWT) several decades ago. It cross breeds with the endangered white-headed duck (*O. leucocephala*) and to date the UK government has spent £4.6 million culling 6200 of the birds in an attempt to eradicate them.

7.6.5 Vandalism and other criminal activity

Unfortunately zoos have always been the target of anti-social behaviour. Louis XIV built a menagerie in the grounds of the Palace of Versailles and it received its first animals in 1665. He opened it to the public, but soon had to restrict access because of the amount of damage that was caused. In the late 1800s, London Zoo allowed visitors to feed its famous African elephant, Jumbo, until they discovered that some people were inserting pins and other objects into the food.

Zoos with walk-through exhibits containing lemurs, monkeys, bats, butterflies and other species often need to post keepers inside the exhibits to ensure no harm comes to the animals. Visitors have been known to catch and kill bats and butterflies. At the Cotswolds Wildlife Park an attempt was made in 2008 to smash through a window in a lion enclosure.

Occasionally, animals are stolen from zoos. In 2009 two Goeldi's monkeys (*Callimico goeldii*) were stolen from Tweddle Children's Animal Farm, in northeast England (Anon., 2009d). In 2008 a seven-year-old boy broke into the Alice Springs Reptile Center in the Northern Territory of Australia and was captured on CCTV killing 13 reptiles and feeding them to a saltwater crocodile (*Crocodylus porosus*) (Tedmanson, 2008).

7.6.6 Keeper safety

Keepers are exposed to a wide range of risks to their health and safety. These include:

- Risk of physical injury due to accident caused by:
  - enclosure design, e.g. deep dry moats, deep water, gates, electricity
  - mechanical machinery, e.g. lifting equipment, cleaning equipment
  - zoo vehicles, e.g. delivery and maintenance vehicles
  - moving and handling heavy objects
  - food preparation equipment.
- Risk of disease:
  - zoonosis – a disease transmitted from an animal to a human
  - infection resulting from injury
  - allergic reaction to animals, bedding, vegetation or animal food.
- Risk of attack or poisoning:
  - physical injury caused by attack, e.g. by a large predator

- accidental or intentional crush injury caused by contact with a large, heavy animal, e.g. elephant, rhino
- poisoning – e.g. from a bite or sting, or from touching the skin of a poisonous animal (see Table 11.6).

In order to control the risk to keepers from animals it is useful to categorise different taxa based on the risk they pose (see Section 7.6.2).

7.6.6.1 Keeper enclosure requirements

Keeper access to enclosures and animal houses needs to be designed so that animals are unable to escape when the keeper enters. This may be achieved by an 'air lock' arrangement of doors whereby the keeper passes through one door – giving access to an enclosed space – and closes it behind him before a second door is opened, giving access to the enclosure itself. It should not be necessary to say that, even with such an arrangement, keepers should not enter enclosures when dangerous animals are present. However, keepers have been killed doing just that, particularly with big cats.

Where large, dangerous animals need to be moved from one space to another mechanical or electrical remotely operated doors may be utilised, preferably in conjunction with CCTV so that a keeper inside the animal house can see, for example, animals in the outside enclosure as the doors are operated (Fig. 7.32). Light levels indoors need to be sufficiently high to allow unrestricted viewing of animals at all times, including at night.

Keepers may need access to a weigh bridge to monitor the growth of large species such as rhinos, elephants and other large mammals. Ideally this should be built into the enclosure when it is first constructed and located in such a position that the animals have to walk over it when moving from one part of the enclosure to another.

7.6.6.2 Attacks on keepers

Keepers are occasionally killed by animals in zoos. This is often the result of an animal that was previously not considered to be dangerous becoming aggressive. Sometimes it is the result of poor working practice, for example a keeper working alone with an animal instead of with a colleague. Divers at the *Blue Planet* aquarium feed fish from inside their tank. The divers work in pairs, one distributing the food and the

Fig. 7.32 Door opening mechanism. It is important that this can be operated from outside the enclosure.

other looking out for the sharks that share the tank. The sharks are fed at a different time from other fishes (Fig. 7.33).

The dangers posed by large marine species should not be underestimated. In February 2010 a trainer at *SeaWorld Park* in Orlando, Florida, was killed when a killer whale (*Orcinus orca*) pulled her into a pool.

Elephant keeper deaths

Elephant keeping is an extremely dangerous profession. The occupation of 'elephant trainer' has the highest fatality rate of any occupation documented by the US Department of Labour Statistics. Based on 600 known elephant trainers in the United States the fatality rate is 333 per 100,000 (USDL, 1997). One study documented 15 elephant-related deaths between 1976 and 1991, approximately one death per year (Lehnhardt, 1991). In the UK, three keepers, at different zoos, were killed by cow elephants within a period of just 20 months in 2000 and 2001.

Fig. 7.33 Shark feeding.

Benirschke and Roocroft (1992) analysed data on 36 Asian elephants responsible for serious accidents involving keepers and concluded that bulls were significantly more dangerous than cows. The mean age of bulls involved in accidents was 18 years and the mean age of cows was 25.3 years. These ages coincide approximately with the age when bulls first come into musth and cows become matriarchs (Kurt, 1995). It has recently been suggested that some – 'low-ranking' – elephant keepers may be at particular risk by virtue of their position in the herd hierarchy (Litchfield, 2005).

7.6.7 Emergency procedures and public safety

Zoos must have emergency procedures in place in case of a fire, animal escape or other emergency which might require evacuation or containment of the public away from a dangerous situation. A sufficient number of staff should be trained in first aid and first aid posts should be appropriately marked as should exit routes. If some parts of the zoo are remote from the main buildings signs at these locations should indicate emergency telephone numbers.

The SSSMZP (2004) states that procedures relating to escapes of animals should include:

- the reporting of each escape to a senior member of staff
- details of how staff should respond to an escape in all circumstances, for example whether visitors are present or absent, whether or not staff are on duty, whether one or many animals have escaped
- procedures in the event of escape including recapture, visitor protection, informing the police and the licensing authority
- visitor control and evacuation of the zoo
- securing the perimeter fence by closing all access points
- the provision of firearms and darting equipment, details of which should be agreed by the zoo operator and the police
- the provision of suitable equipment – including vehicle protection – for the recapture party.

A zoo should have a clear chain of responsibility in the event of an emergency and this should be notified to all staff in writing and displayed on notice boards. An appropriate member of staff should be available at all times to make decisions regarding the use of firearms or darting equipment and the euthanasia of an escaped animal.

Zoos are required by the SSSMZP (2004) to record all escapes and to report any escape from the confines of the zoo to the licensing authority within 24 hours. Risk assessments should be continually reviewed.

The SSSMZP (2004) requires that emergency drills are carried out in the zoo at least four times a year. Sometimes such drills are conducted in such a manner that staff believe they are responding to a real incident. Zoos should be equipped with a public address system to warn visitors about safety issues and staff should be equipped with radios.

7.7 ANIMAL ENCLOSURE REQUIREMENTS

7.7.1 Introduction

Apart from the necessity to provide suitable barriers between animals and people, animal enclosures need to fulfil the basic biological needs of the species. Husbandry guidelines are available for many species and many of these provide detailed specifications for enclosures (Box 7.2).

In order to create an enclosure that allows animals to adapt and cope physiologically and behaviourally to confinement a number of factors must be considered, including:

- the amount of space
- the complexity of the space
- the social environment (group size, group structure, and stocking density)
- the human–animal relationship
- the animal's ability to control and predict events (Tennessen, 1989).

There is a widespread acceptance that animals have certain behavioural needs. However, there is controversy over which behaviours an animal needs to perform and how important specific behaviours are to animal welfare (Jones and McGreevy, 2007).

When animals are confined we limit:

- their physical environment
- their ability to choose companions.

Unfamiliar animals need to establish their relationships to other individuals when they are put together in the same enclosure, often with agonistic behaviour. This can be avoided by using individual housing but this is often impractical and undesirable in a zoo because it removes the animals' ability to express normal social behaviour and form relationships with conspecifics.

If social animals are kept in unnaturally large groups in a confined space, this can disrupt their social behaviour, for example if the social group is too large for the animals to be able to recognise all of the other individuals in the group (D'Eath and Keeling, 2003). The maintenance of a social hierarchy assumes that each of the individuals knows its rank in relation to all of the others.

The lack of ability to control or predict events may cause stress in animals. (See Section 11.9).

7.7.2 Animal safety

The safety of animals, keepers nd visitors must be paramount when enclosures are being designed. Outdated designs may present serious hazards to animals. These may include deep concrete dry moats, poorly designed wet moats, and steel spikes embedded in concrete floors to keep animals away from barriers. Steep slopes, dry moats and deep water can be fatal to some animals. When dry moats were common in zoos elephants and other large mammals regularly fell into them and had to be destroyed.

Fires have led to a number of deaths in zoos. In 1995 a fire in the primate building of the Philadelphia Zoo killed 23 animals, including six lowland gorillas (*Gorilla gorilla*), three orangutans (*Pongo* sp.), four white-handed (lar) gibbons (*Hylobates lar*), and 10 lemurs (Janofsky, 1996). In 2006 three giraffes (*Giraffa camelopardalis*), including a week old calf and its mother, died in a fire in the giraffe house at Paignton Zoo. In 2007 a fire in a building at Indianapolis Zoo killed three turtles, two birds, an armadillo, two rodents, one snake and several hissing cockroaches.

Some injuries to animals are the result of poor visitor behaviour. Visitors sometimes throw food and other items, such as drink cans and batteries, into open enclosures, thereby compromising the safety of the animals. In 2006 a gibbon at Chicago's Lincoln Park Zoo had part of its arm amputated after becoming stuck in fencing while trying to reach food thrown by a thoughtless visitor (Nyhuis and Wassner, 2008).

7.7.3 Size and shape

Although it is impossible for zoos to provide most species with enclosures similar in size to their natural home ranges, it is nevertheless important for zoo designers to consider their space requirements very carefully. Where territorial species are to be kept consideration should be given to the number of individual territories that need to be accommodated. In the wild, home range is not a fixed characteristic of a species and may vary considerably depending upon the availability of resources, especially food. Many predators have

Box 7.2 Exhibit requirements for jaguars (*Panthera onca*) (adapted from Law, 2010).

Design parameters

- Standard cage designs can be used.
- Optimum design would be naturalistic, utilising complex artificial or natural features creating a vertical structure to maximise the available area and decrease animal loading.
- Design should accommodate natural felid behaviour: territoriality, scent marking, defence of territory against conspecifics.
- Design should minimise pressure from visitors.
- Vegetation, rockwork and climbing structures may help to reduce stress.
- Water features are recommended and should be more than 1 metre deep, with shallow areas to encourage play.

Primary containment

- Highest level of security required.
- The top of the containment should be completely enclosed where possible.
- Dry moats should be at least 25 feet (7.6 m) wide and vertical jump walls at least 15 feet (4.6 m) high.
- Mesh or fencing should be supported by cantilevers and hot wires are recommended for open-top exhibits.
- Care should be taken in the landscaping and position of furniture to prevent escape and to prevent access to areas where public, staff or animals may be injured.
- Fence or mesh should be no less than 6-gauge composition. Good results are achieved with 2 × 4 inch (51 × 102 mm) mesh, but 2 × 2 inch (51 × 51 mm) is recommended in keeper work areas. Jaguars may bite or pull flexible mesh risking tooth damage and damage to the mesh itself.

Enclosure size

- Outdoor enclosure: at least 300 square feet (28 m^2), with 50% extra space per additional specimen.
- Indoor enclosure: at least 20 × 15 feet (6.1 × 4.6m), with 50% extra space per additional specimen.
- Minimum exhibit height: 8 feet (2.4 m) but 12 feet (3.7 m) recommended.
- Additional space would be optimal for facilitating introductions and for breeding pairs, to reduce potential aggression.

Enclosure components

- climbing structures of live, dead or artificial trees
- elevated resting sites – artificial snags or ledges – with long viewing distances (at least one per animal)
- artificial or natural rocks to provide visual and auditory barriers
- multi-level complex pathways to reduce stereotypic behaviour
- landscaping that simulates natural cover and promotes walkways and provides shade and escape routes.
- Plant materials must be tested for toxicity before planting.

Night holding

- Each specimen should have its own individual shelf or nest box in a shift cage.
- This should be at least 8 × 8 × 8 feet (2.4 × 2.4 × 2.4 m).
- The design of the shift cages should prevent contact between adjacent incompatible animals.
- Shift doors should be designed to prevent tail injury.
- Doors between individual holding units should be considered to facilitate introductions.
- Isolated birthing dens should be available, with low-light capability, and located so as to minimise disturbance.
- Closed-circuit TV monitoring is advised.

a large home range simply because they require access to a large area in order to obtain sufficient food. Polar bears (*Ursus maritimus*) may have a home range of up to 125,000 km^2 (48,250 square miles). In contrast, the brown-throated three-toed sloth (*Bradypus variegatus*) has a home range of less than 2 ha (5 acres). In some species the territory size varies seasonally. The European robin (*Erithacus rubecula*) has a territory of about 0.55 ha (1.3 acres) in spring and summer but this decreases to about half this size in winter.

In Holland red foxes (*Vulpes vulpes*) have been found to occupy home ranges varying in size from 116 to 880 ha in different habitats (Niewold, 1976). In suburban Oxford, mean fox territories were 44.8 ha, while in a sheep-farming area in the north of England they were up to 1300 ha (Macdonald, 1980). Wild chimpanzees (*Pan troglodytes*) have smaller home ranges in mixed rainforest and larger ranges in woodland forest mosaics and open savannah, so the space they utilise may vary between 5 and 400 km^2 (Kingdon, 1997).

Clubb and Mason (2007) have suggested that the welfare of many carnivores is compromised in zoos as a result of inadequate enclosure sizes. The brown bear (*Ursus arctos*) and snow leopard (*Uncia uncia*) adapt well to captivity, but the polar bear and clouded leopard (*Neofelis nebulosa*) are hard to breed and tend to develop abnormal behaviours. Clubb and Mason claim that for many carnivore species, stereotypy levels and relatively high captive infant mortality rates are related to natural ranging behaviour. They found that animals with large home range sizes had higher rates of infant mortality in captivity than those with smaller ranges. They also found that pacing stereotypies were more likely to occur in species that had a large median daily travel distance in the wild than in those that travelled shorter distances. Clubb and Mason suggest that enclosure designs for carnivores should focus on their ranging behaviours for example by providing more space, multiple den sites, and greater day-to-day environmental variability.

The methodology used by Clubb and Mason has been criticised and zoo officials have insisted that large carnivores in zoos are not routinely suffering as a result of poor enclosures (Randerson, 2003). Their work did not involve experimentation. However, other studies have directly examined the effects of enclosure size on welfare. Li *et al.* (2007) examined the influence of enclosure size and animal density on Père David's deer (*Elaphurus davidianus*) at Dafeng Nature Reserve in China. They compared faecal cortisol levels – an indicator of stress – and behaviour in a group of deer stags kept in a large enclosure at low animal density (200 ha; 0.66 deer/ha) with stags kept in a small display pen with high density (0.75 ha; 25.33 deer/ha). In a second experiment, Li *et al.* compared faecal cortisol levels in a group of 12 stags which were moved from a 100 ha enclosure to a small pen of 0.5 ha. In the first experiment, they found that the frequency of conflict behaviour was higher in the small display pen than in the large enclosure. The animals in the small pen also exhibited higher cortisol concentrations. In the second experiment, no difference was found between the faecal cortisol levels on sampling days, but the mean level was significantly higher on the day after transfer to the small pen than on the day before. The use of cortisol to measure stress is discussed in Section 11.9.1.

Enclosure shape may be important in some species and it may be necessary to construct refuges for subordinate animals or females. Where aggression is likely to occur, circular enclosures or houses remove the possibility of individuals being trapped in corners.

For many species, the quality of the space within an enclosure may be much more important than the quantity. A small but elaborate enclosure may provide much more opportunity for an animal to engage in normal behaviour than a large homogeneous space.

Arboreal species can make better use of a small enclosure than can animals that cannot climb, provided suitable structures are provided. Obviously, the amount of space available to an arboreal animal will depend upon the complexity of the climbing structures. Mesh fencing, although unattractive, can significantly increase the useable volume of a cage for animals such as monkeys and parrots. It is unlikely that an enclosure could be designed where 100% of the volume could be used, except perhaps for some small birds and, of course, small fish kept in an aquarium. Mellen (1997) suggests that for small cats 75% of the enclosure's vertical space should be available, but this appears to be a completely arbitrary figure.

One way of increasing the amount of space available to an animal is to train it so that it may be taken out of its enclosure. Some zoos that keep elephants walk them around the zoo before it opens. Other species, such as tame lemurs, may be walked on leads and introduced to visitors.

Husbandry guidelines and manuals provide minimum or recommended sizes for enclosures. These are often presented as minimum dimensions with a suggestion that larger enclosures would be preferable (Table 7.6).

Table 7.6 Minimum enclosure requirements for elephants, cotton-top tamarins, emperor penguins and flamingos.

Elephants (Elephantidae)	Outdoor enclosure	Indoor enclosure	Source	
Space for first animal	167.2 m^2 (1800 sq.ft.)	Adult cows: 37.2 m^2 (400 sq.ft.) Adult bulls or cows with calves: 55.7 m^2 (600 sq.ft.) each	*AZA Standards for Elephant Management and Care* (2003).	
Each additional animal	83.6 m^2 (900 sq.ft.)	As above		
Cotton-top tamarin (*Saguinus oedipus*)	**Cage size**	**Source**		
Smallest size for a single family group	3 m (length) × 2 m (width) × 2.5 m (height)	Savage, A. (ed.) (1995). *Cotton-top Tamarin Husbandry Manual.* Roger Williams Park Zoo, Providence, RI.		
Emperor penguin (*Aptenodytes forsteri*)	**Pool surface area**	**Pool depth**	**Land surface area**	**Source**
Per bird for first six birds	1.67 m^2 (18 sq.ft.)	1.33 m (4 ft.) for any number of birds	1.67 m^2 (18 sq.ft.)	Penguin Taxon Advisory Group (2005). *Penguin Husbandry Manual*, 3rd edn. AZA.
Each additional bird	0.84 m^2 (9 sq.ft.)		0.84 m^2 (9 sq.ft.)	
Flamingos (Phoenicopteridae)	**Outdoor enclosure**	**Outdoor pool size**	**Indoor enclosure (land and pool)**	**Source**
Per bird	1.4 m^2 (15 sq.ft.) Nesting area: 0.9 m^2 (10 sq.ft.)	Large enough to accommodate all of the flock	1.4 m^2 (15 sq.ft.) Pool depth: 30–60 cm (1–2 ft.)	Brown, C. and King, C. (2005). *Flamingo Husbandry Guidelines.* AZA, EAZA and WWT.

7.7.4 Substrate

The nature of the substrate in an enclosure can affect the welfare of animals. The effect of rubberised flooring on the behaviour of Asian elephants has been examined by Meller *et al.* (2007). They observed six elephants at Oregon Zoo and concluded that the flooring may have provided a more comfortable surface for locomotion and standing resting behaviour. However, the elephants did not choose to use those rooms with rubber floors over those without them, and both normal locomotion and stereotypic locomotion increased on rubber floors. Foot problems are common in elephants kept in zoos (Fowler, 1993).

A study conducted at the California National Primate Research Center found that rhesus macaques (*Macaca mulatta*) kept on a grass substrate spent more time foraging and less time grooming than those kept on a gravel substrate. Increased time spent grooming by macaques kept on gravel may have contributed to a significantly higher hair loss in gravel enclosures (Beisner and Isbell, 2008).

Abrasive substrates are useful for some species. Grit is placed on the floor of some giraffe houses to help in wearing down the animals' hooves. Using perches made of rough materials will help to prevent overgrowth of birds' claws.

The nature of the substrate will determine the extent to which it can assist with enrichment. Bark chippings make an interesting substrate for many primates because they will spend time searching it for hidden food. Some elephant houses have floors covered in

Fig. 7.34 Sand is a good substrate for animals that naturally dig for food and burrow such as aardvarks (*Orycteropus afer*).

a deep layer of sand (e.g. Chester Zoo and Dublin Zoo). This gives the elephants the opportunity to dust bathe indoors and dig for buried food. It also provides a soft surface for them to rest and sleep on. Sand is also a useful substrate for burrowing animals such as aardvarks (*Orycteropus afer*) (Fig. 7.34).

7.7.5 Water

Water is an important component of many enclosures, used as a barrier, for ornamental purposes or as an enrichment, or all three. It may take the form of:

- a moat barrier around the perimeter of the enclosure, or a canal which may act as a waterway for boats
- a large pool for bathing, swimming and drinking, e.g. for aquatic animals such as sea lions, penguins and beavers or for mammals that enjoy using water, e.g. elephants and tigers
- a small pool in a small vivarium
- the medium in an aquarium
- a complex water feature such as an artificial waterfall and stream, e.g. in a leopard enclosure.

Water features are sometimes expensive to manage. Some zoos have built enclosures in which pool water is re-circulated after passing through a reedbed that acts as a filter. However, the water has to be pumped around these systems and they often become blocked with faeces and other materials that fall into the pool. It is essential for zoos to maintain high standards of water quality because it is an important medium for the spread of many diseases and it is also an important habitat for insect vectors of disease. Where animals use the water for drinking care must be taken to prevent contamination with toxic chemicals. In some cases it may be necessary to exclude animals from water used as a barrier with an electric fence.

7.7.5.1 Aquarium water

The water requirements in aquariums vary depending upon the species being kept (Table 7.7). The following factors need to be considered:

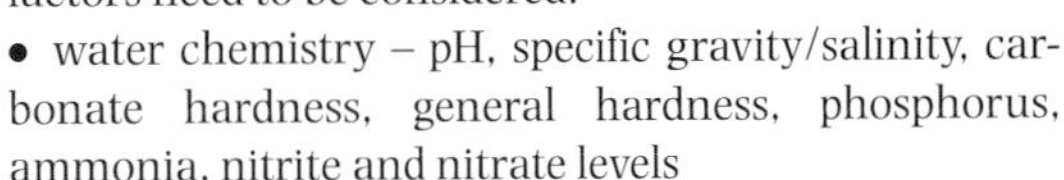

- water chemistry – pH, specific gravity/salinity, carbonate hardness, general hardness, phosphorus, ammonia, nitrite and nitrate levels
- temperature
- light spectrum
- movement (running or still)
- depth (deep or shallow)
- volume (depending upon the size and number of fishes kept).

Water must be filtered. This may be achieved in a number of different ways:

- biological (using a bacterial system)
- mechanical (using sieve-like structures)
- chemical (using zeolite removes excessive ammonia)
- ultraviolet light (destroys harmful pathogens).

All water filtration systems should have some biological component to be effective.

7.7.6 Temperature

Maintaining a suitable temperature is an important consideration, especially where a species is being kept in a zoo located at a completely different latitude or altitude from where it evolved. *Cheetah Rock* at ZSL Whipsnade Zoo contains artificial rock formations which incorporate heat pads on the top and inside the sheltering area (Fig. 7.35).

Electric heat pads may be built into the enclosures of a wide range of species from reptiles to rock hyraxes

Table 7.7 Basic environmental requirements for aquariums.

Habitat	Temperature	Lighting	Salinity
Cool freshwater (simulating temperate riverine environments)	Usually none, may need cooling in hot climates	Broad spectrum UV, controlled to mimic day length	None
Tropical freshwater (simulating tropical riverine environments)	Temperature approximately 22–28°C, thermostatically controlled	Broad spectrum UV, controlled to mimic day length	None
Warm marine (simulating open sea or reef environments)	Temperature approximately 22–26°C, thermostatically controlled	Specific UV lighting, depending upon type of corals, controlled to mimic day length	Species-specific, approximately 1.020–1.025 specific gravity. Requires protein skimmer
Brackish (simulating tidal environments)	Thermostatically controlled heat source	Broad spectrum UV, controlled to mimic day length	Species-specific, approximately 1.005–1.010 specific gravity. Requires protein skimmer

Adapted from Rayers (2009).

Fig. 7.35 An artificial rock in the *Cheetah Rock* exhibit at ZSL Whipsnade Zoo. The rock contains heat pads and also acts as a vantage point from which cheetahs can look beyond their immediate surroundings.

(*Procavia capensis*). Exhibits for other species such as meerkats (*Suricata suricatta*) and tortoises often include suspended heat lamps (Fig. 7.36). Temperature gradients should be created in vivariums so that reptiles and amphibians may self-regulate their body temperature by moving to a suitable position.

Many species need access to shade during the heat of the day. Penguins at Edinburgh Zoo are provided with canopies and water sprays to help them keep cool (Fig. 7.37). Canopies are also often provided for mammals, for example gorillas at *Durrell* and Asian elephants at ZSL Whipsnade Zoo, while others are provided with pools, for example tigers at *Marwell Wildlife* and Asian elephants at Dublin and Chester Zoos.

In the wild, some species make little use of shade. According to Schaller (1972) wild adult lions (*Panthera leo*) often spend the whole day in the open, even when shade is available a short distance away. Of 791 lions

Fig. 7.36 Galapagos giant tortoises (*Geochelone elephantopus*) basking under heat lamps in an indoor enclosure.

observed on the plains of the Serengeti, 88% were found on the open plains and just 12% were resting in kopjes (rocky outcrops). However, cubs are much less tolerant of the sun and must seek shade. Wild lions often choose prominent resting places such as a hillside, rock or branch of a tree. Zoos often provide elevated resting places for their lions, and other big cats, under which they can seek shade if required.

Many mammal species hibernate during the winter. At the Ueno Zoological Gardens, in Tokyo, an exhibit has been created that allows visitors to observe Japanese black bears (*Ursus thibetanus japonicus*) during hibernation (Itoh *et al.*, 2010). Air-conditioning and illumination equipment were installed in the exhibit to simulate the environment of a mountainous habitat in winter. In the wild this species hibernates in tree hollows. As well as providing interest for visitors, the exhibit has allowed researchers to gather information about the behavioural and physiological changes that occur in black bears during hibernation.

7.7.7 Humidity

The control of humidity is unimportant to many species. However, some species may have requirements for high humidity – if they normally inhabit humid tropical areas – or low humidity, if they are desert species. Warm humid atmospheres encourage the growth of fungi and other microbes. Damp conditions can lead to respiratory infections, while low humidity can cause dysecdysis (abnormal shedding of the skin) in reptiles and tail necrosis in rodents. Low humidity has also been implicated in necrosis of the pinnae in the grey short-tailed opossum (*Monodelphis domestica*) (Benato *et al.*, 2010).

Fig. 7.37 Gentoo penguins (*Pygoscelis papua*) at Edinburgh Zoo are kept cool with cold water sprays and metal canopies.

7.7.8 Light

Natural light is desirable in indoor exhibits. The Casson Pavilion at ZSL London Zoo was designed to direct natural light through 'funnels' from above onto the areas where elephants were held while the public areas remained relatively dark.

Animals in vivariums, aquariums and nocturnal houses are generally exposed to artificial lighting and care must be taken to match this to the requirements of the species. The quality and quantity of light can affect the health and behaviour of some species.

Photoperiod can affect biological rhythms and influence behaviours such as hibernation and breeding behaviour in many species. This should be controlled by timers to simulate the natural day length required by the species.

There is some evidence that the use of full spectrum ultraviolet (UV) bulbs diminishes aggression in certain large felids in indoor enclosures (Law, 2010). Many species of reptiles require UV light, without which bone rarefaction (weakening) may occur. Ultraviolet light is made up of UVA, UVB and UVC light. Reptiles require both UVA – which affects activity cycles – and UVB – which is important in vitamin D_3 synthesis. Vivariums should provide a gradient of UV light and shade so that reptiles can self-regulate their exposure. It should be noted that UV light is blocked by glass.

7.7.9 Off-show areas

An off-show area is one to which the public has no access. Such areas may be important for the holding of sick or young animals, or as breeding areas.

Off-show areas do not need to be as complex as those where the animals are exhibited. For example, in an aquarium, they may be simple tanks filled with water and little else. For terrestrial species they may be simple cages or paddocks.

Fig. 7.38 Amur leopard (*Panthera pardus orientalis*). Leopards of all species benefit from the provision of high-level platforms and elaborate climbing structures constructed from tree trunks and rocks.

Some zoos have breeding centres which are physically separate from the public areas. Chester Zoo has a Parrot Conservation Centre to which the public has no access and which is screened off from the rest of the zoo.

7.7.10 Enclosure furniture

The term 'furniture' is used for a wide variety of structures which form part of the exhibit accessible to the animals. This may include:

- logs and branches for climbing, scent marking, scratching, etc. (Fig. 7.38)
- feeding devices (see Section 10.6.3)
- climbing structures, e.g. wooden or steel climbing frames, climbing walls, etc.
- high-level ledges or resting platforms
- waterfalls.

The presence of furniture may significantly affect behaviour. Mallapur *et al.* (2002) studied the effect of enclosure design on space utilisation in singly housed Indian leopards (*Panthera pardus*) in four zoos in Southern India. They concluded that leopards used the edge or edge zone of the enclosure for stereotypic pacing, the rear for resting and the remainder of the enclosure for other activities. Those leopards housed in structurally enriched on-exhibit enclosures (e.g. with sleeping platforms, logs and trees) showed higher levels of activity than those housed in barren enclosures.

When confined, animals have very little control over what happens to them. Increasing the complexity of an enclosure can increase an animal's available choices and help it cope with novel or stressful situations. The provision of pen dividers, retreat areas or vegetation can improve many indicators of animal welfare by creating opportunities for concealment from other animals or from people. Blaney and Wells (2004) found that the welfare of lowland gorillas (*Gorilla gorilla*) can be improved through the use of camouflage net barriers to reduce the animals' sight of visitors.

Carlstead *et al.* (1993) found that leopard cats (*Prionailurus bengalensis*) stressed by the presence of large cats (*Panthera* spp.) exhibited less pacing and reduced cortisol levels when provided with hiding places. They also found an inverse correlation between cortisol levels and exploration.

Furniture in vivariums should include rocks, plants, and branches – of various thicknesses – and should provide a variety of surfaces and textures to aid ecdysis (skin shedding). Hiding places should be provided so the animals can escape from the gaze of visitors and to help to create light and temperature gradients.

Other aspects of the use of furniture in enclosures as enrichment are discussed elsewhere (see Section 10.6.6).

7.7.10.1 Protecting trees

Many enclosures include large trees. For safety reasons, some species capable of climbing must be prevented from doing so by, for example, covering the lower part of the tree trunk with a smooth material and/or attaching an electric fence at a suitable height. Many herbivores such as elephants, giraffes, rhinos and antelopes will damage trees by stripping off the bark. The vascular tissue is just below the surface and if it is removed all of the way round the tree becomes 'ringed' and will die. To prevent this many zoos use electric fences to keep the animals away from the trees or cover the bark with wire mesh or rope to protect it.

7.8 INTRODUCING ANIMALS TO A NEW ENCLOSURE

When animals are introduced into a new enclosure or an old enclosure is extended or modified, it may be some time before they utilise all of the space available.

When a bull pen was added to the outdoor Asian elephant (*Elephas maximus*) enclosure at Chester Zoo some individuals walked into it on the first day it was made accessible. Others would not step beyond the boundary between the new pen and the original enclosure for many weeks, even when they could see other elephants they knew inside. It was three months before all eight were regularly seen in the new area (Rees, 2000a).

Where animals are introduced into a new enclosure which contains glass it is useful to smear the glass with a harmless opaque material to prevent attempts to walk or fly through it, until the animals become acclimatised. Some zoos stick black silhouettes of birds of prey on windows to deter birds from attempting to fly through the glass.

7.9 SEPARATION OF ANIMALS AT NIGHT

Some animals may need to be separated from conspecifics at night for a number of reasons:

- to prevent dominant animals from taking all of the food
- to prevent males from harassing or mounting females in a confined area
- when females are about to give birth and there is a risk of the offspring being attacked or even killed by males or other females.

7.10 HYGIENE

It is essential that the management of animal housing, food preparation and other procedures involving animals includes appropriate hygiene precautions. Animal enclosures inevitably become soiled with urine and faeces from their intended and unintended occupants (e.g. small birds, rodents, etc.) and this clearly needs to be removed. Houses for large mammals often have a sloping floor which allows the free drainage of urine.

Keepers may unintentionally introduce materials into enclosures which may pose a contamination risk. Bedding and substrate materials such as woodchips, shredded paper, wood shavings and bark are widely used in enclosures. Baer (1998) has discussed the contamination risks associated with these and other materials. For example, wood products are likely to have been in contact with wild animals before processing and may therefore harbour infectious diseases, and bedding may be contaminated with pesticide residues and with the fungus *Aspergillus fumigatus*, which causes disease in some birds.

7.10.1 Disinfectants

Disinfectants may be physical or chemical. Physical disinfectants include heat, sunlight and steam. High temperatures may be used, for example, to sterilise veterinary equipment. Chemical disinfectants are widely used to clean food preparation areas, indoor animal housing and other places and objects used by animals and keepers. They are usually used in solution but some are used as a gas to fumigate buildings after a disease outbreak.

Disinfectants may act on microbes in one of three ways:

- as oxidising or reducing agents
- as corrosives or coagulants
- as poisons.

Dirty walls and floors must be washed before applying disinfectant as the dirt may protect the bacteria by covering them up, and it may also react chemically with the disinfectant, rendering it useless. Power hoses are widely used to clean the floors of animal houses but steps must be taken to prevent operatives from exposure to aerosols of fine particles. Such procedures should be carried out when there is no risk of exposing the public to the spray.

Dilution of disinfectants with clean water is generally required, following manufacturers' recommendations. They are more effective if used warm and should be left in contact with surfaces for 10–30 minutes before washing away. Mixing disinfectants may result in the formation of a useless compound which has no disinfecting effect.

The types of disinfectants used may be controlled by legislation for certain circumstances in particular jurisdictions. For example, the Diseases of Animals (Approved Disinfectants) (England) Order 2007 (SI 2007/448) governs the use of disinfectants in England and specifies those approved for use in outbreaks of disease. Disinfectant mats may be located at entrances to sensitive areas such as quarantine areas and nurseries (Fig. 7.39).

Fig. 7.39 A disinfectant mat at the entrance to an animal enclosure. These mats can help to reduce the spread of disease organisms and are especially useful in quarantine areas or where young animals are housed. This mat is located at the entrance to a duckling nursery which is open to the public.

7.11 ENCLOSURE DESIGN AND SUSTAINABILITY

Designers of enclosures should take all aspects of sustainability into consideration including the use of locally sourced natural materials and recycled materials, energy efficiency and water conservation. Energy conservation should be a major consideration in the design of enclosures, especially where tropical species are to be kept in a temperate climate. Some zoos have built water treatment systems into their new exhibits. At Chester Zoo water from the elephant pool is circulated through a reedbed and returned to the pool. Some modern zoo buildings have green roofs as an integral part of their design, for example buildings at the Wildfowl and Wetlands Trust site at Martin Mere (Fig. 7.40), the *Komodo Dragon* exhibit at ZSL London Zoo and the *Cloud Forest* exhibit at *Durrell*. Old railway sleepers and other reclaimed wood can be used to make barriers and enclosure furniture; for example, old fire hose is useful for making hammocks and climbing structures for primates.

Fig. 7.40 A green roof on the flamingo house at Martin Mere (Wildfowl and Wetlands Trust).

7.11.1 Case study: The ZSL's *Rhinos of Nepal* exhibit

The *Rhinos of Nepal* at ZSL Whipsnade Zoo is home to a herd of Asian greater one-horned rhinoceroses (*Rhinoceros unicornis*) and is the Zoological Society of London's first fully 'green' exhibit. The building makes use of natural sunlight and utilises recycled and local materials such as recycled railway sleepers and local sandstone. The exhibit makes use of a Wilo AF 150 rainwater utilisation system to supply water to the rhino pool. The system collects rainwater in a 30,000 litre underground tank, but when there is insufficient rainwater the system automatically draws potable water from the mains supply. Wastewater from the enclosure is filtered through a reedbed system before it drains away. The pool water is heated with a solar thermal and an air heat exchanger which uses 75% less fossil fuel energy than a conventional gas boiler. The recycled water is filtered through a high-tech biological filter, saving 20,000 m^3 of water per year. The exhibit cost £1 million to construct.

7.12 FURTHER READING AND RESOURCES

Some useful specifications for animal enclosures published by the New Zealand government are contained in Anon. (2007).

India's Central Zoo Authority has published a substantial document on barrier design (Gupta, 2008) that is available from its website (www.cza.nic.in).

Information about exhibit design is available from the website of *ZooLex* Zoo Design Organization (www.zoolex.org) and that of Jon Coe (http://joncoedesign.com).

The SSSMZP (2004) provides general guidelines on requirements for enclosures in the UK.

A photographic library of England's listed buildings, including those located within zoos, may be found at www.imagesofengland.org.uk/default.aspx

The *International Zoo Yearbook* volume 34 covered *Aquariums* and volume 29 covered *Horticulture in Zoos*.

Yew, W. (1991). *Noah's Art: Zoo, Aquarium, Aviary and Wildlife Park Graphics*. Quon Editions, Singapore.

7.13 EXERCISES

1. Construct a table of the various types of exhibit barriers and list their advantages and disadvantages.
2. How can art help zoos to convey a conservation message?
3. List the ways that horticulture can enhance the experience of animals living in zoos and that of zoo visitors.
4. How can zoos balance the need to satisfy the biological requirements of the animals they keep with their obligation to protect enclosures of historical and architectural importance?
5. List the ways that zoos have met their legal obligation to make their exhibits and buildings accessible to disabled visitors.
6. Design an enclosure for a named animal species, listing in detail the requirements of the species.
7. Should enclosures always be designed to mimic the natural environment?
8. To what extent is it important to provide animals with large enclosures?
9. Describe the ways in which zoos have attempted to make their enclosures more environmentally friendly.
10. Discuss the advantages and disadvantages of exhibiting animals in drive-through enclosures.

8 NUTRITION AND FEEDING

... it was tremendous to know that these chimpanzees actually ate meat. Previously scientists believed that ... they were primarily vegetarians and fruit eaters. No one had suspected that they might hunt larger animals.

Jane Goodall

Conservation Status Profile

Chinese alligator
Alligator sinensis
IUCN status: Critically Endangered A1c; D
CITES: Appendix I
Population trend: Unknown

An Introduction to Zoo Biology and Management, First Edition. Paul A. Rees.

8.1 INTRODUCTION

Nutrition is the study of the processes by which an organism takes nutrients in from the environment and assimilates them into its body. Most of what is known about animal nutrition comes from studies of domestic and farm animals. Zoos must provide appropriate diets for a wide range of species, and although these diets were initially based on our knowledge of the requirements of domestic species, zoos now recognise that many species have very specific nutritional requirements.

In the 1950s Dr Hans Wackernagel, at the Basel Zoo, was a pioneer in recognising the importance of nutrition in zoo animals. Knowledge of the nutritional requirements of domestic animals was used to formulate diets in-house for many species. In the 1930s H. L. Ratcliffe had begun to create complete diets for animals at the Philadelphia Zoo when few nutritionally complete feeds were commercially available.

Zoos should develop a feeding strategy for each species which provides:

- a nutritionally balanced diet
- a diet that reasonably simulates natural feeding behaviours
- a nutritionally balanced diet that the animal consumes consistently
- a diet that meets all of the above criteria, and is practical and economical to feed (EZNC, 2010).

8.2 THE CONSTITUENTS OF FOOD

Animals need a balanced diet consisting of appropriate quantities of:

- carbohydrates
- lipids
- proteins
- vitamins
- minerals
- fibre
- water.

What constitutes an appropriate quantity depends upon the species and will vary from one individual to another depending upon a variety of factors such as age, sex, physical activity and state of health. Animals also need a sufficient quantity of food to provide for their energy needs. Dietary adjustment may be necessary for female mammals that are pregnant or suckling.

8.2.1 Carbohydrates

Carbohydrates are important energy sources for animals. They include simple sugars such as glucose, lactose, fructose and sucrose, and larger molecules such as starch. Carbohydrates occur as:

- monosaccharides – e.g. glucose
- disaccharides – e.g. maltose
- polysaccharides – e.g. starch, glycogen.

During digestion, enzymes break down starch into maltose, and maltose into glucose. Glucose is the primary source of energy for animals. Carbohydrates act primarily as an immediate energy source during cellular metabolism, but if they are present in excess they are stored in cells as glycogen. Much of the energy stored in carbohydrates is released as heat during metabolism.

Some carbohydrates have structural roles. For example, chitin is a polysaccharide that is a major component of the cuticle of arthropods and occurs in the integuments of many other invertebrates.

8.2.2 Lipids

Lipids include a number of different types of molecules including fats, cholesterol and phospholipids. Fats are triglycerides, consisting of a backbone of glycerol linked to three fatty acids. The type of fat is determined by the types of fatty acids.

Fats are stored in the body as adipose tissue. They act as long-term energy reserves and, in many animals, adipose tissue stored under the skin acts as an insulator, for example in marine mammals. Fats that are liquid at body temperature are called oils.

Animals can synthesise most of the fatty acids they need. However, for most species there are some essential fatty acids that they cannot produce for themselves and which must therefore be present in the diet.

Phospholipids are an important part of the cell membrane. Cholesterol is also important in the structure of biological membranes and in the synthesis of steroid hormones.

8.2.3 Proteins

Proteins consist of subunits called amino acids. They are more complex than carbohydrates and lipids because their structure is determined by the precise sequence of amino acids. In the cell, proteins are constructed from instructions in the DNA contained in the chromosomes. These proteins perform a variety of functions. Some are enzymes and control the metabolic activity of the cell. Others act as hormones or antibodies. Proteins also form most of the structural components of cells, including collagen which strengthens skin, and actin and myosin, which are components of muscle.

Some species have a specific dietary requirement for essential amino acids: lysine, methionine, tryptophane, leucine, isoleucine, phenylalanine, threonine, histidine, valine and arginine. Lysine is important for growth and milk production.

8.2.4 Vitamins

Vitamins are organic molecules that are essential in the diet. They serve a variety of functions, but are not broken down and therefore provide no energy. Vitamins are either lipid soluble (A, D, E and K), and can be stored in body fat, or water soluble (thiamine, riboflavin, niacin, B_6, pantothenic acid, biotin, folic acid, B_{12} and C) and must be ingested frequently because they cannot be stored. Vitamins are required in relatively small quantities and too much may be harmful. This is especially true for the lipid-soluble vitamins because they accumulate in the body.

Animals kept in captivity often suffer from vitamin deficiencies, either because their diet is very restricted or because the naturally occurring vitamins have been destroyed by food preparation techniques.

Some vitamins must be present in the food if the animal is to remain healthy. Others are synthesised within the animal's body. Different species require different vitamins. For example, vitamin C (ascorbic acid) is a vitamin for primates, some birds, fishes and invertebrates. It is not a vitamin for amphibians, reptiles and many birds and mammals, which can synthesise it. Any vitamins produced in the lower part of the alimentary canal only become available if the animal eats its own faeces (coprophagy). Rabbits produce pellets that are rich in the B vitamins.

Deficiency diseases vary between species and a variety of apparently unrelated conditions may be caused by a deficiency in a single vitamin.

8.2.4.1 Vitamin A (retinol)

This vitamin plays an important part in cellular metabolism, vision, bone development and epithelial cell integrity. Deficiency may cause ulceration of the cornea, blindness, xerophthalmia, and cellular changes in the trachea resulting in decreased resistance to infection. Vitamin A is only found in a preformed state in animal tissues, but some precursors, such as carotenes, are synthesised by plants. Carnivores may derive much of their requirement by consuming the livers of their prey where vitamin A is stored. This vitamin is toxic if consumed in excess and supplements should only be provided in small quantities. However, polar bears and pinnipeds contain very high levels in their livers, suggesting that they have a high degree of tolerance.

8.2.4.2 Vitamin D

Vitamin D is concerned with the control of calcium and phosphorus in circulation and the absorption of calcium in the gut. It occurs in two forms, D_2 (ergocalciferol) and D_3 (cholecalciferol). Vitamin D_2 occurs in plant material after irradiation by sunlight, and vitamin D_3 is produced in the skin of animals as a result of exposure to ultraviolet light. Sun-dried hay contains more vitamin D_2 than hay that has been artificially dried. Vitamin D deficiency causes abnormal bone growth (rickets) in young animals and a softening of the bones in adults (osteomalachia). Most vertebrates cannot store vitamin D so signs of deficiency may occur in a few weeks or months. Excess vitamin D may result in bone demineralisation and calcification of the kidney. Species differences occur in the biological activity of the two forms of vitamin D, and there is evidence that New World monkeys use D_3 more efficiently than D_2.

8.2.4.3 Vitamin E (alpha-tocopherol)

Vitamin E acts as an antioxidant and protects cells from oxidative changes caused by free radicals. Deficiency of this vitamin can result in cardiac and skeletal muscle myopathies, anaemia, red blood cell haemolysis, and fat degeneration. Vitamin E is found widely in plants but little occurs in animal tissues. The vitamin E content of foods is influenced by a wide variety of factors including food processing methods, season of harvest and storage conditions. Vitamin E supplements may be required by animals suffering a high parasite load or infection. It may also be useful in animals that receive a diet high in fat, particularly those that are high in polyunsaturated fatty acids. Oversupplementation with vitamin E does not generally cause problems.

8.2.4.4 Vitamin K

Vitamin K occurs naturally in two forms: K_1 which is synthesised by plants and K_2 which is synthesised by microbes. A synthetic form, K_3, is also called menadione. Vitamin K is important in blood clotting. Deficiency is highly unlikely in healthy animals that are fed natural foods or manufactured food products because it is widely distributed and it is synthesised by organisms in the gut. Alfalfa is a good source. If animals are fed intestinally active antibiotics they may need to be given menadione supplements. If these are given in excess it may cause anaemia and porphyrinuria (the presence of porphyrins in the urine).

8.2.4.5 Vitamin B complex

This includes riboflavin, nicotinic acid, pantothenic acid, choline, biotin and thiamine. Most of the B vitamins are widely distributed in plant tissues or animal and microbial tissues. Many act as coenzymes that are important in a range of metabolic processes. Deficiency signs often involve central nervous system problems resulting in convulsions and lack of coordination. Other signs may also occur, such as diarrhoea, anaemia and impaired growth. Herbivores may obtain considerable amounts of B vitamins from the synthetic activity of gut microbes. Vitamin B_1 (thiamine) deficiency is rare but can occur in pinnipeds that consume raw fish as they contain enzymes that will destroy thiamine (thiaminases). Certain plants such as bracken, horsetails and sweet potato leaves contain similar anti-thiamine compounds and the consumption of these plants may cause a secondary deficiency. Deficiency signs include muscular spasms, a staggering gait and loss of appetite. A sudden loss of the righting reflex in saltwater crocodile hatchlings (*Crocodylus porosus*) was corrected by injections of thiamine hydrochloride, after they were found floating or lying on their sides (Boden, 2007). Thiamine deficiency is responsible for the 'star-gazer' phenomenon in newly hatched chicks, in which the individual's head is permanently held back with the beak pointing upwards.

8.2.4.6 Vitamin C (ascorbic acid)

Some species can synthesise vitamin C, others need to absorb it in their diet. Bats, guinea pigs and anthropoid primates cannot synthesise the vitamin. It is important to provide animals with a dietary source of vitamin C in the absence of information regarding the ability of particular species to synthesise it. Even those species that can produce it may not produce sufficient and may need a supplementary dietary source. Vitamin C is important in many metabolic reactions and in the synthesis of the collagen found in cartilage. Deficiency signs include sore joints, bones and muscles, listlessness, abnormal bone growth, anorexia and increased susceptibility to disease. Good sources of vitamin C include citrus fruits and the leafy parts of plants from the cabbage family. It is relatively unstable and deteriorates rapidly in many foods. Vitamin C has low toxicity and high levels may need to be added to manufactured foods. Particular care must be taken to ensure that primates receive adequate vitamin C in the diet.

8.2.5 Minerals

The importance of different elements in the diet varies between species. Mammals need iron to make the blood pigment haemoglobin. However, in molluscs, crustaceans and some spiders the respiratory pigment is haemocyanin, and this contains copper in place of iron. Vertebrates need calcium for their skeletons and birds need it for egg shell production. While these nutrients are essential in trace amounts, most are highly toxic in large concentrations. The nutritional importance of minerals to mammals is summarised in Table 8.1.

Environmental factors other than diet may interact with minerals and vitamins in some taxa. For example, reptiles need adequate exposure to UV light in order to promote the absorption of calcium and the synthesis of vitamin D_3.

8.2.6 Fibre

Fibre plays an important role in the process of digestion and is also a source of energy. Fibre consists largely of cellulose, lignin and hemicellulose obtained from plants. Cellulose and hemicellulose can be digested by microbial fermentation in herbivorous mammals but lignin is almost impossible to digest. Monogastric species – those with simple guts – do not use fibre as a major source of energy. However, primates may digest substantial amounts of fibre. Mammals rely upon symbiotic microbes to provide the cellulase enzymes required to break down cellulose as their guts are unable to secrete them. These enzymes ferment the cellulose and other carbohydrates producing volatile fatty

Table 8.1 Minerals of nutritional importance in mammals.

Mineral	Function	Deficiency/toxicity signs
Calcium and phosphorus	Bone structure and many metabolic reactions, muscle contraction, nerve function, blood clotting and enzyme activation	Required in large amounts and in appropriate ratios during growth and lactation. Bone demineralisation, tetany and death may occur if the food provided has a low calcium: phosphorus ratio, as skeletal stores are depleted. Phosphorus may be an important deficiency in grazers.
Magnesium	Involved in muscle contraction and nerve conduction, synthesis of proteins, fats, carbohydrates and nucleic acids	Vasodilation, convulsions, calcification of soft tissues. Deficiency may occur in ruminants grazing on spring pastures low in available magnesium.
Sodium	Helps to maintain electrolyte balance and osmotic pressure	Salt craving. Deficiencies most likely when lactating, growing or working. Sodium toxicity can occur if not given access to water.
Chloride	Helps to control acid–base balance, catalyses certain enzymes	Deficiencies and toxicities are rare.
Potassium	Helps to control acid–base balance and osmotic pressure	Deficiencies are rare, but may occur in herbivores fed diets low in forage and high in concentrates (as grains are a poor source).
Sulphur	Important in some amino acids and protein structure	Non-ruminants appear to have no dietary requirement. Microbes in the gut of ruminants use dietary sulphur to synthesise some amino acids and B vitamins.
Iron	Important in the structure of haemoglobin, myoglobin and enzymes	Deficiencies are rare. Dietary iron may be important in animals that have suffered blood loss and are anaemic. Many mammalian milks are low in iron and hand-rearing using cow's milk may lead to deficiency.
Copper	Important in connective tissue and melanin synthesis. Component of enzymes that mobilise stored iron	Deficiencies are rare. Hepatic iron accumulation. Deficiency may occur where ruminants are fed on molybdenum-rich soil or if there is excessive dietary zinc.
Iodine	Component of thyroid hormones that regulate tissue metabolism	Deficiency may occur in species feeding on vegetation growing on soil with low iron concentration. Enlargement of thyroid gland, retarded growth. Iodine occurs in high levels in some marine products.
Cobalt	Component of vitamin B_{12}	Required for B_{12} synthesis. Ruminants have a relatively high requirement due to the inefficient production of B_{12} in the rumen, and poor absorption of the vitamin in the small intestine.
Zinc	A cofactor in many metabolic reactions. Involved in wound healing, protein synthesis, immune system	Growth retardation, anorexia, impaired reproduction (especially males). Availability of zinc is low in some plants. Deficiency may occur if there is excess calcium, cadmium or copper in the diet. Excess zinc may interfere with iron and copper absorption and utilisation.
Manganese	Involved in the development of the bone matrix, fat mobilisation and gluconeogenesis	Higher levels required for reproduction than for growth. Deficiency may cause ataxia in newborn, neonatal death, loss of reproductive function, impaired growth, skeletal abnormalities. Some grains contain low levels.
Selenium	Component of glutathione peroxidase which protects cells from destruction by peroxides	Deficiency causes skeletal muscle degeneration, necrosis, calcification and liver pathologies. May occur if animals are fed plants grown on selenium-deficient soils. Selenium toxicity may occur at relatively low levels.

Based on information in Allen and Oftedal (1996).

acids which are then absorbed by the gut and used as an energy source. Species of mammals that rely on microbial fermentation for their nutrition must supply the microbes in their gut with sufficient amounts of fibre or gastric disorders may develop.

8.2.6.1 Browse

The term 'browse' refers to the vegetation that zoos provide for browsing animals; essentially, branches containing leafy material. This provides important nutrients for browsers and also acts as an enrichment, especially if it is located at the level at which it would normally occur in the wild, for example suspended at height for giraffes. Many browsers will spend a great deal of time obtaining food from branches, even when other food, such as hay, is provided at the same time. Elephants will spend hours using their toe nails to break small pieces of bark from tree trunks provided as food and enrichment. Obviously, care must be taken to provide palatable and non-toxic species as food (see Table 11.4).

The browse preferences of colobus monkeys (*Guereza kikuyuensis*) were studied at the Central Park Zoo in New York by Tover *et al.* (2005). They found that the majority of the troop preferred browse that was relatively low in fibre. However, preferences varied between the sexes. Nasturtium (*Tropaeolum majus*), which contained the least lignin, was generally favoured by females, but was the least favourite choice for males.

8.2.7 Water

Some species need access to drinking water while others obtain their water from their food or from metabolic processes. Many species need to evaporate water from their body surface in order to lose heat. These animals need access to large amounts of water when they become heat-stressed during periods of hot weather. Water is also important in providing the medium by which many species excrete electrolytes and waste materials from their kidneys. Nursing mammals may need additional water to assist with milk production. Care should be taken to maintain high standards of hygiene with water vessels as water is important in the transmission of many diseases. Drinking water should be replaced daily.

8.2.8 Commercially produced foods and supplements

The diets of animals living in zoos are unlikely to be identical to what they would obtain in the wild (Fig. 8.1). They may therefore need nutrient supplements to prevent deficiency diseases or during the breeding season.

Mineral requirements may be fulfilled by, for example, providing mineral blocks (Fig. 8.2) or adding minerals to other foods (Table 8.2). Penguins are marine birds and therefore likely to require additional salt if kept in freshwater pools. This can be supplied by putting salt tablets inside their fish.

Mazuri® Zoo Feeds include a very wide range of products for exotic species, some of which are supplements and others of which are intended as the main diet. They include, for example: Mazuri Bear Diet, Mazuri Callitrichid High Fiber Diet, Mazuri Crocodilian

Fig. 8.1 Food pellets for giraffes.

Fig. 8.2 A mineral block.

Table 8.2 Some common nutritional supplements*.
(The products listed here have been developed by specialist manufacturers in collaboration with staff of the International Zoo Veterinary Group.)

Product	Contents	Taxa
Aquavits	Vitamins A, D_3, E, C, B_1, D-pantothenate, ferrous fumarate, inositol, folic acid, choline bitartrate, kelp, biotin	Dolphins, pinnipeds, penguins, pelicans, crocodilians, turtles, sharks
Ferrous gluconate tablets	Ferrous gluconate – an iron supplement	Most species, particularly those fed on an iron-deficient fish-only diet
Salt tablets	Sodium chloride	Marine mammals and birds kept in freshwater
Elasmobranch tablets	Vitamins A, D_3, E, C, B_{12}, B_1 thiamin, B_2 riboflavin, nicotinic acid, B_6 pyridoxine, pantothenic acid, iodine	Sharks and rays fed on frozen fish or invertebrate diets
Nutrazu Aquatic Omnivore Gel	A complete diet: protein, oil, fibre, ash, vitamins A, D, E, cupric sulphate, stabilised vitamin C	Omnivorous aquarium fish and crustaceans (e.g. crabs)
Nutrazu Aquatic Herbivore Gel	A complete diet: protein, oil, fibre, ash, vitamins A, D, E, cupric sulphate, stabilised vitamin C	Herbivorous aquarium fish, and algal grazers
Carnivore calcium	Mineral supplement with optimum calcium/phosphorus ratio for carnivores of 3:1	All carnivores, e.g. cats, birds of prey, monitor lizards
Herbivore calcium	Mineral supplement with optimum calcium/phosphorus ratio for herbivores of 1.8:1	All herbivores, especially reptiles
Cricket Calci-paste	High calcium diet for nutrient loading crickets	Insectivorous animals, e.g. reptiles, birds, small mammals
Emcelle Natural Vitamin E Supplement	Vitamin E solution suitable for mixing with drinking water or liquid diets	Elephants, rhinos, camelids, giraffes and large herbivores
Vitamin D3 and E in oil	Vitamins D_3 and E in coconut oil	New World monkeys – prevents nutritional bone disease. Especially important if deprived of natural sunlight
Zoovet Colour Feed Supplement	Pigment precursors for birds with natural yellow-red feather colouring based on carotenoids: canthaxanthin and beta-carotene	Scarlet ibis, storks, flamingos, pelicans
Zoo-E-Sel	Vitamin E and selenium	Equids, giraffes, tapirs, camels and other grazers when overwintered indoors
Zoovet Electrolyte	A water-soluble electrolyte probiotic powder, containing enzymes, vitamins and minerals	Used for general rehydration, in quarantine situations, transportation, periods of stress and veterinary treatment. Suitable for most species

*Care must be taken in the use of supplements to avoid direct toxicity or nutrient imbalance.

Diet, Mazuri Insectivore Diet, Mazuri Ratite Diet and Mazuri Zebra Pellets. The company also manufactures products for animals that need special diets – such as Mazuri Callitrichid Diabetic Gel and Mazuri Ostrich Breeder (for very highly productive birds) – vitamin and mineral supplements and a milk replacement. Each product is accompanied by a detailed diet sheet which provides information about the ingredients and nutrient content, and mixing and feeding directions.

8.3 ENERGY

In addition to supplying animals with nutrients, food also supplies the energy they need for activity, growth and reproduction. Energy is measured as kilojoules (kJ) or calories (Cal). Endotherms (mammals and birds) need more energy per unit mass than ectotherms (all other taxa) because they need to maintain a constant body temperature.

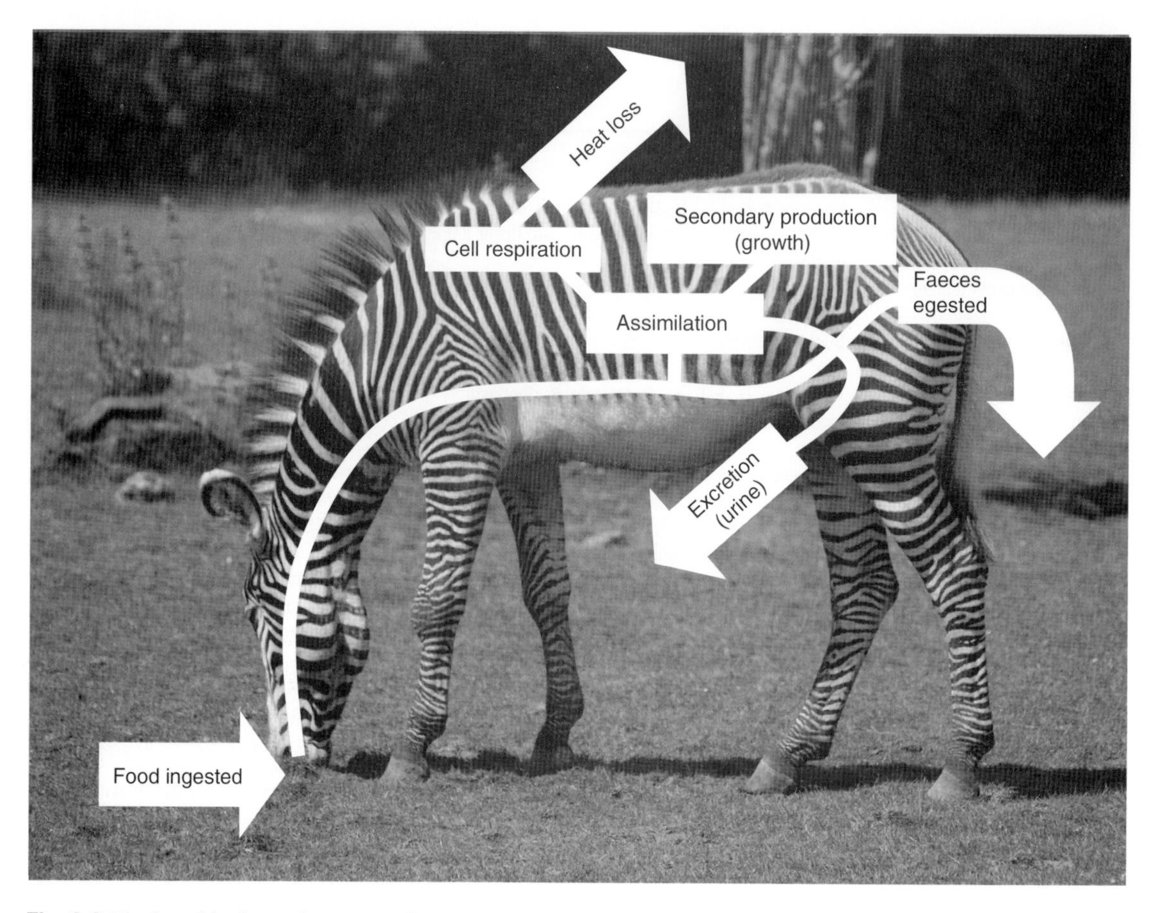

Fig. 8.3 The fate of food eaten by an animal.

The rate at which animals in a food chain accumulate dry mass (biomass) or energy is called secondary production. It is measured as the dry mass laid down (or the energy therein) per unit area per unit time.

8.3.1 Energy budgets: the fate of the energy in food

Some of the food eaten by an animal will pass through its gut and be egested as faeces, without ever having entered the animal's body. The remainder is absorbed across the gut wall (assimilated) (Fig. 8.3). The fate of this assimilated food can be expressed as an equation in terms of either energy or dry mass:

$$F = E + R + U + P$$

where F = food eaten, E = faeces egested, R = expenditure on cellular respiration, U = urine and P = secondary production (growth).

This formula can be rearranged to calculate secondary production:

$$P = F - E - R - U$$

In a large herbivorous mammal much of the assimilated energy may be used in respiration, simply to keep the animal alive. The remainder may be used to produce new cells (for growth, tissue repair, or reproduction).

Table 8.3 Comparison of energy budgets of selected ectotherms and endotherms (arbitrary units based on a food intake of 100 units of energy).

Taxon	Feeding mode	Egested (*E*)	Assimilated	Respired (*R*)	Net production (*P*)	*R*/*P* ratio
Grasshopper	Herbivore (ectotherm)	63	37	24	13	1.8
Wolf spider	Carnivore (ectotherm)	8.2	91.8	57	27	2.1
Perch	Carnivore (ectotherm)	16.5	83.5	61	22.5	2.7
Cow	Herbivore (endotherm)	60	40	39	1	39
Owl	Carnivore (endotherm)	15	85	85	0	High

Adapted from King (1980).

The energy costs of maintaining a constant body temperature are massive, especially in cold climates. Tropical mammals kept outside in zoos in temperate regions may need additional food in order to cope with the energetic demands of heat loss. A comparison of energy budgets in ectotherms and endotherms is provided in Table 8.3.

Endotherms have high respiration rates and must keep their body temperatures high. This makes them poor converters of food into biomass compared with ectotherms. Sedentary animals need less energy from their food than active animals. If they eat more food than they need they may put on excessive weight. Obesity is a problem in some zoo animals because they do not need to expend energy searching for food; for example, some zoo elephants are considered to be obese compared with their wild counterparts (Harris *et al.*, 2008) (see Section 8.10.2).

Food quality may affect growth rates in zoo animals. Rich and Talent (2008) examined the effect of prey species on the growth parameters of hatchling western fence lizards (*Sceloporus occidentalis*). The lizards were fed house cricket nymphs (*Acheta domesticus*) or mealworm larvae (*Tenebrio molitor*). The lizards grew well on both diets but those fed on crickets consumed a slightly higher percentage of their body mass per day than those fed on mealworms. The lizards fed on mealworms ingested considerably more metabolisable energy, had higher food conversion efficiencies, higher daily mass gains and a greater total growth in mass than those fed on crickets.

8.3.2 Assimilation efficiency

Gross assimilation efficiency (GAE) is a measure of how well an animal utilises it food. It is calculated by measuring the amount of food ingested and the amount of faeces egested (both as dry mass) over several days, and applying the following formula:

Assimilation efficiency (%)

$$= \frac{\text{Food ingested - Faeces egested}}{\text{Food ingested}} \times 100$$

Gross assimilation efficiency depends upon the efficiency of the gut and the digestibility of the food. It is also known as apparent dry matter digestibility. Some herbivores have very inefficient guts and must eat a great deal of food to compensate for this. If their food is of poor quality they must spend a high proportion of their time eating.

Carnivores have high protein diets that are easily digested and absorbed. Consequently they typically assimilate over 80% of the energy in their diets. Herbivores have diets that contain a great deal of cellulose from plant cell walls. As a result their assimilation efficiencies are much lower. Ruminants have symbionts in their guts which digest cellulose for them, so they may assimilate up to 40% of the energy in their food. Elephants are not ruminants and their guts are much less efficient than this. African elephants (*Loxodonta africana*) have a GAE of approximately 22% (Rees, 1982a). In other words, about 78% of what they eat passes out as faeces (Fig. 8.4).

Knowledge of the GAE of an animal can help to make estimates of carrying capacities in the wild, because it is possible to calculate food consumption from the production of faeces if the GAE is known. For example, Rees (1982a) calculated the carrying capacity for elephants in Tsavo National Park, Kenya from GAE estimates made using elephants at Knowsley Safari Park.

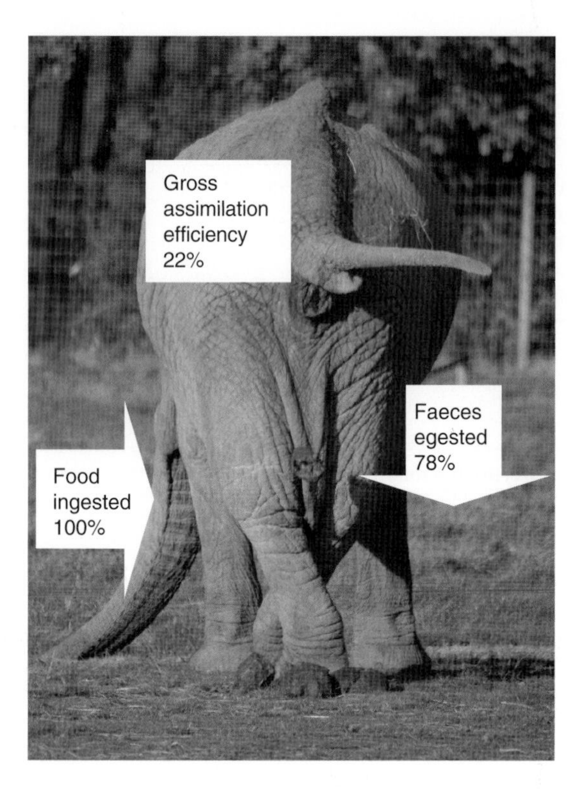

Fig. 8.4 Gross assimilation efficiency in the African elephant (*Loxodonta africana*). (Based on data in Rees, 1982a.)

Table 8.4 Examples of feeding modes based on food type.

Feeding mode	Food
Carnivore	Animals
Insectivore	Insects
Molluscivore	Molluscs
Piscivore	Fish
Avivore	Birds
Spongivore	Sponges
Herbivore	Plants
Folivore	Leaves
Frugivore	Fruit
Graminivore	Grasses
Nectarivore	Nectar
Palynivore	Pollen
Omnivore	Animals and plants
Fungivore	Fungi
Scavenger	Carrion
Cannibal	Conspecifics
Detritivore	Decomposing material

8.4 FEEDING MODES AND BEHAVIOURS

The process of adaptive radiation has produced a diverse range of feeding types and feeding mechanisms within the Animal Kingdom. The taking in of food from the environment by an animal is called ingestion. Some species filter the water they inhabit and extract small organisms such as phytoplankton and zooplankton as food. These animals are referred to as filter feeders (e.g. some whales and many small marine and freshwater organisms). Most terrestrial vertebrates are bulk feeders and obtain their nutrients by consuming all or part of another organism.

Some animals feed on a very wide range of prey and are 'generalists' with respect to their diet, while others have evolved to specialise on a very specific type of food. Most people are familiar with the ecological classification of animals into carnivores, herbivores, omnivores and detritivores, but this is simplistic, and in reality these types may be further divided into narrower specialisms (Table 8.4). However, these terms mask the fact that some species may be carnivores at one time of the year and herbivores at another. Some animal diets are complex and, in the wild, will vary with the availability of food items. In mammals, the dentition is a useful indicator of the type of food taken (Box 8.1).

In addition to the classification of feeding modes by the type of organisms taken, there are many other terms that describe very specialised modes of feeding, for example coprophagy (faeces), osteophagy (bones), xylophagy (wood), trophallaxis (regurgitated food), oophagy (eggs) and paedophagy (young animals). Some of these terms are useful but, from a practical point of view, zoos need to know exactly what to feed their animals and the quantities they require.

Some animals obtain some of their nutrients by absorbing dissolved organic material across their body surface. Such animals require a large surface area relative to their body volume, so are usually either small or long and thin. Many aquatic invertebrates, such as sponges, obtain at least part of their nutrition in this way.

Many animals acquire some nutrients from endosymbionts (algae or bacteria) that live inside their bodies. Some sponges, corals and clams obtain nutrients from photosynthetic endosymbiotic algae (zooxanthellae). Some mammals obtain vitamins B_{12} and K from bacteria living in their large intestine.

8.4.1 Coprophagy

Coprophagy is the eating of faeces. It is natural for some species to do this. Some young mammals, for example pigs, dogs, non-human primates and elephants, eat the faeces of their parents when they are

Box 8.1 Mammalian dentition and dental formulae.

Animals like crocodiles and alligators, and most non-mammalian vertebrates, possess teeth that all have essentially the same shape and function. This type of dentition is referred to as homodont. However, most mammals have heterodont dentition, i.e. they possess different types of teeth, each with a specific function. This may be described in a dental formula which indicates the number of each type of tooth in the upper and the lower jaw on one side of the skull. The tooth types are abbreviated with their initial letter:

I = incisors   C = canines   P = premolars   M = molars

The giraffe (*Giraffa camelopardalis*) has no incisors or canines in its upper jaw and a long gap, called a diastema, between the incisors and canines and the cheek teeth in the lower jaw (Fig. 8.5a).

Giraffe dental formula: I 0/3   C 0/1   P 3/3   M 3/3 = 32

Tigers (*Panthera tigris*) possess all four types of tooth in the upper and the lower jaw, including four very large incisors (Fig. 8.5b). In the Carnivora the last upper premolar and the first lower molar on each side of the jaw are called 'carnassials'. These teeth function like scissor blades to slice the food and crush bones.

Tiger dental formula: I 3/3   C 1/1   P 3/2   M 1/1 = 30

(a)

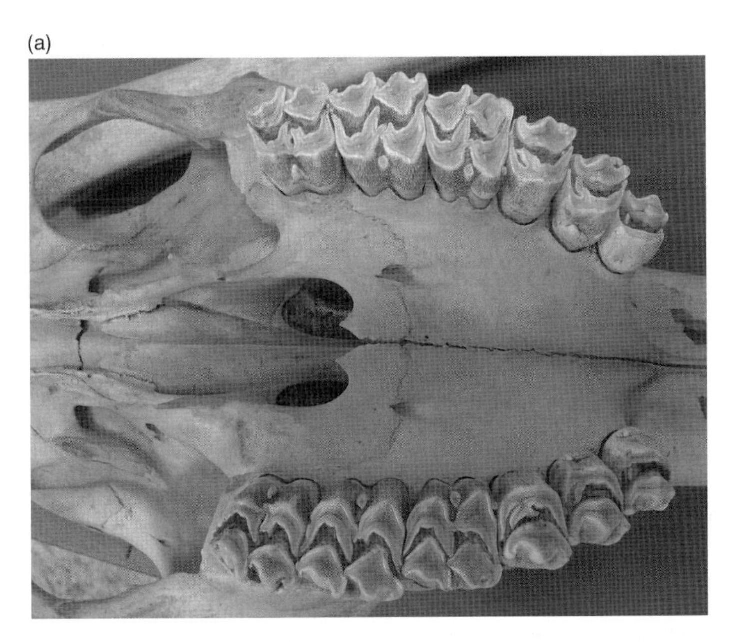

Fig. 8.5 (a) Premolars and molars in the upper jaw of a giraffe (*Giraffa camelopardalis*).

(continued)

Box 8.1 (*cont'd*)

(b)

Fig. 8.5 (Cont'd) (b) Skull of a tiger (*Panthera tigris*). Note the large carnassial teeth.

very young. This helps to populate their guts with the bacteria necessary for digestion. Others animals, for example great apes, may develop coprophagy as a self-stimulatory response to living in captivity (Stevenson, 1983).

8.5 DIGESTIVE SYSTEMS

Very small animals absorb food materials over their body surface but larger animals need a digestive system to acquire the food and a circulatory system to transport it to the tissues. The digestive systems of vertebrates exhibit considerable variation between taxa. However, the various components must perform essentially four functions:

- acquisition (ingestion) and mechanical breakdown of the food into small pieces
- digestion of the food by enzymes
- absorption of the food into the body
- removal of undigested material (egestion).

Vertebrates obtain their food using their mouths (sometimes assisted by hands or trunks) and most then break down the food using teeth. Some chew (masticate) their food before swallowing it, but others – for example crocodiles – swallow it in large pieces with little chewing. Others – ruminants – chew their food, swallow it, then chew it again before swallowing it a second time (see Section 8.5.1).

In all vertebrates the digestive or alimentary system is essentially a coiled tube (gut) leading from the mouth to the anus through which food passes. Associated with this tube are a number of auxiliary organs, particularly the liver and pancreas, which provide some of the enzymes and other chemicals required to digest the food. A simple vertebrate gut consists of:

- a mouth leading to a buccal cavity containing teeth which provide a means of mechanically breaking up food while some digestive enzymes are added
- a stomach which stores the food temporarily while it is mixed with more enzymes which continue chemical digestion
- an intestine where additional enzymes are added and where absorption of digested food occurs along with the reabsorption of water
- a rectum which temporarily stores faeces before they are released through the anus to the outside.

The gut of a bird consists of:

• the mouth which carries a horny bill
• the oesophagus and crop – a distensible sac-like extension of the oesophagus, used to store food
• the proventriculus and gizzard – the equivalent of the stomach
• the small intestine – equivalent to the duodenum, jejunum and ileum in mammals
• the caeca – two blind-ending tubes where the small intestine runs into the large intestine which contain bacteria (one or both are absent in some species)
• the large intestine (rectum) – responsible for most of the water reabsorption and forming the faeces into discrete boli (small balls)
• the cloaca – the combined opening to the outside of the bird's digestive, excretory and reproductive system.

The gizzard of grain-eating birds contains stones which assist in the mechanical breakdown of the food. Some species regurgitate indigestible food in pellets, for example owls and some fish-eating species.

The digestive systems of mammals vary considerably from species to species. The basic structure consists of a mouth, buccal cavity and oesophagus leading to a stomach, a small intestine (consisting of a duodenum, jejunum and ileum), followed by a large intestine (colon) leading to a rectum which expels waste via an anus. The caecum is a blind-ending tube which branches off from the first part of the large intestine and terminates in the appendix.

Herbivores that feed on cellulose that is partly digested by colonic bacteria have a long small intestine and an unusually large colon. Carnivores usually have a short, narrow colon with almost no caecum. The stomach in primates varies in complexity depending upon the diet of the species.

8.5.1 Ruminants and hindgut fermenters

Animals that eat poor quality food have essentially two options:

• eat huge quantities of food and process it quickly and inefficiently, or
• eat less food and process it slowly and more efficiently.

The food of ungulates is highly fibrous and contains a great deal of cellulose. Ungulates may be divided into two groups based on the functioning of their digestive systems: hindgut fermenters and ruminants.

Perissodactyls (horses, rhinos and tapirs) are hindgut fermenters. They possess relatively simple guts and the food is completely digested in the stomach. It then passes to the large intestine and the caecum where microbes ferment the cellulose.

In ruminants the food is retained for much longer in a more complex digestive system. Food passes to the first stomach (rumen) where it is fermented by microbes. It is then regurgitated to be chewed a second time and mixed with saliva. At this stage the food is referred to as 'cud' and ruminants are said to 'chew the cud'. Next, the food is swallowed a second time. It bypasses the rumen and passes directly to the second stomach chamber, the reticulum. Bacteria pass with the food to the omasum (third stomach) and then the abomasum (fourth stomach), where digestion is completed. Nutrients are absorbed in the small intestine and some additional fermentation and absorption occurs in the caecum.

There are two suborders of ruminants:

• Ruminantia – the deer, moose, elk, reindeer, caribou, antelopes, giraffe, bison, cow, sheep, goat and relatives. These animals possess a four-chambered stomach (Fig. 8.6).
• Tylopoda – camels, llama, alpaca and vicuna. These animals do not possess an omasum.

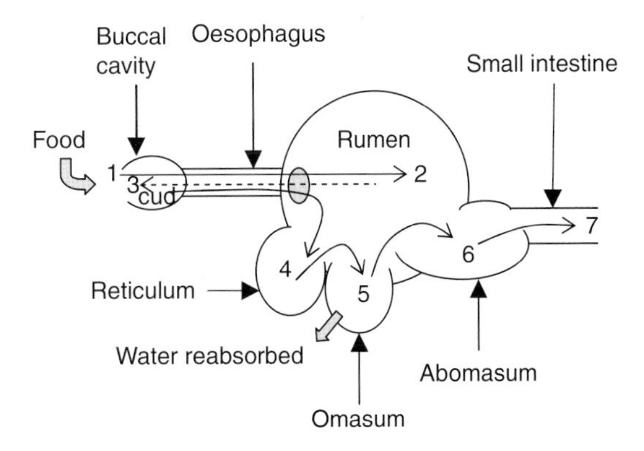

Fig. 8.6 The movement of food through the ruminant gut. Food is taken in through the mouth (1), chewed, and swallowed. Initially it passes to the rumen (2). It is then regurgitated back into the mouth (3) as 'cud' and is chewed again. The second time it is swallowed the bolus of food passes to the reticulum (4). From here it moves to the omasum (5) and then the abomasum (6) before entering the small intestine (7).

In a typical equid (a hindgut fermenter) it takes around 48 hours for food to pass through the gut, and about 45% of the cellulose is utilised. In a typical antelope (a ruminant), the food takes around 80 hours to pass through the gut but around 60% of the cellulose is used. Elephants are hindgut fermenters. They need to consume a large quantity of food and have evolved to process this quickly rather than use the slower process of rumination. Their guts are simple and the food passage time is between 21 and 46 hours in Asian elephants (*Elephas maximus*) (Rees, 1982a). The largest ruminants are hippopotamuses (*Hippopotamus amphibius*). However, they have a relatively fast food passage time and do not chew the cud.

It takes a long time for a ruminant to process low quality food, but a simple gut allows a much faster throughput. So, evolution has not produced very large ruminants because they would not be able to process enough food to support their size. Very small herbivores need more food per unit of body weight compared with larger herbivores. However, they have not evolved rumination because they can spend time selecting better quality food.

8.6 WHAT SHOULD ZOOS FEED THEIR ANIMALS?

Keepers experienced in keeping particular taxa know which foods to provide for their animals. This, however, does not mean that all keepers of a particular taxon agree on optimum diets. For many species standard husbandry guidelines are available which contain information on appropriate and inappropriate foods. However, Dierenfeld (1996) has recognised that aviculturalists, herpetologists, wildlife rehabilitators and others have developed successful diets for particular species or that work in particular circumstances. Zoos cannot hope to provide their animals with the foods they would normally utilise in the wild, but they can attempt to provide the equivalent nutrients.

Food preferences may be established by offering animals a range of foods. This method has been widely used by ecologists attempting to establish the feeding habits of wild animals. A useful starting point is to offer foods likely to be available in their natural habitat. Of course, the problem for zoos is that they inevitably keep animals that eat foods that are not available to zoos so they are forced to develop a nutritionally equivalent diet (Table 8.5). It should also be appreciated that animals kept in captivity will often select foods that they would not eat in the wild (Fig. 8.7).

Care must be taken to avoid food items which might be toxic. The inability of some species to break down particular toxins – due to a lack of the appropriate enzymes – makes certain plants toxic to particular species (see Table 11.4).

For some species, keepers may have preconceptions about what types of foods an animal might eat. For example, it used to be commonly believed that rodents

Table 8.5 A zoo diet for aardvarks (*Orycteropus afer*).

Per pair of animals:
1 apple (peeled)
1 banana (peeled)
1 orange (peeled)
1 carrot
200 g cooked mince
100 g mealworms
1.5 kg Baker's (commercial feed)
1 kg Vitalin muesli
5 litres water (*ad lib*)
60 g Vionate (vitamin/mineral supplement)
2 calcium tablets
2 pinches of vitamin E

Source: Blackpool Zoo.

Fig. 8.7 A beaver (*Castor fiber*) with a corn cob. Many species will eat food in captivity that would not form part of their diet in the wild.

Box 8.2 Nutrition in captive keas (*Nestor notabilis*) (based on Pullar, 1996).

Keas are opportunists and will eat a variety of foods. Breeders disagree regarding the optimum diet. However, most animals offered a variety of foods are likely to select those that will provide the nutrients they need at the time.

Foods commonly fed to captive keas include:

- Pellets or crumbles that are specially formulated to supply the total nutrient requirements of parrots
- Sunflower seeds, peanuts, walnuts in limited amounts*
- Oats, maize and blue peas and other seeds (soaked or dry)
- Cheese – up to 25 g per bird three times per week (may be increased to 75 g per day for birds rearing chicks)
- Fruit biscuits
- Fresh green food – sow thistle (puha), flowering dandelion, and chickweed. Alternatively silverbeet and celery. Birds with chicks should be offered fresh green food every day
- Fruits, especially apples and oranges
- Carrots
- 'Natural foods' – coprosma berries, wild grasses and flowers, hawthorn berries
- Cuttle fish may be given to supplement calcium intake.

Birds rearing chicks may be offered a 'nectar mix' made from rolled oats, 'Complan', sugar (or glucose), wholemeal bread and a suitable vitamin supplement (see Pullar (1996) for recipe).

*Feeding excessive quantities of high protein feeds (e.g. seeds, nuts, meat) can cause obesity and gout.

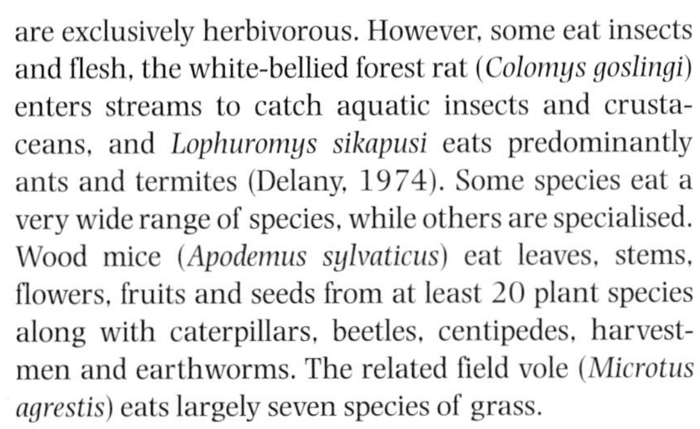

are exclusively herbivorous. However, some eat insects and flesh, the white-bellied forest rat (*Colomys goslingi*) enters streams to catch aquatic insects and crustaceans, and *Lophuromys sikapusi* eats predominantly ants and termites (Delany, 1974). Some species eat a very wide range of species, while others are specialised. Wood mice (*Apodemus sylvaticus*) eat leaves, stems, flowers, fruits and seeds from at least 20 plant species along with caterpillars, beetles, centipedes, harvestmen and earthworms. The related field vole (*Microtus agrestis*) eats largely seven species of grass.

Some wild animals have diets that vary seasonally. Wild wood mice in an English woodland eat seeds for most of the year but switch to a diet of mainly arthropods in May and June (Watts, 1968). Many wild carnivores are opportunists and will eat the most abundant prey organisms. This sometimes means that the same species may have widely differing diets in different parts of its range. Feral domestic cats (*Felis catus*) in temperate areas feed predominantly on small birds and rodents, but in desert areas they take lizards.

Vester *et al.* (2008) examined nutrient digestibility and faecal characteristics in five captive felid species (bobcats (*Lynx rufus*), jaguars (*Panthera onca*), cheetahs (*Acinonyx jubatus*), Indochinese tigers (*P. tigris corbetti*) and Siberian tigers (*P. t. altaica*)) fed on a beef-based raw diet (Nebraska Brand® Special Beef Feline, North Platte, NE). They found putrefactive compounds in the faeces that have been linked to disease states in humans and suggest that care must be taken to manage life-long gut health bearing in mind the longevity of these animals in captivity. They also found variations in digestibility of the diet between species and concluded that it may not be appropriate to feed all species of cat the same raw meat carnivore diet and that meat sources should be matched to the wild diet.

Husbandry guidelines exist for many species and may be used to determine the appropriate quantities of foods to offer particular species (Box 8.2). These guidelines will have been drawn up by keepers, nutritionists and others with many years of experience. However, for some species guidelines are not available and the appropriate quantities may need to be established by trial and error, possibly by offering a series of small feeds during the day until the animals stop feeding. If animals leave some of the food offered each day the likelihood is that they have been offered too much. In some situations offering too much food may have hygiene implications; for example, too much fish thrown into a penguin pool may result in contamination of the water with rotting fish.

The amount of food required by an individual must satisfy its energy and its nutrient requirements. This may vary depending upon:

- Age – Youngsters need more energy and nutrients (for growth) than adults of a similar size.
- Sex – In polymorphic species males are often larger than females so need more food.
- Health – Dietary requirements may change in sick animals, for example if an animal is diabetic or has a deficiency disease.
- Activity level – Active animals require more energy than inactive animals.
- Environmental temperature – Homiotherms need more energy than poikilotherms of the same size in order to maintain their body temperature.
- Females need more – and possibly different food – if pregnant or suckling/feeding her young.

Care must be taken to include food used in enrichment devices in the total diet.

In the United States, the Animal and Plant Health Inspection Service (APHIS) lays down basic feeding requirements for a range of species, without specifying exactly what food should be provided (9 CFR Ch.1 (1-1-09 Edition)):

> **§ 3.105 Feeding**
>
> (a) The food for marine mammals shall be wholesome, palatable, and free from contamination, and shall be of sufficient quantity and nutritive value to maintain all of the marine mammals in a state of good health. The diet shall be prepared with consideration for age, species, condition, size, and type of marine mammal being fed. Marine mammals shall be offered food at least once a day, except as directed by veterinary treatment or professionally accepted practices.

8.6.1 Requirements of specific taxa

Care must be taken to establish the specific needs of particular taxa. Some primates deteriorate quickly in the absence of food or water. Most New World monkeys eat more or less continuously during the day. They will not feed without water, and many species will

Table 8.6 Bamboo species cultivated by San Diego Zoo to meet the nutritional needs of giant pandas (*Ailuropoda melanoleuca*).

Bambusa beechyana
B. glaucescens 'Fernleaf'
B. glaucescens 'Alphonse Karr'
B. oldhamii
B. textilis
B. tuldoides
B. ventricosa
B. vulgaris 'Vittata'
Fargesia fungosa
Phyllostachys aurea
P. aureosulcata
P. bambusoides
P. bissetti
P. nigra
P. vivax

Source: www.giantpandaonline.org/captivemanagement/nutrition_articles/bamboocult.htm (accessed 30 June 2010).

Fig. 8.8 Magellanic penguins (*Spheniscus magellanicus*) feeding in water, as they would in the wild.

show adverse effects if deprived of food for just a day. If their access to water is disrupted squirrel monkeys can dehydrate and develop hypoglycaemia in less than 24 hours (Abee, 1985).

Secure specialist food supplies are critical to some species. San Diego Zoo grows its own *Eucalyptus* plants to feed its koala population (*Phascolarctos cinereus*) and bamboo to feed its giant pandas (*Ailuropoda melanoleuca*) (Table 8.6), and the National Zoo in Washington DC grows bamboo for its giant pandas, red pandas (*Ailurus fulgens*) and Asian elephants (*Elephas maximus*).

After an earthquake struck Sichuan Province in China in May 2008 some giant pandas had to be evacuated from the panda breeding centre because local people were unable to collect bamboo.

8.7 HOW SHOULD FOOD AND WATER BE PRESENTED?

Where possible, food and water should be presented to each species as it would be in the wild (Fig. 8.8). For many species it may be quite appropriate to provide food in simple dishes (Fig. 8.9). However, if an animal is arboreal, and normally feeds in the trees, food should be presented off the ground (Fig. 10.15). Animals that browse at high level, for example giraffes, should be provided with branches suspended at head level. This is especially important for animals that will be released into the wild as there is considerable evidence that some species have difficulty locating suitable food after release if they have been captive-reared and fed in an inappropriate manner.

Fig. 8.9 Feeding keas (*Nestor notabilis*) at Orana Park, New Zealand. (Courtesy of Bethan Shaw.)

Fig. 8.10 Chimpanzees (*Pan troglodytes*) both compete for, and share, food.

The manner in which food is presented to animals can be an important enrichment to their daily routine. For example, a scatter feed – small items of food thrown into the enclosure – at an unpredictable time will initiate foraging and feeding activity if food is not available all of the time.

There is evidence that young animals learn to forage from adults. Diamond and Bond (1991) found that wild fledgling keas (*Nestor notabilis*) discovered little food for themselves and most new food sources were excavated by adults. Social factors were important in the acquisition of foraging expertise in different ways at different stages of development.

The SSSMZP (2004) lays down standards for the provision of water and food. It should be presented in an appropriate manner for the species and should be of appropriate nutritive value, quality, quantity and variety. When supplying food the following characteristics of the animals should be considered:

- size
- condition
- physiological status
- reproductive status
- health status.

Species that require water at all times should have unrestricted access to clean water. Receptacles of food and drink should be placed so as to minimise the risk of contamination by animals, wild birds, rodents and other pests, and they should be regularly cleaned. Uneaten food should be removed sufficiently regularly to maintain hygiene. Self-feeding devices should be checked to ensure they are functioning properly.

The social behaviour of the species should be considered when supplying food and water. Feeding and drinking receptacles should be designed and placed in such a way as to allow them to be accessed by all of the individuals in a group. Animals may experience intraspecific or interspecific competition – if kept in multi-species exhibits – for food (Fig. 8.10). Young, old or subordinate individuals may have their access to food restricted by stronger individuals if too few feeding points are provided or if these are inaccessible because of their position.

Most species kept in zoos are fed daily. However, some zoos will only feed their big cats every other day or have regular 'starve' days in an attempt to simulate the position in the wild. The feeding frequency of predators is determined in part by the abundance of prey and their hunting success. This varies between and within species. Schaller (1972) estimated that lions (*Panthera leo*) living on the plains of the Serengeti ate once in every 2 to 2.5 days and in woodlands about every 3 to 3.5 days on average. McLaughlin (1970) found that solitary cheetahs (*Acinonyx jubatus*) killed every two or three days while cheetahs with cubs killed gazelles at a rate of almost one per day.

It is generally considered inappropriate and unethical to feed live vertebrate prey to animals, and the SSSMZP (2004) recommends that this should only be done in exceptional circumstances and then only under direct veterinary supervision. However, live insects such as locusts or crickets are commonly fed to insectivorous animals such as lizards, and this can be an important enrichment for them.

Fig. 8.11 A vending machine for bird food at Martin Mere (Wildfowl and Wetlands Trust).

8.8 FEEDING BY VISITORS

Some zoos allow visitors to feed some of their animals. This should be done on a selective basis and the zoo should control which animals are fed, and the type and quantity of food given. This is usually achieved by selling small bags of grain or other suitable foods at an appropriate point. The animals that may be fed are often domestic animals such as goats and sheep kept in a petting zoo or similarly separated from the main collection. The Wildfowl and Wetlands Trust sells bags of food suitable for the species in its wildfowl collection via vending machines (Fig. 8.11). Where feeding by the public is allowed it is essential to provide hand-washing facilities with anti-bacterial soaps.

8.9 FOOD PREPARATION AND STORAGE

Food should always be prepared under hygienic conditions. In many zoos, food preparation areas are located within individual animal facilities themselves, thereby reducing the risk of contamination when it is transported to the enclosures. Sometimes these areas are on show to visitors (Fig. 8.12). Staff must observe strict personal hygiene standards while preparing food

Fig. 8.12 Food on display to visitors before it is fed to monkey species at Chester Zoo.

Fig. 8.13 Keepers preparing food for their animals.

Fig. 8.14 The hay barn at Blackpool Zoo.

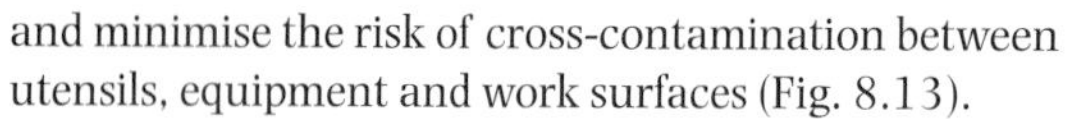

and minimise the risk of cross-contamination between utensils, equipment and work surfaces (Fig. 8.13).

Inappropriate storage of food may lead to deterioration in its quality. Food should be protected from dampness and contamination by fungi. It should also be protected from contact with and contamination by insects, birds, rodents and other pests. Perishable food and drink should be refrigerated unless they are supplied fresh each day. Containers used to store food should not be used for anything else. Some zoos have sufficient space to store large quantities of food on their own premises (Fig. 8.14).

8.10 NUTRITIONAL PROBLEMS

8.10.1 Deficiency diseases

A deficiency disease is one which is caused by the absence from the diet of some substance or element that is essential to health. The essential element may be a mineral or vitamin, or a protein or amino acid. Although starvation results from the lack of sufficient food, it is not a deficiency disease.

Deficiency diseases are highly variable in nature. Some are relatively simple, while others are extremely complex. Where they involve a vitamin the condition is often called avitaminosis, and the vitamin concerned is specified. Some deficiency diseases may be caused by one or several factors. For example, anaemia may be caused by an iron deficient diet, a lack of the trace elements copper and cobalt – which aid the assimilation of iron – or a deficiency of folic acid, vitamin B_6 or vitamin B_{12}. Calcium deficiency may result in osteomalachia or rickets. This occurs in intensively bred species such as cockatiels (*Nymphicus hollandicus*) and African grey parrots (*Psittacus erithacus*). Some other deficiency diseases and their causes appear in Table 8.1 and Section 8.2.4.

8.10.2 Obesity and assessment of body condition

Obesity is a problem in some animals living in zoos. It can occur for a number of reasons including lack of exercise and excessive feeding. In some species dominant animals may acquire more than their fair share of the food provided by virtue of their dominant position in the group.

Obesity is associated with increased health risks (e.g. cardiovascular disease, diabetes and high blood pressure) and decreased activity in monkeys and apes. Many captive chimpanzees (*Pan troglodytes*) subjectively appear to be overweight. Body mass index (BMI) was found by Videan *et al.* (2007) to be a good measure of relative obesity for male chimps but not for females, and abdominal skinfold measurements were determined to be a good tool for determining and monitoring obesity. Staff at the Primate Foundation of Arizona were able to reduce weight in several obese female chimps by 5–7% over the course of one year by monitoring and reducing calorific intake and monitoring BMI and skinfold measurements.

Elephants tend to be less active in captivity than they would in the wild. Harris *et al.* (2008) found that 17.1% of 76 elephants in UK zoos were obese in the opinion of their keepers, although no measurements were made. Wemmer *et al.* (2006) have developed a method of measuring the body condition of Asian elephants (*Elephas maximus*). They used visual assessment to assign numerical scores to six different regions of the body which were then totalled to give a numerical index ranging from zero to 11. This index may be useful to monitor the body condition of elephants in zoos.

8.10.3 Calorific restriction, longevity and disease

Experiments with a number of species have shown that restricting calorific intake can have a number of beneficial effects including a reduction in the incidence of disease and an increased life span. A 20-year study of rhesus macaques (*Macaca mulatta*), at the Washington National Primate Research Center, found that calorific restriction delayed the onset of age-associated pathologies. Life was extended in those monkeys whose diet was restricted – compared with control fed animals – and they experienced a reduced incidence of diabetes, cancer, cardiovascular disease and brain atrophy (Colman *et al.*, 2009).

8.10.4 Hand-rearing and nutrition

Infant mammals need milk. However, milk composition is highly variable between species. The milk of aquatic mammals such as whales and seals is rich in fats compared with the milk of land mammals (Table 9.2). Cow's milk contains little iron and its use in hand-rearing may lead to iron deficiency if supplements are not provided. Zoologic® Milk Matrix is a line of six different milk replacement products that may be used alone or blended together to closely match the composition of the milk produced by various species.

8.11 SOURCES OF DIETARY INFORMATION

8.11.1 EAZA Nutrition Group

The EAZA Nutrition Group was established to improve communication, education and research concerning zoo animal nutrition within zoos in Europe. It produces a number of nutrition books that contain the scientific contributions to the European Zoo Nutrition Conferences and other publications. The group's mission is:

> To promote and support nutrition in zoological institutions as an essential component of their conservation mission.

8.11.2 The USDA National Nutrition Database

This database is produced by the US Department of Agriculture and provides information on the composition of foods and dietary supplements available in the United States. The National Nutrient Data Bank is a repository of information about more than 130 nutrients for over 7000 foods.

8.11.3 Zootrition®

Zootrition® is dietary management software that provides zoo managers with a means of comparing the nutritional content of specific food items and calculating the overall nutritional composition of diets. This allows the identification of potential nutrient deficiencies and toxicities. The software was developed by St Louis Zoo with support from WAZA. It contains information about a wide range of foods and has a facility for zoos to enter information about others. The software features include:

- a feeds database of over 3000 feedstuffs
- an energetics calculator which allows the calculation of the daily energy requirement of a variety of mammals, birds and reptiles based on body weight, feeding habits, activity levels and physiological state
- a nutritional recommendations database which contains details of the nutritional requirements for a wide range of species
- the capacity to create diet records from the feeds database and record feeding schedules and preparation details
- the ability to compare different feedstuffs with animal nutrient requirements
- a reference library which includes information on nutrient toxicities, feed handling and storage, enrichment, etc.

8.12 FURTHER READING AND RESOURCES

The National Research Council Committee on Animal Nutrition of the US National Academy of Sciences produces a series of publications which provide details of the nutrient requirements of most species of domestic and laboratory animals.

The USDA National Nutrition Database is available online at www.ars.usda.gov/main/site_main.htm?modecode=12-35-45-00

The EAZA Nutrition Group's webpages provide numerous sources of information about animal nutrition (www.eaa.net/activities/Pages/Nutrition.aspx) including details of Zoo Animal Nutrition books, abstracts of papers presented at the European Zoo Nutrition Conferences, and EAZA Nutrition Group Newsletters.

More information about Zootrition® is available from www.zootrition.org

The *International Zoo Yearbook* covered nutrition and feeding in volume 6, *Nutrition of Animals in Captivity*, volume 16, *Principles of Zoo Animal Feeding*, and volume 39, *Zoo Animal Nutrition*.

8.13 EXERCISES

1. Investigate the dietary requirements of a species of your choice.
2. Discuss the variety of ways that food may be presented to zoo animals.
3. Calculate the gross assimilation efficiency of an African elephant from the data below:
   Dry weight of food consumed in 7 days = 145 kg
   Dry weight of faeces produced in 7 days = 112 kg.
4. Why are there no really small or really large ruminants?
5. Discuss the differences in the manner in which ruminants and hindgut fermenters digest cellulose.
6. Explain the importance of the main types of foods in the nutrition of zoo animals.
7. What is a deficiency disease and how may it be prevented?

9 REPRODUCTIVE BIOLOGY

In certain circumstances an individual may leave more adult offspring by expending care and materials on its offspring already born than by reserving them for its own survival and further fecundity. A gene causing its possessor to give parental care will then leave more replica genes in the next generation than an allele having the opposite tendency.

William Hamilton

Conservation Status Profile

Blue-throated macaw
Ara glaucogularis
IUCN: Critically Endangered A2bcde; C2a(ii)
CITES: Appendix I
Population trend: Decreasing

An Introduction to Zoo Biology and Management, First Edition. Paul A. Rees.

9.1 INTRODUCTION

An understanding of the behavioural and physiological changes associated with reproduction is an important part of the knowledge required to provide appropriate husbandry for breeding animals. Before mating can occur in many species individuals must first select and then court a mate.

9.2 REPRODUCTIVE BEHAVIOUR

9.2.1 Choosing a mate

Individuals of some species will not simply mate with any member of the opposite sex: they must have the opportunity to choose their own mate. This may not be a problem in animals kept in relatively large social groups. However, in small groups there may be insufficient choice for pairing to occur.

Mate choice may affect reproductive success. Pinyon jays (*Gymnorhinus cyanocephalus*) that choose partners which are similar in weight, age and bill length produce more offspring than those that do not. Reproductive success in barnacle geese (*Branta leucopsis*) is better when pairs comprise a large and a small individual, and in willow ptarmigan (*Lagopus lagopus*) when individuals of a pair are of similar social standing. Spoon *et al.* (2006) have discussed the importance of the behavioural compatibility of mates in parenting and reproductive success in cockatiels (*Nymphicus hollandicus*).

When the silverback gorilla (*Gorilla gorilla*) – 'Bobby' – died at ZSL London Zoo in December 2008 staff had to find a suitable new mate for their females. In August 2009 keepers gave the three females – 'Mjukuu', 'Effie' and 'Zaire' – a photograph of 'Yeboah', a male from Boissière du Doré zoo in France, prior to his arrival at London. Yeboah was given photographs of the three females. Keepers hoped that these photographs would help the animals recognise each other when they eventually met (Anon., 2009e).

9.2.2 Courtship

There is no clear relationship between phylogeny and the complexity of courtship behaviour. Some insects and crabs have elaborate courtship behaviour, while sheep exhibit little courtship. Reproductive behaviour is energy consuming and it places the participants in a vulnerable position so it must be made efficient. Courtship helps to do this.

Courtship has a variety of purposes:

- It may synchronise the reproductive behaviour of the male and the female. Sometimes the mutual performance of complex courtship patterns confirms that both partners are receptive. In some species the female is made receptive by watching the male's display. In sticklebacks synchronisation occurs as a result of a mutually stimulating sequence of behaviours which forms a reaction chain. Courtship may be important in inducing ovulation or egg production in birds. The displays of a male bird may induce a female to lay eggs. The whooping crane (*Grus americana*) does not ovulate until a male has performed a courtship dance around her, jumping in the air and flapping his wings. These cranes can be induced to produce eggs by humans performing the dance after which the eggs are artificially inseminated.
- It prevents wasteful mating with the wrong sex. Courtship signals tend to be species- specific. However, hybrids between closely related species occur in the wild and in captivity. The ten-spined stickleback (*Pungitius pungitius*) has a courtship similar to the three-spined stickleback (*Gasterosteus aculeatus*), but the male nuptial colours are different: black in the ten-spined and red in the three-spined. The colours only appeal to females of the correct species.
- It may help to reinforce the pair bond. This is important if both parents are needed to rear the offspring.
- Some ritualised courtship patterns may allow the partner to assess the quality of a future mate. For example, some bird species make symbolic presents of nesting material. Australian bowerbirds (Ptilonorhynchidae) build a shelter called a bower which is used only for courtship. Bowers are made of twigs and other plant material. Some are shaped like small wigwams; others are saucer-shaped with a 'maypole' in the centre. In some animal species (e.g. grouse, hares, some antelope species) the males carry out communal displays on special areas of ground called leks.
- Courtship displays may be necessary to reduce aggressive behaviour in the male, and sometimes in the female.

Many elements of courtship behaviour are ritualised forms of submissive or aggressive gestures. Sometimes juvenile behaviours, such as food begging, are carried out as ritualised courtship. Courtship preening is observed in many duck species and the movements often serve to emphasise bright markings on the wings.

Fig. 9.1 Roan antelope (*Hippotragus equinus*) courtship. The male is exhibiting the 'leg beat' behaviour.

In some finches males bow to females during courtship in ritualised beak-wiping movements.

In many antelope species there is an elaborate courtship sequence. In the early stages the male stimulates the female to urinate – by licking and nuzzling her genital region – and then tests her urine to assess her reproductive condition. He smells the urine and flehms. Flehmen is characterised by raising the head, turning back the lips and wrinkling the nose, and breathing is stopped briefly.

Variation occurs between taxa and the different patterns of behaviour have been described in detail by Ewer (1968). In the Tragelaphinae (spiral-horned antelopes) the sequence consists of the male following the female; he comes alongside while driving her forwards, lays his head on her back and then copulates with his neck in contact with her back. In some antelope species the male checks to see if the female will allow him to mount by raising a foot and tapping her hind leg. This is called the 'leg beat' (or laufschlag). Fig. 9.1 illustrates this behaviour in roan antelope (*Hippotragus equinus*).

9.2.3 Mating systems

A very wide variety of mating systems exists within the Animal Kingdom and it is important that zoos are aware of the types of systems that naturally occur within the species they are attempting to breed.

- Promiscuity – Male mates with many females. No bonds form, e.g. in insects.

- Monogamy – One male pairs with one female (for at least one breeding period).
- Polygamy – There are two types:
  - Polygyny – One male mates with many females. This occurs in environments where resources are patchy and males control the patches.
  - Polyandry – Many males mate with one female. This occurs where the female is larger than the male or for some reason controls the resources needed by the male. It is common in plover-like birds.

In many social species some males will not be able to find female partners and they remain on the periphery of the group, sometimes forming bachelor herds. Generally speaking, as many males are born as females; this results in zoo populations containing surplus males in those species where males hold harems of females, as in many antelope species. Some zoos keep all male groups as a resource for breeding programmes. Knowsley Safari Park keeps a bachelor herd of bongos (*Tragelaphus eurycerus*) which is constantly changing as individuals are transferred between Knowsley and other zoos in Europe for breeding purposes.

9.3 THE PHYSIOLOGY OF REPRODUCTION

An understanding of reproductive physiology is important if populations of zoo animals are to be properly managed. The reproductive state of an animal may be monitored by recording changes in its behaviour and the cyclic changes in the level of hormones circulating in the blood. Such monitoring allows keepers and vets to prepare for births and to vary diets depending on the changing nutritional needs of pregnant individuals.

9.3.1 Reproduction in mammals

All mammals reproduce by internal fertilisation. All mammalian taxa are viviparous – give birth to miniature adults – apart from monotremes, which lay eggs. Marsupials give birth to poorly developed young which continue their development in the female's pouch after birth. The following account refers largely to placental mammals.

9.3.1.1 The male reproductive system

Male mammals possess two testes which contain seminiferous tubules. The tubules produce sperm in a process called spermatogenesis, which are stored in the epididymis on each testis until ejaculation. During ejaculation the sperm are carried by muscular contractions down a tube, the vas deferens, to the urethra of the penis. Sperm are released in a fluid called semen which is made up of components from the Cowper's gland (a lubricating mucus), the seminal vesicles (a viscous fluid containing fructose as an energy source), and the prostate gland (a fluid that maintains a suitably alkaline environment).

Sperm are placed in the female's genital tract by a process known as intromission. To achieve this, the penis needs to be erect. The process of erection is achieved by erectile tissue in the penis filling with blood, assisted, in most mammals, by a bone.

The production of sperm is stimulated by follicle-stimulating hormone (FSH). Interstitial cell-stimulating hormone (ICSH) – which is called luteinising hormone in females – stimulates the interstitial cells (Leydig cells) of the testes to produce androgens. Androgens, such as testosterone, are steroid hormones that stimulate the development and functioning of the male reproductive organs and control the development of secondary sexual characteristics. They also stimulate muscle growth. FSH and ICSH are both produced by the anterior pituitary gland (at the base of the brain) and their secretion is stimulated by gonadotrophin-releasing-hormone (GnRH) from the hypothalamus in the brain.

9.3.1.2 The female reproductive system

The reproductive organs in female mammals are the two ovaries. These produce eggs (oocytes) – by a process known as oogenesis – which are released during ovulation. Each ovary contains a number of primary oocytes. After ovulation the oocyte is moved down the oviduct by cilia – microscopic hair-like structures – towards the uterus. Fertilisation occurs in the oviduct and embryonic development begins. The fertilised egg (zygote) then implants in the wall of the uterus to complete its development. This period is referred to as gestation and varies widely between mammals from a few days to many months and is positively correlated with the weight of the adult animal (Table 9.1; Fig. 9.2).

In primates the uterus has a single chamber which has evolved for the development of a single offspring at a time. Most other mammals have a bicornate (Y-shaped) uterus with two horns in which litters develop.

Ovulation and the subsequent events are under the control of hormones secreted by various endocrine

Table 9.1 Reproductive parameters of selected mammals.

Species	Oestrous cycle (days)	Oestrus (hours/days)	Gestation (days)	Litter size
Spider monkeys (*Ateles* spp.)	24–27	2 days	200–232	1
Chimpanzee (*Pan troglodytes*)	36	6.5 days	202–261	1, rarely 2
Snow leopard (*Uncia uncia*)	15–39	2–12 days	90–103	1–5
Bongo (*Tragelaphus euryceros*)	21–22	3 days	282–287	1
Asian elephant (*Elephas maximus*)	22	4 days	615–668	1, rarely 2
Black rhinoceros (*Diceros bicornis*)	17–60	6–7 days	419–478	1
Beavers (*Castor* spp.)	14	10–12 hours	100–110	1–9 (usually 2–4)
Golden hamster (*Mesocricetus auratus*)	4	27.4 hours	16–19	2–16 (mean 9)

Source: Nowak (1999).

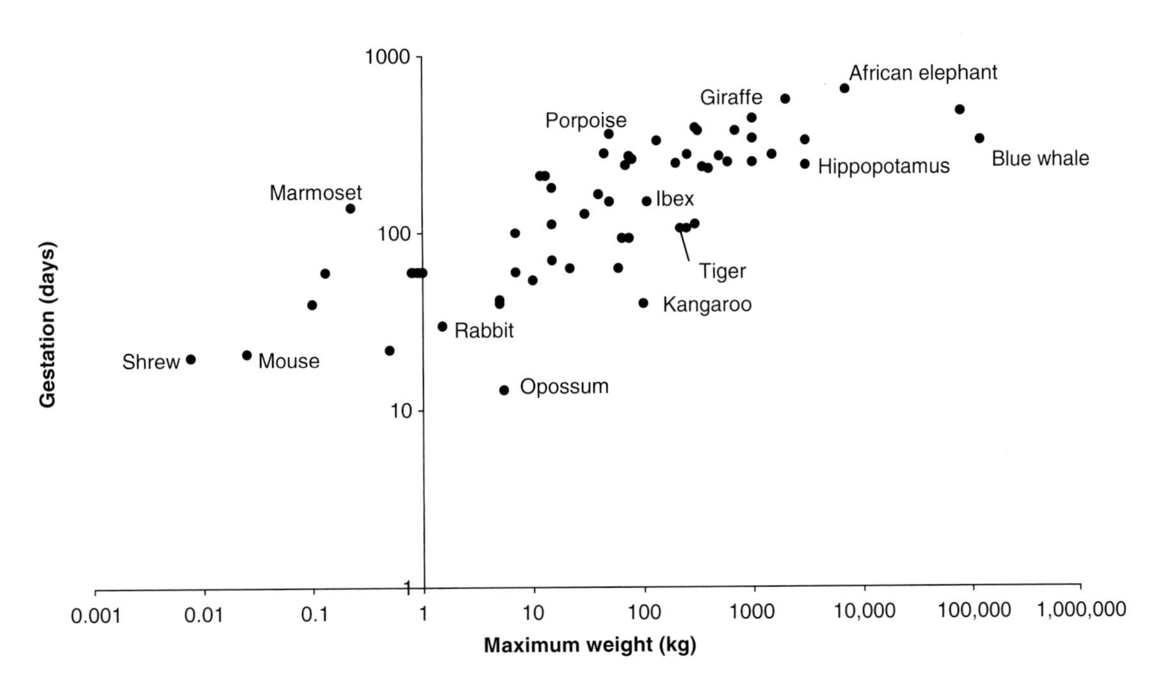

Fig. 9.2 The relationship between maximum body weight and gestation in mammals. (*Source*: Rees, unpublished data.)

glands. In female mammals the anterior pituitary gland releases two gonadotrophins, follicle-stimulating hormone (FSH) and luteinising hormone (LH), under the influence of gonadotrophin-releasing hormone (GnRH) from the hypothalamus. Follicle-stimulating hormone stimulates the development of a primary oocyte within a follicle (Fig. 9.3). When the follicle is mature it is called a Graafian follicle. The Graafian follicle migrates towards the surface of the ovary and ruptures, releasing the egg into the abdominal cavity and it is then picked up by the ciliated fimbriae at the end of the oviduct. Luteinising hormone triggers ovulation and stimulates the development of the corpus luteum in the ovary. This is produced from the remains of the ruptured Graafian follicle. Both the follicle and the corpus luteum function as endocrine glands. The follicle produces oestrogen. The corpus luteum produces oestrogens and progesterone. Both of

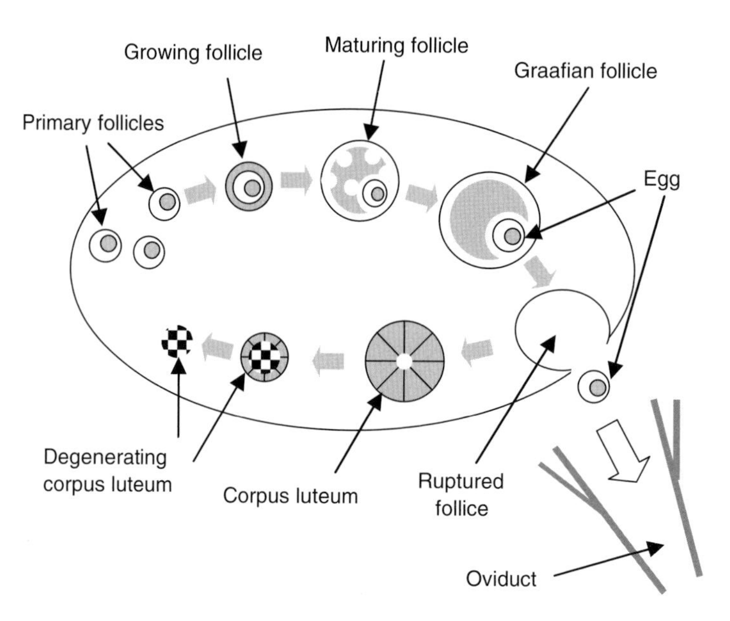

Fig. 9.3 Mammalian ovary with developing follicle. (Adapted from Harris, 1992.)

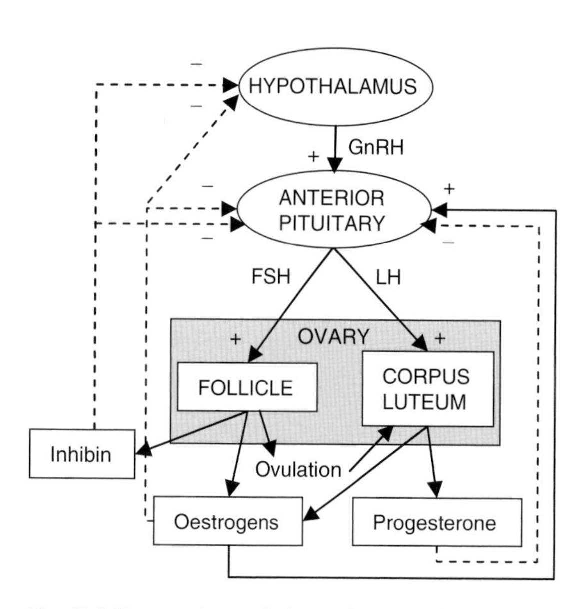

Fig. 9.4 Hormonal control of reproduction in female mammals, illustrating the role of gonadotrophin-releasing hormone (GnRH), follicle-stimulating hormone (FSH) and luteinising hormone (LH). Stimulatory (+) and inhibitory (–) interactions between the hormones are indicated. (Adapted from Harris, 1992.)

these steroid hormones prepare the endometrial lining of the uterus for the implantation of the zygote. In addition, oestrogens also promote the development of the secondary sexual characteristics and appropriate reproductive behaviours.

Oestrogens and progesterone have important feedback effects on the hypothalamus and the anterior pituitary. As the follicle ripens, the oestrogens produced from this – and later from the corpus luteum – inhibit FSH secretion, thereby preventing the development of other oocytes. Oestrogens also stimulate the secretion of LH, causing ovulation. After ovulation, the progesterone secreted by the corpus luteum inhibits LH secretion. Another hormone, inhibin, is produced by the ovary and this inhibits the production of FSH. Oestrogens and inhibin act directly on the anterior pituitary and also on the hypothalamus, inhibiting the secretion of GnRH (Fig. 9.4).

In many female mammals GnRH is released by the hypothalamus in a regular cycle, thereby producing a cyclicity in the release of eggs. In most species of mammals the females become sexually receptive at the time of ovulation. At this time they are said to be in oestrus or 'in heat'. These mammals exhibit an oestrous cycle. In humans and gorillas the females

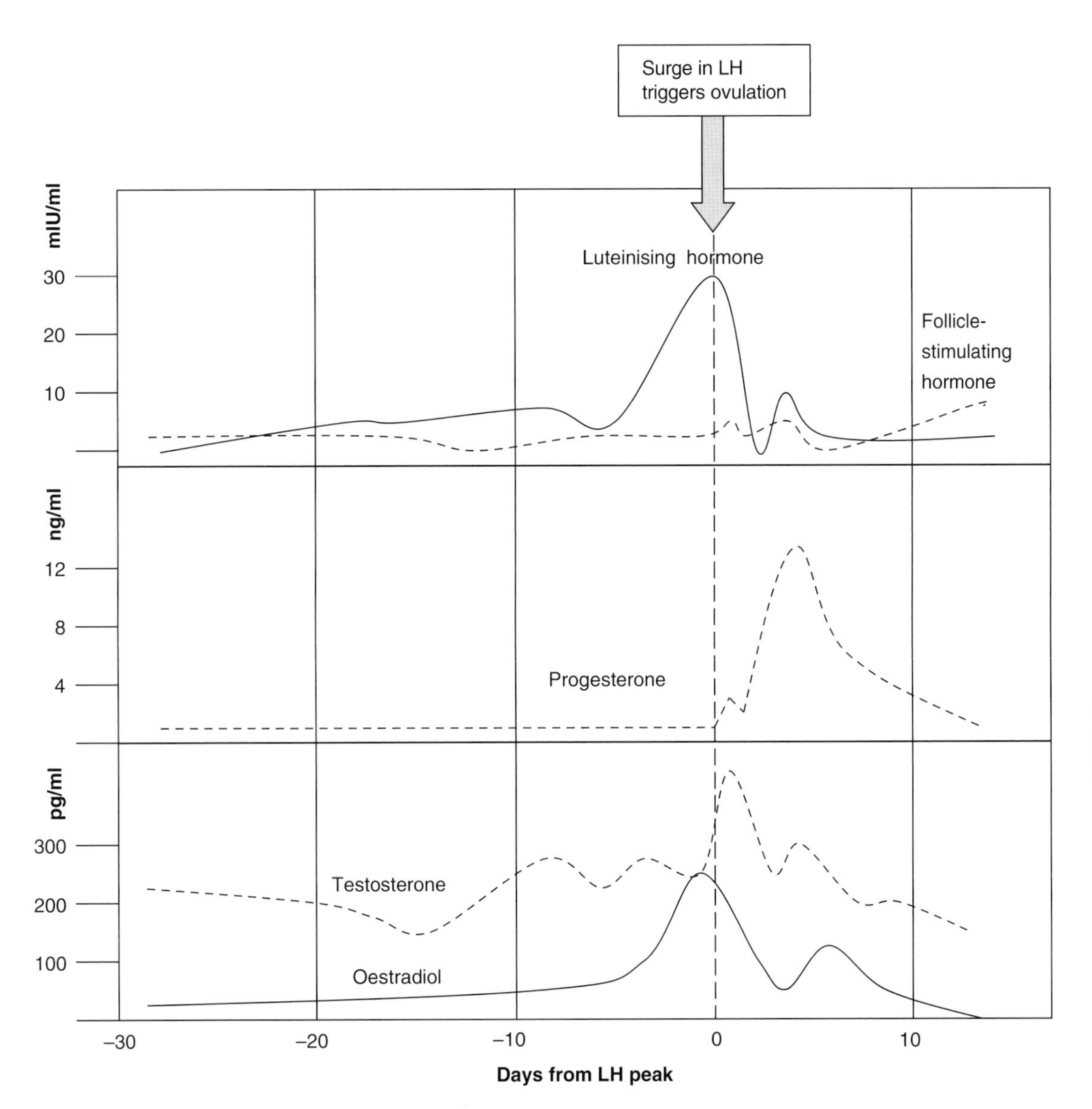

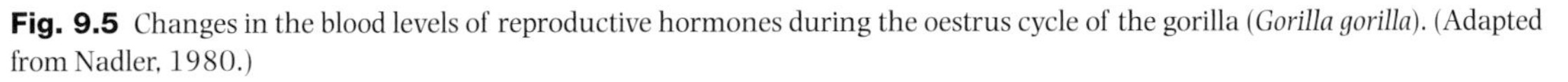

Fig. 9.5 Changes in the blood levels of reproductive hormones during the oestrus cycle of the gorilla (*Gorilla gorilla*). (Adapted from Nadler, 1980.)

are sexually receptive regardless of whether or not an oocyte is available to be fertilised (Fig. 9.5). In primates the decline in the levels of oestrogens and progesterone as the corpus luteum degenerates results in the sloughing off of some of the endometrium in a menstrual discharge. Consequently, the cycle in these species is call the menstrual cycle. In some species pheromones released in the urine cause a synchrony of the oestrous cycles of females living together (Kashiwayanagi, 2003).

At the beginning of the oestrous cycle oestrogen and progesterone levels are low so FSH secretion is not inhibited. As FSH levels rise a new follicle ripens in the ovary and an oocyte develops within it. The follicle produces oestrogens which trigger a surge in LH production which results in ovulation and the formation of the corpus luteum. The latter continues to produce oestrogens and progesterone which ends LH secretion. The oestrogens and progesterone stimulate the development of the endometrium in the uterus. If conception does not occur, the corpus luteum degenerates, the levels of these hormones fall and a new cycle begins.

In some mammal species there is no oestrous cycle. Instead, ovulation is triggered by external stimuli. This external stimulation may be seasonal, ensuring that young are born at a particular time of year when food supply and other conditions are favourable. Reflex (or induced) ovulation occurs in some species, and the females only ovulate when stimulated by copulation. This mechanism ensures that time and eggs are not wasted, and occurs in a number of taxa including rabbits and camelids.

Reflex ovulation may be the result of physical stimulation or the result of a chemical mechanism. An ovulation-inducing factor has recently been documented in the seminal plasma of llamas and alpacas, and it has been demonstrated that this, rather than physical stimulation of the genital tract, is the trigger for ovulation (Ratto *et al.*, 2005). Pheromones in rat urine can induce reflex ovulation in female rats in the absence of copulation and mounting (Kashiwayanagi, 2003).

9.3.1.3 Delayed implantation

Delayed implantation or embryonic diapause causes the embryo to remain in a state of dormancy prior to implantation. This can extend gestation by up to a year. There are two types of embryonic diapause. Facultative diapause is usually associated with metabolic stress. If copulation takes place when a female is still suckling an existing offspring, implantation may be delayed until the individual is weaned. This type occurs in some marsupials, rodents and insectivores. Obligate diapause allows the mother to delay birth until the environmental conditions are optimal. This type occurs in some pinnipeds, mustelids, ursids and armadillos.

9.3.1.4 Pregnancy

If conception occurs, the fertilised egg (zygote) divides and develops into a cluster of cells called a blastocyst. The outer layer of this, the trophoblast, forms a layer called the chorion. This secretes a hormone called chorionic gonadotrophin (CG) which enters the bloodstream of the mother and prevents the corpus luteum from degenerating. Oestrogens and progesterone continue to stimulate the development of the endometrium.

The chorion produces finger-like processes (chorionic villi) which embed themselves in the endometrium shortly after fertilisation. The structures develop into the foetal part of the placenta. The placenta allows exchange of nutrients, metabolic wastes, respiratory gases and other materials between the blood supplies of the mother and the foetus. The placenta eventually takes over most of the production of oestrogens and progesterone and the level of CG declines. These hormones stimulate the further development of the uterus and promote the development of the mammary glands, preparing them for milk production (lactation). The presence of progesterone in the mother's bloodstream inhibits lactation until after birth.

9.3.1.5 Parturition

Towards the end of gestation the smooth muscle of the uterus (myometrium) begins to contract. These contractions gradually become stronger and more frequent under the influence of the hormone oxytocin, which is secreted by the posterior pituitary gland. This process ultimately culminates in birth (parturition).

Oxytocin may be administered by a vet to induce labour or accelerate parturition, if a pregnant animal is clearly in distress and struggling to give birth. It may also be given to initiate the release of milk.

9.3.1.6 Lactation

The process of milk production by mammals is called lactation. It is controlled by the release of the hormone prolactin from the anterior pituitary gland. As long as suckling occurs regularly prolactin continues to be released because nerve endings in the nipples prevent the hypothalamus from secreting the hormone dopamine that would otherwise inhibit its production. These nerve impulses also stimulate the release of oxytocin which causes the discharge of milk from the nipple ('milk let-down'). In monotremes, females have no nipples and the milk is secreted onto hairs where it is licked off by the young. Milk composition in mammals is highly variable in composition (Table 9.2).

Table 9.2 Milk composition in selected mammals (g per litre).

Taxon	Fat (%)	Protein (%)	Lactose (%)	Ash (%)	Total solids (%)
Antelope	1.3	6.9	4.0	1.3	25.2
Bear, polar	31.0	10.2	0.5	1.2	42.9
Bison	1.7	4.8	5.7	0.96	13.2
Camel	4.9	3.7	5.1	0.7	14.4
Cow, Jersey	5.5	3.9	4.9	0.7	15.0
Dolphin	14.1	10.4	5.9	—	30.4
Elephant	15.1	4.9	3.4	0.76	26.9
Human	4.5	1.1	6.8	0.2	12.6
Kangaroo	2.1	6.2	Trace	1.2	9.5
Monkey	3.9	2.1	5.9	2.6	14.5
Rat	14.8	11.3	2.9	1.5	31.7
Seal, grey	53.2	11.2	2.6	0.7	67.7
Whale	34.8	13.6	1.8	1.6	51.2

Adapted from various sources, including Jensen (1995).

9.3.2 Reproduction in birds

All birds are oviparous (egg-laying), produce cleidoic (self-contained) eggs, and fertilisation is internal. Male birds have two testes that produce sperm. Females usually possess two ovaries and two oviducts when very young, but later only the left ovary and left oviduct become functional. The outlet for eggs or sperm is called the cloaca in both sexes. A bird's testes or ovary greatly increase in size during the breeding season. Their small size during the rest of the year lightens the load for flight.

9.3.2.1 Mating and egg production

In male birds the sperm is evacuated through the cloaca. In most species the cloacae of both sexes must be everted and brought into contact for sperm to be passed to the female. However, some 'primitive' bird groups (ducks, storks, ostriches, rheas, emus and cassowaries) have an erectile penis on the ventral wall of the cloaca.

The female's ovary produces ova released by the process of ovulation. Immediately after ovulation each ovum consists of a yolk and germinal spot which will form the embryo, if fertilisation occurs. As it passes down the oviduct it becomes covered with albumen and then enclosed in two keratin membranes. These later become separated at one end of the egg to form an air pocket. The egg then enters an enlarged area of the oviduct ('uterus') where a protein matrix is deposited on its surface. A shell then crystallises on this matrix, made mostly of calcium carbonate. The fully formed egg is eventually released via the cloaca. Soft-shelled eggs are sometimes produced as a result of a nutritional deficiency.

Respiratory gases are exchanged through pores in the shell, assisted later by the allantois and chorion (two membranes). Water also evaporates through the pores and the air pocket expands to fill the space created by the lost water.

Occasionally a bird may be unable to discharge an egg because it has become stuck in the oviduct. Such a bird is described as being 'egg-bound' and will typically stand straining and pressing, in obvious discomfort. A small dose of liquid paraffin may successfully dislodge the egg.

Female birds lay a clutch of eggs which may vary in size from one to 20 eggs. Clutch size may depend upon:

- species
- age of the mother
- physical condition of the mother
- food availability
- mortality rate of the species
- size of the brood patch.

The number of clutches produced per year varies between species and is usually between one and three. In many species, fecundity can be increased by removing the eggs from the nest as they are laid. The practice is widely used in poultry farming and by people who harvest eggs from wild birds (see Section 12.7.4).

Eggs range in size from 150 mm × 130 mm in the ostrich (*Struthio camelus*) to about 9 mm long in hummingbirds. The size of the egg is generally proportional to the size of the bird that produces it. However, smaller birds lay large eggs in proportion to their body weight. For example, hummingbirds (Trochilidae) lay eggs that are 15% of their body weight, while ostriches lay eggs that are only 2% of their body weight. Kiwis lay only one large egg that can be 25% of their body weight.

Egg shape often reflects the nest environment. For example, cliff-nesting birds lay eggs that are pointed at one end so that they roll in a tight circle and not off the cliff. Egg colour varies within and between species, and may act as camouflage. The last eggs to be laid in a clutch are often paler in colour than those laid earlier, and have fewer spots. Food may also have an effect on egg colour.

Nearly all birds sit on their eggs to keep them at the best temperature for growth in a process known as incubation. This ranges from 10 days to more than two-and-a-half months, depending on the species. Either one sex or both sexes incubate, again depending on the species. During incubation the eggs are turned so they warm evenly and develop properly. This turning process twists thick strands of albumen (the chalaza) which hold the yolk (and embryo) in the centre of the egg.

Many species possess a brood patch. This is an exposed patch of skin with many blood vessels to provide extra heat for the eggs during incubation. When embryonic development is complete the chick breaks through the shell. In some species the chick possesses an egg-tooth on the outside of the upper beak to assist in hatching, which is subsequently shed.

There are two types of chicks:

- Precocial – These are born in an advanced state (e.g. gulls, ducks, ostriches, fowls). They hatch covered with down, have considerable fat reserves, their eyes are open, and they can immediately run or swim to escape predators. They are able to leave the nest immediately they are referred to as 'nidifugous'.
- Altricial – These are born in a retarded state of development (e.g. passerines, birds of prey, woodpeckers, toucans) blind and helpless, with no feathers and no food reserves. They are completely dependent upon their parents for food and protection. Most chicks are of this type and are referred to as nidicolous.

Mothers of altricial young expend less energy by laying a smaller egg and incubating it for a shorter time than do precocial mothers. However, they spend more energy providing the greater care required by altricial chicks after hatching.

9.3.2.2 Hormonal control of reproduction in birds

Reproduction in birds is coordinated by hormones similar to those in mammals. Gamete production and ovulation are triggered by gonadotrophins (FSH and LH) produced by the anterior pituitary gland. Prolactin stimulates the production of 'pigeon's milk' by the parent's crop in pigeons and doves. Reproductive behaviours are stimulated by oestrogens (from the ovary) and androgens (from the testes) whose secretion generally follows an annual rhythm which is synchronised by day length. Many of the same hormones coordinate other aspects of biology that must be synchronised with reproduction. For example, testosterone triggers the partial moult to breeding plumage and promotes aggression and territorial behaviour, while prolactin is involved with the timing of migration.

9.3.3 Reproduction in reptiles

Most reptiles reproduce by laying shelled eggs which are fertilised internally. Both sexes possess a cloaca. However, in all male reptiles except Tuataras (*Sphenodon punctatus*) it is modified to form a penis. In tortoises and crocodiles it is a simple single structure, but in

Table 9.3 Clutch sizes and mode of reproduction in selected reptile species.

Species	Mode of reproduction	Clutch size/number of young
European grass snake (*Natrix natrix*)	Oviparous	12–20 eggs
Adder (*Vipera* spp.)	Ovoviviparous	6–20 young
Green lizard (*Lacerta viridis*)	Oviparous	*c.*10 eggs
Viviparous lizard (*L. vivipara*)	Ovoviviparous	*c.*10 young
Chameleon (*Chamaeleo bitaeniatus*)	Ovoviviparous	*c.*6 young
Gharial (*Gavialis gangeticus*)	Oviparous	*c.*40 eggs

Fig. 9.6 A Eurasian black vulture (*Aegypius monachus*) incubating an egg in a nest built on an artificial cliff.

snakes and lizards it is double, forming two hemipenes. Tuataras mate by pressing their cloacae together.

After mating the sperm travel up the oviduct and fertilise the ova as they ripen. This is sometimes months or years after copulation. After fertilisation the eggs travel down the oviduct, acquiring yolk, albumen (in many species) and then a shell. The shell is membranous or only slightly calcareous.

A clutch contains between 10 and 60 eggs, depending on the species (Table 9.3). The incubation period depends upon temperature and can last up to 12 or 15 months in the Tuatara. Like birds, young reptiles possess an egg-tooth to assist with hatching.

The common chameleon (*Chamaeleo chamaeleon*) is oviparous and lays eggs from which the young later hatch. The related *C. bitaeniatus* is ovoviviparous. In this species the fertilised eggs are retained in the oviduct until hatching, without any nutritional support from the mother. The young then appear as miniature adults. In some species there is a placenta.

9.3.3.1 Case study: Parthenogenesis in Komodo dragons

Komodo dragons (*Varanus komodoensis*) normally reproduce sexually. However, Watts *et al.* (2006) identified parthenogenetic offspring produced by two female Komodo dragons that had been kept at separate institutions and isolated from males, using genetic fingerprinting. Many zoos keep only females of this species, and males are often moved between collections for breeding purposes. Watts *et al.* suggest that perhaps the sexes should be kept together to avoid triggering parthenogenesis, which will decrease genetic diversity within the captive population.

9.3.4 Provision of denning and nesting sites

Many mammals seek shelter when about to give birth so it may be useful to provide denning sites within an enclosure. Alternatively, some animals should be confined in separate quarters from the rest of their social group near the time of birth and when caring for young offspring. This is especially important for species that are kept in zoos where the climate might pose a risk to pregnant animals or neonates (e.g. tropical mammals kept in zoos in temperate areas), and situations where the intervention of staff may be necessary.

Zoos need to provide appropriate nest boxes and other types of nesting places for birds. For some species, for example the Eurasian black vulture (*Aegypius monachus*), it may be necessary to provide artificial cliff ledges (Fig. 9.6). Where animals are free-ranging they may choose inappropriate sites for reproduction.

Free-ranging wildfowl may nest on paths and even on walls and hedges where they may be disturbed by visitors.

Some species may be particularly sensitive to disturbance immediately before and after birth. This may require animals to be kept off-show and adjustment to normal keeper routines such as enclosure cleaning. The use of closed-circuit television in denning areas and nest boxes allows keepers and vets to monitor animals without disturbance and may also reduce visitor disappointment when animals are off-show if TV screens are available in public areas.

9.3.5 Assessing reproductive status

Reproductive status in mammals may be assessed using a number of methods (Hodges, 1996):

- Invasive:
  - laparotomy (incision in the abdominal wall)
  - laparoscopy (inspection via a small incision using a lens/miniature camera)
  - blood sample (hormone analysis).
- Requiring capture or restraint:
  - vaginal smear
  - ultrasonography
  - saliva sample (hormone analysis).
- No contact required:
  - behaviour observation
  - urine sample (hormone analysis)
  - faecal sample (hormone analysis).

In mammals, the time of ovulation may be determined by detecting a fall in oestrogen and a rise in progesterone in the blood. Tests for pregnancy use antibodies to detect the presence of chorionic gonadotrophin in urine, a hormone produced by the developing zygote following fertilisation of the egg. Pregnancy in gorillas (*Gorilla gorilla*) and orangutans (*Pongo* spp.) can be detected using human pregnancy tests and these animals may be trained to urinate on pregnancy test sticks.

Ultrasound scans can be used to confirm pregnancy in many species. In giant pandas this is difficult because the embryo is extremely small. Cheung *et al.* (2010) have demonstrated the use of radiography and ultrasonography as non-invasive methods for the study of reproductive cycles in captive Asian yellow pond turtles (*Mauremys mutica*) in Taiwan. Radiography was used to monitor clutch size and ultrasonography was used to measure the growth in ovarian follicles. Egg-shell images were detected on the sixth or seventh day after ovulation.

Fig. 9.7 A female Asian elephant (*Elephas maximus*) donating blood from a vein in her ear which was later centrifuged to extract the plasma. This contains antibodies and was used to make up bottles of feed for a calf born to another cow a short time later, when he was unable to suckle immediately after birth.

Birds' eggs may be tested for embryos by a process known as 'lamping'. This simply involves shining a bright light through the egg thereby producing a silhouette of the developing chick.

Experienced keepers may know when their animals are in oestrus by changes in their appearance or behaviour. There are several reliable signs in the yak (*Poephagus grunniens*): standing to be mounted by a male, raising the tail, frequent urination, swelling of the vulva and congestion of the vulvar mucosa. Elephants raise their tails stiffly during pregnancy and the frequency of this behaviour appears to increase as they near parturition.

Many other behaviours are associated with reproduction. Females of some species, for example bears and hyenas, may exhibit denning behaviour. A female giant panda almost always undergoes a pseudopregnancy if she ovulates and fails to conceive. She sleeps a lot, builds a bamboo nest and exhibits the changes in hormone levels associated with pregnancy, but does not produce an offspring.

9.3.6 Preparing for a birth

Occasionally keepers have to make detailed preparations for a birth. This may involve producing a flow chart to predetermine what should happen under different circumstances, for example if the offspring is seriously injured or rejected by the mother. This reduces the need for debate amongst the staff immediately after the birth itself.

A young mammal needs to suckle from its mother to obtain antibodies from her milk. These antibodies may be obtained from the blood plasma of an adult conspecific – by centrifugation – in advance of the birth, and bottle fed to the infant if it is rejected by its mother (Fig. 9.7).

9.4 FURTHER READING

Further reading relevant to this chapter appears in Chapter 12 (Section 12.11).

9.5 EXERCISES

1 Discuss the problems and decisions faced by keepers when an animal rejects it offspring.
2 Describe and explain the hormonal changes that occur during the oestrous cycle of a mammal.
3 How are hormone assays used to monitor pregnancy in a mammal?
4 What preparations might keepers and vets make for the birth of an elephant?

10 ZOO ANIMAL BEHAVIOUR, ENRICHMENT AND TRAINING

Clearly one of the most urgent problems in the biology of zoological gardens arises from the lack of occupation of the captive animal.

Heini Hediger

Conservation Status Profile

Andean (spectacled) bear
Tremarctos ornatus
IUCN status: Vulnerable A4cd
CITES: Appendix I
Population trend: Decreasing

An Introduction to Zoo Biology and Management, First Edition. Paul A. Rees.

10.1 INTRODUCTION

Manning (1972) has defined behaviour as including:

> ...all those processes by which an animal senses the external world and the internal state of its body, and responds to changes which it perceives.

The main function of the nervous system is to produce behaviour. We may attempt to explain this behaviour in terms of neurophysiology or we may simply study the behaviour itself. It is possible to study and to manipulate the behaviour of animals without fully understanding the underlying physiological mechanisms. Keeping animals in captivity inevitably restricts the range of behaviours that they are able to exhibit. A knowledge of behaviour is essential if we are to understand many of the factors which affect the welfare and reproduction of animals living in zoos and which may therefore adversely affect breeding programmes.

Biography 10.1 **Heini Hediger (1908–92)**

Professor Heini Hediger was a Swiss zoologist who conducted pioneering work in animal behaviour and was responsible for the concept of 'flight distance' in animals. He is considered to be the 'father of zoo biology' and was once the Director of Zurich Zoo. Hediger published a number of books on the biology of zoo animals including *Studies of the Psychology and Behaviour of Captive Animals in Zoos and Circuses* (1955), *Wild Animals in Captivity: An Outline of the Biology of Zoological Gardens* (1964) and *Psychology and Behaviour of Animals in Zoos and Circuses* (1969), and co-authored *Born in the Zoo* (1968) and *Man and Animal in the Zoo: Zoo Biology* (1970).

Biography 10.2 **Konrad Lorenz (1903–89)**

Lorenz was an Austrian ethologist who studied the relationship between instinct and learned behaviour, particularly in birds, in which he described the phenomenon of imprinting. Lorenz is considered to be one of the founders of ethology. In 1973 he shared the Nobel Prize with Nikolaas Tinbergen and Karl von Frisch. Lorenz published many influential books including *King Solomon's Ring* and *On Aggression.*

10.2 THE DEVELOPMENT OF THE STUDY OF ANIMAL BEHAVIOUR

In the second half of the last century there were two main schools of animal behaviour:

- the behaviourists (experimental psychologists), based in the USA (notably B. F. Skinner, J. B. Watson and E. L. Thorndike)
- the ethologists, based in Europe (notably Konrad Lorenz and Nikolaas Tinbergen).

The experimental psychologists performed laboratory experiments involving mazes and 'Skinner boxes' to train animals. They were almost exclusively interested in learning and ignored the role of the natural environment. They concentrated on a small number of animal types, principally rats, mice, pigeons and monkeys.

The ethologists studied the behaviour of animals in their natural environment. This school of behaviour developed in Europe and in 1973 Lorenz, Tinbergen and Karl von Frisch received the Nobel Prize in Physiology or Medicine:

> ...for their discoveries concerning organisation and elicitation of individual and social behaviour patterns.

Ethologists studied a range of species and were not particularly interested in the process of learning. They concentrated on studying adaptive behaviour, reproduction, instinct and motivation. Ethologists were particularly interested in the role of evolution in the development of behaviour: the survival value of behaviour.

10.2.1 Tinbergen's four questions

Tinbergen (1963) published an important paper in which he outlined the four major questions in the study of animal behaviour, the 'four whys':

1 Causation – What causes an animal to exhibit a particular behaviour? What mechanisms underlie the behaviour? What is its motivation?
2 Development – How did a particular behaviour develop during the lifetime of the animal, i.e. what is the ontogeny of the behaviour?
3 Function (survival value) – Why does an animal exhibit a particular behaviour? How does it help it to survive? What are the consequences for the animal's fitness?
4 Evolution – How did evolution produce a particular behaviour?

When a scientist undertakes a study of a particular behaviour it is important to consider which of the four questions above is being addressed. In a zoo context we would probably be more interested in the causation, development and function of behaviour and less interested in its evolution. For example, understanding the cause of an abnormal behaviour might help a zoo to prevent it and understanding how a normal behaviour develops might help a zoo to provide appropriate conditions for normal development. Understanding the function of behaviours is important in appreciating the relationship between animals and their environment, and the possible consequences of releasing poorly adjusted animals back into the wild.

Biography 10.3

Nikolaas Tinbergen (1907–88)

Tinbergen was a Dutch-born zoologist who specialised in the study of instinctive behaviour. He was one of the founders of ethology and shared the Nobel Prize with Konrad Lorenz and Karl von Frisch in 1973. Tinbergen is credited with originating the four questions that should be asked about any animal behaviour: How is it caused? How does it develop? How did it evolve? and What is it for? He helped to found the Serengeti Research Institute in Tanzania and his published works include *The Study of Instinct* and *The Herring Gull's World.*

10.2.2 Measuring behaviour

Animal behaviour is a continuous activity but for the purposes of scientific study we must break it up into small units. This requires us to define discrete behaviours and divide time up into relatively small units.

Data may be collected by focusing on a single animal at a time for an extended period (focal sampling) or by examining a group simultaneously at discrete points in time. Many zoo studies use instantaneous scan sampling, in which a whole group of animals is observed at fixed points in time (e.g. every five minutes) and the behaviour of each animal at that instant is recorded. Martin and Bateson (1993) have described the most widely used methods in behaviour studies.

Using instantaneous scan sampling it is possible to calculate the activity budget of an animal – the amount of time it spends on each activity, such as feeding, sleeping, walking, and so on (Fig. 10.11). It also allows us to examine correlations between environmental variables and behaviours, for example the relationship between temperature and activity (Fig. 10.12). Many zoo studies have measured the difference in the amount of time spent performing different behaviours before and after the introduction of an enrichment device, for example a feeding device.

10.2.3 Ethograms

An ethogram is essentially a list of all the behaviours that a species exhibits. Standard ethograms for individual species do not exist so each study produces its own. The result is that different scientists may use different definitions of behaviours, making it difficult to compare studies that examine the same phenomena. In some studies scientists may only study a subset of all possible behaviours so their ethogram may only list, for example, aggressive behaviours or social behaviours, depending upon the purpose of the study. A partial ethogram for the serval (*Leptailurus serval*) is given in Table 10.1.

10.3 THE BEHAVIOUR OF ANIMALS IN ZOOS

10.3.1 The mental state of animals

Early studies of behaviour, particularly those of psychologists like Skinner, Thorndike and Watson, developed a paradigm in which only observable and quantifiable events were considered. It was believed that animal learning could be explained

Table 10.1 Partial ethogram for the serval (*Leptailurus serval*).

Behaviour	Description
Stalking	Prey-directed, concealed, with or without movement, ears focused and directed towards a particular object or area, standing or slowly moving.
Rest	Stationary, eyes closed or partially closed, animal lying down, settled with head up or down.
Alert	Eyes open, awake, aware and obviously attentive to its surroundings.
Social grooming	Licking, biting, rubbing serval.
Body contact	Head, neck, body rubbing or tail contact.
Bared teeth	Animal showing teeth, may be accompanied by hissing, spitting or other aggressive vocals.
Eating	Swallows, chews, does not simply hold in mouth, also includes grass.
Sex/courtship	Sex, courtship and any mating behaviour, all active copulatory behaviour.

Adapted from Geertsema (1985).

by conditioned responses to stimuli and instincts. However, the results of more recent studies are difficult to explain without acknowledging the importance of mental states in animals. Long-term field studies of species such as chimpanzees (Goodall, 1971, 1986), lions (Schaller, 1972) and elephants (Moss, 1989) have revealed complex societies and the work on language acquisition in apes (e.g. Savage-Rumbaugh and Lewin, 1994) has been instrumental in revealing the cognitive abilities of animals. These studies have influenced the way scientists think about animal welfare. If we accept that animals have complex mental states it follows that they may also have complex psychological needs.

10.3.2 What is normal behaviour?

One of the 'Five Freedoms' requires that animals should be free to express normal behaviour. But what is normal behaviour?

There are over 4800 species of mammals. We know very little about the behaviour of most of these species in the wild. We tend to know most about large common species (e.g. elephants, lions, chimpanzees) and relatively little about smaller and rarer species (e.g. many small rodents). We know even less about the majority of birds, reptiles and amphibians, and very little indeed about the behaviour of most fishes and invertebrates.

Some species have been very well studied in the wild. These include the apes and many other primates. However, we cannot assume that closely related species behave in a similar way. For example, bonobos (*Pan paniscus*) are far more sexually active than chimpanzees (*P. troglodytes*) and show much higher frequencies of homosexual behaviour (Bagemihl, 1999). If we assumed that bonobos in captivity should behave like chimpanzees we would have to conclude that captive bonobos exhibit an abnormally high level of homosexual sexual activity. But the difference is species-specific and nothing to do with captivity.

Zoo visitors can often be heard saying that particular zoo animals are inactive because they are 'bored' or even 'sad' (Box 10.1). This perception depends upon their expectations and, for most people, their only experience of natural animal behaviour is what they see on the television. Film makers do not make films about resting animals, so our perception tends to be that animals spend much of their time feeding, fighting, courting, mating, migrating or engaging in some other activity that is interesting to watch. For most species, this is far from the truth.

In order to establish whether or not a zoo animal is behaving normally we need to define what 'normally'

Biography 10.4

Jane Goodall (1934–)

Dr Goodall is a British zoologist who is a world authority on chimpanzees. She is the author of many books including *In the Shadow of Man* and *Through a Window*. Goodall established the longest continuous field study of chimpanzees in the Gombe Stream National Park in western Tanzania in 1960 and founded the Jane Goodall Institute in 1977 to support this research. Goodall holds a PhD from the University of Cambridge but never studied for a first degree. She is in very large part responsible for changing our attitude towards the treatment of chimpanzees in captivity.

means. 'Normally' could mean the absence of abnormal or unnatural behaviour. Such behaviour is discussed later in Section 10.5. But if 'normal' means 'as it behaves in the wild', this is difficult to establish. For many species, it is impossible to compare the behaviour of wild individuals with that of individuals kept in zoos because suitable studies do not exist. In addition, individuals of the same species may behave differently depending upon their sex, age, group size, food availability, temperature and many other aspects of their ecology. In the wild, the amount of time an animal spends engaged in different activities depends upon the environment. When food is scarce an individual may spend more time foraging or searching for prey than when it is common; on hot days it may spend more time resting than on cold days.

Visitors, and even keepers, are unlikely to see relatively uncommon behaviours. In Asian elephants copulation lasts a few seconds. An adult bull at Chester Zoo mounted a cow on 16 out of 45 days of a study spanning 10 months. In other words, there was about a one in three chance of a visitor who attended the zoo on just one of these days being in the zoo when mating was occurring. On average, a visitor (or anyone else) would have had to observe these elephants for about 12 hours to see one mating (Rees, 2004a).

Meerkats (*Suricata suricatta*) communicate the presence of predators to each other using alarm calls (Fig. 10.1). Captive-born meerkats use the same repertoire of alarm calls as wild individuals and can recognise potential predators through olfactory cues. Hollén and Manser (2007) found that they could distinguish between the faeces of potential predators (carnivores) and non-predators (herbivores) and concluded that this ability may have been retained in captive animals because of the recency of relaxed selection on these populations. However, it is clear that some captive-born animals have lost – at least temporarily – some of the behaviour necessary for survival in the wild.

Biography 10.5

George B. Schaller (1933–)

Dr George Schaller is a zoologist and conservationist who is renowned for his detailed field studies of a number of iconic species including giant pandas, mountain gorillas and snow leopards. He has held a number of posts at universities in the United States, including Stanford and Johns Hopkins, and was formerly Director of the New York Zoological Society's International Conservation Program. He is currently a Senior Conservationist at the Wildlife Conservation Society and has been instrumental in protecting important wildlife areas in the USA, China, Brazil, Pakistan and South East Asia.

Schaller's books include *The Serengeti Lion*, *The Year of the Gorilla* and *The Deer and the Tiger*.

Lions (*Panthera leo*) living in zoos are often inactive, and Haas (1958) found that zoo lions sleep for 10–15 hours per day. However, wild African lions are also inactive on average for 20–21 hours per day (Schaller, 1972). The best time to view a lion in a zoo is at feeding time.

African elephants (*Loxodonta africana*) kept on grass in a safari park spent 75% of the day feeding (Rees, 1977). However, Asian elephants (*Elephas maximus*) kept on bare soil and supplied with hay and other food spent just 34.5% of their time feeding (Rees, 2009a). The time wild elephants spend feeding is variable, but African elephants have been recorded feeding for 75.5% (Wyatt and Eltringham, 1974) and Asian elephants for over 90% of the time (McKay, 1973). Hutchins (2006) has argued that wild elephants exhibit so much variation in their behaviour that comparisons with zoo elephants are of limited value. Nevertheless, it must be true that if elephants living in zoos are deprived of continuous access to food they cannot spend long periods feeding, and this is what many large herbivores naturally do.

10.3.3 Social behaviour

Social behaviour is the behaviour exhibited by animals towards others of their species. An understanding of a species' social behaviour is essential if a zoo is to keep it in a suitable enclosure and in a social group of appropriate size and composition (Table 10.2).

Social interaction is important for some species. In the wild many species such as wildebeest, weaver birds and starlings live in huge social groups apparently devoid of social structure. However, some species form relatively small, permanent groups, for example wolves and lions.

Fig. 10.1 Meerkats (*Suricata suricatta*) watching an aircraft flying overhead, perhaps concerned that it might be a predator.

Box 10.1 Why do some animals look sad or bored?

Zoo visitors often comment that animals look bored. But is this really the case or are they just doing what they might do in the wild? 'Boredom' is a human concept and often visitors simply see animals resting or exhibiting other normal behaviour (Fig. 10.2).

Sadness is also a human concept and we tend to decide that a person is sad when they have a particular facial expression, with the corners of the mouth turned down. In humans, a sad expression is associated with discomfort, pain, helplessness, loss or bereavement. Many psychologists believe that sad emotion faces are low intensity forms of crying faces which can be observed early in newborn babies.

Primate faces are similar in basic structure to ours, so it is natural that we should expect similar facial expressions to represent the same emotions. This, however, is not the case. Howler monkeys have a naturally sad-looking face, to us, but not to other howler monkeys. Some zoos erect signs to explain that their monkeys are not sad.

There is considerable research interest in the evolution of facial expressions as a means of communication. Some studies on chimpanzee facial expressions have been conducted at the Yerkes National Primate Center in Atlanta (Parr and Waller, 2006).

(*Continued*)

Box 10.1 (*cont'd*)

Fig. 10.2 A yawning polar bear (*Ursus maritimus*).

Fig. 10.3 A 'sad-looking' black howler monkey (*Alouatta caraya*).

Table 10.2 Social organisation in selected mammal species in the wild.

Species/subspecies	Group size/social organisation
Chimpanzees (*Pan troglodytes*)	Groups of 15–120. Composition variable: all male groups, single mothers with offspring, mixed groups.
Western lowland gorilla (*Gorilla gorilla gorilla*)	Groups of 5–12. Small harems dominated by single adult male.
Hamadryas baboon (*Papio hamadryas*)	Basic unit: one male with one or more females (a harem). Related harems associate in 'clans', which converge to form 'bands'. Several hundred animals may gather at feeding or resting sites to form a regional 'troop'.
Black rhinoceros (*Diceros bicornis*)	Basic unit: female with her young. Adult females may form a temporary association. Adult males generally solitary.
Eland (*Taurotragus oryx*)	Juvenile assemblies of up to 50 provide the nucleus for female herds. May form temporary congregations of up to 1000. Male herds occur of up to six or seven. Larger male groups are very temporary.
Scimitar-horned oryx (*Oryx dammah*)	Almost always seen as herds of 10 or more.
Giraffe (*Giraffa camelopardalis*)	Social units are temporary. Mixed sex groups of up to 50 occur. The only stable associations are between mother and offspring.
Asian elephant (*Elephas maximus*)	Family groups of adult females and their young. Males solitary or associate in bull groups. May aggregate in much larger associations.
Meerkats (*Suricata suricatta*)	Pack of around 10–30, including several breeding pairs.

Partly based on data in Kingdon (1997).

One result of this stability of group composition is that members of the same group are able to recognise one another allowing the development of complex relationships, such as dominance hierarchies and cooperative behaviour. Animals which have evolved this degree of sophisticated behaviour should be allowed to live in naturalistic social groups if they are kept in zoos. One problem that may occur when groups are first established is that natural groups generally contain at least some close relatives. They are not random assemblages of unrelated individuals. It may only be possible to achieve natural groups in zoos several generations after they are established if the groups are initially comprised of individuals from a variety of sources.

One of the problems created by captive breeding programmes is that social groups are frequently broken up in order to prevent inbreeding. In many species close relatives often live together in the same social group. In captivity young are often separated from their families at a relatively early age in order to provide new genes for a social group held by another zoo. In the long run there is a risk that this may inadvertently produce socially inadequate animals, especially among animals like primates which have a complex social life and rely heavily on learning from older animals for their survival in the wild.

10.3.3.1 Social structure and dominance systems

Animals exhibit a wide variety of social structures. They may be solitary, social, colonial or eusocial (i.e. they live in a group which exhibits a division of labour with some reproductive and some 'worker' individuals, e.g. termites, ants, bees and naked mole rats (*Heterocephalus glaber*)).

Animals in zoos should be kept in social groups appropriate to their species. Zoo associations often publish recommended group sizes for particular species, but zoos may not always adhere to these. In a survey of 194 zoos, which held a total of 495 Asian elephants and 336 African elephants, 20% of the elephants were found to be living alone or with a single conspecific (Rees, 2009b). The mean group size was 4.28 animals, and many zoos did not comply with the minimum group size requirements of their regional association.

Repeated interactions between pairs of members of a social group often result in the establishment of a dominance hierarchy in which individuals are ranked in a pecking order. In a simple linear hierarchy animal A dominates animal B, B dominates all others except A, animal C dominates all but A and B, and so on.

Fig. 10.4 Barbary macaques (*Macaca sylvanus*) grooming. In primate species grooming helps to establish and maintain a group hierarchy.

Some species live in permanent social groups which consist of a breeding pair and several other animals ranked in a dominance hierarchy, which all help to defend the territory and feed the young. These species are referred to as cooperative breeders and include grey wolves (*Canis lupus*), lions (*Panthera leo*), olive baboons (*Papio anubis*), meerkats (*Suricata suricatta*), Florida scrub jays (*Aphelocoma coerulescens*) and Princess of Burundi cichlid fish (*Lamprolous brichardi*).

In most species of cooperative breeders dominance systems are based on age, with older individuals dominating younger individuals. The ranks of individual animals in these systems are usually established when they are young, often by play fighting. In other species, particularly among males, dominance rank depends upon competitive ability, so individuals gain dominance as they reach their prime and then lose dominance as they age, as in male baboons. In some animals, females inherit their rank from their mother. This occurs in hyenas (*Crocuta crocuta*), baboons and many other primates. Close relatives within each matriline support each other in contests with females from other matrilines. Matrilines are ranked in order. In many primate societies social hierarchies are maintained by grooming, with subordinate animals grooming dominant individuals (Fig. 10.4).

10.3.3.2 Intra specific aggression, density and stress

Studies of a population of voles (*Microtus pennsylvanicus*) in the field found a positive correlation between adrenal size (and thus activity) and population size, suggesting that population size was being regulated to some extent by the effect of social interactions via the endocrine system (Christian and Davis, 1966).

Increasing population density often leads to increased aggression, a phenomenon first described by Calhoun (1962) in rats (*Rattus norvegicus*). Some species are able to adjust to crowding by adopting a 'coping strategy'. Sannen *et al.* (2004) compared the behaviour of a colony of bonobos (*Pan paniscus*) during winter, when they were housed in a heated indoor hall (spatially crowded), and summer, when they also had access to an outdoor island (control). They found that spatial crowding increased aggression and that the bonobos applied a coping strategy which involved increased grooming.

Enclosures may need to be designed and operated to keep aggressive individuals apart. This may mean keeping them in separate enclosures, or, where this is not possible, allowing individuals into the outside enclosure on alternate days so that aggressors are never kept together. Some bachelor gorilla groups (*Gorilla gorilla*) are successfully managed by keeping the animals together during the day but separating them at night (Williams, 2006).

Coe *et al.* (2009) have discussed design considerations for facilities housing bachelor groups of gorillas. The breeding success experienced in zoos is likely to produce an increased need to form bachelor groups of this species in the future. Managing escalating aggression between younger males and silverbacks may require intervention by keepers, so easy access is required to on- and off-exhibit facilities. Coe *et al.* suggest that facilities for bachelor groups should include design features that reduce contact aggression, increase affiliation, allow safe outlets for species-specific behaviour and provide visual barriers and escape routes.

Enclosure designs and management need to be species-specific. Sometimes keeping animals together reduces aggression. When a male-dominated group of pig-tailed macaques (*Macaca nemistrina*) was given access to two rooms instead of one, part of the group was able to remain out of sight of the male. His loss of control resulted in a threefold increase in female aggression (Erwin, 1979).

Even animals that have previously been kept together without incident may become aggressive towards each other and it is not unknown for individuals from some species, for example big cats and bears, to kill their mates.

Keepers need to be aware of the possible effects of population density on the success of breeding programmes. Population density can affect the reproduction and survival of individuals (Chitty, 1960). Christian (1963) has suggested that the intensity and form of social interactions within an animal group depend upon population density. Social interactions modulate the function of the endocrine system via neurosecretory pathways. Endocrine changes in turn modify fecundity, fertility and disease resistance causing populations to decrease when density is high and increase when it is low.

Captive amphibians and reptiles often establish dominance hierarchies, although they are less evident in free-ranging animals (Alberts, 1994). Subordinates may have reduced access to food or may be more restricted in their use of space than dominant individuals. Subordinates may also fail to reproduce (Evans and Quaranta, 1951). The housing of male Goliath frogs (*Conraua goliath*) – a highly territorial species – in close proximity (i.e. with direct visual and auditory contact) resulted in the death of all subordinate males (Hayes *et al.*, 1998). Growth rates in hatchling snapping turtles (*Chelydra serpentina*) were lower in those raised in groups than in those raised in isolation (McKnight and Gutzke, 1993) and high density had the effect of depressing growth rates and elevating plasma corticosteroid levels in juvenile American alligators (*Alligator mississippiensis*) (Elsney *et al.*, 1990).

10.3.4 Cannibalism and infanticide

Cannibalism has been recorded in a number of animals, from insects to mammals. In the social insects cannibalism helps to regulate colony size and serves to recycle valuable nutrients within the colony. Colonies of all of the termite species investigated thus far eat their own dead and injured. Cannibalism is also seen in the preying mantis (in which species the female often eats the male after mating), ladybird larvae, scorpionflies, krill, crows, gulls, mice, rats and alligators. Schaller (1972) observed lions killing and then eating cubs in the Serengeti during territorial disputes. Cannibalism has also been observed in primates. Hamadryas baboons

(*Papio hamadryas*) will occasionally kill and eat the young of another pair. In the Entellus monkey (*Presbytis entellus*), a species of langur, Hrdy (1977) has observed young being killed as a result of fighting between a resident male, who holds a harem of females, and usurping males from peripheral male groups.

Cannibalism of larger individuals on smaller conspecifics occurs in amphibian and reptile populations. There is evidence that this is a common density-dependent regulatory mechanism in free-ranging populations rather than an aberrant behaviour (Simon, 1984).

There is a high incidence of male infanticide and feticide in captive plains zebras (*Equus burchellii*), most often when a new male is introduced into the herd (Pluháček and Bartoš, 2005). Feticide is the abortion of a foetus due to harassment of the mother or forced copulation by an adult male of the same species. After conception, the sooner a new male arrived in the herd, the less likely it was that the female's offspring would survive. Infanticide does not appear to have been documented in wild populations of plains zebra.

Records of male infanticide among ungulates are rare compared with rodents, primates and carnivores and they are more common among odd-toed ungulates than among even-toed ungulates (Pluháček and Bartoš, 2005).

In a zoo environment elephants sometimes kill neonates. However, cows that kill a neonate may go on to become perfectly good mothers to subsequent calves (Rees, 2001c). This may be the result of unfamiliarity. Elephants that are not raised in breeding herds may never see a calf let alone experience a birth.

10.3.5 Home range, territoriality and enclosure size

The home range of an animal has been defined by Burt (1943) as:

> ...that area traversed by an individual in its normal activities of food gathering, mating, and caring for the young.

He did not consider occasional sallies outside the area to be part of the home range. A territory is any defended area. An animal may have a home range and a territory; for example, gulls have a small territory around the nest but a larger home range within which they forage for food. The sizes of home ranges and territories

Table 10.3 Home ranges of selected mammals in the wild.

Species	Home range	
	Minimum	Maximum
Cheetah (*Acinonyx jubatus*)	50 km²	1000 km²
White rhinoceros (*Ceratotherium simum* (♀))	4 km²	12 km²
De Brazza's monkey (*Cercopithecus neglectus*)	5 ha	10 ha
Barbary macaque (*Macaca sylvanus*)	25 ha	1200 ha
Giraffe (*Giraffa camelopardalis*)	5 km²	> 654 km²

From Kingdon (1997).

in nature are not fixed and vary with changes in the environment.

When designing an animal enclosure it is useful to consider the size of the normal home range of the species (Table 10.3). However, it would be unrealistic, and in many species unnecessary, to attempt to replicate this. Many species range over very large areas in the wild. In East Africa, wildebeest (*Connochaetes taurinus*) migrate between the Serengeti in Tanzania and the Masai Mara in Kenya. This annual cycle is driven by the need to find sufficient food and water. Where food and water are plentiful, for example Ngorongoro crater, the wildebeest do not migrate and consequently have a much smaller home range.

In predatory species, territories are often found to be large when prey density is low and small when prey densities are high. The winter feeding territories of the Northern harrier (*Circus cyaneus*) varied from 4 to 125 ha depending upon the availability of mice (Temeles, 1987). In the Serengeti, hyenas (*Crocuta crocuta*) 'commute' to areas of high prey density in order to find food for themselves and their young. The distances they are forced to travel vary seasonally as they must move with the migrating herds of wildebeest and other species.

In territorial species it is essential to provide adequate space for the individuals the enclosure is intended to house. In highly territorial species it may be necessary to keep individuals in separate enclosures in order to prevent competition for space. Space requirements clearly vary between species. The small size of an enclosure does not necessarily compromise the welfare of an animal, and in some cases the quality of the environment may be much more important than the overall size.

Group size and home range size of carnivores in the wild may be constrained by patterns of resource dispersion (the 'resource dispersion hypothesis') (MacDonald, 1983). But animals kept in captivity may have all of their resource needs (food, shelter, etc.) fulfilled in a relatively small space, thereby allowing large groups to be maintained within a relatively small home range.

The small size of enclosures does not automatically lead to the development of abnormal behaviours. Solitary marmosets (*Callithrix* spp.) did not develop stereotypic behaviours when raised in small laboratory cages as readily as did other primates under similar conditions (Berkson *et al.*, 1966). Captive-born rattlesnakes suffered no behavioural deficiencies when reared in small transparent plastic boxes compared with those raised in larger terrariums (Marmie *et al.*, 1990).

Zoo animals may become very attached to the spaces they occupy in zoos. When *Hurricane Andrew* destroyed exhibit barriers in the Miami Metrozoo in Florida, many animals remained in their territories (Kreger *et al.*, 1998). When their enclosure was enlarged at Chester Zoo, it was over three months before all members of a herd of eight elephants fully utilised the newly available space (Rees, 2000a).

10.3.6 Introducing animals to an existing group

Integrating new individuals into an existing social group may be problematic where the species is extremely territorial, has a ridged hierarchical structure, or contains individuals that are particularly aggressive. Initially, it may be necessary to separate animals with a fence – through which they can see and smell each other – in order to judge their compatibility before allowing physical contact.

Aggression may be a problem when newborn animals are introduced into a social group. When a bull Asian elephant (*Elephas maximus*) calf, 'PoChin', was born at Chester Zoo he was initially kept indoors with his mother and a single allomother. Then, over a period of days, he was introduced to the other adult cows and two older calves (one of each sex). During this period he was kicked a number of times by the other animals. When he was finally allowed in the outdoor enclosure and introduced to his father he was kicked to the ground when he stayed too close to a cow that the adult bull was attempting to court. Notwithstanding these incidents, PoChin was eventually fully accepted by all members of the herd.

When large felids are to be introduced into the same enclosure for the first time a vet with a tranquilliser dart gun should be present as a precaution. Fights between cats can be fatal even if the animals know each other. An adult male snow leopard (*Uncia uncia*) at a British zoo recently inexplicably killed an adult female he had mated on a previous occasion.

10.3.6.1 Case study: Integrating meerkat groups

Dixon (2005) has described two methods of integrating meerkat (*Suricata suricatta*) groups used at Auckland Zoo. This species is highly territorial so aggression between the groups was expected. In the first method, two groups were kept in a sand and rock exhibit divided by a rock wall, with two PVC tunnels connecting the two sides. At the beginning of the process, one group was housed in each side and access to the tunnels was changed with time as follows:

- Both tunnels were blocked at both ends with wooden caps.
- Wooden caps were replaced with plastic caps containing holes for two weeks, thereby allowing some visual access.
- Plastic caps were replaced with transparent Perspex caps to allow greater visual access.
- A cap from one end of the tunnel was removed allowing close visual access, i.e. individuals from the two groups could approach the cap from different sides.
- After 10 days small holes were drilled in the Perspex caps, allowing the animals to smell and hear each other.
- Non-breeding animals were then removed from the two groups.
- Tunnel caps were removed completely so the animals could freely associate.

The whole process took three months.

On a separate occasion a different integration procedure was used. This time the meerkats were distracted and confused in various ways:

- Males were rubbed with talcum powder so that they smelt the same.
- Rhino dung and sacks of citronella oil were added to the enclosure.
- Males were introduced before being given a large scatter feed, in an attempt to create a bond through foraging.

Fig. 10.5 An aggressive response of a male chimpanzee (*Pan troglodytes*) to the presence of visitors. This chimp repeatedly rubbed a box across the window and then ran into the enclosure, grabbed a handful of bark from the floor, climbed to the top of the enclosure and then threw the bark through the mesh roof at visitors.

- Log positions were changed.
- Cardboard boxes were added to the enclosure.
- All animals were denied access to the nest box to avoid the possibility of individuals being cornered.

Both of these methods appeared to be successful in minimising aggression. However, it must be appreciated that these were not controlled, scientific studies, but practical methods designed by keepers, which appeared to produce the desired result.

10.4 THE EFFECT OF HUMANS ON THE BEHAVIOUR OF ANIMALS IN ZOOS

10.4.1 The effect of visitors on behaviour

When it was common to see visitors feeding zoo animals many species such as bears, apes, monkeys and elephants spent much of their time 'begging' for food by extending a hand or trunk, or, in the case of bears, standing up on their hind legs. The presence of visitors may have unexpected effects on animals. Shy animals like okapis (*Okapia johnstoni*) and maned wolves (*Chrysocyon brachyurus*) may hide out of sight, gorillas (*Gorilla gorilla*) are intimidated by eye contact, mandrills (*Mandrillus sphinx*) are intimidated by the colour red and gibbons by visitors who imitate their calls and appear to threaten their territory.

Some species appear to enjoy human contact, or are at least curious about humans, even when there is no possibility of obtaining food from them. Orangutans (*Pongo* spp.), chimpanzees (*Pan* spp.), monkeys and lemurs may sit close to windows observing visitors as they observe them.

Some species may display aggressively at visitors (Fig. 10.5). Other species may be rather more amenable to encounters with visitors (Fig. 10.6).

Some species will 'entertain' themselves by throwing objects at visitors. Chimps may throw faeces and other

Fig. 10.6 Humboldt penguins (*Spheniscus humboldti*) meeting visitors at South Lakes Wild Animal Park, where they are allowed to help at feeding time.

materials. A bull Asian elephant (*Elephas maximus*) who was previously at Chester Zoo used to dig up stones from his enclosure and throw them at visitors with his trunk with surprising accuracy.

A number of scientific studies have indicated that the presence of visitors may have adverse effects on the behaviour of certain species. As a result, some zoos have reduced the extent to which animals are exposed to visitors at viewpoints by obscuring windows and screening off the perimeters of enclosures (Fig. 7.15).

Apes may learn behaviours from visitors, for example waving and sticking their tongues out. Chamove *et al.* (1988) studied 15 species of primates and found that when visitors were present the animals were less affiliative, more active and more aggressive. These changes were particularly marked in arboreal species, especially smaller species. The effects were reduced by 50% by lowering the height of the spectators. In the same study, a group of mandrills (*Mandrillus sphinx*) exhibited a linear increase in attention to visitors, in activity and in stereotypic behaviour as visitor numbers increased. The researchers concluded that visitors are a source of stressful excitement rather than a source of enrichment for primates.

Studies of the effect of visitors on zoo animals have shown that characteristics such as visitor presence, density, activity, size and position are associated with behavioural, and to a lesser extent physiological, changes (Davey, 2007). However, although researchers have concluded that in some cases the effects on the animals are positive (enriching) and in other cases they are negative (undesirable), the extent to which visitors affect animals' welfare is unclear. Davey suggests that in order to draw confident conclusions about visitors' effects more studies are needed using, among other things, a wider range of animal groupings and measures of stress. He emphasises the need for zoo staff to be aware of possible visitor effects.

10.4.2 Effect of keepers on behaviour

There have been relatively few studies of the effects of keepers on the behaviour of animals in zoos, although there are a number of studies of the effects of humans on farm and domesticated animals (e.g. Hemsworth *et al.*, 1996; Rushen *et al.*, 1999a and b).

Even when feeding by visitors is not allowed many species associate keepers with food and will become excited if they approach. Elephants may even perform a 'greeting ceremony' normally reserved for other elephants – vocalising loudly, flapping their ears and urinating – when one of their keepers enters their enclosure. In some zoos keepers are allowed a great deal of contact with the animals (Fig. 10.7).

Visitors may come into close contact with animals in a Children's Zoo or a walk-through exhibit. Sometimes animals in these situations exhibit undesirable behaviours towards visitors. Lemurs will jump into children's pushchairs and farm animals will pursue visitors for food. Anderson *et al.* (2004) studied the effect of the close presence of a keeper on the undesirable behaviours exhibited by African pygmy goats (*Capra hircus*) and Romanov sheep (*Ovis aries*) in a contact yard at Zoo Atlanta. They defined undesirable behaviours as aggressive, avoidance and escape behaviours directed at humans, that is, behaviours which were incompatible with the purpose of gaining a positive experience through contact with the animals. Contrary to what was expected, the researchers found that the close presence of a keeper increased the rate of undesirable behaviour. In addition, they found that the rate of undesirable behaviour increased as the rate of visitor touches increased.

Jensvold (2008) has suggested that the relationship between chimpanzees (*Pan troglodytes*) and their keepers can affect their welfare. In a study conducted at Zoo Northwest Florida in Gulf Breeze, Jensvold showed that when keepers communicated with chimpanzees using typical chimp behaviour and vocalisations, the chimps engaged in more friendly behaviours (e.g. play) than when the keepers communicated with human speech and human behaviours. The chimps also interacted more with keepers when they used chimp behaviours.

Fig. 10.7 A keeper with a cheetah (*Acinonyx jubatus*). The behaviour of keepers towards animals may have a significant effect on their welfare. (Courtesy of Bethan Shaw.)

Handling animals may cause stress in some species. This is discussed in Section 11.13.

10.5 STEREOTYPIC BEHAVIOUR

Captive animals often develop stereotypic behaviours that are rarely observed in wild or free-ranging animals (Boorer, 1972). A stereotypy is any movement pattern that is performed repeatedly, relatively invariant in form, and has no apparent function or goal (Ödberg, 1978).

Stereotypic behaviour may take a number of forms, including:

- swaying from side to side or rocking backwards and forwards
- walking the same route around a pen or cage
- pacing back and forth along a fence line (Fig. 10.8)
- repeatedly walking in circles
- head-bobbing
- bar-biting (Fig. 10.9)
- neck-twisting
- tongue-playing

Fig. 10.8 A tiger (*Panthera tigris*) pacing along the fence line of its enclosure. Note that there is a concrete path immediately inside the fence which reduces wear to the ground. Pacing along fence lines is a common stereotypy in large predators.

Fig. 10.9 A giraffe (*Giraffa camelopardalis*) bar-biting, a common stereotypy in this species.

- excessive grooming
- coprophagia
- vomiting
- self-mutilation.

Stereotypic behaviour is very common in some species kept in zoos. A study of 214 giraffe (*Giraffa camelopardalis*) and 29 okapi (*Okapia johnstoni*) living in 49 US zoos found that almost 80% performed at least one type of stereotypic behaviour (Fig. 10.10) (Bashaw *et al.*, 2001).

Stereotypies have a wide range of origins and proximate causes (Mason, 1991). Some occur when an animal is consistently unable to reach a particular goal by performing appetitive behaviour, unable to reach a desired place, or unable to escape from disturbance (Carlstead, 1996). Although stereotypic behaviour is sometimes considered to be an indicator of poor welfare, it is unclear whether or not the performance of such behaviour reduces the level of stress or aversion experienced. Attempts to correlate the performance of stereotypic behaviour with stress levels in elephants have been inconclusive. A study of three captive African elephants (*Loxodonta africana*) found that differences in mean cortisol levels did not clearly correspond with the expression of swaying (Wilson *et al.*, 2004).

Rushen (1993) has discussed the possibility that stereotypies help animals to cope with adverse environments. However, after examining a number of studies that appear to support a 'coping' hypothesis, he concludes that more research is necessary and he offers alternative explanations to some of the conclusions drawn. Furthermore, he states that since

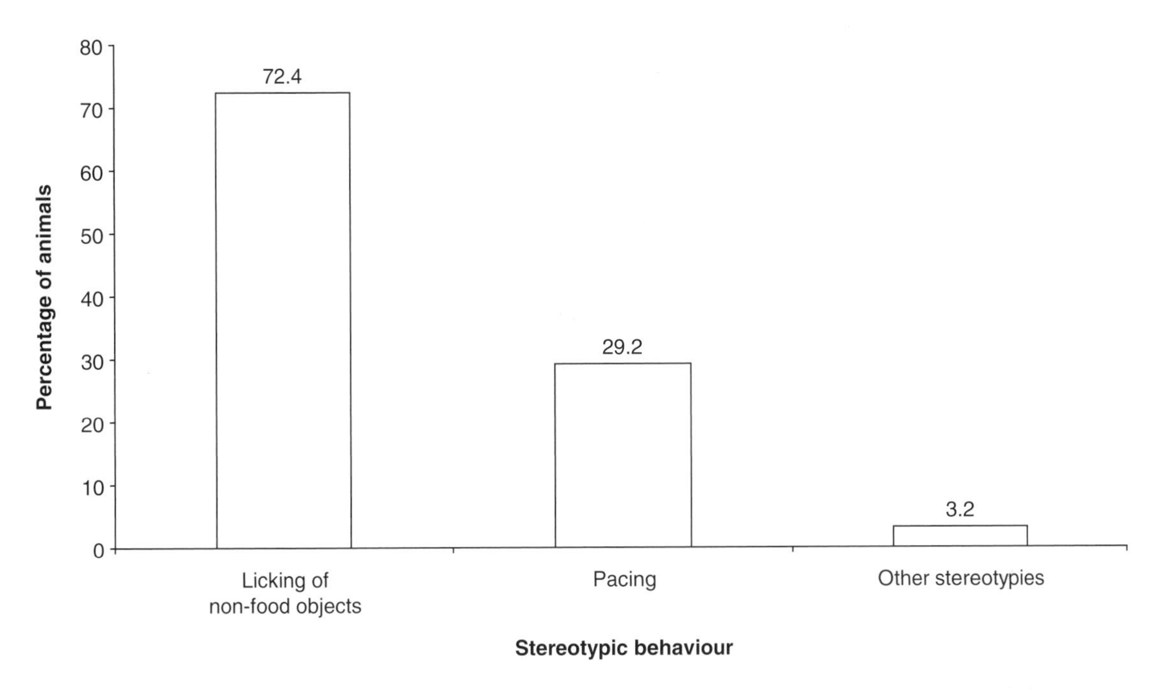

Fig. 10.10 Percentage of giraffes (*Giraffa camelopardalis*) and okapis (*Okapia johnstoni*) that exhibited stereotypic behaviours ($n = 257$). (Based on data in Bashaw *et al.*, 2001.)

different forms of stereotypic behaviour are likely to have different causes, it is a mistake to assume that all stereotypies are a response to stress. Some individuals may be predisposed to develop stereotypic behaviour. Schwaibold and Pillay (2001) have shown that stereotypic behaviour is genetically transmitted in the African striped mouse (*Rhabdomys pumilio*).

Mason (1991) has discussed the relationship between stereotypies and suffering in detail. She concludes that 'Like a scar, a stereotypy tells us something about past events', and that it suggests that a behaviour pattern has been repeatedly elicited, probably in an environment that demanded little variation in its performance. For example, a female Asian elephant (*Elephas maximus*) at Chester Zoo walked in a large anticlockwise circle, sometimes for many hours each day (Rees, 2009a). She had previously been kept in a circular pen in another zoo. Although the presence of stereotypic behaviour is not necessarily an indication of poor welfare, Mason asserts that it often develops in environments that independent evidence shows cause poor welfare.

Some repetitive behaviours may be discouraged by simple interventions. For example, if a monkey repeatedly walks the same route along a series of branches attached to its cage the pattern may be broken by rearranging the branches.

10.5.1 Early deprivation and stereotypic behaviour

In a zoo environment animals are sometimes removed from their mother earlier than would occur naturally and sometimes they are weaned prematurely. Many stereotypic behaviours appear to be related to deprivation early in life. In some cases weaning promotes the rapid emergence of stereotypic behaviours. The 'source behaviours' from which these develop may include:

- natural mother/offspring behaviours
- escape behaviours
- exploration behaviours
- the early expression of normal adult behaviours.

Weaning may lead to persistent changes in motor control, motivational state and temperament that may have an effect on the later development of stereotypic behaviour. These effects might not be apparent until some time after weaning has occurred (Latham and Mason, 2008).

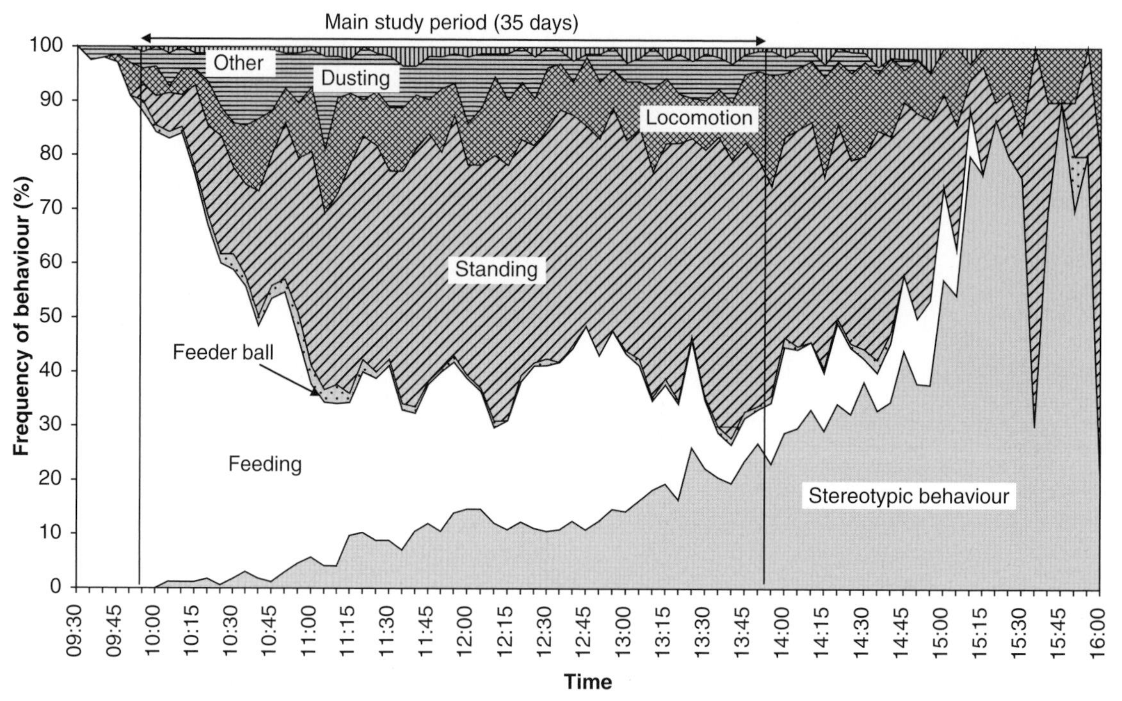

Fig. 10.11 Activity patterns in five Asian elephant (*Elephas maximus*) cows. (Reproduced with permission. Originally published in Rees, 2009a. © 2009 Wiley-Liss). Outside of the main study period data were collected on <35 days.

10.5.2 Case study: Some factors affecting stereotypic behaviour in Asian elephants

In elephants, stereotypic behaviour typically consists of repetitive pacing or rhythmic head movements (Rees, 2009a). Studies of stereotypic behaviour in elephants have frequently examined short-term phenomena, for example the effects of changes to a social group (Schmid *et al.*, 2001), the response to a particular environmental enrichment technique (Wiedenmayer, 1998; Stoinski *et al.*, 2000), or the effects of pregnancy (Szdzuy *et al.*, 2006). Other studies have examined the effects of different husbandry methods (Schmid and Zeeb, 1994; Schmid, 1995; Gruber *et al.*, 2000), behavioural sleep (Tobler, 1992), nocturnal behaviour (Brockett *et al.*, 1999; Weisz *et al.*, 2000; Wilson *et al.*, 2006), seasonal variation (Elzanowski and Sergiel, 2006) and the relationship between stereotypic behaviour and serum cortisol levels (Wilson *et al.*, 2004).

Harris *et al.* (2008) found that 54% of the 77 elephants kept in UK zoos exhibited stereotypic behaviour and more than a quarter of them stereotyped for more than 5% of the day. However, this study was based on a relatively short period of observation and there may be considerable variation in the amount of time individual elephants spend stereotyping. Short sample periods may result in gross over- or underestimates of behaviours.

In a study conducted over 35 days at Chester Zoo, five adult cow Asian elephants (*Elephas maximus*) exhibited stereotypic behaviour with frequencies ranging from 3.9% to 29.4% of all observations, with considerable variation exhibited between days and between individuals (Rees, 2009a). Feeding was the dominant activity early in the day but this was replaced by stereotypic behaviour as the day progressed (Fig. 10.11). The frequency of stereotypic behaviour was consistently higher on cold days than on warm days, at all times of the day (Rees, 2004b) (Fig. 10.12) and elephants spent more time stereotyping during the winter months than during the summer months (Rees, 2009a). Elephants prone to stereotypic behaviour showed an increase in the mean frequency of this behaviour as maximum daily temperature fell. Stereotypic behaviour increased

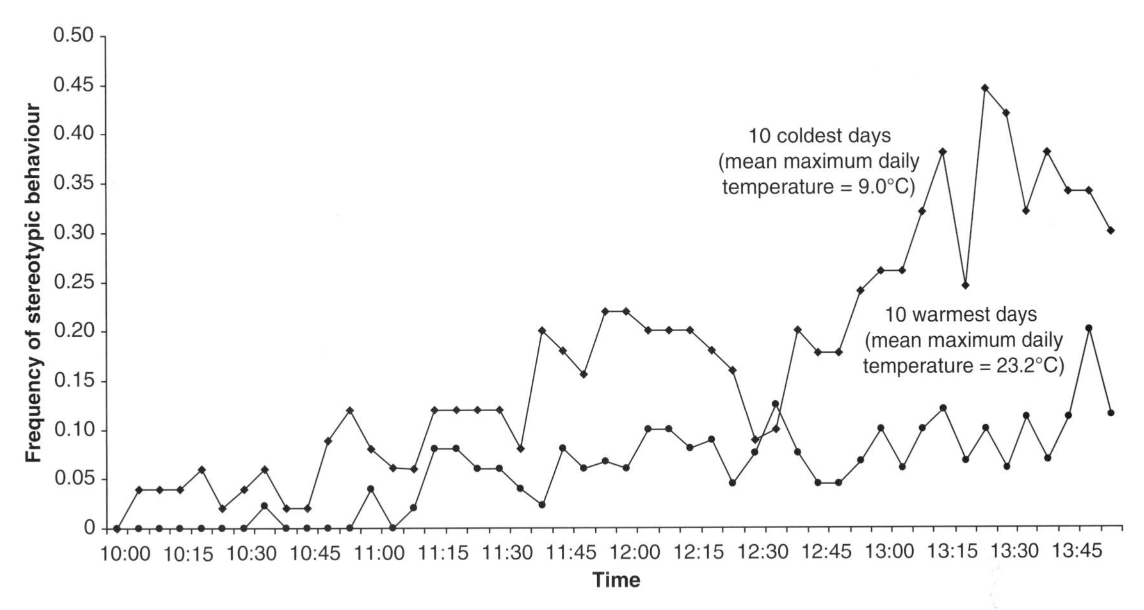

Fig. 10.12 The frequency of stereotypic behaviour exhibited by five Asian elephant (*Elephas maximus*) cows on cold and hot days. (Reproduced with permission. Originally published in Rees, 2004b. © 2003 Elsevier Ltd.)

throughout the day reaching a maximum at around 14:00 hours (when data collection ceased). This may have been indicative of a progressive increase in appetitive behaviour caused by hunger. Later in the day the elephants spent more of their time near the entrance to the elephant house waiting to return to their indoor quarters where they were fed in the afternoon (Rees, 2009a).

Stereotypic behaviour occurs mainly prior to feeding time in many captive mammalian species when individuals are motivated to perform food acquisition behaviours (Rushen, 1984; Carlstead *et al.*, 1991). The physical thwarting of attempts to move to their primary daily food source and the predictable time of feeding may both have been important in affecting the frequency of stereotypies in the elephants at Chester. Within this herd, stereotypic behaviour in adult cow elephants was negatively correlated with the frequency of feeding behaviours (Fig. 10.13). Elephants were inactive (i.e. exhibited behaviours other than locomotion) for between 70.1% and 93.9% of the time. Creating more opportunities for elephants to exhibit foraging behaviour and the introduction of greater unpredictability into management regimes, especially

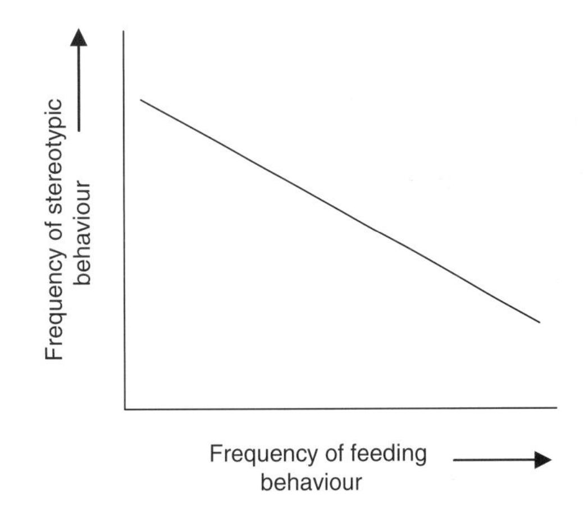

Fig. 10.13 The relationship between feeding behaviour and stereotypic behaviour in Asian elephants (*Elephas maximus*). (Adapted from Rees, 2009a.)

feeding times, may reduce the frequency of stereotypic behaviour and increase general activity levels.

In the Chester Zoo elephant herd the frequency of stereotypic behaviour was a function of differences

between individual animals, time of day, time of year, temperature and food availability.

10.6 ENVIRONMENTAL ENRICHMENT

Zoos have developed a wide range of devices and techniques in an attempt to enrich the environments of their animals.

10.6.1 What is environmental enrichment?

The Behavior Advisory Group (now the Behavior Scientific Advisory Group) of the AZA has defined enrichment as:

> ...a process for improving or enhancing zoo animal environments and care within the context of their inhabitants' behavioral biology and natural history. It is a dynamic process in which changes to structures and husbandry practices are made with the goal of increasing behavioral choices available to animals and drawing out their species appropriate behaviors and abilities, thus enhancing their welfare.
>
> AZA (2009)

The accreditation standards of the AZA require all AZA-accredited institutions to have 'a formal written enrichment program that promotes species appropriate behavioral opportunities' (Standard 1.6.1; AZA, 2010b).

The AZA recommends that the enrichment programme be based on current information in biology, and should include:

- goal-setting
- planning and approval process
- implementation
- documentation/record keeping
- evaluation
- subsequent programme refinement.

This sequence has been given the acronym SPIDER (see Box 10.2).

A more concise definition of environmental enrichment has been suggested by Shepherdson (1998):

> ...an animal husbandry principle that seeks to enhance the quality of captive animal care by identifying and providing the environmental stimuli necessary for optimal psychological and physiological well-being.

The term 'behavioural enrichment' is frequently used synonymously with environmental enrichment (e.g. Markowitz, 1982). However, the latter term is to be preferred as enrichment may confer benefits other than changes in behaviour, such as improved reproductive success.

Enrichment may consist of devices, techniques and practices that keep animals occupied and increase the range and diversity of behavioural opportunities. These may include:

- feeding devices
- hiding food or freezing it in blocks of ice
- using natural vegetation and substrates in enclosures
- designing an enclosure to maximise behavioural opportunities
- training.

Apart from improving well-being in captive animals, environmental enrichment is an important conservation tool because it can:

- improve reproductive success by providing an appropriate social and physical environment for successful reproductive behaviour and parental care
- provide the developmental environment necessary for the growth of behaviourally normal animals
- promote the maintenance of the species-typical behaviours necessary for survival in the wild.

Enrichment may also have an important educational role because it may increase animal visibility and it may allow visitors to observe and learn about natural behaviours. Of course, this may not be the case where animals are provided with toys or frozen blocks of food as enrichment.

Enrichment can have a number of benefits for animals. It may:

- promote the expression of natural (species-appropriate) behaviours
- provide mental stimulation via a complex environment
- provide an opportunity for the animal to respond to environmental conditions using its evolutionary adaptations
- increase opportunities for exercise
- provide increased control of the environment, thereby reducing stress
- provide appropriate learning opportunities.

Box 10.2 The SPIDER framework for assessing the effectiveness of husbandry, training or enrichment.

SPIDER is a model framework for assessing the effectiveness of behavioural husbandry and may be used to assess the effectiveness of a training or enrichment programme (Fig. 10.14).

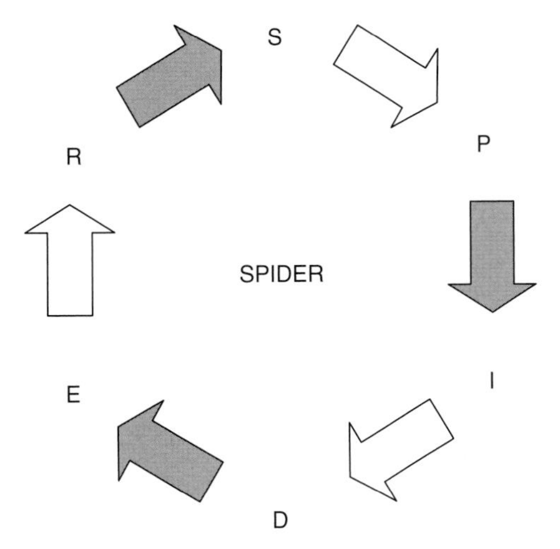

Fig. 10.14 SPIDER.

S – Setting goals

Identify the desired behaviours, and how these may be affected by the natural and individual history of the animals.

P – Planning

Develop a plan outlining the steps involved in training the behaviour.

I – Implementing

Implement the training plan.

D – Documenting

Document information about how the training sessions are progressing.

E – Evaluating

Evaluate trends in behavioural changes over time.

R – Readjusting

Readjust the training plan if this is necessary after reviewing progress.
(See Colahan and Breder, 2003.)

Although the use of environmental enrichment in zoos is currently very popular, and, in some jurisdictions, required by law or as a condition of accreditation by a zoo organisation, it is not a new concept. Some 60 years ago, Hediger (1950) recognised the inadequacy of most zoo environments. London Zoo was using an automatic device to feed its seals (*Halichoerus* spp.) by the 1960s (Morris, 1960). This carried fish around the pool before they were released to be caught by the chasing seals. Markowitz pioneered the 'behavioural engineering' approach to enrichment – based on operant conditioning techniques – at Portland Zoo (now Metro Washington Park Zoo) in the 1970s. He devised a food dispenser for white-handed gibbons (*Hylobates lar*) which dispensed food when they brachiated high in the enclosure (Markowitz, 1982).

10.6.2 The concept of 'behavioural needs'

Animals may be motivated to perform some behaviours in the absence of any need to do so. Domestic cats will continue to perform prey-catching behaviour towards live prey even after they have received adequate food (Leyhausen, 1979) and rats and pigeons will 'work' for food (by operating a lever or pecking at a light) even when food is freely available (Neuringer, 1969).

Under some circumstances, preventing animals from performing these appetitive behaviours may be stressful. This has led to the concept of 'behavioural needs' and the suggestion that the performance of behaviours observed in the wild should be used as a benchmark for evaluating the psychological well-being of captive animals. This is problematic for several reasons:

- For many species we have little or no information about how they behave in the wild.
- For those species that have been studied in the wild it is clear that there is a great deal of variation in behaviour within and between wild populations.
- It is unlikely that the absence of some behaviours in captive animals is an indicator of poor welfare, e.g. predator-avoidance behaviours.

10.6.3 Food and feeding as an enrichment

Some of the most effective enrichment techniques involve searching for hidden food. This may be explained by the fact that animals are strongly motivated to seek information from their environment, and this results in high levels of exploratory behaviour (Inglis and Fergusson, 1986). Shepherdson *et al.* (1993) observed reductions in abnormal behaviour in leopard cats (*Prionailurus bengalensis*) and a fishing cat (*Felis viverrina*) when they were required to search for their food.

Enrichment options must take into account group dynamics in socially housed animals as some techniques and devices may have negative consequences for some individuals. For example, feeding devices may be monopolised by dominant animals so that subordinate individuals receive no benefit. A large metal feeder ball at Chester Zoo was often monopolised by a single bull Asian elephant (*Elephas maximus*) to the extent that on one occasion he spent more than 56% of his time interacting with it (Rees, 2009a).

10.6.3.1 Food placement

Simple changes to the method of feeding may act as an enrichment. Scatter feeds (throwing small pieces of food over a wide area of the enclosure) will increase the amount of time some animals, for example primates, spend foraging and feeding.

Suspending food out of reach will make animals work for their food, for example hanging meat from the top of a cage so that small cats have to jump to reach it. Markowitz (1982) found that servals became more animated and more interesting to visitors when fed 'flying meatballs' attached to a rope or rod swung over the heads of the cats. Herbivores that normally feed in the trees should have their food placed above the ground to simulate this (Fig. 10.15).

At South Lakes Wild Animal Park keepers feed their lions (*Panthera leo*), tigers (*P. tigris*) and jaguars (*P. onca*) by fixing pieces of meat to the top of tall wooden telegraph poles (Fig. 10.16). Visitors observe the animals climbing for their food from high wooden walkways that provide unobscured views over the tops of the fences. This feeding method provides a spectacle for visitors and exercise for the big cats.

Some species appear to be prepared to spend time trying to obtain food which is difficult to reach even when other food is freely available. When the Asian elephants (*Elephas maximus*) at Chester Zoo were held behind a dry moat some of the adults regularly kneeled down in order to reach grasses and other plants growing in the moat when they could simply have picked up food which had been put out in their enclosure. Others would stand on their hind legs in order to reach plants growing on top of a high wall.

10.6.3.2 Carcass feeding

Feeding whole animal carcasses to large predators can be a useful feeding enrichment. Big cats will drag carcasses around and spend long periods of time dismembering them (Fig. 10.17). However, such sights may be unacceptable to many zoo visitors and, in some zoos, can only be practised when visitors are absent. Where small prey items are involved, for example rats, fish and young chicks, the sensibilities

Fig. 10.15 A ring-tailed lemur (*Lemur catta*) feeding from a suspended branch.

of visitors are less likely to be offended (Fig. 10.18). A useful alternative employed by some zoos is a prey mannequin which can be 'killed' by a large predator and dragged around.

10.6.3.3 Puzzle feeders and other feeding devices

Any device that makes it more difficult for an animal to find or obtain its food is likely to increase the amount of time it spends foraging and feeding. However, care must be taken to include all of the food made available during the day in its total diet to avoid over-feeding.

Zoos and their keepers have been extremely inventive in producing devices capable of making animals work for their food. These range from machines that will drag food along the ground or around a pool for predators to chase and 'kill', to puzzle boxes that require primates to perform a series of operations in the correct sequence in order to obtain a food reward.

Wells and Irwin (2009) studied the effect of three feeding devices (food-filled baskets, polyvinyl chloride tubes and frozen ice pops) on the behaviour of moloch gibbons (*Hylobates moloch*) at Belfast Zoo and concluded that such devices offered an effective enrichment for this species. Many other studies of feeding enrichment have reported beneficial effects on behaviour (Reinhardt and Roberts, 1997).

Puzzle feeders Puzzle feeders are devices that contain food which can only be accessed by the animal learning to use the appropriate behaviour. For example, puzzle balls will dispense small items of food through small holes if they are rolled into the correct position (Fig. 10.19); apes can obtain food from holes in artificial

Fig. 10.16 A tiger (*Panthera tigris*) collecting food from the top of a tall pole at South Lakes Wild Animal Park, Cumbria.

Fig. 10.17 Lions (*Panthera leo*) feeding on cow carcasses.

Fig. 10.18 A ground hornbill (*Bucorvus abyssinicus*) feeding on a dead rat.

termite mounds using thin sticks (Fig. 10.20). Puzzle feeders are sometimes located at the rear of artificial structures like tree trunks or termite mounds, and therefore out of sight of visitors, while others are on full view in the enclosure (Fig. 10.20). 'Kong' toys are strong rubber hollow chew toys into which food can be inserted. They are available in a range of colours and sizes and were originally designed for domestic dogs.

The use of puzzle feeders is most often associated with primates and elephants because they require manipulation with hands or trunks. However, they have also been used for other species, including octopuses. Brady *et al.* (2010) studied the effect of enrichment on giant Pacific octopuses (*Enteroctopus dofleini*) at Cleveland Metroparks Zoo, Ohio, and found that enrichment – including prey puzzles – increased activity and behavioural diversity and reduced the amount of time spent on resting and locomotion.

Other feeding devices include:

Burlap bags

This is a bag made of sacking. It may be filled with hay and food treats and then tied closed. The animal has to get into the bag and sort through the hay to find the treats.

Feeder tyres

Plastic feeder tyres are commercially available but many keepers make their own from old car tyres. They are useful for bears, primates, elephants and other species. Animals obtain small pieces of food from small holes in the sides of the tyre.

Popsicles

A popsicle is a large block of ice containing small pieces of food, usually fruit or vegetables. Very large popsicles may take some time to produce and yet be very quickly disposed of by a large animal like an elephant that can crush the ice with ease using its foot. When made from blood for carnivores these ice blocks are known as 'bloodsicles'.

Feeder tubes

This is a tubular feeder which has small holes drilled in its surface through which small pieces of food fall when the tube is rolled or manipulated. The tube can be easily suspended. One end of the tube has a screw cap through which the feeder is filled. Feeder tubes are useful for encouraging play and problem-solving in primates, elephants and other animals. Plastic feeder tubes are commercially available but a simple device is not difficult to make.

Programmable insect feeders

A sequential programmable insect dispenser that blows living insects into a cage (or into the air) for feeding animals has been described by Gans and Mix (1974). These devices are useful for lizards and are a substitute for live prey.

Programmable food dispensers

'On Time Wildlife Feeders' are commercially available feeders originally designed to be used by hunters to attract game animals. They are either battery- or solar-powered

Fig. 10.19 A red-tailed guenon (*Cercopithecus ascanius*) attempting to extract food from a puzzle ball.

Fig. 10.20 An artificial termite mound feeder for gorillas (*Gorilla gorilla*).

and fitted with programmable clocks which control the release of food onto a spinning plate which throws the food out in a circular pattern around the feeder.

10.6.4 Auditory and olfactory enrichment

Okapi (*Okapia johnstoni*) are particularly shy and sensitive animals. At *Marwell Wildlife* visitors are asked to be quiet when they visit the Okapi Breeding Centre and in their indoor quarters the animals are played calming music. Recent experiments have shown that many zoo-housed species (gorillas (*Gorilla gorilla*), black-footed cats (*Felis nigripes*), elephants) gain welfare advantages from auditory and olfactory stimulation (Wells and Egli, 2004; Wells *et al.*, 2006).

Olfactory enrichment may take the form of predator or prey scents, novel odours or pheromones. Some scents are available commercially (e.g. chicken, fish, honeysuckle and catnip). Some keepers put the faeces of other species in their enclosures (e.g. the faeces of antelope in a lion's enclosure). *Feliway*® is a commercially available synthetic feline facial pheromone – used by cats to mark their territory – which may act as an olfactory enrichment in felid enclosures.

10.6.5 Toys

Many species, ranging from parrots to elephants, benefit from the provision of toys. Large unsuspended and suspended tyres or balls (Fig. 10.21) are useful toys for a wide range of large mammals, from great apes and elephants to tigers and polar bears. A '30 inch Safari Ball' is commercially available and useful for tigers and bears. The ball is hollow and may be filled with water or sand. If left empty it will float on water. Plastic kelp

Fig. 10.21 A large plastic ball used as an enrichment for lions (*Panthera leo*). Note the scratch marks.

is available as an enrichment for dolphins or pinnipeds. Head butting posts have been used for bison (*Bison* sp.) and heavy sacking bags may be suspended as butting objects for smaller species like babirusa (*Babyrousa* spp.). Orangutans (*Pongo* spp.) will play with old clothing, using it to cover their heads as they would use large leaves in the wild. The gorilla (*Gorilla gorilla*) in Fig. 10.22 wrapped himself up in a blanket on a cold day.

10.6.6 Enrichment and enclosure design

Many enclosures have design features which may act as important enrichments for animals, particularly the substrate and the furniture.

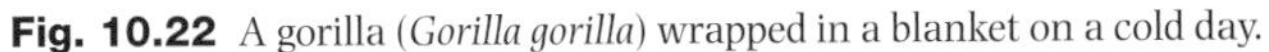

Fig. 10.22 A gorilla (*Gorilla gorilla*) wrapped in a blanket on a cold day.

Fig. 10.23 Elephants meeting visitors on the lawn at Bristol Zoo *c*.1960. (Courtesy of Peter and Enid Rees.)

Substrate The use of a suitable substrate may act as a significant enrichment for some animals. For example, if bark is used as the substrate in a monkey cage it will provide more interesting opportunities for foraging than a concrete or tiled floor.

Ropes and fire hose Ropes and lengths of old fire hose can provide useful opportunities for climbing in arboreal species such as monkeys and apes.

Hammocks Hammocks can provide elevated resting places for animals like primates and bears. They are sometimes constructed from plaited fire hose because of its strength.

Climbing frames Where possible it is important that climbing structures move in a natural fashion, like trees, so the animals can get used to a moving environment. This can be achieved by suspending branches and other materials from ropes and elastic materials. Arboreal animals that have been reared in artificially static environments may have difficulty in adapting to moving around in real trees.

Elevated resting sites These often consist of platforms supported on wooden posts. They are particularly useful for big cats, bears and great apes. Apart from providing an opportunity to climb they may also allow the animals to see beyond the confines of their own enclosure.

10.6.7 Television and computers

The behaviour of primates in zoos may be enriched by the use of television, video and computer 'games'. Bloomsmith and Lambeth (2000) showed videotapes of chimpanzees, other animals and humans to 10 chimpanzees (*Pan troglodytes*). The chimps watched the monitor for 38.4% of the time available and individually housed chimps watched it more than those kept in a social group. The chimpanzees' behaviour was not altered extensively by watching the videos, but it did occupy a significant proportion of their activity budget, suggesting that it may be a useful enrichment.

10.6.8 Conspecifics and other species as an enrichment

For social species there can be no doubt that the presence of conspecifics acts as an important enrichment. Social animals spend a considerable amount of time interacting with each other and without such interaction they would inevitably have little to do for much of the day. For an elephant kept on its own there is no substitute for another elephant (Rees, 2000b).

Some zoos may allow species to reproduce in order to produce offspring that will enrich the lives of their parents and the rest of the social group. Some of these offspring may eventually be euthanised if there is insufficient space for them and they cannot be transferred to another zoo.

Multi-species exhibits can generate a number of enriching experiences for the species concerned. Different species may choose to associate with each other – for example antelope species – or they may have to spend time avoiding each other and competing for food. The relationship between African elephants (*Loxodonta africana*) and hamadryas baboons (*Papio hamadryas*) kept in a multi-species exhibit is described in Section 7.5.4.

10.6.9 Other sources of enrichment

Some zoos allow visitors to interact with some of their animals. Honolulu Zoo in Hawai'i brings its Asian elephants out of their enclosure to meet the public. Such encounters with elephants were once commonplace in zoos – as were elephant rides – but few zoos now consider them safe (Figs 10.23 and 10.29).

Box 10.3 An example of simple enrichment for primates.

At the National Zoo in South Africa keepers have added enrichment to their management plan for hamadryas baboons (*Papio hamadryas*) (Fig. 10.24). The following furniture was added to their enclosure:

- Hammocks made of fire hose
- Feeding tables
- Wood stumps with hiding places
- Areas of woodchips.

Keepers also changed the way they were providing food by:

- Presenting food whole instead of chopped
- Wrapping food in newspaper
- Putting food in pine cones
- Hiding food in the woodchip area
- Placing or hiding food in new areas of the exhibit.

Activity was monitored before and after the introduction of enrichment. The changes to the management of the baboons appeared to increase activity (especially foraging), reduce stereotypic behaviour and reduce aggression. This example was subsequently used in a training programme to show other zoo staff that enrichment programmes do not have to be complex or time-consuming (Cloete *et al.*, 2008).

Fig. 10.24 Male hamadryas baboon (*Papio hamadryas*).

Lemurs are popular animals for visitor encounters in a free-ranging environment, and some zoos will allow the public to feed them with food they supply. Domesticated animals are kept by some zoos in a Children's Zoo or Pets' Corner. Some of these animals undoubtedly seek interactions with people (see Section 10.4).

Some zoos provide elephants and chimpanzees with the opportunity to paint. Blackpool Zoo sells paintings created by its elephants to raise money for elephant conservation. Participating in research, training for animal shows and veterinary examination may also be enriching activities for some species. For example, some types of research require active participation by the subjects and may provide them with mental stimulation (e.g. cognitive studies of apes).

10.6.10 Does enrichment improve welfare?

The purpose of providing enrichment for animals kept in zoos is to improve their welfare. In particular the intention is generally to increase the time spent engaged in species-typical behaviours or goal-directed behaviours and reduce the amount of time spent engaged in abnormal and stereotypic behaviours (Box 10.3).

In a recent meta-analytical review, Shyne (2006) examined the effects of enrichment on stereotypic behaviour in mammals kept in zoos reported in 54 published studies. The species examined were mostly primates (mostly gorillas and chimpanzees), bears, felids (mostly big cats), seals and walruses, canids, giraffes and okapis. Shyne concluded that 'enrichment substantially reduces the frequency of stereotypic behavior exhibited by mammals living in zoo environments'. However, she recognises that 'in zoo practice, all methods of enrichment may not continue to stimulate the animals and reduce negative behaviors'. Most of the studies she examined collected data immediately after enrichment was introduced, but most did not report whether or not the enrichment was still effective after data collection stopped. Clearly food puzzles are likely to continue to be stimulating for carnivores, but pieces of novel furniture (e.g. a tree branch lying on the ground) may not. It is important that keepers monitor the use of enrichment devices and recognise when they lose their novelty value.

10.6.10.1 Enrichment as a health benefit

Much of the benefit gained from enrichment is behavioural, but sometimes it may have an obvious physical benefit. When perches at several heights were added to the cages of group-housed squirrel monkeys (*Saimiri sciureus*), males showed a decrease in the propensity to develop the foot and tail ulcers associated with floor contact, and overall morbidity decreased as a result of reduced aggression (Williams *et al.*, 1988).

Although enrichment devices may confer behavioural and health benefits, in spite of their initial interest in novel stimuli, many animals will eventually habituate to the enrichment provided by zoos (see, e.g., Bloomsmith and Lambeth, 2000). When a device does hold the attention of an animal over an extended period of time there is a risk that it may develop repetitive behaviour which amounts to being stereotypic. Is repeatedly operating a device to obtain a food reward any less of a stereotypy than any other repetitive behaviour?

10.7 ANIMAL TRAINING

Zoos train their animals for a variety of reasons including:

- to perform for the entertainment of visitors
- to illustrate normal behaviour
- as an enrichment
- to assist veterinary examination or treatment
- to participate in research.

10.7.1 Principles of training

The basic method of training animals is the same regardless of its purpose. Training is achieved by a type of learning called operant conditioning. When a particular behaviour is performed and the consequences of the behaviour are reinforced, for example by receiving a reward, the animal is more likely to repeat the behaviour in the future. Positive reinforcement may take the form of a small food reward or it may simply be verbal praise. In order for the desired behaviour to be performed on demand the trainer must teach the animal to associate a command with the desired behaviour and the subsequent reward (Box 10.4).

10.7.2 Animal shows

Historically zoos have used trained animals to entertain visitors. Some zoos still do. The good animal shows include individuals performing natural behaviours, for example:

- birds of prey giving flying displays
- ground hornbills (*Bucorvus abyssinicus*) 'catching' artificial food (plastic frogs) (Fig. 10.28)
- sea lion feeding
- penguin feeding
- shark feeding in aquariums.

Some of these activities require training, others do not. Some animal shows may have a conservation message. Asian elephants at ZSL Whipsnade Zoo push over an artificial villagers' hut and knock over a hinged tree to demonstrate the damage they do in the wild. Other shows may be used to demonstrate

Box 10.4 Clicker training.

Clicker training is a reward-based system which allows a variety of species, from parrots to primates, to be trained to perform desirable behaviours. These might include, for example, presenting a part of the body for examination by a keeper or vet, or approaching the side of a cage for medication.

The clicker is a simple metal or plastic box that makes a distinctive 'click' sound when a thin metal plate is depressed (Fig. 10.25). With some clickers the volume/tone of the sound can be altered. This is important if an animal is particularly sensitive to the sound.

Fig. 10.25 A clicker.

Animals can be trained to perform a particular behaviour by rewarding that behaviour with a small piece of food (positive reinforcement). However, when this is done without the use of a clicker there may be a delay between the performance of the desired behaviour and the receipt of the reward (Fig. 10.26). The animal may then not know exactly which behaviour is being rewarded. The clicker allows the trainer to mark the behaviour precisely with a click before the reward is presented, so the animal learns to expect the reward after hearing the click.

Training an animal to touch a target stick

Initially it is important to make the association between the click and a food reward (Fig. 10.27). Press the clicker once and give a treat immediately. Repeat until the animal knows a treat will follow a click.

1 Put a small amount of a tasty, smelly food on the end of the stick.
2 Extend the target stick and offer it to the animal.
3 When he touches the stick, click once (never repeatedly) and reward with food.
4 Offer the stick again and repeat step 3 above.

(*Continued*)

Box 10.4 *(cont'd)*

5 Once he understands he needs to touch the stick to get the reward, move the stick further away so he has to take a step forward to reach it.
6 When he touches the stick, click and reward.
7 Gradually move the stick further away and repeat step 6.

This training may now be extended by varying the amount of time between the click and the delivery of the reward. This is important because it will not always be possible to give the reward immediately after the click. The click acts as a 'bridge' that links the reward with the behaviour.

Once the animal reliably delivers the behaviour every time the stick is produced it may be possible to associate the behaviour with a cue or a command such as 'touch'.

Some simple rules for clicker training

1 You must reward every time you click even if you do it accidentally or in error. Remember if you make a mistake it is not the animal who is at fault.
2 Only use the clicker for training.
3 Only click once between rewards.
4 Keep training sessions short.
5 Only teach one new behaviour in any session.

Fig. 10.26 Target training a sea lion.

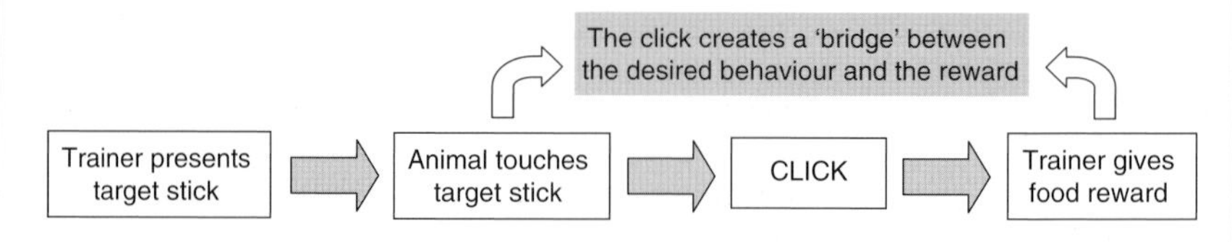

Fig. 10.27 Creating a bridge during clicker training.

Fig. 10.28 A ground hornbill (*Bucorvus abyssinicus*) 'catching' a plastic frog during a public display.

an animal's intelligence or physical prowess, for example:

- macaws solving puzzles
- coatis climbing ropes
- dolphin shows
- feeding tigers by suspending food high up on a pole (Fig. 10.16).

The worst types of show present animals merely for entertainment with no educational message, such as macaws riding miniature bicycles or elephant orchestras. In the past animals were widely used to give rides to visitors. This is relatively rare now but some zoos still give elephant rides (Fig. 10.29).

10.7.3 Veterinary examinations and research

Some keepers train animals to cooperate in veterinary examinations and procedures, and in maintaining hygiene, for example oral examination and teeth-cleaning in sea lions and washing elephants and inspecting them for foot abscesses and other problems. These behaviours may be incorporated into animal shows.

Animals can become accustomed to repeated non-aversive procedures such as weighing or taking blood samples. Giraffes (*Giraffa camelopardalis*) have been trained to enter a restraint device voluntarily (Wienker, 1986). However, animals do not habituate to procedures that are very aversive such as some transport stress.

Great apes can be trained to drink from a bottle, a skill that may help in giving medication. At *Disney's Animal Kingdom*, a number of primate species – including gorillas (*Gorilla gorilla*), mandrills (*Mandrillus sphinx*) and gibbons – are trained to perform a wide range of behaviours (Colahan and Breder, 2003). These behaviours may be divided into three categories, depending upon their purpose:

- Husbandry – e.g. presenting the face, foot, arm, hand, knee, open mouth, tongue.
- Veterinary – e.g. accepting an ear thermometer, injection, oral medication, wound-cleaning, stethoscope and ultrasound examination.
- Research – e.g. semen collection and urine collection.

10.7.4 Is training enrichment?

For some species there can be little doubt that training acts as an enrichment to their lives in a zoo. Many species, such as elephants, sea lions and macaws, appear to enjoy taking part in animal shows and clearly do so for very small food rewards. However, zoos vary greatly in their attitudes towards animal shows. Some will train their animals to do tricks, others will only train their animals to exhibit natural behaviour and still others will not use their animals in shows at all.

10.8 FURTHER READING AND RESOURCES

The Behavior Advisory Group (now the Behavior Scientific Advisory Group) of the American Zoo and Aquarium Association (AZA) has published ethograms for a wide range of species using a website administered by Lincoln Park Zoo at www.lpzoo.org/ethograms/about.html

The Shape of Enrichment website provides access to a wide range of articles on enrichment techniques and the Proceedings of the International Conference on Environmental Enrichment (www.enrichment.org).

Fig. 10.29 An elephant ride at Colombo Zoo, Sri Lanka. Such sights used to be common in American and European zoos before concerns about safety led to their demise.

A number of companies produce enrichment devices aimed at the zoo market. Many of these are well made and designed, but even those that are relatively simple tend to be expensive. Examples include Sanctuary supplies (www.sanctuarysupplies.com.au) and Aussiedog (www.aussiedog.com).

Bolhuis, J. J. and Giraldeau, L.-A. (eds.) (2005). *The Behaviour of Animals: Mechanisms, Function, and Evolution*. Blackwell Publishing Ltd., Oxford.

Kleiman, D. G., Thompson, K. V. and Baer, C. K. (eds.) (2010). *Wild Mammals in Captivity: Principles and Techniques for Zoo Management*, 2nd edn. University of Chicago Press, Chicago, IL.

Shepherdson, D. J., Mellen, J. D. and Hutchins, M. (eds.) (1998). *Second Nature: Environmental Enrichment for Captive Animals*. Smithsonian Institution Press, Washington DC and London.

Young, R. J. (2003). *Environmental Enrichment for Captive Animals*. Wiley-Blackwell, Oxford.

10.9 EXERCISES

1. What is stereotypic behaviour?
2. Describe the measures that keepers may take to reduce stereotypic behaviours in a range of named species.
3. What is environmental enrichment?
4. 'The best form of enrichment for a social animal is other animals of the same species.' Discuss.
5. Why do keepers train animals?
6. Discuss the role of reinforcement in animal training.

11 ANIMAL WELFARE AND VETERINARY CARE

The concept 'normal' is one of the most difficult things to define in the whole of biology but at the same time it is unfortunately as indispensable as its counterpart, the concept 'pathological'.

Konrad Lorenz

Conservation Status Profile

Scimitar-horned oryx
Oryx dammah
IUCN status: Extinct in the Wild
CITES: Appendix I
Population trend: Extinct in the wild

An Introduction to Zoo Biology and Management, First Edition. Paul A. Rees.

11.1 INTRODUCTION

Much of what we know about animal health and welfare comes from studies undertaken on farm and domestic animals. However, a great deal of information is available on certain exotic species, for example elephants, many primates, bovids, birds and fishes, many of which have been kept in captivity in large numbers for many years. Other taxa are less well understood simply because they are kept by so few zoos and are not kept as pets.

A disease may be defined as an impairment of normal physiological function that affects an animal. Diseases produce characteristic symptoms and signs. The signs of a disease are the physical evidence of its presence; the symptoms are what a patient experiences. Inevitably, vets must rely on the signs an animal exhibits in order to make a diagnosis since animals have no means of communicating how they feel.

Diseases may be divided into two types:

- non-infectious diseases.
- infectious diseases

Non-infectious diseases are diseases which are not caused by a pathogen and cannot be passed from one animal to another. They may be caused by the environment (e.g. stress), dietary deficiencies (e.g. lack of vitamin C), or they may be inherited. Many diseases may have a genetic basis but be modified by the environment.

Diabetes mellitus is a condition in which there is excessive glucose in the blood (hyperglycaemia). This is usually caused by the inability of the pancreas to produce sufficient insulin and causes thirst, weight loss, and recurrent infections. In severe cases it may result in kidney damage and blindness. Overweight animals are more likely to develop diabetes than those of normal weight. Diabetes is common in monkeys, especially when they have been provided with a high sugar diet.

Animals in zoos may suffer from a very wide range of infectious diseases. They are caused by bacteria, viruses, fungi, parasites and other infective agents. Some diseases affect a narrow range of species, while others may attack a wide variety of taxa. For example, rabies is widely transmissible between mammals and birds, but feline leukaemia virus (FeLV) affects only felids.

The signs caused by a particular disease may vary depending upon the infected species. For example, a rabid dog shows no fear of water but a human infected with rabies shows extreme hydrophobia. Some animals may die of rabies without exhibiting any signs whatsoever.

11.2 PREVENTATIVE MEDICINE

Diagnosis and treatment of animals is generally more difficult in zoo animals than in domestic species. It is, therefore, extremely important that zoos practise continuous surveillance of their animals wherever possible. If preventative medicine is practised effectively it will reduce the necessity for clinical interventions. It should include surveillance of individual animals and monitoring of entire social groups.

Components of a preventative medicine programme should include:

- regular faecal examinations
- treatment for parasites
- health screening procedures
- vaccinations
- quarantine of new arrivals
- necropsy (autopsy) of dead animals
- a pest control programme.

When animals are anaesthetised for any reason there is an opportunity to perform routine health checks of teeth, skin and so on, and to perform simple procedures such as hoof trimming.

It is impossible, in a work of this size, to provide a comprehensive account of all of the diseases that may affect zoo animals. The following descriptions are intended only as an introduction to some common and important infections.

11.3 INFECTIOUS DISEASES

Infectious diseases are those that are caused by microorganisms and may be passed from animal to animal (sometimes from species to species).

11.3.1 Disease-causing organisms

Diseases may be caused by a number of different types of organisms which are referred to as pathogens:

- bacteria – e.g. brucellosis
- viruses – e.g. rabies
- fungi – e.g. chytridiomycosis
- protozoa – e.g. trypanosomiasis
- prions – infectious agents made primarily of protein, e.g. bovine spongiform encephalopathy (BSE).

Each disease has its own characteristic signs.

11.3.2 Signs of disease

Routine inspection of animals by keepers is essential for the maintenance of good health. Keepers should be aware of any obvious changes in their animals such as:

- Smell – Does the breath smell, indicating poisoning, the presence of tooth decay or an abscess?
- Temperature – Is the body temperature abnormally high, indicating the presence of infection?
- Gait – Is gait abnormal, suggesting an injury or disease of the nervous system?
- Posture – Is the animal maintaining an unusual posture?
- Faeces – Does the animal have diarrhoea? Are the faeces an unusual colour?
- Urine – Does it have an abnormal smell or colour?
- Behaviour – Is the animal inactive or overactive? Is it engaging in repetitive behaviours?
- Feeding – Has there been a change in appetite?
- Skin condition – Does the skin smell, indicating the presence of a poison? Does the skin look abnormal? Are lesions or parasites present?
- Eyes – Are the eyes clear or is conjunctivitis or a discharge evident? Does the animal have impaired vision?

11.3.3 Body temperature

A rise in body temperature above the normal range accompanies almost all cases of acute infectious disease. Normal body temperature varies between taxa (Table 11.1). Homiotherms (mammals and birds) control their own body temperature physiologically and tolerate a relatively narrow range of temperature, whereas the body temperature of poikilotherms (reptiles, amphibians, fishes and invertebrates) depends upon on the ambient temperature.

The normal course of a disease involves a steady rise in temperature, followed by a steady fall. If the temperature subsequently increases again, this suggests a relapse. A sudden rise and fall in temperature is abnormal. A fluctuating temperature which shows little tendency to return to normal usually indicates an active focus of disease such as an abscess.

A fall in temperature may be the result of starvation, collapse, coma or a substantial loss of blood. It also occurs in some forms of kidney disease.

Table 11.1 Normal average body temperatures in selected homiothermic animals.

Taxon	Temperature °C	Temperature °F
Elephant (Elephantidae)	36.4	97.5
Grey kangaroo (*Macropus major*)	36.4	97.5
Koala (*Phascolarctos cinereus*)	36.4	97.5
Chimpanzee (*Pan troglodytes*)	37.2	99.0
Camels (Camelidae)	37.5	99.5
Serotine bat (*Eptesicus serotinus*)	38.1	100.6
Black bear (*Ursus americanus*)	38.3	100.9
Bison (*Bison bison*)	39.0	102.2
Small birds	42.5	108.5

Adapted from data in Burton (1970) and Boden (2007).

11.3.4 Some important infectious diseases

Species kept in zoos may suffer from a very wide range of diseases. Some of these are restricted to a narrow range of species, while others may affect many different taxa including man. The examples given here are intended only as an illustration of the diversity of types of disease encountered in zoos and are not intended to be comprehensive. In addition, some of the diseases which may specifically affect birds and monkeys are listed in Tables 11.2 and 11.3 respectively.

11.3.4.1 Bacterial diseases

Anthrax

Anthrax is caused by the bacterium *Bacillus anthracis*. It is found in mammals, especially herbivores, and is usually fatal. The bacterium may form highly resistant spores which may live in the soil for 10 years or more and still be capable of infection. Spores have been found in bone meal, wool, hides, feeds and blood fertilisers,

Table 11.2 Some common diseases of birds.

Ascaridiasis	Bumble-foot	Candidiasis
Capillariasis	Coccidiosis	Crop impaction
Dry gangrene of the feet	Fatty tumours	Feather cysts
Giardiasis	Hepatitis	Osteomalachia
Pacheco's disease	Papilloma (warts)	Pneumonia
Pox	Prolapse of the cloaca	Pseudotuberculosis
Psittacosis	Rickets	Trichomoniasis

and they may be carried to the soil surface by earthworms. Infection may be spread by ingesting spores in food or water, through a cut or by inhalation.

Three forms of anthrax are recognised. In peracute (very acute) cases animals may be found dead without having shown any signs. Acute cases exhibit a raised temperature, rapid pulse, 'blood-shot' eyes, and cold feet and ears, followed by prostration, unconsciousness and death. In subacute (moderately acute) cases the animal may linger for up to 48 hours with a very high temperature and laboured respiration. Swellings may occur in the neck and lower chest. A 'carbuncle' may form at the site of infection if infected through the skin. Early administration of antibiotics may be effective.

Tuberculosis (TB)
This is a contagious disease caused by the bacterium *Mycobacterium tuberculosis*. It affects a wide range of mammals (including humans), birds, reptiles and fishes and is characterised by the formation of nodules or tubercles in almost any tissue or organ. Infection may be through the respiratory system, digestive tract, a wound, contaminated feed, infected dung or by sexual contact. Treatment in domesticated animals is generally not attempted, but in zoo animals TB is sometimes treated with para-aminosalicylic acid. Control measures include good hygiene, good ventilation and good feeding.

Brucellosis
Brucellosis is caused by infection with bacteria from the genus *Brucella*. Signs often take the form of undulating fluctuation of temperature ('undulant fever'). It may also cause abortion, arthritis, infertility and a range of other signs. Brucellosis may affect a wide range of animals including goats, sheep, cattle, horses, dogs, foxes, deer, poultry, cetaceans and humans. The harbour porpoise may carry *B. maris*; hares *B. suis*; and *B. abortus* has been found in deer, foxes, waterbuck (*Kobus ellipsiprymnus*) and rodents. Care must be taken in handling and disposing of aborted foetuses, foetal membranes and discharges.

Actinomycosis (lumpy jaw)
This occurs in many taxa including dogs, pigs, birds, reptiles and humans. It is caused by an anaerobic bacterium, *Actinomyces bovis*, and probably only becomes pathogenic by invading tissues through a wound. It commonly occurs when the permanent cheek teeth are erupting. Typically, lesions occur on the cheeks, pharynx and the jaws. Swelling in bone and other tissue may cause interference with mastication, swallowing or breathing depending on the location of the lesion. Antibiotics may be an effective treatment.

Leptospirosis
This is caused by infection with bacteria from the genus *Leptospira* which are found in surface water. It commonly occurs in cattle, horses, pigs, sheep, dogs and humans. It has also been found in wild mammals, including mice, hedgehogs, voles and shrews. Spread may be partly via contamination of pasture with the urine of infected animals. Leptospires can be inhaled and can penetrate intact mucous membranes and abraded skin. Signs may include generalised illness, jaundice, kidney failure, fever, abortion and death. Treatment is by antibiotics (especially streptomycin) and vaccines are available. *Leptospira icterohaemorrhagiae* causes jaundice in dogs and Weil's disease in humans.

Psittacosis (parrot fever)
Psittacosis is caused by the bacterium *Chlamydia psittaci* and results in severe respiratory illness in members of virtually all bird species, especially those of the parrot family (Psittaciformes), and in humans. Active infection is often triggered by stress. Infected birds may exhibit listlessness, diarrhoea, conjunctivitis, sinusitis and respiratory signs. The condition may be fatal. Treatment is with tetracycline or doxycycline in the feed.

Bumble-foot
This is a condition of the feet of birds in which an abscess forms in the soft tissue between the toes. It may be caused by the penetration of a sharp object and the resulting abscess may cause lameness. The abscess usually contains *Staphylococcus* but other micro-organisms may be involved, including *Brucella abortus*. Treatment involves opening the abscess and evacuating the pus. Lack of vitamin A may make birds more susceptible to infection.

11.3.4.2 Viral diseases

Rabies
This is an inoculable contagious disease caused by a Lyssavirus. It affects virtually all mammals and occasionally occurs in birds. It is almost always fatal in humans. It is transmitted by bites or scratches and there is also a risk of infection from contamination of wounds or eyes by the saliva. Rabies causes derangement of the nervous system, a change in temperament and, eventually, paralysis. Signs may occur as early as the ninth day after being bitten but may also only

appear after several months. The principal wild animal vectors are foxes, wolves, jackals, coyotes, badgers, martens, skunks, mongooses and bats. Foxes are highly susceptible to infection and their vaccination has been very successful at controlling rabies in Western Europe. Rabies is a notifiable disease in most countries. In 1965 a leopard (*Panthera pardus*) cub imported into Scotland from Nepal was found to have rabies. It was originally held in quarantine at Edinburgh Zoo but died after being transferred to the Royal (Dick) Veterinary College in Edinburgh (Sharp and McDonald, 1967).

Rift Valley fever (enzootic hepatitis)
This disease is caused by a bunyavirus that is transmitted by mosquitoes. It causes necrosis of liver cells and abortion. The disease affects cattle, sheep, horses, donkeys, goats, buffaloes, camels and humans. It occurs mostly in Africa, but there is concern that it may spread through the Mediterranean countries and the Middle East. A live vaccine is available.

Foot and mouth disease (FMD)
Foot and mouth disease is caused by an aphthovirus. It is highly contagious and can affect all cloven-hooved species. Infected animals have small fluid-filled blisters (vesicles) in the mouth and on the feet. Females may also have vesicles on the skin of the udder or teat. The virus is present in the vesicles and the fluid which is released when they burst. When there are lesions in the mouth the virus spreads via saliva. It is also spread in faeces, urine and from lesions in the feet, and it is excreted in milk. Foot and mouth disease may be spread by wind, watercourses, people, vehicles and migratory birds. The disease is transmissible to humans but infection is usually mild. Vaccination of livestock is practised in some countries but many control the disease by slaughtering. Outbreaks of FMD in Britain in recent years have resulted in the temporary closure of deer parks and zoos.

Avian influenza (fowl plague)
Avian influenza is caused by the virus *Myxovirus influenzae*. It mainly affects domesticated fowl but also occurs in ducks, geese, turkeys and most common wild birds. Affected birds may die suddenly. Signs include an elevated temperature, drooping wings and tail, fast and laboured respiration, and lack of movement. The bird may tuck its head under a wing while squatting on its breast. Oedema of the head and neck are common. Vaccines are available. Severe acute respiratory syndrome (SARS) is a form of avian influenza that is transmissible to humans and may be fatal.

Newcastle disease
Newcastle disease is a paramyxovirus that affects birds and may be contracted by humans. It may be spread by the wind. The disease may cause a reduction in egg production and soft-shelled eggs. Infected birds may exhibit respiratory signs (breathing difficulties) or nervous signs (e.g. paralysis of wings or legs), but rarely both. In mild cases the only sign may be black diarrhoea. Live and inactivated vaccines are available.

West Nile virus
This causes an infection mainly in wild birds, especially corvids. Infected birds may die and literally fall out of the sky. The virus is related to yellow fever and Japanese encephalitis viruses, and is transmitted by mosquitoes. West Nile virus can infect primates, including humans.

11.3.4.3 Disease caused by prions

Bovine spongiform encephalopathy (BSE)
This is caused by a prion (a self-replicating protein) which causes spaces to develop in the brain tissue. BSE was first recognised in 1986. The disease has been called 'mad cow disease' as it affects the nervous system in cattle causing them to become hypersensitive to noise, frightened and aggressive. The head is lowered and they exhibit an abnormal gait with hind limb swaying. At rest, muscle twitching can be seen. BSE has been transmitted to cattle because BSE-infected cow products were used to make meat and bone meal. Meat and bone meal is now banned from all animal feeds. BSE has been recorded in several antelope species in zoos. There may be a link with feline spongiform encephalopathy as some big cats in zoos developed this after eating bovine heads before BSE was identified. The disease has been transmitted to monkeys experimentally.

11.3.4.4 Fungal diseases

Ringworm
Ringworm is a contagious disease caused by fungi (e.g. *Trichophyton* spp., *Microsporum* spp., *Oidmella* spp.) which live on the surface of the skin or in the hairs of infected areas. It appears as patches of raised, dry, crusty skin, often more or less circular in form, where the hairs have fallen out, and scales and scabs have formed. It can affect a number of taxa of mammals and birds. Treatment is by oral administration of griseofulvin or topical application of, for example, natamycin or enilconazole.

Chytridiomycosis (chytrid fungus)
Since 1999, declines in amphibian populations around the world have been linked to the emergence of a new species of fungus called *Batrachochytrium dendrobatidis*. This organism, a chytrid fungus, was first discovered in a museum specimen of the frog *Xenopus laevis* from 1938 collected in South Africa. It is thought to have spread around the world when international trade in this species began in the 1930s (Weldon et al., 2004). The fungus causes a disease called chytridiomycosis and is often fatal. It spreads through water and through contact between amphibians. Mortalities have been reported from many zoos including facilities in Japan, Australia, the USA and Europe (including the UK). The most cost-effective treatment is the application of itraconazole administered as a bath. Strict quarantine precautions must be taken to avoid transferring the disease between collections or to a collection from the wild. Some animals can carry the disease without showing any signs, so strict testing for infection with chytrid fungus is essential.

11.3.4.5 Parasitic diseases

Protozoan diseases

Trypanosomiasis Trypanosomiasis is the name given to a group of diseases caused by flagellated protozoans of the genus *Trypanosoma* which are found in the bloodstream (Fig. 11.1). One of these diseases, African trypanosomiasis, is transmitted by the tsetse fly (*Glossina*). The disease is usually chronic but acute cases occur and mortality rates may be high. Signs include intermittent fever, anaemia and loss of condition. Lymph nodes are often enlarged. A chancre (hard swelling) occurs at the site of the insect bite and is the first sign of infection. Drugs are used for both prophylaxis and treatment. American trypanosomiasis (Chagas disease) occurs in South and Central America and is transmitted to animals and people by blood-sucking triatomid bugs. Other vectors include rats, mice, foxes, ferrets and vampire bats.

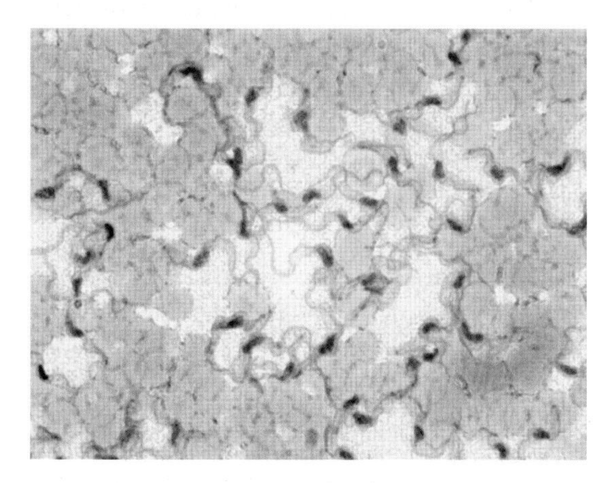

Fig. 11.1 Blood smear showing infection with *Trypanosoma brucei rhodesiense* which causes sleeping sickness.

White spot
This is a parasitic disease of fish – particularly members of the carp family – in which white cysts occur all over the body. It is caused by the protozoan *Ichthyophthirius multifiliis*. The organism lives at the bottom of ponds from where it releases the infective stage into the water. Treatment is by application of zinc-free malachite green to the water.

Disease caused by roundworms and platyhelminths

Fluke disease This is infestation with *Fasciola hepatica*, the common liver fluke of sheep. It is found in most herbivorous animals including pigs, goats, cattle, horses, rabbits, hares, beavers, kangaroos, elephants and humans. It is generally found in the bile ducts of the liver, but also occurs in other organs, causing anaemia and hepatitis. The intermediate hosts are various species of snails (*Limnaea* spp.). Cercariae may be ingested with water or when encysted on grass. Control may involve the use of anthelmintics on infected animals and land drainage to control snails. Chemicals can be used to kill snails. In Zambia ducks have been used to clear snails from flooded farmland (Boden, 2007).

Schistosomiasis (bilharziosis)
This is a disease caused by infestation with a platyhelminth from the genus *Schistosoma*. It generally lives in the portal and mesenteric veins and infects a wide range of animals including cattle, sheep, camels, water buffalo, horses, donkeys, dogs and humans. Infestation may be fatal. Transmission is via water and the intermediate hosts are snails. Molluscicides such as copper sulphate may be used to treat pasture and drugs such as praziquantel may be effective in treating infected animals.

Hydatid disease
This disease is caused by a small tapeworm, *Echinococcus granulosus*, in its cystic larval stage. The usual hosts are the dog and fox, but it can affect cattle, sheep, horses, wallabies and other species, including

humans. Eggs released in the faeces of infected individuals are eaten by grazing animals. Infection may also occur by drinking contaminated water or by exposure to wind-blown eggs. Swallowed eggs hatch in the intestines and migrate to the liver in the blood. Some remain there, forming hydatid cysts, while others may form cysts in the lungs, spleen, kidney, bone marrow cavity or the brain. Routine worming of animals is essential for the control of *Echinococcus*.

Toxocariasis
This is an infection caused by roundworms from the genus *Toxocara*. It can occur in dogs, foxes, cattle and humans. Transmission is through resistant eggs in faeces which may survive in soil for long periods (Fig. 11.2). Infection may result in impaction of the bowel and kidney damage. Infection in humans can cause tumours to develop in the lung, liver, eyes and brain. Several anthelmintics are effective against the adult worms.

Other common diseases caused by parasites

Toxoplasmosis Toxoplasmosis is a disease of most warm-blooded animals, including humans. It is caused by a coccidian parasite, *Toxoplasma gondii*. Cats are a particularly important vector. Cats and other carnivores may become infected by ingesting cystozoites within cysts in the muscles of their prey, or from oocytes present in feline faeces. Infection may cause a variety of signs, depending upon the species, including abortion, perinatal mortality, coughing, distressed breathing, diarrhoea

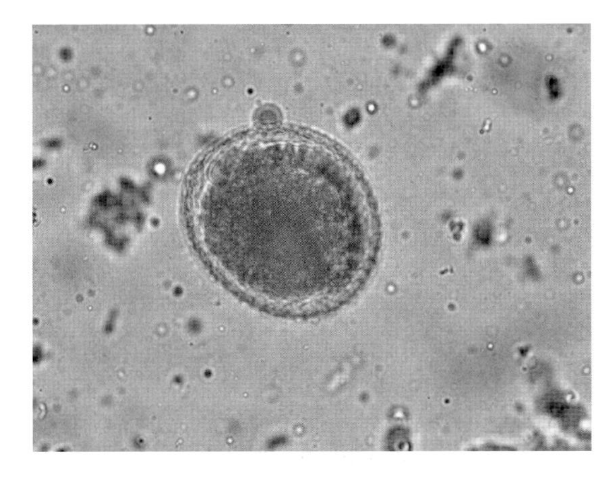

Fig. 11.2 *Toxocara* sp. ovum. This parasite causes toxocariasis in a number of species and is transmitted via microscopically small eggs in faeces.

and encephalitis. Pregnant women should avoid contact with infected faeces and other sources of infection.

Warbles
Warbles are swellings about the size of marbles on the backs of animals, caused by the larvae of various species of warble flies (*Hypoderma* spp.). They are found in cattle, horses, goats, reindeer, deer and other species. The condition commonly occurs in young animals, causing a loss of condition and sometimes even death. If a number of larvae in warbles are crushed death may be caused by anaphylactic shock. Some larvae may migrate through the body and damage the spinal cord. Larvae may be killed with a systemic insecticide.

Mange
Mange is a contagious skin disease caused by mites. The mites lay their eggs in the skin and the resulting larvae cause intense irritation. Attempts by the infected animal to relieve the discomfort cause damage to the skin. Different species of mite cause different types of mange: sarcoptic (scabies), psoroptic, chorioptic and demodectic. Treatment may involve the use of ivermectin, amitraz or doramectin. Treatment may be difficult where mites have penetrated deep within the skin.

11.3.5 Zoonoses

A zoonosis is a disease that is communicable between animals and humans. Such diseases may be transmitted by bacteria, viruses or other microbes, or by parasites. They include rabies, anthrax and other potentially fatal infections. Zoo procedures and facilities should be designed to minimise the risk of animals passing diseases to staff and visitors, and people passing diseases to animals.

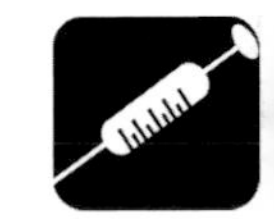

11.3.5.1 Primates and zoonoses

Primates suffer from a wide range of infectious diseases (Table 11.3), some of which are zoonoses. For example,

Table 11.3 Some infectious diseases of monkeys.

Dysentery (caused by *Shigella*)	Herpes simian B virus	Infectious hepatitis
Kyasanur fever	Leptospirosis	Malaria
Marburg disease	Measles	Monkey pox
Phycomycosis	Pneumonia	Rabies
Simian foamy virus	Simian haemorrhagic fever	Simian sarcoma virus
Toxoplasmosis	Tuberculosis	Yellow fever

simian foamy virus can be transmitted from apes to humans, measles is readily transmitted from humans to New World monkeys, and night monkeys may contract *Herpes simplex virus* (Weller, 1994). Staff should not work with night monkeys if they have cold sores and staff with family members who have measles should not work with cebids. Anyone who has been vaccinated against measles should not have contact with cebids for at least two weeks after vaccination (ILAR, 1998).

Primates in modern zoos are often exhibited behind glass or separated from the public by moats and other substantial physical structures to avoid contact. During veterinary examinations and handling staff must take particular care to avoid infection (in either direction) by wearing suitable protective clothing, face masks and other equipment.

Surveillance of zoo animals is essential if precautions are to be taken to prevent the spread of disease. In 2007 the first confirmed outbreak of callitrichid hepatitis/lymphocytic choriomeningitis virus (LCMV) was reported from a zoo in the UK (Masters *et al.*, 2007).

The Great Ape Health Monitoring Unit (GAHMU)
The GAHMU is a network of researchers from different disciplines who are concerned about disease in great apes. Among other things, they are studying emerging disease in great apes and they have created a Great Apes Emergency Task Force made up of experienced vets to deal with outbreaks of disease in the wild.

11.4 DENTAL PROBLEMS

Veterinarians have to deal with a wide range of dental problems in zoo animals. Primates often suffer from tooth decay as a result of eating too many sugary fruits. Dietary changes may alleviate this problem. Periodontal disease is the most prevalent pathology of captive mammals. This is disease that affects one or more of the periodontal tissues: the alveolar bone, periodontal ligament, cementum and gingivae (gums).

Many species can be trained to open their mouths on command so that keepers and vets may examine their teeth and oral cavity. It is sometimes, but not always, possible to detect the presence of an oral abscess from the odour it produces.

Some species, such as felids, may damage their teeth chewing fencing and other containment structures. Female Asian elephants (*Elephas maximus*) may possess small 'tushes' instead of tusks. They often rub these against doors and other structures and may damage them or even snap them off. If they fracture longitudinally infection may reach the pulp, the living part of the tooth. Some vets routinely cut off most of the tush as a preventative measure.

Specialist expertise is sometimes required to deal with dental problems. Harley Street dentist Peter Kertesz has performed dental operations on a wide range of species from elephants and snow leopards (*Uncia uncia*) to gorillas and pandas. He founded *Zoodent International* in 1985. Kertesz is dental consultant to the Zoological Society of London and the International Zoo Veterinary Group and is the author of *A Colour Atlas of Veterinary Dentistry and Oral Surgery* (Kertesz, 1993).

11.5 CONGENITAL PROBLEMS

Some species suffer from significant congenital problems that may affect their survival in zoos, or if they are ultimately released to the wild. For example, Amur leopards (*Panthera pardus orientalis*) exhibit heart murmurs which may be indicative of a genetic disorder and could compromise the recovery of populations in the wild.

Multiple ocular coloboma (MOC) is a congenital eye malformation that occurs in snow leopards and some other species, including humans. The malformation affects the upper eye lid, retina and optic nerve, but the cause is not fully understood (Barnett and Lewis, 2002). There may be a genetic link or the condition may arise in offspring following a nutritional deficiency or other problems during pregnancy.

Woolly monkeys (*Lagothrix* spp.) show a high frequency of elevated blood pressure, and veterinary procedures should take this into account (ILAR, 1998).

11.6 PEST CONTROL

Pest control is essential in zoos because many pest species are capable of transmitting disease to zoo animals. A successful pest control programme should be continuous and requires a concerted effort by zoo staff. It is essential to minimise the number of places where pest species may be harboured within a zoo and to implement a variety of mechanical control methods (e.g. traps) and chemical controls (poisons) (Fig. 11.3). Care must be taken in the choice of control methods, and the choice and storage of poisons, to reduce the risk of secondary poisoning of zoo animals.

Fig. 11.3 A rat trap. It is important to keep rats and other predators out of animal enclosures as they may kill other animals, destroy eggs and transmit parasites and diseases.

Some very common pest species are a threat to zoo animals. Cockroaches act as intermediate hosts for gut parasites of primates and birds, and rodents can transmit *Francisella tularensis*, *Listeria* spp., *Salmonella* spp. and *Leptospira* spp.

Predators such as domestic cats and dogs and wild carnivores such as foxes can cause significant damage to zoo collections through predation and by the transmission of diseases such as rabies, parvovirus and canine distemper. It is important that perimeter fencing is designed to exclude these animals. It may also be necessary to enclose completely aviaries to exclude pigeons, starlings and other birds that may carry avian diseases. In addition, wild birds may consume contaminated food and transmit disease organisms via their faeces.

11.7 POISONING

Poisons may enter the body by being swallowed or inhaled, through a wound or sometimes through broken skin. A substance may act as a poison for one species and yet have no discernible effect on another species. This generally depends in part on whether or not the species possesses an enzyme capable of detoxifying the poison.

Often substances which are essential in the diet may act as poisons if ingested in large quantities. Excessive amounts of some vitamins may be fatal. Care must be taken when feeding vitamin-rich foods to animals.

Apparently innocuous substances may be hazardous for some species. Cattle calves have been fatally poisoned in the UK as a result of feeding on waste chocolate. Heart failure may have been caused by the theobromine content (Boden, 2007).

Care must be taken in sourcing meat for carnivores. Horse meat containing sodium pentobarbitone has been unknowingly fed to tigers resulting in a sedative effect and sometimes death.

11.7.1 Poisoning by plants

Many plants that may be present in zoos are toxic to animals, including rhododendrons, yew, lupins, laburnum and laurel (Table 11.4). For this reason, garden trimmings should never be fed to animals. Care must be taken to ensure that paddock animals do not have access to poisonous plants in pasture and that such plants are not present in hay or silage.

Ragwort (*Senecio jacobaea*) contains pyrrolizidine alkaloids which cause liver damage, and is poisonous to a wide range of animals including horses, sheep, cattle, deer and hares. Bracken (*Pteridium aquilinum*) contains thiaminase, an enzyme which causes a thiamine deficiency in some species. It also contains carcinogens and chemicals which may depress bone marrow function. Poisoning by ragwort or bracken may be fatal. Common charlock, also known as wild mustard (*Brassica sinapis*), may be dangerous to livestock after the seeds have formed pods if eaten in large quantities.

11.7.2 Poisoning by chemicals

Poisoning may result from the careless use of paints, weed killers, and insecticides. Even small quantities of lead paint may be toxic to animals. Birds may be killed by inhaling poisonous vapours. Birds have died after being taken into a newly painted room and also after inhaling polytetrafluorethylene given off from the overheating of a non-stick pan. In separate incidents, birds and a dog died from inhaling acrolein given

Table 11.4 Selected plants which have parts that are poisonous to some animals.

Common name	Scientific name	Poison parts
Alder buckthorn	*Fragula alnus*	Berries
Cherry laurel	*Prunus laurocerasus*	Leaves
Privet	*Ligustrum vulgare*	Berries
Yew	*Taxus baccata*	Foliage and wood
Daffodil	*Narcissus* spp.	Bulbs and leaves
Foxglove	*Digitalis purpurea*	All parts
Honeysuckle	*Lonicera periclymenum*	Berries
Wood anemone	*Anemone nemorosa*	All parts
Potato	*Solanum tuberosum*	Tubers (when green)
Tulip	*Tulipa* spp.	Bulbs and foliage
Rhubarb	*Rheum* spp.	Foliage
Black nightshade	*Solanum nigrum*	All parts
Deadly nightshade	*Atropa bella-donna*	All parts
Bracken	*Pteridium aquilinum*	Foliage
Male-fern	*Dryopteris filix-mas*	Foliage

Source: Blackpool Zoo.

off by overheated cooking fat (Boden, 2007). Birds may be poisoned by the zinc in galvanised wire fencing and by lead in paint and some plastics. Some wood preservatives can cause hyperkeratosis (thickening of the skin) in some species.

11.8 MEASURING WELFARE

Most of the research on animal welfare has been conducted on animals kept on intensive farms and on laboratory animals kept in cages. Concern about the welfare of animals kept in these conditions gave rise to philosophers' interests in animal rights and the ethics of keeping animals in captivity (see Chapter 6). More recently the attention of both philosophers and scientists has turned to animals kept in zoos.

11.8.1 Measures used to provide evidence of welfare

Animal welfare is concerned with the 'quality of life' of an individual. A number of different variables may be used to assess animal welfare (Fraser and Weary, 2005). These may be divided into three categories:

- Biological functions
- Affective states
- Natural living.

Biological functions include variables that would indicate that the animal is functioning normally from a physiological point of view. Indicators of poor welfare would be:

- increased levels of stress hormones
- a reduction in the competence of the immune system
- increased incidence of disease and injury.

Indicators of good welfare would be:

- a high survival rate
- a high growth rate
- high reproductive success.

The affective states of an animal include indicators of pain and distress. Indicators of poor welfare would be:

- behavioural signs of fear, pain, frustration, etc.
- physiological changes considered to reflect fear, pain, etc.
- behavioural signs of aversion or learned avoidance.

Indicators of good welfare would be:

- behavioural signs of contentment or comfort
- performance of behaviours thought to be pleasurable (e.g. play)
- behavioural signs of approach/preference.

Finally, satisfactory welfare may require that animals can live relatively natural lives, indicated by the performance of natural behaviours. Good welfare would be indicated by the performance of normal behaviours. Poor welfare would be indicated by:

- behavioural or physiological indicators of thwarted natural behaviour
- the performance of abnormal behaviour.

Using different measures of welfare may lead to similar conclusions. Elephants dust bathe more on hot days than on cold days (Rees, 2002b). Allowing elephants access to soil or sand may be good for their welfare on hot days because:

- It may help to protect them from stress caused by the heat (a biological functioning criterion).
- They could otherwise suffer from the heat (an affective state criterion).
- Dusting is a natural behaviour (a natural living criterion).

However, sometimes the three approaches to assessing welfare produce conflicting conclusions. An elephant may exhibit a great deal of stereotypic behaviour and yet be a very successful breeder.

In many cases there is no correct way of weighing the different indicators of welfare status and the scientific assessment of welfare must include a conceptual framework where values play an important role. In some cases zoos must decide whether or not to exhibit an animal that appears to be in poor condition. The chimpanzee (*Pan troglodytes*) in Fig. 11.4 has a large fatty growth which vets have decided not to remove, even though its presence gives the animal an unusual appearance. Instead the zoo displays signs explaining that the lump does not cause the animal any significant problems.

Increased awareness of animal welfare in zoos has led to improvements in non-invasive and non-contact methods of monitoring physiological functions. For example, Suzuki *et al.* (2009) have described a long-term respiration monitoring system which used microwave radar to provide information about respiratory movements in a hibernating black bear (*Ursus thibetanus japonicus*) at Tokyo Zoo.

Fig. 11.4 A female chimpanzee (*Pan troglodytes*) with a benign fatty growth. Zoo authorities sometimes have to make difficult decisions about whether or not to operate on sick animals. In this case the zoo opted not to intervene. Instead it displays a sign explaining the chimp's condition to visitors.

11.8.2 Communication and welfare

Some types of communication can be used to assess welfare. For example, the squealing of animals during castration is likely to be an honest signal. Failure to recognise the importance of communication behaviour may adversely affect welfare in captive animals. Many rodent species mark their cages with urine and faeces. If these marks are removed by excessive cleaning females may become so distressed that they may cannibalise their litters. Infant rodents emit ultrasonic calls when distressed. These are similar to sound emissions made by video monitors and other equipment, and this noise may adversely affect the welfare of adult rodents and their young (McGregor, 2005).

11.9 PSYCHOLOGICAL STRESS

Threatening situations occur in the wild and in captivity and evoke similar physiological responses. An animal that cannot cope with this may suffer stress. Animals under chronic stress may become extremely ill and die. Chronic stress can cause animals to develop gastric ulcers and tumours in the pituitary gland. Animals have not evolved adaptive mechanisms to deal with chronic conflicts and in the wild they would normally have the opportunity to escape. The containment and management of zoo animals must take this into account.

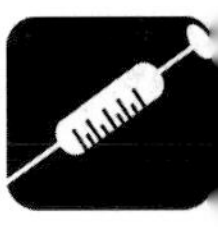

Fear is a very strong stressor. Psychological stress is fear stress and may be caused by:

- restraint
- handling
- transportation
- contact with people
- exposure to novel situations or objects.

The amount of stress experienced by an animal in any particular situation will depend upon a complex interaction between:

- previous experience
- genetic factors.

A wild rat which is unable to escape from the territory of a dominant resident male may die after several hours of intermittent attacks even if it is not wounded (Barnett, 1964).

An animal that has been trained and habituated to a particular method of restraint may have baseline cortisol levels and be behaviourally calm. In contrast, an individual that has not been trained or habituated may exhibit raised cortisol levels – an indicator of stress – when similarly restrained (Grandin, 1997).

Mental stress, anxiety and frustration can adversely affect the health of animals. It can increase susceptibility to disease and infection by parasites, inhibit reproduction, and reduce milk production. In extreme cases stress can result in death.

There is no generally accepted definition of the term 'stress.' It may refer to a physiological response (e.g. the response to cold or heat). It may also refer to a behavioural response (e.g. the social response to overcrowding, or the causes of pathology). The physiological response is called 'general adaptation syndrome' (Selye, 1936). This has three stages:

1 Flight or fight response – An acute activation of the sympathetic nervous system and the adrenal medulla occurs resulting in the secretion of catecholamines.

2 Resistance phase – Activation of the neuroendocrine system occurs: the HPA axis (hypothalamic-pituitary-adrenal). Adrenocorticotrophic hormone (ACTH) is secreted by the pituitary gland in the brain. This stimulates the release of a number of glucocorticoid hormones from the adrenal cortex, especially cortisol. These stimulate the conversion of amino acids into glucose to provide energy. Other pituitary hormones may be released that inhibit growth and suppress reproduction (e.g. growth hormone, prolactin, thyroid-stimulating hormone, and gonadotrophins). The secretion of sex steroids is decreased.

3 Final stage – If adaptation to the stressor does not occur or it is not removed, gastric ulceration may occur and there may be a lowering of immunological function.

Siberian tigers (*Panthera tigris altaica*) at Bucharest Zoo developed gastroenteritis due to a failure to adapt to new quarters and as a result of persistent building noise (Cociu *et al.*, 1974). However, giant pandas (*Ailuropoda melanoleuca*) exposed to noise resulting from the demolition of an adjacent exhibit in the Smithsonian National Zoo did not appear to experience decreased welfare, although their activity budgets and stress levels were affected and their responses were individual-specific (Powell *et al.*, 2006).

If emotional arousal is avoided when an animal is exposed to an environmental change that might cause stress the HPA axis is not activated. In rhesus macaques, if the temperature of the room is suddenly increased by 15°C there will be an increase in circulating corticosteroid levels. However, if the temperature is raised 1°C per hour to 15°C above normal there is no such effect (Mason, 1971). Such experiments show that the hormonal changes that occur in stressful situations depend upon subjective emotional experience. One of the most obvious chronic stressors for a zoo animal is the inability to escape or avoid fearful situations. The presence of zoo visitors may act as a chronic stressor and may prevent breeding in some species. However, the quality of space may be more important than the quantity in reducing stress. The animal may only need to perceive that it can retreat to safety rather than actually withdrawing (Hediger, 1950). Uncertainty about the actions of keepers and vets may cause stress. Chronic stress may lead to depression, lethargy and even death in some animals.

Animals that have been deprived of a diversity of stimuli may respond badly in novel situations. Christian and Radcliffe (1952) reported 14 zoo animal deaths that resulted when animals that had been kept in small indoor cages were subjected to the stress of transfer to a new cage or disturbance by workers. In all cases the adrenal cortex had atrophied indicating an inability to respond to an extreme stimulation and a novel situation.

Carlstead (1996) has emphasised that the behavioural responses of wild mammals in captivity to aversive stimuli are diverse, idiosyncratic and situation-specific. An individual that is crouching in the corner of its cage may be experiencing more stress than one in the same situation that is bounding from wall to wall (Duncan and Filshie, 1980). Behaviour alone is not a good guide to the amount of stress an animal is experiencing. Young squirrel monkeys (*Saimiri sciureus*) which were separated from their mothers and placed in a novel environment exhibited increased plasma cortisol levels and behavioural signs of stress such as increased vocalisation and activity. When separated from their mothers, but placed with conspecifics in a familiar environment, they showed few signs of distress but plasma cortisol levels were still high (Levine, 1983).

Many zoos have screened-off enclosures where the animals were previously exposed to public gaze, and new exhibits have been designed to reduce the exposure of animals (Figs 7.15 and 7.27). Many older orangutan exhibits include indoor accommodation where visitors view the animals through large windows.

Some zoos have covered these windows with bamboo screens containing 'letterbox' windows to reduce the amount of disturbance caused by people.

11.9.1 Measuring stress

Stress in mammals can be measured using non-invasive methods by analysing the levels of cortisol found in serum, saliva or urine. Cortisol is a stress-response hormone that reflects the activity of the HPA axis. As stress increases cortisol levels increase. Faecal cortisol metabolites have also recently been identified as an index of stress, and Bayazit (2009) has suggested that measurement of these metabolites in intensively managed species may be useful to identify factors which affect animal well-being in captivity. In particular it may be useful in determining which factors may have an adverse effect on reproduction, for example housing and handling procedures.

11.9.2 Flight distances

The 'flight distance' of an animal is the shortest distance that it is prepared to tolerate between itself and a human before it flees. This will vary between individuals of a species. Some individuals will be tamer than others. 'Tameness' may be defined as having no flight response with respect to man (Hediger, 1950).

Older animals may be more difficult to tame than young ones because they may have had more negative experiences with people. Older wild-caught moose may never adapt to captivity and may die of heart failure as a result of stress.

In the wild flight distances vary between species and within a species, dependent upon the degree of exposure to people. In the Serengeti National Park, Nyahongo (2007) has shown that the flight distances-from a slowly driven vehicle-of impala (*Aepyceros melampus*), Thomson's gazelle (*Gazella rufifrons*), topi (*Damaliscus lunatus*), zebra (*Equus burchellii*) and wildebeest (*Connachaetes taurinus*) were typically less than 50 m in the central Serengeti but more than 150 m in the western corridor which was frequented by fewer visitors. This has implications for safari parks where visitors drive through 'reserves' in their own vehicles. Often animals may come very close to vehicles or even come into physical contact with them, so visitors need to be kept moving to prevent damage to cars and to the animals.

11.10 LONGEVITY OF ANIMALS IN ZOOS

There has been considerable discussion about the longevity of animals in zoos. Zoos often claim that animals live longer, healthier lives in zoos than they do in the wild, while some animal welfare groups claim the exact opposite. The truth is probably that the situation varies between species, and also between zoos.

A species' mortality profile in captivity is likely to differ significantly from that in the wild. Animals in captivity have the benefits of veterinary care, a lack of predators and a reliable food supply. However, for some species a number of factors may be important in adversely affecting survival rates in captivity, including:

- injuries sustained from exhibits (Leong *et al.*, 2004)
- poor adaptation to captivity or to a zoo's climate (Karstad and Sileo, 1971; Gozalo and Montoya, 1991)
- high levels of obesity in some species (Taylor and Poole, 1998)
- higher perinatal mortality caused by inbreeding (Wielebnowski, 1996)
- efficient spread of disease in the close quarters of captivity, within and between species (Ward *et al.*, 2003; de Wit, 1995).

Kohler *et al.* (2006) used the ISIS database (see Section 13.3.1) to construct life tables for 51 species, mostly mammals. In total they examined 35,229 individual animals present in zoos between the beginning of 1998 and the end of 2003. They included apes, small primates, carnivores, hoofstock, kangaroos, crocodilians, ratites and raptors. In most of the groups studied female survivorship significantly exceeded that of males above age five years. They found no significant difference between the mortality of wild-born and captive-born animals.

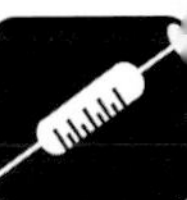

In the wild Burchell's (plains) zebra (*Equus burchellii*) has a mean expectation of life of about nine years. This reflects the high juvenile mortality rate of perhaps 50%. In captivity this species may live to the age of 40 years (Nowak, 1999). In his analysis of zoo bear populations, Kitchener (2004) noted that 32% of captive polar bears were more than 20 years old, whereas only 3% were this age in a well-studied wild population in Hudson Bay, Canada (Ramsey and Stirling, 1988). Gorillas in zoos had a maximum longevity of 33.4 years in the 1960s (Jones, 1962) but this had increased to 54 years by 30 years later (Nowak, 1991). Over the same period the longevity of Goeldi's monkeys (*Callimico goeldii*) increased from just 2.3 to 17.9 years.

However, reduced longevity is an issue for some species. A recent study of elephants living in zoos suggests that they live much shorter lives than those in the wild. After analysing data from 4500 animals Clubb *et al.* (2008) concluded that elephants in zoos have half the median life span of conspecifics in protected populations in range states. However, the survivorship curves produced in this analysis cannot take into account the effect of the improved conditions that now exist in many modern zoos and it will be many years before it will be possible to assess the effect of these improvements on survival.

Table 11.5 Some common diseases of elephants.

Multiple abscess (often on the feet)	Anthrax	Blackleg
Botulism	Elephant pox	Enzootic pneumonia
Foot and mouth disease	Influenza	Myiasis
Parasitic gastroenteritis	Pasteurellosis	Rabies
Salmonellosis	Schistosomiasis	Stephanofilarial dermatitis
Tetanus	Trypanosomiasis	Tuberculosis

11.11 OLD AGE

Zoos are increasingly having to deal with the problems of old age in their animals. These problems include:

- diseases of old age, such as arthritis
- lack of a direct contribution to breeding programmes
- long-term provision of accommodation thereby competing for space needed for animals of breeding age.

Kitchener (2004) studied the prevalence of dental and skeletal pathologies in captive bears aged at least 15 years. He identified a wide range of problems, including fused vertebrae in around 55% of individuals and broken canine teeth in over 70%.

Although old animals may be a burden to some zoos, in some species older animals may have a useful function within a social group, for example post-reproductive cow elephants may act as allomothers to the calves of younger animals.

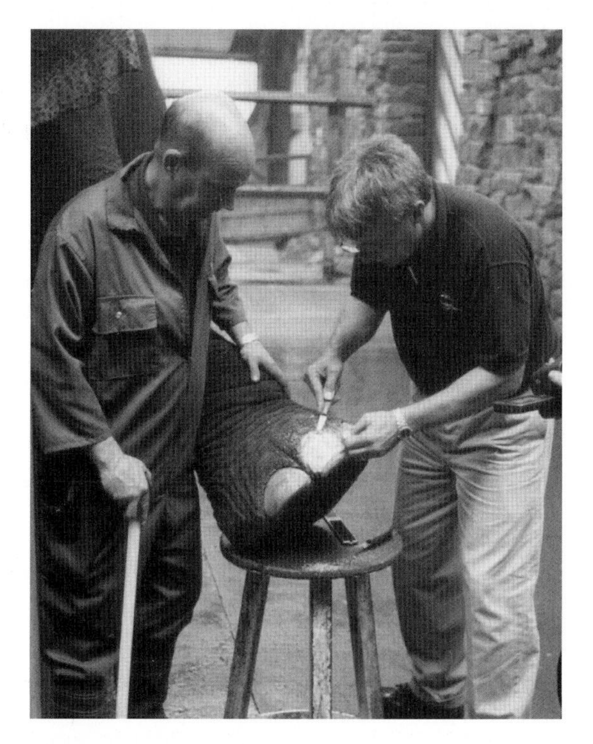

Fig. 11.5 Elephant foot care. Elephants in zoos are prone to foot problems.

11.12 CASE STUDY: DISEASES IN ELEPHANTS

Elephants suffer from a wide range of diseases (Table 11.5). In captivity, they are particularly susceptible to foot infections (Fig. 11.5). Elephants are frequently moved between zoos for breeding purposes and in order to disperse juveniles from their natal group. The risks associated with the movement of elephants between collections have increased since the appearance of new fatal herpesviruses in the zoo population (Montali *et al.*, 1998; Rickman *et al.*, 1999). One is fatal for African (*Loxodonta africana*) and the other for Asian elephants (*Elephas maximus*). The disease affects predominantly young elephants, and has been recorded in North America, Europe and Israel. Rickman *et al.* (2000) have suggested that otherwise healthy African elephants may have been the source of the herpesvirus that causes death in both species. This has implications for the management of zoo elephants, and the European Endangered Species Programme (EEP) species coordinators (Dorresteyn and Terkel, 2000) have recommended that Asian and African elephants should not be kept in mixed groups.

Herpesviruses are not the only threats to captive elephants. Tuberculosis has also emerged as a disease of concern. Mikota *et al.* (2000) isolated *Mycobacterium tuberculosis* from 18 of 539 elephants (3.3%) in North America between August 1996 and May 2000. The recent outbreak of foot and mouth disease, which resulted in the closure to the public of zoos in the UK, illustrates the potential susceptibility of zoo elephants to disease that may originate outside the zoo community. One advantage of a dispersed zoo population is that it cannot be destroyed by localised disease outbreaks. Artificial insemination may reduce the need to move animals between zoos, although even semen could potentially transmit disease.

11.13 ANIMAL HANDLING, CAPTURE AND TRANSPORTATION

Handling animals poses risks to keepers and other zoo staff, and to the animals themselves. Staff may sustain physical injuries, contract diseases or, with some species, be poisoned via bites or other physical contact. The animals may suffer injury or be subjected to considerable stress, depending upon the methods used.

11.13.1 Handling

When handling animals keepers should eliminate stress as much as possible. Restraining and washing an animal may cause stress. Following an oil spill, Magellanic penguins (*Spheniscus magellanicus*) that were rescued and washed exhibited elevated levels of corticosterone compared with those that were not oiled, or oiled but not washed (Fowler *et al.*, 1995).

Studies of farm animals have illustrated the effects of stress on physiology. Fear responses caused by rough, unskilful or inconsistent handling by people can interfere with the basic endocrine processes that control reproduction and growth (Hemsworth and Coleman, 1998). By comparing responses to 'rough' and 'gentle' handling, Rushen *et al.* (1999a) have shown that fear of people affects the milk yield, behaviour and heart rate of cows during milking.

Some animals are particularly susceptible to the stress associated with restraint and handling, for example spider monkeys (*Ateles* spp.) and woolly monkeys (*Lagothrix* spp.) (ILAR, 1998).

11.13.2 Staff safety

Whenever large or dangerous animals need to be handled there is a potential risk to the zoo staff involved (Table 11.6). Appropriate risk assessments should be undertaken beforehand to ensure that staff are aware of the risks and that these are minimised, and staff need to be supplied with appropriate equipment (Fig. 11.6). A risk assessment should include consideration of the risk of escape.

In a survey of vets working for the Ministry of Agriculture and the Institute for Research on Animal Diseases in the UK, 397 out of the 563 respondents (70.5%) reported injuries resulting from animal

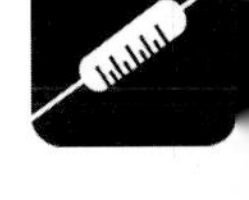

Table 11.6 Some examples of poisonous animals.

Species	Origin	Source of toxin
Poison arrow frogs (*Dendrobates* spp. and *Phyllobates* spp.)	South and Central America	Skin
Komodo dragon (*Varanus komodoensis*)	Komodo, Indonesia	Bite
Gila monster (*Heloderma suspectum*)	Mexico and SW USA	Bite
King cobra (*Ophiophagus hannah*)	India and SE Asia	Bite
Brazilian wandering spiders (*Phoneutria* spp.)	South and Central America	Bite
Stonefish (Synanceidae)	Indo-Pacific oceans	Fin spines
Puffer fishes (*Tetraodon* spp.)	Red Sea and Indo-Pacific region	Bite, skin
Australian sea wasp or box jellyfish (*Chironex fleckeri*)	Seas around Australia	Sting cells
Blue-ringed octopuses (*Hapalochlaena* spp.)	Coasts of Australia and parts of SE Asia	Bite
Cone shells (*Conus* spp.) (Gastropoda)	Tropical and subtropical seas	Harpoons
Flower sea urchin (*Toxopneustes pileolus*)	Indo-Pacific	Spines and pedicellaria (small 'pincers')
Solenodon (*Solenodon* spp.) (Rodentia)	Caribbean islands	Bite

Fig. 11.6 A keeper handling a snake with a snake hook.

handling (Boden, 2007). Handling by keepers in zoos has resulted in a number of serious injuries and fatalities particularly as a result of contact with big cats and elephants.

11.13.2.1 Elephant handling: free contact vs. protected contact

Elephants have killed more keepers than any other species. Increasingly, modern zoos are adopting a zero-handling or protected contact approach to managing their elephants to protect their staff and improve conditions for their animals. In the free contact system of management keepers work directly with elephants, entering their enclosures and making regular physical contact with them. The protected contact system requires that there must be a barrier – a protected contact wall – between the keeper and the elephant at all times.

Protected contact requires elephants to be trained using trial-and-error learning. This begins with a response that subsequently becomes attached to a novel stimulus. For example, the response required might be to stand in a particular position near a fence. The stimulus that the trainer wishes to attach to this response is a target (a striped or coloured area) at the end of a long pole held near the elephant's head. The correct behaviour – standing near the target – is rewarded by positive reinforcement: praise and a small piece of food. Using this method an elephant may be trained to present its feet, ears and other parts of its body for inspection by keepers and vets via ports in the protected contact wall (Fig. 11.7). Handling elephants using protected contact reduces stress for the elephants as they participate voluntarily and largely removes the need to chain them for washing, veterinary examination and treatment. Roocroft (2009) has emphasised the importance of zoos developing similar protocols and using similar equipment when handling elephants so that elephants and keepers can move between institutions with ease.

11.14 TRANSPORTATION

Zoo animals may need to be captured and moved between locations for a number of reasons:

- Initial transportation to a zoo following capture in the wild
- Transportation between zoos to facilitate a captive breeding programme
- Transportation between enclosures within the same zoo
- Transportation to a veterinary hospital for treatment.

Some animals are small enough or tame enough to be captured and transported without difficulty. Others may be large, dangerous, difficult to catch or particularly susceptible to the stress involved in capture and transportation. Some animals may need to be immobilised in order to facilitate transportation while others may be caught in a trap or persuaded to enter a crate.

Immobilising animals

Large animals may be immobilised using a projectile hypodermic syringe containing an anaesthetic fired from a gun, crossbow or blowpipe. The same equipment may be used to administer antibiotics and vaccines and is particularly useful when dealing with dangerous animals. Dart guns use a compressed air discharge system and are only effective up to a distance of about 36 m (40 yards). In some jurisdictions the use of dart guns may be controlled by law. In the UK their use is restricted by s.5 of the Firearms Act 1968.

Large animals may be immobilised by administering Immobilon (M99), a neuroleptanalgesic (i.e. it combines sedation with analgesia (pain relief)).

Fig. 11.7 Target training an elephant for protected contact.

It consists of etorphine hydrochloride combined with acepromazine. It acts as a respiratory depressant and slows the heart rate. The effects of Immobilon are reversed by injecting Revivon (M5050 or diprenorphine hydrochloride). Etorphine hydrochloride was first synthesised by Bentley and Hardy in 1963 and is chemically related to morphine. When given subcutaneously, it is 1000 to 80,000 times more potent than morphine as an analgesic. Immobilon can cause catatonia at very low dose levels; 5.0 mg may immobilise a rhino.

Immobilon is fatal to humans. If accidental injection occurs the antidote, Narcan (naloxone), must be administered quickly. An assistant should always be available to do this if necessary. A vet without access to the antidote died within 15 minutes after accidental self-inoculation with Immobilon. The eyes should be protected and gloves should be worn as even skin contamination has required hospitalisation (Boden, 2007).

Appropriate sites for the administration of Immobilon with a projectile syringe in the ostrich (*Struthio camelus*) are illustrated in Fig. 11.8. These birds are extremely dangerous when angry and could easily disembowel a human with a kick.

Tranquillisers may have unexpected effects on animals that appear to make a full recovery. Male bighorn sheep (*Ovis canadensis*) fight over females by head-butting and kicking. Sheep tranquillised with ketamine or xylazine lose their fighting skills and consequently slip down the dominance hierarchy (Pelletier *et al.*, 2004).

11.14.1 Trapping and crate training

Some small animals may be captured using nets. However, others may need to be caught in traps or persuaded to walk into transportation cages or crates (Fig. 11.9). Being captured in a trap may be extremely stressful for an animal. Once captured, placing a cover over the trap may calm the animal. Many mammals will respond to the closing of a trap door by running and colliding with it. They may then charge repeatedly at both ends of the trap and scrape at the base of the

Fig. 11.8 Suitable darting sites for an ostrich (*Struthio camelus*): the thigh and the lumbo-sacral area.

Fig. 11.9 California sea lion (*Zalophus californianus*) with transport cage.

door. This may result in physical damage to the animal. Four out of seven male feral cats (*Felis catus*) sustained bleeding noses when captured as part of a neutering project. None of the five females and three kittens captured sustained damage (Rees, 1982b).

Many animals will initially be extremely suspicious of traps placed in their enclosures. Most will habituate to their presence eventually and, if the door is left open, will enter the trap to obtain food left on the floor. An animal may be trained to enter a crate by allowing it to feed in the crate with the doors open (Fig. 11.10). Once an animal is accustomed to entering a crate it can be trained to adapt to crate movements during transportation by gently rocking the crate while it is inside. It may also be useful to accustom an animal to noise prior to transportation by, for example, using a radio.

11.14.2 Transport stress

Transportation can cause stress in animals. Research on farm animals has demonstrated that they will choose to avoid noise and vibration if possible and do not habituate to very aversive stimuli. Studies of cattle during transportation have shown that cortisol levels did not decrease with experience when they were subjected to repeated truck journeys during which they fell down (Fell and Shutt, 1986). Pigs held in a noisy vibrating pen – which simulated the effects of transportation – learned to press a panel switch with their snouts to obtain relief from the vibration for 30 seconds and kept the machine immobile for 70–80% of the time. The pigs switched the machine off more often when the speed of vibration was increased and when they had eaten a large meal prior to the test. Aversion to the conditions did not diminish with time (i.e. they did not habituate to the conditions) and pigs exposed only to the noise did not learn to operate the switch (Boden, 2007).

The stress of long travel may cause transit tetany in ruminants and, rarely, horses. This is mainly the result of hypocalcaemia (low blood calcium). Signs range from excitability to lack of muscular coordination and staggering as calcium levels fall. The animal sweats, breathing rate increases, gait becomes stiff, the tail is raised and spasms similar to tetanus occur. Eating and drinking may become impossible; recumbency, coma and death may follow. The animal should be kept quiet and given a mineral replacement solution.

Animals kept in poor conditions may respond very badly to transportation. Pigs reared in impoverished intensive farming conditions often die during transportation to the slaughterhouse as a result of contact with unfamiliar pigs and the stress associated with extreme changes in surroundings (Carlstead, 1996).

A number of studies have examined means of alleviating psychological and physiological stress during

Fig. 11.10 A Barbary macaque (*Macaca sylvanus*) feeding in an open trap. Baiting traps with food will help to habituate animals to them, making them easier to catch when necessary.

the transportation of zoo animals. Dembiec *et al.* (2004) simulated transport by relocating five tigers (*Panthera tigris*) in a small transfer cage. They found that even short-term transportation procedures can cause significant increases in immune-reactive (IR) cortisol concentration in tigers but that levels remained elevated longer in naïve individuals (9–12 days) than in those that had previously experienced the procedures (3–6 days). The naïve tigers also exhibited a greater intensity of behavioural indicators of stress, for example faster pacing. This study suggests that prior exposure to elements of the transport procedure may lead to a degree of habituation and therefore reduced stress.

When marine mammals are transported there is a risk of damage to internal organs, especially the lungs, if their weight is not supported in water. Dolphins are frequently transported on mattresses. Suzuki *et al.* (2008) have demonstrated that the use of high-performance mattresses – similar to those used in human nursing care to prevent pressure ulcers – provides cardio-pulmonary benefits to Indo-Pacific bottlenose dolphins (*Tursiops aduncus*). They recorded lower breathing rates, lower heart rates and higher exhaled CO_2 concentrations compared with using standard mattresses.

11.14.3 IATA Live Animal Regulations

The International Air Transport Association publishes regulations for the safe transport of live animals. It specifies minimum requirements for the safe international transport of animals and indicates the precautions that should be taken on the ground and in the air by airlines, cargo agents and persons responsible for the care of animals (IATA, 2006). These regulations are enforced by the European Union and many other countries for the import and export of live animals, and have been adopted by CITES as their official guidelines. Some husbandry manuals provide useful information on animal transportation (e.g. Jolly, 2003; Box 11.1).

11.15 QUARANTINE

The term 'quarantine' refers to a period of time that an animal is detained in isolation to prevent the spread of contagious disease. The term also refers to the place where the animal is detained. Ideally, this should be a purpose-built facility. However, very large animals such as elephants and giraffes may have to be quarantined within the main collection due to their size.

Box 11.1
Transportation requirements for giraffes (*Giraffa camelopardalis*) (based on Jolly, 2003).

Transport by road and sea

- Giraffe is secured in a transport crate and loaded onto a semi-trailer truck – usually a low-loader because of the animal's height – by crane and secured.
- Travel route should be planned in advance with a support vehicle and possibly a police escort.
- Power lines may need to be raised to allow access.
- Giraffe travelling by container ship should be accompanied by keeping or veterinary staff.
- Ample food needed for long sea journeys. Giraffe may become seasick.
- Giraffe should be quarantined during transport.

Transport by air

- Only young giraffe may be transported by air.
- Planning should begin early because giraffe calves grow quickly (7–13 cm per month in their first year).
- Maximum height restriction in aircraft is approximately 3 m. This must include the frame of the crate and the metal aircraft loading pallet. This leaves an internal crate measurement of about 2.7 m.
- The take-off and landing trajectory of the aircraft must be kept at a low angle to avoid sudden pressure changes as this may cause the giraffe to lose consciousness.

Transport crate requirements

- For air transport giraffe crates must comply with IATA Live Animal Regulations.
- The crate must be sturdy enough not to warp or twist, but must not be excessively heavy.
- The crate should have a roof or a roof frame with a canvas cover.
- The crate should be large enough for the giraffe to lay down and get up again during transport.
- The crate design should allow the provision of food and water, and medical treatment, through a number of hatches.

Stress

Stress can be reduced by:

- Crate training prior to transportation
- Avoiding excessive noise, lights and unfamiliar people
- Stabilising temperature and avoiding cold draughts by covering the crate with tarpaulins.

11.15.1 National and international responsibilities for animal disease control

Each country has its own quarantine regulations relating to the arrival of animals at ports of entry. These regulations change from time to time in order to deal with threats from specific diseases from animals from particular parts of the world. The World Organisation for Animal Health (Office International des Epizooties (OIE)) informs member countries of animal disease outbreaks and sets international standards for animal health protection. The responsibility for quarantine varies between countries. In the UK, quarantine is the responsibility of the Quarantine Section of Animal Health, an executive agency of DEFRA. In the United States, the responsibility falls to the Animal and Plant Health Inspection Service (APHIS) of the US Department of Agriculture, and in Australia it is the responsibility of Biosecurity Australia, a unit within the Biosecurity Services group of the Department of Agriculture.

11.15.2 Purpose and procedures

Animals entering a zoo must undergo quarantine. This serves two purposes:

- It prevents the spread of disease to other animals in the zoo, by allowing time for signs of disease to manifest themselves.
- It allows staff an opportunity to undertake a thorough physical examination, diagnostic testing and vaccination against disease. Testing might include:
    - radiographs
    - serological tests

- ○ haematological tests
- ○ clinical chemistries
- ○ tests for parasites.

Quarantine also provides an opportunity, where necessary, to apply a method of future identification to an animal entering a collection and to create its medical records.

Quarantine facilities should be designed to:

- allow the safe handling of animals
- allow the proper cleaning of cages, enclosures and shipping crates
- exclude vermin and potential disease vectors.

Ideally, quarantine facilities should be staffed by dedicated keepers to prevent transmission of disease to animals in the main part of the zoo. These staff should be skilled at recognising signs of disease and able to monitor food intake and faecal characteristics.

11.15.3 Notifiable diseases

Some diseases are so serious, either because they threaten human health or because they are of great economic importance, that they must, by law, be notified to the state veterinary authorities in the countries where they occur. These 'notifiable diseases' vary from country to country. The diseases that are notifiable in the UK are listed in Table 11.7.

Table 11.7 Diseases which are notifiable in the UK.

African horse sickness	African swine fever
Anthrax	Aujeszky's disease
Avian influenza (bird flu)	Bluetongue
Bovine spongiform encephalopathy	Brucellosis (*Brucella abortus*)
Brucellosis (*Brucella melitensis*)	Classical swine fever
Contagious agalactia	Contagious bovine pleuro-pneumonia
Contagious epididymitis (*Brucella ovis*)	Contagious equine metritis
Dourine	Enzootic bovine leukosis
Epizootic haemorrhagic virus disease	Epizootic lymphangitis
Equine infectious anaemia	Equine viral encephalomyelitis
European bat Lyssavirus (EBLV)	Foot and mouth disease
Glanders and Farcy	Goat pox
Lumpy skin disease	Newcastle disease
Paramyxovirus of pigeons	Pest des petits ruminants
Rabies (classical)	Rift Valley fever
Rinderpest (cattle plague)	Scrapie
Sheep pox	Swine vesicular disease
Teschen disease (porcine enterovirus encephalomyelitis)	Tuberculosis (bovine TB)
Vesicular stomatitis	Warble fly
West Nile virus	

Source: DEFRA (2010).

11.15.4 Vaccinations

Vaccination programmes vary between zoos. Hinshaw *et al.* (1996) suggest that the programme for any particular institution should consider:

- the types of species present
- the history of infectious disease problems in the species and the facility
- the potential risks and benefits of each type of vaccine
- up-to-date information on the susceptibility of various species to diseases
- the safety and efficacy of the vaccines available
- recommended vaccination intervals.

Vaccines can be divided into two general types:

- 'Live' or 'modified live' vaccines – These are attenuated strains of the infectious organism that, once injected, multiply in the body of the animal causing an immune response but not clinical disease.
- 'Killed' vaccines – These contain infectious agents that have been inactivated by heat or chemicals so that they cannot cause infection. These vaccines may not produce as strong an immune response as that caused by a live vaccine.

In the past only dead virus vaccinations have been used. However, some modified live vaccines are now considered safe to use in selected species. Care must be taken in using modified live vaccines as they can be fatal in some species.

Where rabies is endemic a zoo should consider vaccinating all mammals that are housed outdoors against this disease. Vaccination of carnivores is particularly important because of their susceptibility to a range of diseases including:

- canine distemper
- canine parvovirus
- canine infectious hepatitis
- leptospirosis
- feline panleukopenia
- feline rhinotracheitis
- feline calcivirus.

It is common to vaccinate the great apes against:

- polio
- tetanus
- measles.

They may also be vaccinated against influenza and streptococcal infections.

11.16 FURTHER READING AND RESOURCES

A list of notifiable diseases in the UK is available from the DEFRA website at www.defra.gov.uk/foodfarm/farmanimal/diseases/atoz/notifiable.htm. The DEFRA website also contains information about zoonoses (www.defra.gov.uk/foodfarm/farmanimal/diseases/atoz/zoonoses/index.htm).

The OIE website contains a more comprehensive list of animal diseases and zoonoses: www.oie.int/eng/maladies/en_alpha.htm

A Zoonosis CD-ROM is available from the American Association of Zoo Keepers.

The *International Zoo Yearbook* volume 14 was entitled *Trade and Transport of Animals*, and volume 41 covered *Animal Health and Conservation*.

The following is a list of useful textbooks:

Atkinson, C. T., Thomas, N. J. and Hunter, D. (eds.) (2008). *Parasitic Disease of Wild Birds*. Wiley-Blackwell, Oxford.

Ballard, B. M. and Cheek, R. (eds.) (2010). *Exotic Animal Medicine for the Veterinary Technician*, 2nd edn. Wiley-Blackwell, Oxford.

Fowler, M. E. (2008). *Restraint and Handling of Wild and Domestic Animals*, 3rd edn. Wiley-Blackwell, Oxford.

Fowler, M. E. and Mikota, S. K. (eds.) (2007). *Biology, Medicine and Surgery of Elephants*. Wiley-Blackwell, Oxford.

Fowler, M. E. and Miller, R. E. (eds.) (2007). *Zoo and Wild Animal Medicine: Current Therapy: 6*, 6th edn. Saunders Elsevier, St Louis, MO.

Girling, S. J. and Raiti, P. (eds.) (2004). *BSAVA Manual of Reptiles*, 2nd edn. Wiley-Blackwell, Oxford.

IATA (2006). *IATA Live Animal Regulations*, 33rd edn. International Air Transport Association.

Ladds, P. (2009). *Pathology of Australian Native Wildlife*. CSIRO Publishing, Collingwood, Victoria, Australia.

Samuel, W. M., Pybus, M. J. and Kocan, A. A. (eds.) (2001). *Parasitic Diseases of Wild Mammals*, 2nd edn. Manson Publishing Ltd., London.

Thomas, N. J., Hunter, D. B. and Atkinson, C. T. (eds.) (2007). *Infectious Diseases of Wild Birds*. Wiley-Blackwell, Oxford.

Voevodin, A. F. and Marx, P. A. (2009). *Simian Virology*. Wiley-Blackwell, Oxford.

Vogelnest, L. and Woods, R. (2008). *Medicine of Australian Mammals*. CSIRO Publishing, Collingwood, Victoria, Australia.

West, G., Heard, D. and Caulkett, N. (eds.) (2007). *Zoo Animal & Wildlife Immobilisation and Anesthesia*. Blackwell Publishing Ltd., Oxford.

11.17 EXERCISES

1. Discuss the benefits and disadvantages of the various methods of handling zoo elephants.
2. Describe how you would prepare a small antelope for transportation to another zoo in a crate.
3. How may transport stress be alleviated in animals?
4. Explain why it is important for zoos to implement a programme of preventative veterinary medicine.
5. What should zoos do with very old animals?
6. How should zoo enclosures be designed in order to minimise the risk of transfer of infectious diseases between animals and other animals, and animals and people?
7. Discuss the importance of quarantine to zoos.
8. How can stress affect zoo animals?
9. Describe the measures a zoo should take to reduce the possibility of infection by parasites and disease transmission from pests.

Part 3

Conservation

Zoos have evolved from being more or less random collections of exotic animals held in poor conditions and places of entertainment into sophisticated conservation NGOs. This has partly come about as a result of changes in legislation such as the EU Zoos Directive, but in many cases zoos themselves had already begun to make this transition. Zoos may perform a conservation function in many ways. The remaining chapters consider their roles in captive breeding (including record keeping), education, *in-situ* conservation and reintroduction. It also considers the behaviour of zoo visitors.

12 COLLECTION PLANNING AND CAPTIVE BREEDING

> Most zoos assumed that there was an unlimited supply of exhibits in the wild. No one seemed to suspect that a time might come when that supply might be in danger of exhaustion. So it was not uncommon for big zoos to send out expeditions to look for rare creatures that had seldom if ever been seen before in captivity.
>
> Sir David Attenborough

Conservation Status Profile

Humboldt penguin
Spheniscus humboldti
IUCN status: Vulnerable A2bcde, 3bcde, 4bcde; C1, 2b
CITES: Appendix I
Population trend: Decreasing

An Introduction to Zoo Biology and Management, First Edition. Paul A. Rees.

12.1 INTRODUCTION

Zoos have become largely self-sufficient in animals by developing captive breeding programmes. For many species these programmes have been extremely successful. The rise of the animal rights movement and increasing concern about the international loss of biodiversity have resulted in zoos focusing their breeding efforts on threatened and endangered species. It was, therefore, inevitable that increasing numbers of zoos would join breeding programmes for the relatively small number of rare species that were held in captivity in sufficient numbers to make these programmes viable.

Biography 12.1

Gerald Durrell (1925–95)

Gerald Durrell was a successful animal trader, author and television presenter who established his own zoo on Jersey in the Channel Islands. He worked for a short time as a keeper at Whipsnade Zoo and then began making animal collecting trips to West Africa. He opened the Jersey Zoological Park in 1959. The zoo was dedicated to conservation from the outset and specialised in keeping, breeding and reintroducing endangered species. Durrell was extremely influential in persuading zoos to refocus their efforts on endangered species and to become involved in field conservation. The Durrell Wildlife Conservation Trust now works to protect critically endangered animal and plant species in 16 countries.

12.1.1 Where do zoo animals come from?

Modern zoos acquire their animals in one of three ways:

- By birth in the zoo itself
- By loan purchase or donation from a third party, often another zoo
- Exceptionally from the wild

In the past zoos were free to mount expeditions to remote parts of the world and return with almost any species they chose. The writings of Gerald Durrell (1976) and Sir David Attenborough (2003) describe many collecting trips to Africa, South America, South East Asia and elsewhere which were mounted in the first half of the 20th century to collect animals for Jersey Zoo, London Zoo and others. Zoos wanted to exhibit as wide a range of species as possible and it was not uncommon for them to exhibit single individuals of a species because that was all they could obtain.

In the 1800s the Purchase of Animals Committee was responsible for procuring new animals for the Zoological Society of London. In 1865 it paid the Jardin des Plantes, Paris the equivalent of £30,000 for the African elephant Jumbo (Chambers, 2007) (see Box 3.1).

In the early part of the last century it was easy to purchase a very wide variety of animals in London. *Chapman's*, of Tottenham Court Road, styled itself the *Premier Livestock Emporium of Europe* and, in an advertisement in 1924, claimed to stock 5000 small birds, monkeys, chimpanzees, zebras, various antelopes, lions, tigers, elephants, lizards, iguanas and snakes. The animals were kept at their menagerie in Barnet (Inskipp and Wells, 1979).

In 1872 Belle Vue Zoo, in Manchester, bought an elephant for £680 (over £43,500 now) from Wombwell's Menagerie No. 1 in Edinburgh, and in 1946 the zoo purchased two giraffes for £250 (over £7400 now) each (Nicholls, 1992). A letter sent to the proprietor of Belle Vue in 1919 from the World's Zoological Trading Company Limited which accompanied a catalogue and price list of animals offered for sale is reproduced in Fig. 12.1.

In 1912 Bristol Zoo bought an Asian elephant for £82.10*s* (around £6000 now) and in 1930 it purchased a gorilla – 'Alfred' – for £350 (about £16,000 now) (Warin and Warin, 1985). At the time, Alfred was the only gorilla in captivity in the UK. In 1969 London Zoo's most valuable animal was a giant panda ('Chi-Chi') which was considered to be worth £12,000 (£148,000 now). Discussing the cost of purchasing animals in the mid-1980s, Warin and Warin (1985) suggested that a rare species like an okapi, a gorilla or a white tiger (*Panthera tigris*) might cost £20,000–30,000.

Occasionally, animals are given by one country to another as a symbol of their friendship. China gave pairs of giant pandas (*Ailuropoda melanoleuca*) to American President Richard Nixon for the National Zoo in 1972, President Pompidou of France in 1973

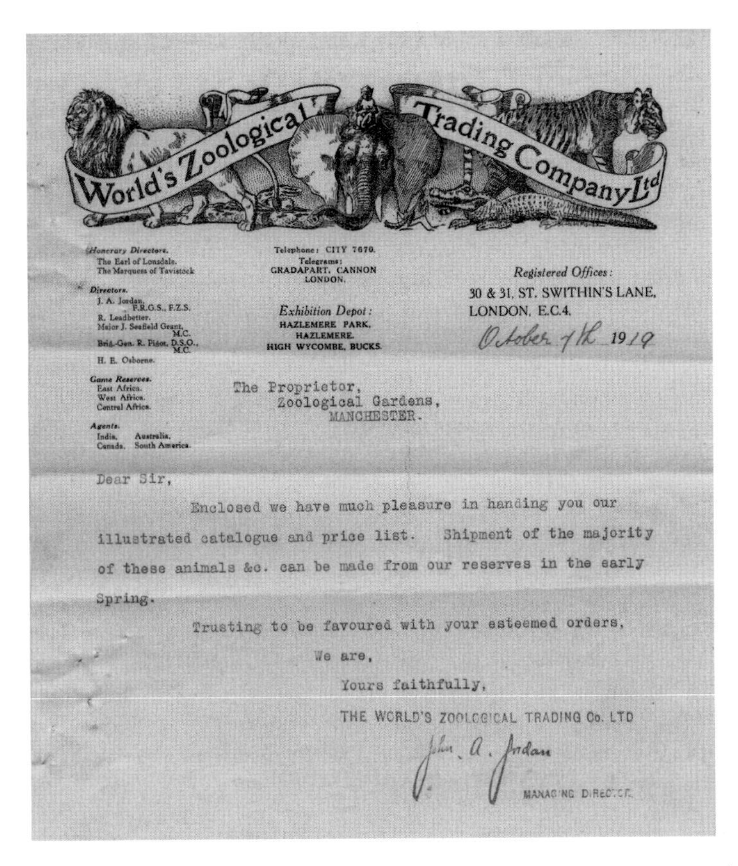

World's Zoological Trading Company Ltd

Honorary Directors.
The Earl of Lonsdale.
The Marquess of Tavistock

Directors.
J. A. Jordan, F.R.G.S., F.Z.S.
R. Leadbetter.
Major J. Seafield Grant, M.C.
Brig.-Gen. R. Pigot, D.S.O., M.C.
H. E. Osborne.

Game Reserves.
East Africa.
West Africa.
Central Africa.

Agents.
India. Australia.
Canada. South America.

Telephone: CITY 7670.
Telegrams: GRADAPART, CANNON LONDON.

Exhibition Depot:
HAZLEMERE PARK,
HAZLEMERE.
HIGH WYCOMBE, BUCKS.

Registered Offices:
30 & 31, ST. SWITHIN'S LANE,
LONDON, E.C.4.

October 7th 1919

The Proprietor,
Zoological Gardens,
MANCHESTER.

Dear Sir,

Enclosed we have much pleasure in handing you our illustrated catalogue and price list. Shipment of the majority of these animals &c. can be made from our reserves in the early Spring.

Trusting to be favoured with your esteemed orders,

We are,

Yours faithfully,

THE WORLD'S ZOOLOGICAL TRADING Co. LTD

John A. Jordan

MANAGING DIRECTOR.

Fig. 12.1 A letter dated 1919 from the World's Zoological Trading Company Limited to the proprietor of Belle Vue Zoo offering animals for sale. (Courtesy of Chetham's Library, Manchester.)

and Prime Minister Edward Heath, for London Zoo, in 1974. When she was Prime Minister of India, Indira Gandhi gave an Asian elephant to Honolulu Zoo as a present from the children of India to the children of Hawai'i and in 2008 Russian Prime Minister Vladimir Putin received a Siberian tiger cub for his 56th birthday which he later gave to a zoo in the Krasnodar Territory of south Russia.

Occasionally animals from zoos – or even circuses – that have closed down may be dispersed to zoos that are prepared to take them. Some zoos will also take unwanted animals from members of the public or those that have been confiscated by customs officials. The alternative may be destruction. However, others will only take in species that fit in with their collection plan. Small reptiles and other small animals are clearly less of a problem than large mammals, and may be kept off-show.

Many zoos contain individual animals that are on loan from other zoos for breeding purposes. In 1984 China began a long-term giant panda lending programme to foreign zoos, which pay to house the animals. In 2008 eight pandas were sent to the Beijing Olympics where they attracted 2.1 million visitors during their 10-month stay. In 2010 six pandas were sent as ambassadors to the Asian Games in Guangzhou. They were temporarily housed at a safari park near the athletes' village.

Nowadays, since the advent of international law controlling trade in endangered species in the 1970s, modern zoos obtain most of their animals from breeding, or from other collections, and it is relatively rare for animals to be taken from the wild. Many modern zoos are now net producers of animals. Dublin Zoo became famous for producing lions (*Panthera leo*). Its first cubs were born in 1857. In the next 30 years 141 cubs were produced and Dublin began supplying lions to zoos all over the world. By 1957 the zoo had bred well over 500 lions (Schomberg, 1957). Most animals in zoos today have been captive bred. Olney *et al.* (1994) noted that 362 of the 633 gorillas held in zoos had been bred in captivity. However, not all species have proved capable of producing self-sustaining zoo populations, for example Asian elephants (*Elephas maximus*) (see Section 12.5.5).

12.1.2 How do zoos decide which species to keep?

Modern zoos have procedures for deciding which species to add to their animal collections. In the past, zoos obtained whatever specimens were available. There are no detailed records of the original animals kept at Bristol Zoo when it opened in 1836, but early press reports refer to 'a young tigress, a leopard, a camel, two sloth bears, a fine antelope and many worthy specimens of the feathered tribe' (Warin and Warin, 1985). As this description illustrates, in these early days it was not unusual for zoos to keep single individuals of a species.

In recent years there has been a trend amongst zoos towards the keeping of fewer species in larger numbers. This is the result of zoos focusing on the development of breeding programmes for a small number of rare species rather than attempting to exhibit as many species as possible. In 1959 London Zoo kept 290 different mammal species (Jarvis and Morris, 1960). By 2006 this had fallen to just 62 species (Fisken, 2007). Over the same period mammal species fell from 133 to 65 at Paignton Zoo (UK), and from 120 to 66 at Chester Zoo (UK). An analysis of 11 major zoos in the UK shows a decrease in the mean number of mammal species kept from 93 in 1967 to 58 in 2003, while over the same period the total number of mammal specimens increased from 4145 to 6020 (Rees, unpublished data). This analysis is based on annual returns made to the *International Zoo Yearbook* by the zoos in London, Whipsnade, Belfast, Chester, Bristol, Edinburgh, Paignton, Twycross, Jersey (now *Durrell*), Dudley and the Welsh Mountain Zoo (Fig. 12.2).

Modern zoos keep a wider range of taxa than was previously the case. In the past, zoos kept relatively few, if any, invertebrates. Of the 11 zoos analysed, only

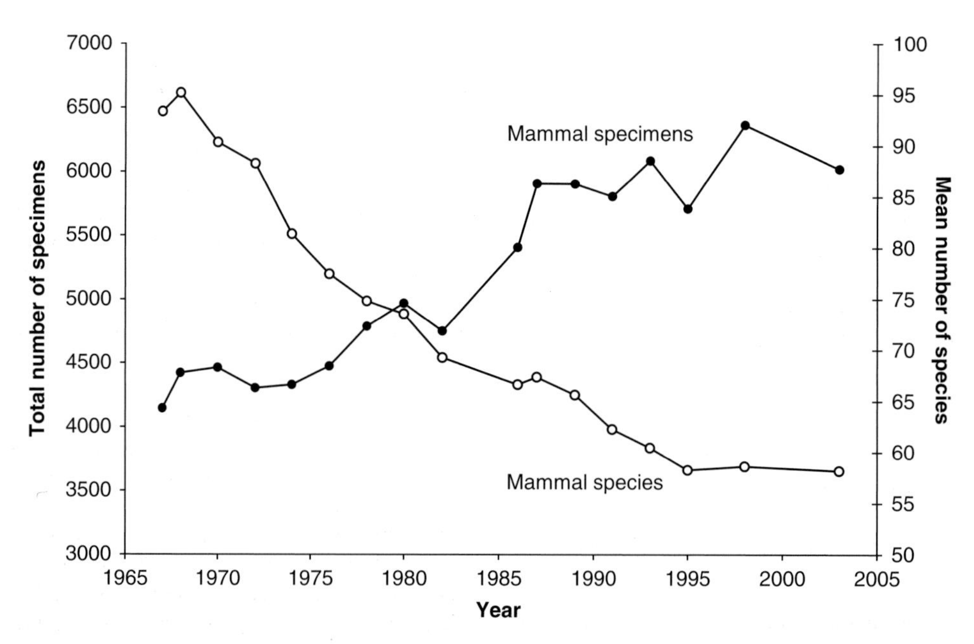

Fig. 12.2 Changes in the mean number of mammal species and the total number of mammal specimens in 11 UK zoos (1967–2003) (see text for details). (*Source*: Rees, unpublished data.)

three kept invertebrates in 1967 but by 2003 only one zoo kept none.

Zoos keep particular species of animals for a variety of reasons:

- Because they are popular with the public – the ABC species, e.g. A is for anteater, B is for bear, etc.
- For historical reasons because the zoo has kept them for a long time
- Because they were added to the collection many years ago and live for a long time
- As ambassadors for conservation
- For entertainment, e.g. animal shows and rides
- Because they can be handled safely by the public
- For educational purposes
- As models to help keepers develop husbandry skills for rarer related species, e.g. meerkats may act as models of rarer mongoose species
- Because they are very rare and part of a captive breeding programme.

Of course, many species may meet several of the criteria listed above.

12.1.2.1 Ambassador species

Some of the animals kept in zoos clearly have little conservation value and zoos claim that their function is to act as 'ambassadors' for their species or for conservation in general (Rees, 2004c). They include individuals:

- from common species that are not endangered in the wild
- from rare species or subspecies for which no cooperative breeding programme exists, perhaps because there are too few individuals in captivity, e.g. forest elephants (*Loxodonta africana cyclotis*)
- from rare species or subspecies that have reproductive potential but are not involved in a cooperative breeding programme, e.g. a tiger (*Panthera tigris*) whose genes are over-represented in the zoo population
- that have no reproductive potential because of their age, behaviour problems, or disease, e.g. a solitary Asian elephant (*Elephas maximus*) in the post-reproductive phase of its life.

12.1.2.2 Are white lions conservation white elephants?

Some zoos have actively sought to attract visitors by acquiring genetic rarities. In April 2004, the West Midlands Safari Park imported four white lions into the UK, at a cost of £250,000. The safari park assigns the subspecific status *Panthera leo krugeri* to these animals in its literature, although this is the subspecies recognised from the Transvaal, regardless of coat colour. White lions are colour variants and do not represent *per se* a particular subspecies or other taxon. They are affected by a recessive genetic condition known as leucism which causes a loss of normal pigmentation in the skin and fur, but not the eyes, lips or paw pads.

White lions in Philadelphia Zoo are reputed to have earned the zoo $1 million in a single year (Croke, 1997). Zoos know that unusual animals will increase visitor numbers. In the three months following the acquisition of two giant pandas by Chiang Mai Zoo in Thailand the zoo received over 400,000 visitors and was predicting a doubling of visitor numbers in 2004 from 600,000 to over 1.2 million (Anon., 2004). The pandas were described by the zoo as 'animal friendship ambassadors' from the Chinese government.

The United Nations Convention on Biological Diversity, 1992 (Art. 2) includes 'diversity within species' as an important element of biodiversity that should be conserved. Subspecies of big cat should clearly be conserved by zoos where possible. However, it seems unlikely that Article 2 of the Convention was intended to include every possible colour variant of a species.

Collection planning is an important part of zoo management. The type of species that zoos keep depends upon the nature of the zoo. Those which are primarily commercial operations keep species that are particularly attractive to the public. Aquariums keep large fish such as sharks, rays and electric eels because they are likely to attract more visitors than cichlids and minnows. Drive-through safari park facilities keep many large herbivore species such as giraffe, wildebeest, zebra, eland and ostrich which generally have a low conservation value but create the illusion of driving through an African national park. Zoos whose primary mission is conservation concentrate on keeping endangered and threatened species but these too may be at least partly selected on the basis of their appeal to the public. Even mission-led zoos need to attract visitors if they are to survive. Good zoos plan their collections and make a great effort to justify the keeping of each species they hold to themselves and to the public. However, zoos cannot ignore the fact that some species are more popular with visitors than others (Moss and Esson, 2010).

12.2 COLLECTION PLANNING

If zoos are to maintain sustainable populations of threatened species they must manage captive groups as metapopulations, that is, as if the social groups held by each zoo were part of a larger population – a population of populations. This is only possible if zoos work cooperatively and make the interests of their own collection subordinate to the interests of the species they are helping to conserve.

Most zoos are managed as independent entities and as such determine their own priorities. There is constant tension in zoos between space limitations and the need to provide a diverse range of exhibits. Although zoos may endeavour to work cooperatively with regard to breeding programmes, zoo directors cannot be forced to send animals to other zoos or to create accommodation for particular species.

Lees and Barlow (2008) have listed a number of challenges facing collection planning in Australasia:

- Lack of zoo space so zoo populations are too small and unstable
- The slow process of replacing long-lived taxa with higher priority taxa
- The difficulty of acquiring good stock from zoos overseas
- Delays in funding for planned exhibit alterations and expansions
- Loss of husbandry skills resulting in poor species programme performance and/or lack of coordination
- Changes in government import and export regulations which can result in the unexpected isolation of zoos.

In Australian zoos, a number of antelope species became demographically extinct as a result of a 2001 suspension of hoofstock quarantine protocols. Suspension of protocols for the importation of zebras after a recent outbreak of equine influenza in the Australian domestic horse industry prevented the movement of animals within the region to establish genetically optimal pairings. Many of the problems facing zoos in Australasia are also faced by institutions in other regions.

Lees and Barlow (2008) have suggested that collection planning could be improved by:

- Establishing more effective processes within individual zoos for collection planning
- Maximising the contribution that reproductively competent animals can make
- Enhancing, improving and evaluating the role of species coordinators
- Maximising the conservation value of zoo populations by managing them at a global rather than a regional level.

12.2.1 Studbooks

Planned breeding in zoos depends upon the existence of studbooks. A studbook documents and tracks the pedigree and demographic history of individual animals in a breeding programme, and it is maintained by a studbook keeper. The purpose of a studbook is to facilitate the efficient genetic management of a captive population. A studbook contains information about the identity of individual animals, their parents, their location, their origin (whether wild-caught or transferred from a zoo), dates of transfer between different zoos, cause of death and other information that may be useful.

The first studbook was established for horses, and published in 1791: *The General Studbook for Thoroughbred Horses*. The first studbook for a wild species was begun in 1923 for the European bison (*Bison bonasus*) using a card index, and first published in 1932 (Tudge, 1991). The next species for which a studbook was created was Père David's deer (*Elaphurus davidianus*) (Fig. 12.3). In 1967 the responsibility for international studbooks was assumed by the IUCN and WAZA which formed a joint committee. This became the Captive Breeding Specialist Group which is now the Conservation Breeding Specialist Group.

International studbooks are listed in the *International Zoo Yearbook* (Table 12.1). In addition to the international studbooks many other studbooks are maintained on a regional basis. Responsibility for the keeping of studbooks falls to staff working in individual zoos, so a particular zoo may keep the studbooks for a number of species. For example, Prague Zoo keeps the following studbooks:

- International studbook of the Przewalski's horse (*Equus ferus przewalskii*)
- International studbook of the Cuban boa (*Epicrates angulifer*)
- International studbook of the Cuban ground iguana (*Cyclura nubila*)
- International studbooks of the Malaysian giant turtles (*Orlitia borneensis*) and giant Asian pond turtles (*Heosemys grandis*)
- European studbook of the tiger cat (*Leopardus tigrinus*).

A studbook keeper may impose restrictions when allowing a zoo to keep a species for the first time. For

Fig. 12.3 Père David's deer (*Elaphurus davidianus*) was saved from extinction in the wild by captive breeding.

example, the studbook keeper for the African hunting dog (*Lycaon pictus*) may require a zoo to demonstrate that it can keep a small bachelor group successfully and securely before allowing it to keep a breeding pack. Hunting dogs are experts at digging under enclosure fences to escape.

Studbooks are essential tools for collection planning. Collection planning occurs at three levels:

- Institutional
- Regional
- Global.

In 2003 WAZA adopted a procedure for the establishment of inter-regional programmes called global species management programmes which may concern a number of species for which international studbooks exist. Institutional and regional collection planning are discussed below.

Table 12.1 Examples of international studbooks, with dates of official approval. (In 2009 there were 181 international studbooks for species and subspecies.)

Common name	Scientific name	Date of approval of studbook
European bison	*Bison bonasus*	1932
Père David's deer	*Elaphurus davidianus*	1957
Przewalski's horse	*Equus ferus przewalskii*	1959
Okapi	*Okapia johnstoni*	1966
Arabian oryx	*Oryx leucoryx*	1966
Black rhinoceros	*Diceros bicornis*	1966
Tiger	*Panthera tigris*	1967
Gorilla	*Gorilla gorilla*	1967
Bonobo	*Pan paniscus*	1967
Orangutan	*Pongo pygmaeus*	1967
Golden lion tamarin	*Leontopithecus rosalia*	1970
Asiatic lion	*Panthera leo persica*	1971
Red-crowned crane	*Grus japonensis*	1971
Giant panda	*Ailuropoda melanoleuca*	1976
Chinese alligator	*Alligator sinensis*	1982
Drill	*Mandrillus leucophaeus*	1985
Cheetah	*Acinonyx jubatus*	1987
Spix's macaw	*Cyanopsitta spixii*	1988
Puerto Rican crested toad	*Peltophryne lemur*	1990
African wild dog	*Lycaon pictus*	1991
Spectacled bear	*Tremarctos ornatus*	1992
Rodriguez fruit bat	*Pteropus rodricensis*	1992
Komodo dragon	*Varanus komodoensis*	1995
Partula snails (all species)	*Partula* spp.	1996
Koala	*Phascolarctos cinereus*	1996
Scimitar-horned oryx	*Oryx dammah*	1996
Fossa	*Cryptoprocta ferox*	2000
Maroon-fronted parrot	*Rhinchopsitta terrisi*	2001
Giant otter	*Pteronura brasiliensis*	2003

Source: *International Zoo Yearbooks*.

12.2.2 Taxon Advisory Groups (TAGs) and Taxon Working Groups (TWGs)

Taxon Advisory Groups (TAGs) are groups of individuals who work together to advance knowledge of the husbandry, nutrition, veterinary care and other aspects of the conservation of a particular taxon and disseminate this to others. They are also involved in decisions about what species should be kept in zoos (Fig. 12.4). EAZA currently has 42 TAGs (Table 12.2).

In the UK, BIAZA now operates combined Taxon Working Groups (TWGs) which have replaced TAGs. These reflect the range of taxa held by zoos in Britain and Ireland:

- Mammal (MWG)
- Bird (BWG)
- Terrestrial Invertebrate (TIWG)
- Reptile and Amphibian (RAWG)
- Aquarium (AWG)
- Plant (PWG)
- Native Species (NSWG).

12.3 INSTITUTIONAL PLANNING

An Institutional Collection Plan (ICP) describes which species a particular zoo intends to keep. ICPs are required by national and regional zoo associations as

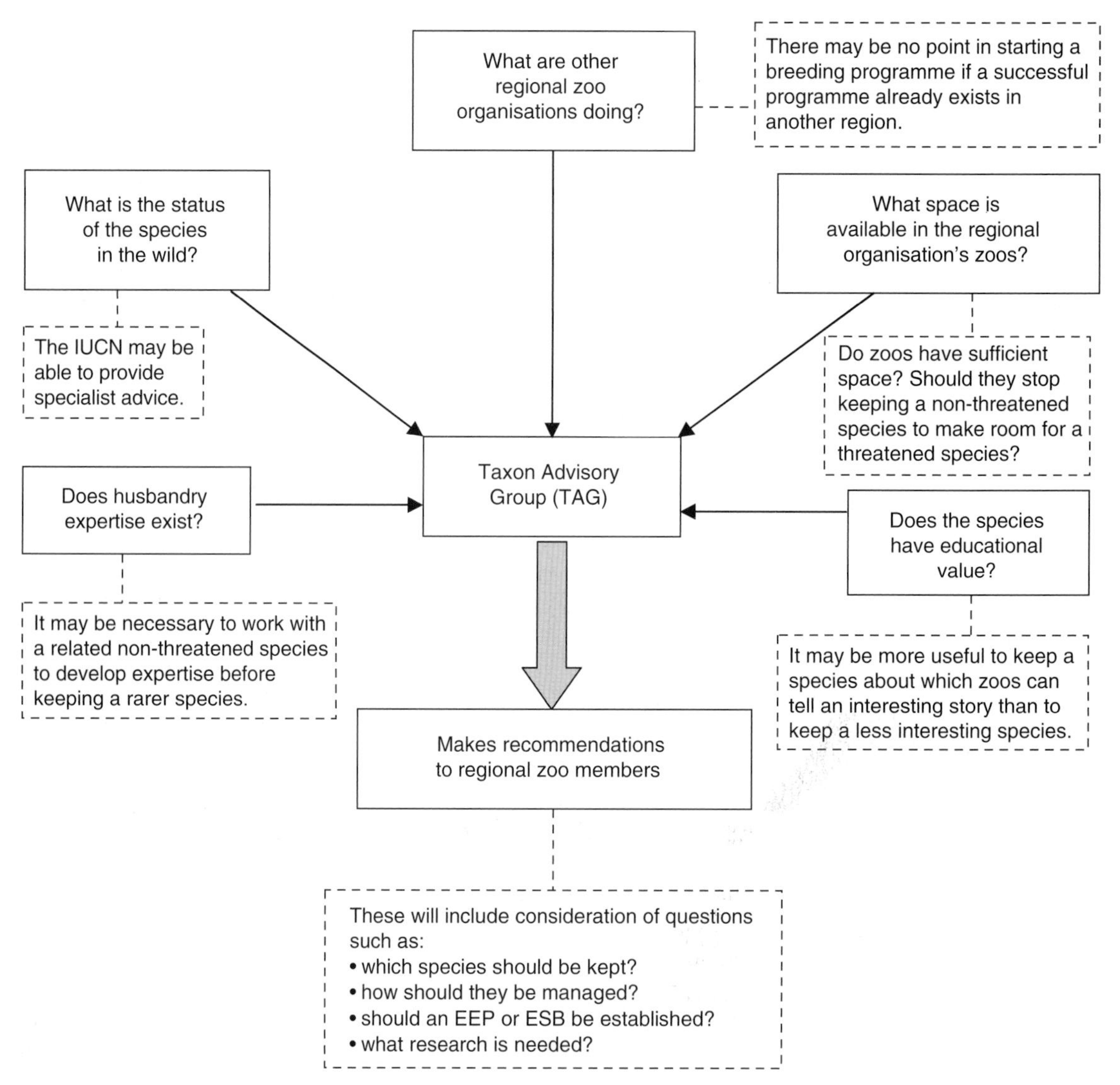

Fig. 12.4 How do TAGs decide which species to keep?

Table 12.2 EAZA Taxon Advisory Groups (TAGs) (July 2010).

Terrestrial Invertebrate	Fish and Aquatic Invertebrate
Amphibian	Reptile
Ratite	Penguin
Pelecaniformes	Ciconiiformes and Phoenicopteriformes
Waterfowl	Falconiformes
Cracid	Galliformes
Gruiformes	Charadriiformes
Pigeon and Dove	Parrot
Owl	Toucan and Turaco
Hornbill	Passeriformes
Monotreme and Marsupial	Prosimian
Callitrichid	Cebid
Old World Monkey	Gibbon
Great Ape	Small Mammal
Canid	Bear
Small Carnivore	Felid
Marine Mammal	Elephant
Equid	Rhino
Tapir and Hippo	Pig and Peccary
Cattle and Camelid	Deer
Antelope and Giraffe	Sheep and Goat

Source: EAZA (2010).

a condition of accreditation and must be updated at specified intervals.

In North America, the AZA requires ICPs to be updated at least every five years and requires that all species and individual animals which are being considered for residence in a collection are evaluated (AZA, 2010c). The ICP should include a statement of justification for all species and individuals in the institution's planned collection. The ICP requires that all species in, and/or being considered for, residence are evaluated with regard to a range of criteria (Fig. 12.5).

12.4 REGIONAL PLANNING

12.4.1 Introduction

A Regional Collection Plan (RCP) is essentially a list of species that zoos within a particular region (accredited to a particular regional zoo organisation) should keep, along with details of the level at which they should be managed and an assessment of how much space should be devoted to each species. RCPs help individual zoos to plan their own collections and select species necessary to fulfil the conservation goals of the

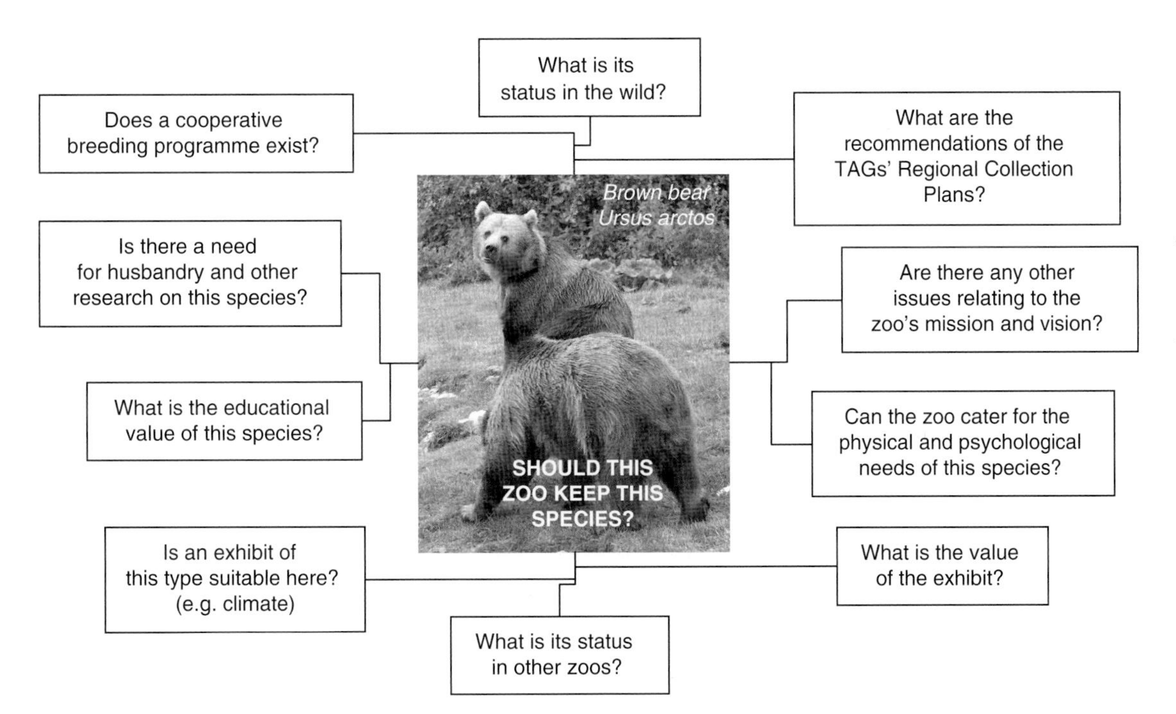

Fig. 12.5 Institutional planning: how does a zoo decide which species to keep? (Based on information in AZA, 2010c.)

Table 12.3 Numbers of species described and classified as threatened by the IUCN in 2010.

Taxon	Described species	Number of threatened species (IUCN categories CR, EN and VU)*
Mammals	5,490	1,143
Birds	9,998	1,223
Reptiles	9,084	467
Amphibians	6,433	1,895
Fishes	31,300	1,414
Subtotal	62,305	6,142
Invertebrates	1,305,250	2,669
Total	1,367,555	8,811

*CR, Critically Endangered; EN, Endangered; VU, Vulnerable.
Source: IUCN Red List Version 10.1 (Table 1). www.iucnredlist.org/documents/summarystatistics/2010_1RL_Stats_Table_1.pdf (accessed 1 July 2010).

regional zoo organisation. The RCP process helps individual zoos to gain access to a wide range of expertise in other institutions.

In deciding if a particular species should be kept, the RCP process should consider a variety of factors, including:

- the mission statement of the TAG
- the recommended function of the taxon (education, conservation, etc.)
- space requirements
- target population sizes
- management criteria
- conservation status of the taxon.

In 2010 the IUCN listed 8811 species of animals as 'threatened' (Table 12.3). Zoos can only hope to establish captive breeding programmes for a relatively small proportion of these (Table 12.4).

Table 12.4 Regional conservation breeding programmes (at December 2009).

American Association of Zoos and Aquariums (AZA)	
Studbooks	484
Population Management Plans (PMPs)	334
Species Survival Plans (SSPs)	110
Taxon Advisory Groups (TAGs)	46
Scientific Advisory Groups	13
Conservation Action Partnerships	4
European Association of Zoos and Aquaria (EAZA)	
Taxon Advisory Groups (TAGs)	41
European Endangered Species Programmes (EEPs)	164
European Studbooks (ESBs)	157
The Zoo and Aquarium Association (formerly the Australasian Regional Association of Zoological Parks and Aquaria (ARAZPA)) – Australasian Species Management Program (ASMP)	
Taxon Advisory Groups (TAGs)	14
Conservation Programs with direct links to *in-situ* conservation	35
Programs releasing *ex-situ* bred animals to the wild in Australia	22
Population Management Programs for sustainable zoo populations	47
Husbandry Research Programs	9
African Association of Zoos and Aquaria (PAAZAB) – African Preservation Programme (APP)	
Studbooks	22
Japanese Association of Zoos and Aquariums (JAZA)	
Managed threatened species or subspecies	64

Source: WAZA (2009).

12.4.2 Europe

12.4.2.1 European Endangered Species Programmes (EEPs) and European Studbooks (ESBs)

Intensive population management of zoo animals began in Europe in 1985 with the establishment of European Endangered Species Programmes (EEPs). Each EEP has a species coordinator who is responsible for collecting information. A Species Committee makes recommendations about which individuals should be used for breeding and the exchange of individual animals between zoos. The organisation of EAZA breeding programmes is illustrated in Fig. 12.6. The role of EAZA TAGs is to develop RCPs for the species that are kept in European collections. As of August 2010 EAZA listed 130 EEPs for mammals. In addition it listed 37 EEPs for birds, seven for reptiles and just two for invertebrates (EAZA, 2010).

European Studbooks (ESBs) have a studbook keeper who manages the population to a lesser extent than the species coordinator. ESBs are kept for a further 176 taxa: one invertebrate, seven fish, two amphibians, 17 reptiles, 68 birds and 81 mammals.

12.4.2.2 European Studbook Foundation (ESF)

The European Studbook Foundation is a non-profit organisation which has the following goals:

- Conservation of reptiles and amphibians in captivity, especially endangered species, by building and maintaining genetically viable populations in captivity
- Management of European studbooks
- Management of breeding programmes which are genetically sound
- Cooperation with reintroduction programmes
- Collection, compilation and publication of knowledge about husbandry and breeding (programmes) for reptiles and amphibians.

12.4.3 United Kingdom

In Britain and Ireland captive breeding programmes were run as the Joint Management of Species Programme (JMSP) until 2006. These were overseen by a special group of experts set up by BIAZA called the Joint Management of Species Committee (JMSC) and smaller, more specifically focused Taxon Advisory Groups (TAGs). BIAZA TAGs have now become Taxon Working Groups (TWGs) (see Section 12.2.2). BIAZA member collections are now encouraged to join the European breeding programmes for their species.

12.4.4 North America

Cooperative breeding programmes in North America are managed by the AZA and are called Species Survival Plan® (SSP) Programs and Population Management Plan Programs. These are managed by the AZA Wildlife Conservation Committee through the work of AZA Taxon Advisory Groups.

12.4.4.1 AZA Taxon Advisory Groups (TAGs)

The AZA currently has 46 TAGs. Some are for large taxa, for example felids, parrots and amphibians, while others are for much smaller taxa, for example tapirs, elephants and rhinoceros.

The function of an AZA TAG is to:

- Examine the conservation needs of an entire taxon (e.g. a species, genus, family, etc.)
- Develop recommendations for population management and conservation
- Develop an action plan that identifies essential goals, scientific investigations and conservation initiatives required to best serve *ex-situ* and *in-situ* populations
- Develop a Regional Collection Plan for the optimal management of *ex-situ* populations (AZA, 2010d).

12.4.4.2 AZA Regional Collection Plan (RCP)

The RCP describes a list of species recommended for management in AZA institutions and the appropriate management level. TAGs develop RCPs to assist institutions in planning their individual collections to ensure that the animal management and conservation goals of the AZA and the individual institutions are achieved.

12.4.4.3 AZA Population Management Levels

Taxon Advisory Groups make population management recommendations for each taxon which help to inform the RCP. The management levels used are:

- Species Survival Plan Populations – these focus on species that require the greatest conservation effort and the AZA requires members to cooperate in breeding them.
- Population Management Plan Populations – these focus on species that require less rigorous conservation effort.

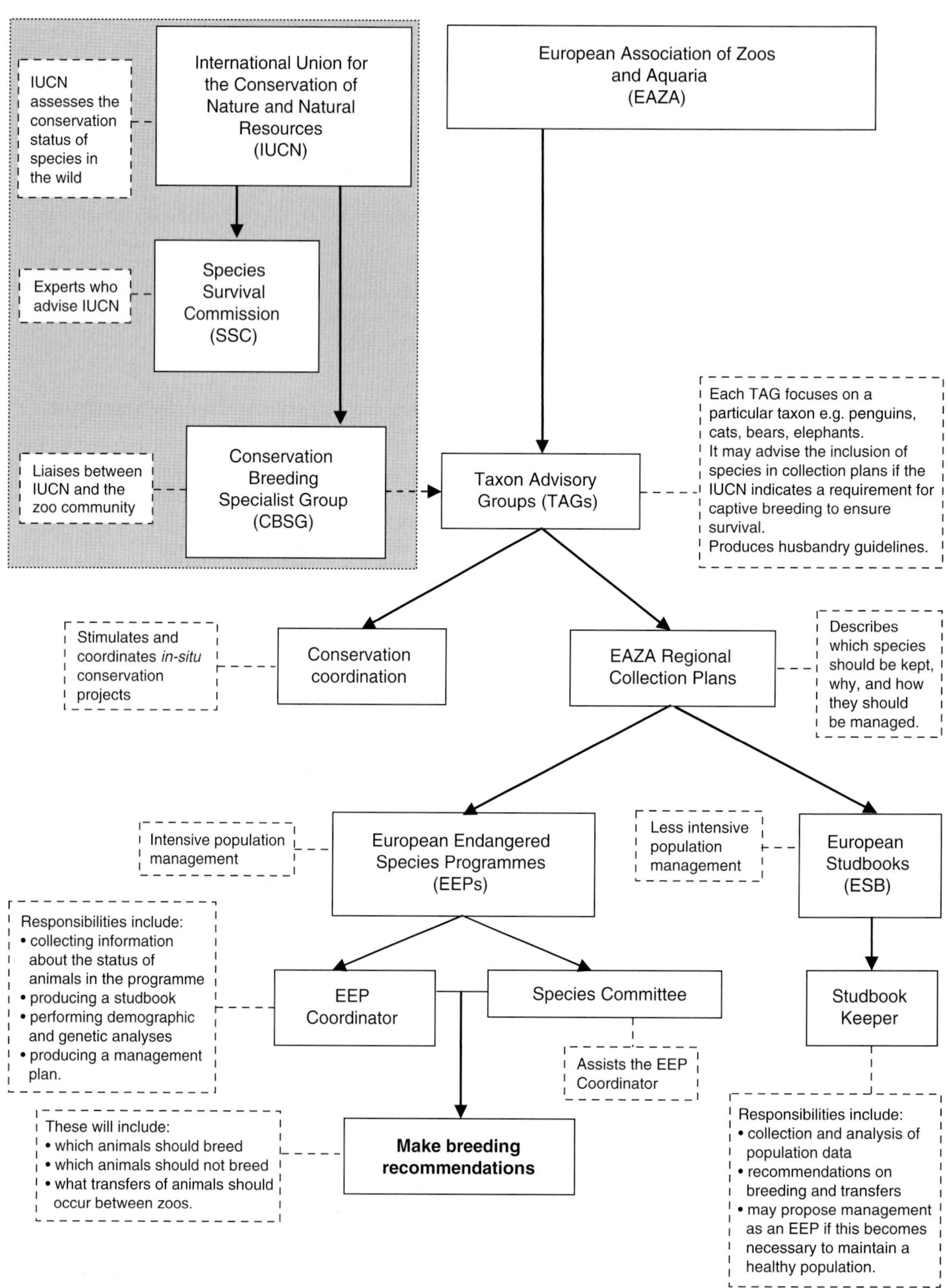

Fig. 12.6 The organisation of EAZA breeding programmes.

• Display, Education and Research Populations – populations which do not require genetic or demographic management and are useful for research, conservation education and display.
• Government Program Populations – these focus on species that are managed by state or federal programmes (e.g. the manatee).
• Phase-Out Populations – species that are being phased out of AZA-accredited institutions by a moratorium on breeding.
• Phase-In Populations – species that are being phased into AZA-accredited institutions and will be transferred to another population management level as appropriate.
• Not Recommended – taxa not currently kept by AZA institutions and not recommended to be phased in.

12.4.4.4 Species Survival Plan® Programs

The mission of an AZA Species Survival Plan (SSP) Program is:

> to manage and conserve a select and typically threatened or endangered, *ex situ* species population with the required cooperation of AZA-Accredited Zoos and Aquariums, Certified Related Facilities, and Approved Non-Member Participants.

These SSP species are often 'flagship species' (Table 12.5).

Each SSP Program is responsible for producing a Master Plan. This identifies population management goals and recommendations to ensure the sustainability of a healthy, genetically and demographically diverse population. Each Program is managed by a TAG and is responsible for managing its studbook and communicating with Program Leaders and Institutional Representatives (IRs). Sometimes differences of opinion may arise between SSP recommendations and individual AZA-accredited institutions. The AZA Animal Management Reconciliation Policy is designed to assist parties in resolving these differences.

Each SSP Program is composed of:
• an SSP Coordinator
• a Management Group (appointed from AZA members)
• expert advisors.

SSP Coordinators must be employed by an AZA-accredited institution and often act as the studbook keeper for the same species. All of the administrative information required to operate a SSP Program is

Table 12.5 Species Survival Plan Programs of the American Association of Zoos and Aquariums (at June 2010).

Mammals
Koala, Queensland
Tree Kangaroo, Matschie's
Wallaby, Yellow-Footed Rock
Leopard, Amur
Leopard, Clouded
Leopard, Snow
Lion
Tiger
Cheetah
Jaguar
Ocelot
Cat, Black-Footed
Cat, Fishing
Cat, Sand
Nocturnal Prosimian
Tamarin, Cotton-Top
Tamarin, Golden Lion
Tamarin, Golden-Headed Lion
Marmoset, Geoffroy's
Monkey, Goeldi's
Sifaka
Lemur, Ring-Tailed
Lemur, Ruffed
Eulemur
Guenon
Colobus
Mangabey
Macaque
Langur
Baboon
Monkey, Spider
Gibbon
Chimpanzee
Bonobo
Gorilla, Western
Orangutan
Elephant
Anteater, Giant
Wolf, Maned
Wolf, Mexican Gray
Wolf, Red
Fox, Fennec
Fox, Island (Northern subspp.)
Fox, Swift
Dog, African Wild
Ferret, Black-Footed
Otter, Asian Small-Clawed
Bear, Sun
Bear, Sloth

(*Continued*)

Table 12.5 *(Cont'd)*

Bear, Polar
Bear, Andean
Panda, Giant
Panda, Red
Bongo, Eastern
Gazelle, Addra and Mhorr
Gazelle, Slender-Horned
Gazelle, Speke's
Oryx, Arabian
Oryx, Scimitar-Horned
Addax
Deer, Burmese Brow-Antlered
Deer, Philippine
Okapi
Zebra, Grevy's
Horse, Asian Wild
Peccary, Chacoan
Pig, Visayan Warty
Babirusa
Hippopotamus
Hippopotamus, Pygmy
Rhinoceros, Black
Rhinoceros, Indian
Rhinoceros, White
Tapir, Baird's (Central American)
Tapir, Malayan
Bat, Rodrigues Fruit
Whale, Beluga
Birds
Kiwi, North Island Brown
Toucan
Mynah, Bali
Kingfisher, Micronesian
Ibis, Waldrapp
Condor, California
Condor, Andean
Vulture, Eurasian Black
Crane, White-Naped
Crane, Wattled
Crane, Red-Crowned
Hornbill
Duck, White-Winged Wood
Bustard, Kori
Cockatoo, Palm
Parrot, Thick-Billed
Peafowl, Congo
Penguin, African
Penguin, Humboldt
Pigeon, Pink
Rail, Guam
Chicken, Attwater's Prairie
Reptiles
Dragon, Komodo
Iguana, Fiji Island Banded
Iguana, Rock
Tortoise, Burmese Star
Tortoise, Radiated
Boa, Virgin Islands
Snake, Louisiana Pine
Rattlesnake, Aruba Island
Rattlesnake, Eastern Massasauga
Crocodile, Cuban
Alligator, Chinese
Amphibians
Frog, Panamanian Golden
Toad, Puerto Rican Crested
Toad, Wyoming
Fishes
Cichlids, Lake Victoria (13 spp.)
Invertebrates
Snail, Partula (4 spp.)
Beetle, American Burying

Adapted from AZA (2010e).

contained in a *Species Survival Plan Program Coordinator Handbook*. The primary functions of an SSP are summarised in Fig. 12.7.

12.4.4.5 Population Management Plan Programs

The mission of an AZA Population Management Plan (PMP) Program is:

> to manage and conserve select *ex situ* species populations with voluntary cooperation.
>
> AZA (2010f)

The emphasis here is on the *voluntary* nature of the cooperation, compared with the *required* cooperation expected in SSPs. There are currently over 300 PMP Programs responsible for developing recommendations for the breeding and transfer of individual animals. Each PMP is managed by the corresponding TAG and administered by a PMP Manager who also acts as the studbook keeper for the species. The operation and responsibilities of the PMP are

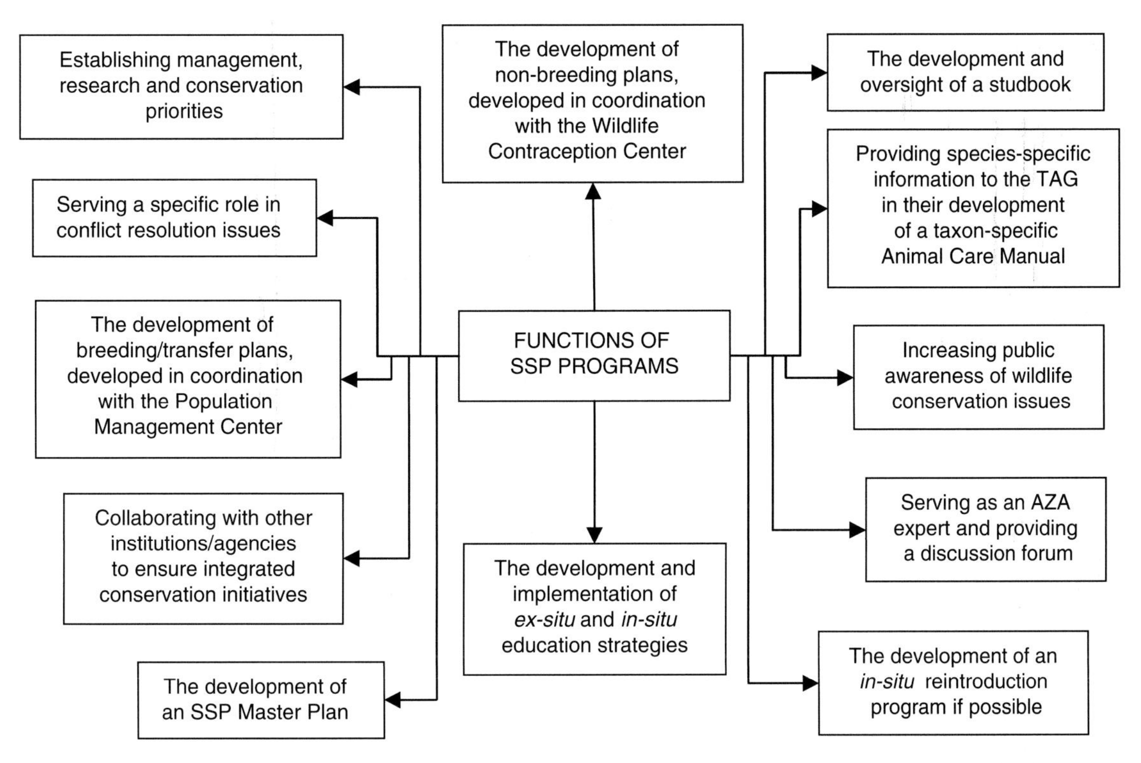

Fig. 12.7 The primary functions of a Species Survival Plan Program. (Based on information in AZA, 2010g.)

described in the *PMP Manager Handbook*. These are broadly similar in nature to those of SSPs (e.g. the development of a studbook, Population Management Plan, Animal Care Manual, etc.).

12.4.5 Australasia

The Zoo and Aquarium Association (formerly the Australasian Regional Association of Zoological Parks and Aquaria) coordinates the Australian Species Management Program (ASMP) across 73 institutions. Under the ASMP there are 14 TAGs and more than 100 programs. Some TAGs focus on individual taxa (e.g. primates, carnivores, marine mammals) while others are concerned with a geographical faunal group (e.g. New Zealand Fauna). Each TAG must produce a Strategic TAG Action Plan which must be updated annually.

Some rare species are not included in captive breeding programmes, perhaps because zoos have insufficient space or because there are too few individuals in zoos to create a viable captive population. For other species there is duplication between regions and the same species may be the subject of a SSP in North America and a EEP in Europe (Table 12.6).

12.5 DEMOGRAPHY OF ZOO POPULATIONS

A population may be defined as a group of organisms of the same species, present at the same geographical location at the same point in time; for example, the population of starlings in London in May 2010, the population of tigers in China in 1986, or the global population of African lions in 1995.

Zoos may need to consider populations of a species at a number of different levels:

- the individual zoo population
- the regional zoo population
- the global zoo population.

Studying the demography of zoo populations helps us to understand:

Table 12.6 Felids whose captive populations are managed in EEPs, ESBs and SSPs (at July 2008).

Species/ subspecies	Scientific name	Breeding programme EEP	ESB	SSP
Cheetah	*Acinonyx jubatus*	•		•
Fishing cat	*Prionailurus viverrinus*	•		•
Pallas cat	*Otocolobus manul*	•		•
Sand cat	*Felis margarita*	•		•
Rusty-spotted cat	*Felis rubiginosus*		•	
Black-footed cat	*Felis nigripes*	•		•
Eurasian lynx	*Lynx lynx*		•	
Clouded leopard	*Neofelis nebulosa*	•		•
Snow leopard	*Uncia uncia*	•		•
Lion	*Panthera leo*			•
Asian lion	*Panthera leo persica*	•		
Siberian tiger	*Panthera tigris altaica*	•		•
Indochinese tiger	*Panthera tigris corbetti*			•
Sumatran tiger	*Panthera tigris sumatrae*	•		•
Sri Lanka leopard	*Panthera pardus kotiya*	•		
Amur leopard	*Panthera pardus orientalis*	•		
Persian leopard	*Panthera pardus saxicolor*	•		
North Chinese leopard	*Panthera pardus japonensis*	•		
Jaguar	*Panthera onca*	•		•
Asian golden cat	*Catopuma temminckii*	•		
Ocelot	*Leopardus pardalis*			•
Margay	*Leopardus wiedii*	•		
Oncilla	*Leopardus tigrinus*		•	
Geoffroy's cat	*Oncifelis geoffroyi*	•		

• their future potential for growth or decline in numbers
• their potential to experience a loss of genetic diversity.

Populations have characteristics that individuals cannot possess, for example:
• a birth rate
• a death rate
• an immigration rate
• an emigration rate
• a density
• a distribution.

Measurement of some of these characteristics helps us to understand many basic aspects of the biology of a species. It also helps us to compare populations of the same species at different points in time, in different geographical locations, or in different zoos. Box 12.1 is an analysis of some of the population data available for the chimpanzee (*Pan troglodytes*) group at Chester Zoo.

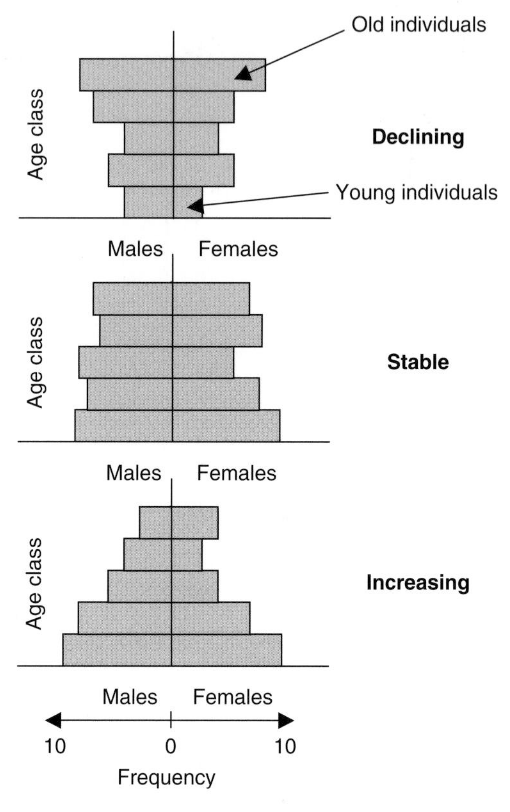

Fig. 12.8 Age structure of theoretical populations which are declining, stable and increasing.

12.5.1 Age structure

The age structure of a population is a representation of the relative numbers of animals of different ages. It may be illustrated in the form of a graph in which horizontal bars indicate the number of males and females of a particular age and individuals are assigned to an age class (e.g. 1–2 years, 3–4 years, etc). The shape of the graph tells us whether the population is growing, stable or declining (Fig. 12.8).

Box 12.1 The Chester Zoo chimpanzees 1932–2006 (by Louise Oates, University of Salford).

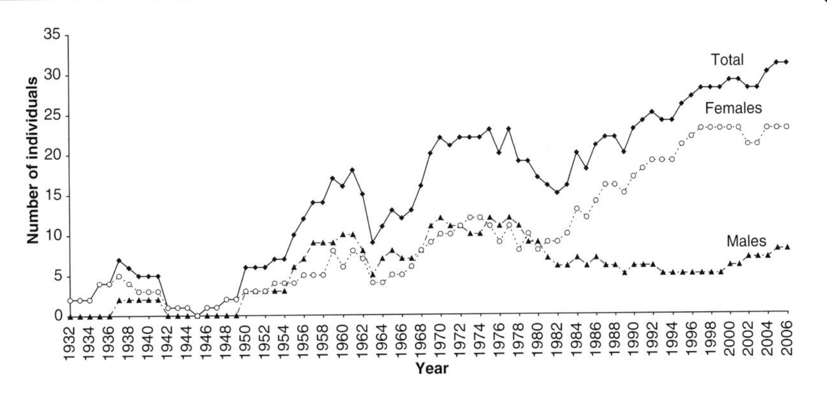

Fig. 12.9 Changes in the Chester Zoo chimpanzee population size and composition, 1932–2006.

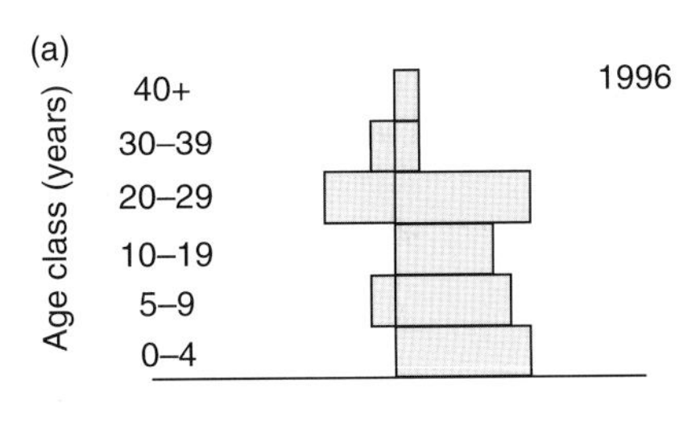

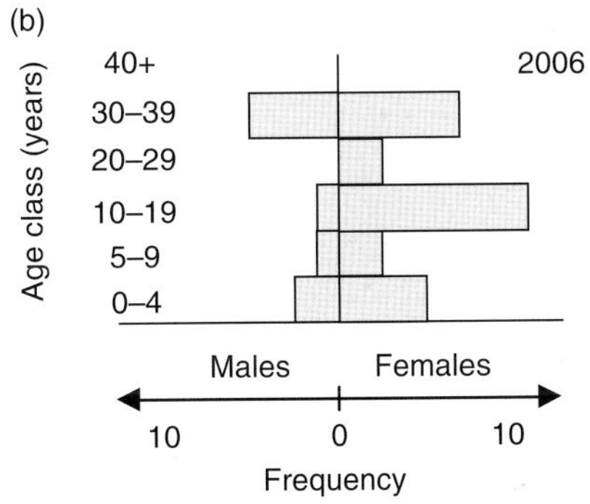

Fig. 12.10 The age structure of the Chester Zoo chimpanzee population in (a) 1996 and (b) 2006.

Records of the chimpanzee (*Pan troglodytes*) group at Chester Zoo began in 1932, when the zoo held just two female chimpanzees. Initially the population was sustained solely by imports, with the group increasing to seven individuals by 1938. The onset of the Second World War and the associated heavy food rationing affected the zoo to such an extent that the entire chimpanzee group died. Following the war a single female was imported and Chester Zoo has held chimpanzees ever since. The population has fluctuated greatly, showing a steady growth over approximately 20 years (Fig. 12.9).

Breeding in the Chester population

The first births in the group were in 1956, but both individuals died in the first year of their lives. Then, in 1958, a female was born and she survived for three years. The population has since increased significantly and, in 2006, numbered 32 individuals. The size of the group has fluctuated markedly with increases mainly due to births and decreases due to a combination of deaths and transfers to other zoos. By 2006 there had been 61 successful births, and to regulate the breeding the zoo often uses temporary contraception. However, the population shows a relatively high percentage of infant deaths, with 45.5% of births resulting in the death of the infant within one year of birth and because of this the population appears to be aging. Some males have contributed heavily to the population, with 'Bolden', 'Boris' and 'Friday' siring many young.

The population in 2006

In 2006, the zoo held 32 chimpanzees and as these are housed together, the group may be as representative of a wild population as is possible in captivity. The oldest individual in the group was Boris, a wild-born male chimpanzee of 39 years, while the youngest was 'Donna', a female born into the group in 2005. Two animals in the group were wild born, compared with four in 1996 (Fig. 12.10). Accordingly, one day, assuming no immigration from elsewhere, all the chimpanzees at Chester Zoo will have been born in captivity.

Box 12.1 *(Cont'd)*

The family relationships within the group are complex and have been reconstructed using ARKS records from the zoo (Fig. 12.11).

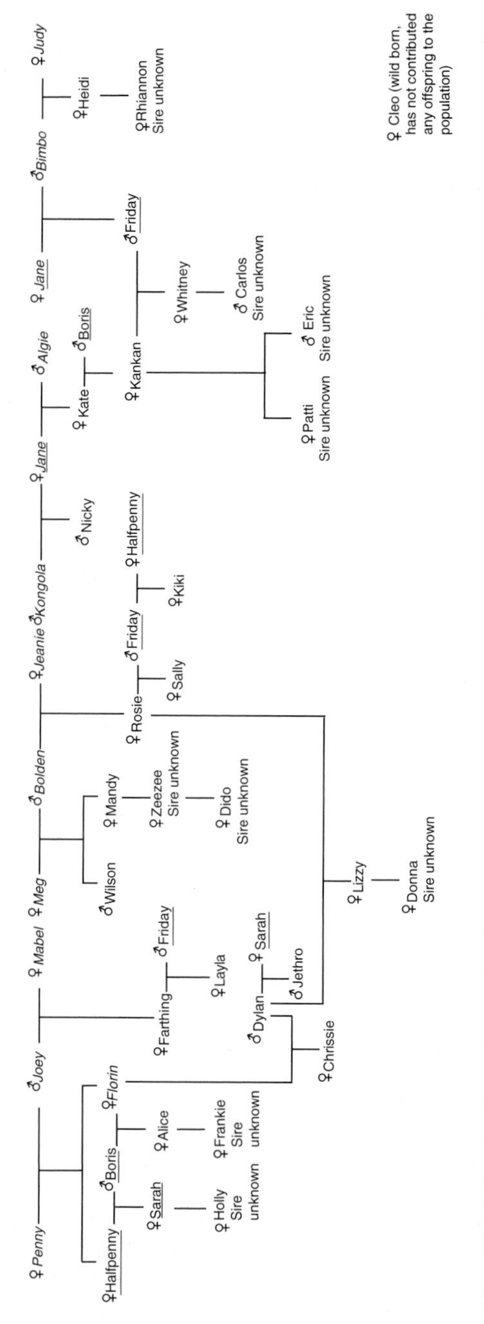

Fig. 12.11 Family relationships in the Chester Zoo chimpanzee population, 2006. Individuals in italics were not present in the group in 2006. Underlined individuals appear in the diagram more than once.

12.5.2 Sex ratio

The sex ratio of a population is the ratio of males to females. In zoo populations the number of each sex is generally expressed in the form:

10.20.2

This means:

10 males, 20 females, 2 unknown sex

In this example, the ratio of females to males is 2:1. In addition there are two individuals of unknown sex that are likely to be juveniles. The sex ratio has an important effect on a population's capacity to maintain genetic diversity (see effective population size below, Section 12.6.2.1 and Fig. 12.16).

The appropriate sex ratio for a zoo population varies between species. The social structure of some species is such that an appropriate social group should contain one adult male and a small group of adult females and their young (e.g. elephants and many antelope species). In other species a 1:1 ratio of the sexes would be appropriate (e.g. bird species that form monogamous pairs). Some zoos may keep bachelor groups of particular species as a resource for other zoos within a regional breeding programme, for example gorillas (*Gorilla gorilla*) or bongos (*Tragelaphus euryceros*).

12.5.3 Life tables

A life table is a mortality schedule for a population. It indicates the death rate experienced by each age class: the age-specific mortality rate.

There are two types of life table:

- A static life table is constructed by counting the number of individuals of each age present in a population at a particular point in time. It assumes that mortality patterns do not change over time so the existing population reflects mortality rates within different age groups in the past.
- A dynamic life table is constructed by following the fate of a cohort of individuals: a group of animals born at approximately the same time. The age at which each animal dies is recorded and this data is used to calculate mortality rates for each age class.

Life tables for zoo animals are usually based on historical data and are therefore static life tables.

Table 12.7 A theoretical example of a life table for a species with a maximum life span of less than five years.

Age (years) x	Survivors at start of age class x l_x	Deaths between age class x and $x + 1$ d_x	Age-specific death rate q_x (d_x/l_x)
0	87	87 − 51 = **36**	36/87 = **0.41**
1	51	51 − 46 = **5**	5/51 = **0.10**
2	46	46 − 20 = **26**	26/46 = **0.57**
3	20	20 − 2 = **18**	18/20 = **0.90**
4	2	2 − 0 = **2**	2/2 = **1.00**
5	0	—	—

Separate life tables are usually – but not always – constructed for each sex and consist of a series of columns (Table 12.7). For the purposes of drawing a survivorship curve, the values of l_x (survivors) are often adjusted so that the first value (for age 0) is 1000 or 1.0.

12.5.4 Survivorship curves

A survivorship curve is a graphical representation of the mortality rates calculated by a life table (Fig. 12.12). It is drawn by plotting the values of l_x against each age class (x). The curve describes the mortality pattern with time and allows us to compare:

- different populations of the same species
- the same species in captivity and in the wild
- males and female mortality in the same species.

Survivorship curves for lions, tigers and cheetahs in zoos are shown in Fig. 12.13. The data for these curves has been adapted from the study of Kohler *et al.* (2006), who used ISIS data to examine mortality in a wide range of species. A recent study of the welfare of elephants in zoos compared survivorship curves in captive and wild populations (Clubb *et al.*, 2008) (see Section 11.10).

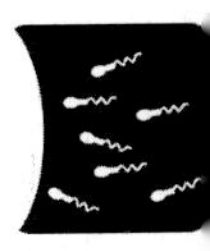

12.5.5 Demographic extinction

Some zoo populations are described as being 'demographically extinct'. In effect this means that the zoo population is not self-sustaining because it contains too few reproductive individuals. The only solution to this is to import more animals from the wild, if they exist.

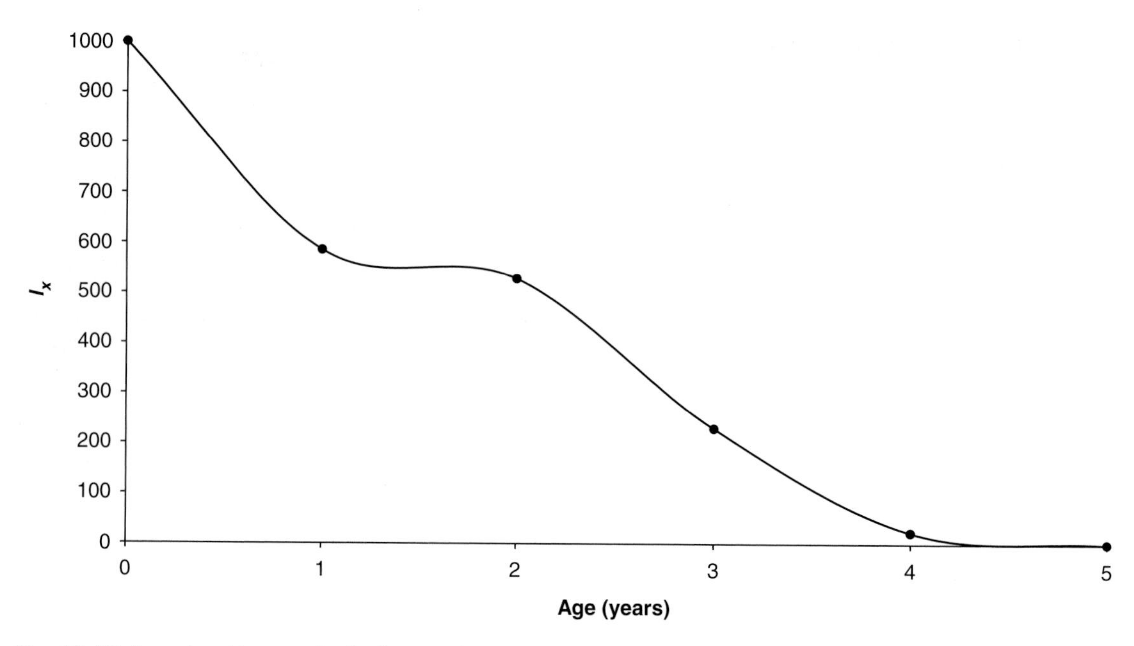

Fig. 12.12 A survivorship curve for the theoretical population described in Table 12.7. (The starting population has been adjusted to 1000 individuals.)

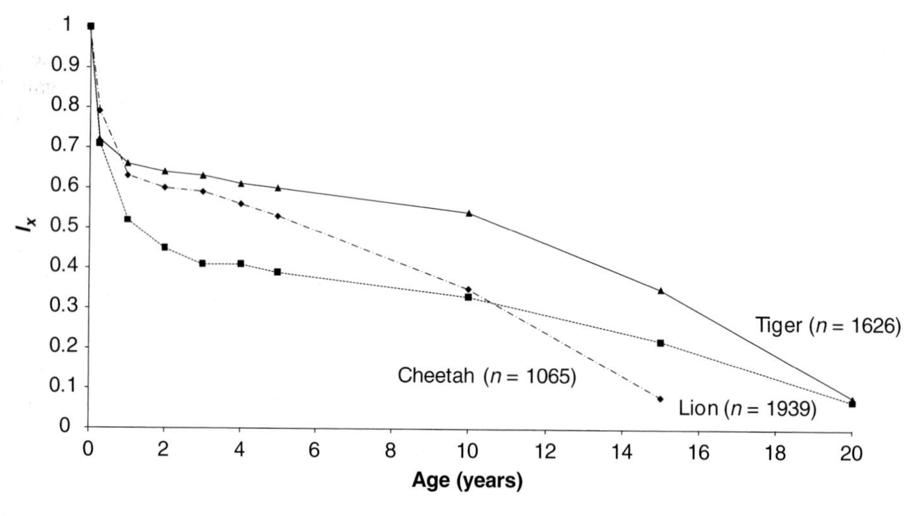

Fig. 12.13 Survivorship curves for three felid species in zoos. (Based on data in Kohler *et al.*, 2006.)

Wiese (2000) analysed data from the North American Asian Elephant Studbook using *Population Management 2000* software, and predicted that the Asian elephant (*Elephas maximus*) population of North America would drop to just 10 individuals in 50 years and be demographically extinct unless birth rates increased drastically or importation of elephants increased. The taking of animals from the wild to supplement zoo populations – especially elephants – is highly controversial, and zoos need to take great care in explaining the reasons for this to the public (Hutchins and Keele, 2006).

12.6 GENETIC MANAGEMENT

Biography 12.2

Gregor Johann Mendel (1822–84)

Mendel was an Augustinian monk who founded the science of genetics after conducting extensive experiments with pea plants. Mendel presented the results of his experiments at meetings of the Brünn Natural History Society in 1865. His ideas were largely ignored by the scientific community until the early 20th century. Mendel's laws of segregation and the independent assortment of characters now form the basis of modern genetics.

12.6.1 Introduction

Genetics is the scientific study of heredity. It owes its origins to a monk called Gregor Mendel who discovered the rules by which characteristics are inherited by studying pea plants (*Pisum sativum*) (Mendel, 1866).

The reproductive organs of animals are called gonads. The sex cells are called sperm in males and ova (singular: ovum) in females. Collectively they are referred to as gametes. In sexually reproducing animals, the gametes carry half of the genes from the mother and half from the father on chromosomes called autosomes. In many species, they also carry one sex chromosome from each parent. When a sperm fertilises an ovum a zygote is formed which contains the full complement of chromosomes.

Somatic (body) cells contain both chromosomes of a pair and are referred to as being diploid. Sex cells (gametes) contain only one chromosome from each pair and are haploid. When a sperm cell fertilises an egg cell two haploid cells (n) fuse to form a diploid zygote ($2n$), where n is the number of pairs of chromosomes. The chromosome number is the total number of chromosomes in the somatic cells and varies between species (Table 12.8).

The sex chromosomes determine the sex of the individual and are responsible for sex-linked characteristics. In many species there are two sex chromosomes but in some the situation is more complex (Table 12.9). The autosomes carry the genes for all of the other characteristics.

Heritable characteristics are transmitted from one generation to another by genes. Genes are sections of deoxyribonucleic acid (DNA) located on chromosomes. Different forms of a gene are known as alleles. A gene may have two or more alleles. The alleles which form a pair are carried on separate chromosomes and pass to different sex cells (sperm or ova) as a result of meiosis.

Table 12.8 Chromosome numbers in selected animal species.

Species	Chromosome number
Carp (*Cyprinus carpio*)	104
Horse (*Equus calibus*)	64
Honeybee (*Apis mellifera*)	56
Human (*Homo sapiens*)	46
Orangutan (*Pongo pygmaeus*)	44
Rhesus monkey (*Macaca mulatta*)	42
Rat (*Rattus norvegicus*)	42
Alligator (*Alligator mississippiensis*)	32
Frog (*Rana pipiens*)	26
Aardvark (*Orycteropus afer*)	20
Fruit fly (*Drosophila melanogaster*)	8

Chromosome number is the total number of chromosomes in somatic (body) cells, including the sex chromosomes (i.e. the diploid number). Adapted from Beck *et al.* (1991).

They then recombine at random when a sperm fertilises an ovum (Fig. 12.14). One allele may be dominant to another, recessive, allele. For example, in tigers, white forms occasionally occur when two recessive alleles for this condition are inherited (Box 12.2; Fig. 12.15).

All of the alleles possessed by an individual make up its genotype. Some of these alleles will be recessive so may not be exhibited in the animal. The exhibited characteristics are referred to as the phenotype. A tiger carrying one recessive allele for the white colour is phenotypically normal, in relation to its colour – because of the presence of one dominant allele which suppresses its effect – but carries an allele for the white condition as part of its genotype. Many inherited characteristics, such as size, are controlled in a much more complex way than this, with many genes being involved which interact with the environment. For example, an individual animal that has inherited genes which should

Table 12.9 Sex determination in a range of taxa.

Sex determination systems:
XX – XY: the male is the heterogametic sex (XY).
ZZ – ZW: the female is usually the heterogametic sex (ZW).
XX – XO: there is only one type of sex chromosome.
('O' indicates the absence of a sex chromosome).

Taxon	Sex chromosomes	
	Male	Female
Mammals	XY	XX
Duck-billed platypus*	5 pairs of XY	5 pairs of XX
Birds	ZZ	ZW
Some lizards	XY	XX
Other lizards and many snakes	ZZ	ZW
Some turtles, lizards and crocodilians	Genes that control sex are temperature-dependent in a critical period during incubation.	
Snakes	ZZ	ZW
Guppies (domestic strain)	ZZ	ZW
Guppies (wild strain)	ZW	ZZ
Drosophila (fruit fly)	XY	XX
Grasshoppers	XO (i.e. one X only)	XX
Butterflies	ZZ	ZW
Hymenoptera (bees, wasps, ants, etc.)	No sex chromosomes. Males hatch from unfertilised eggs and are haploid. Females hatch from fertilised eggs and are diploid.	

**Ornithorhynchus anatinus* 10 sex chromosomes instead of the usual 2 possessed by other mammals (Grützner, 2004).

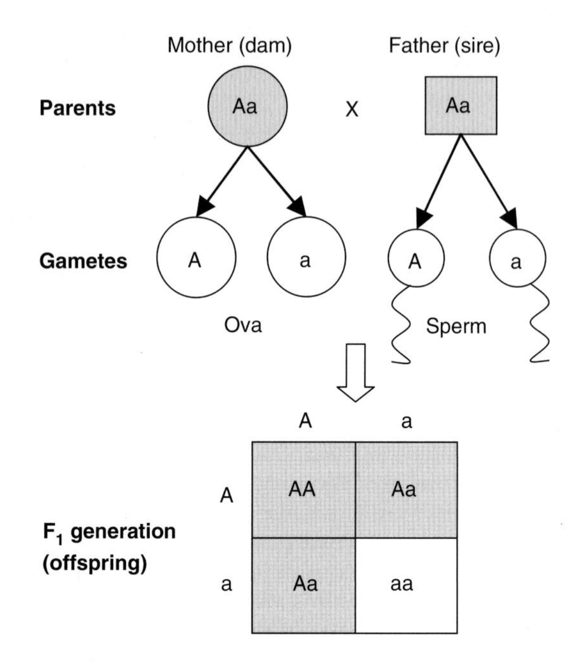

Fig. 12.14 A monohybrid cross between two individuals who are heterozygous for gene A.

make it tall could have its height stunted by the effects of poor nutrition.

In order to record and manage breeding between animals in captive populations zoos keep studbooks which record the genetic relationships between the individuals (see Section 12.2.1).

12.6.2 How many individuals do we need to maintain genetic diversity?

Genetic management is essential to the long-term viability of captive breeding programmes. Each breeding animal only passes on half of its genes to each of its offspring so some loss of genetic diversity is inevitable, especially in very small populations. A population size of 500 animals is often quoted as necessary to maintain genetic diversity (Tudge, 1991) (Table 12.10). However, space in zoos is at a premium so this is an unrealistically high number for many large species.

It is more realistic to plan to conserve 90% of the variation in a population for 200 years and it is possible to calculate how large the population size needs to be to achieve this by considering the proportions of males and females available and by considering the generation time of the species. When generation time is short, a large population is needed to conserve genetic diversity because some is lost each time breeding occurs.

12.6.2.1 Effective population size

In assessing the future breeding potential of captive animals the actual population size is less important than the effective population size. The effective size of a population is a measure of how well the population maintains genetic diversity from one generation to the next (Ballou and Foose, 1996).

Small populations lose genes as a result of chance events. For example, the last two individuals possessing the gene for blue eyes might be killed in a storm. This

Box 12.2 The genetics of white tigers.

White tigers (*Panthera tigris*) are not albino individuals. The white colour is conferred by a recessive allele which can only be expressed if an individual possesses two copies of it: one inherited from its mother and the other from its father (Fig. 12.15a). If the dominant allele (for normal coat colour) is represented by 'W' and the recessive allele (for white coat colour) by 'w', parents that are both heterozygous for this gene have a one in four chance of producing a white cub (Fig. 12.15b).

The individuals that are heterozygous for the gene (Ww) appear normal but carry the allele for white coat colour and may pass it on to the next generation. In the wild it is relatively rare for white tigers to be seen because two heterozygotes or two white tigers would have to mate.

Although white tigers are a genetic rarity there is at least some evidence that they are able to survive in the wild. Seventeen white individuals were shot in the wild by hunters between 1907 and 1933 (Croke, 1997), and white tigers were once considered 'not unusual' in north and east central India (Macdonald, 1984). All wild white tigers have been of the Bengal subspecies (*P. t. tigris*).

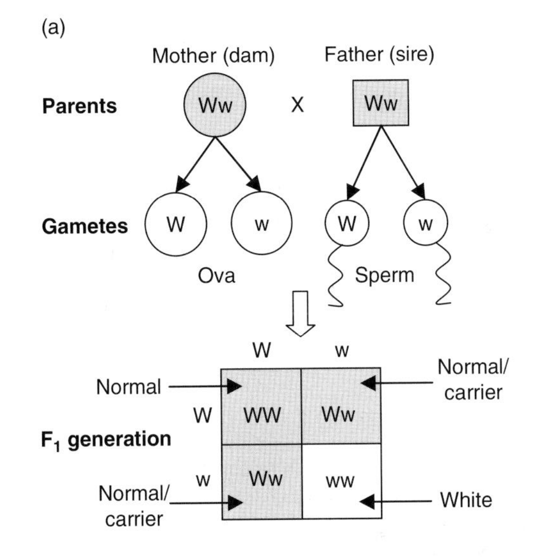

(b)

Fig. 12.15 (a) The results of a cross between two normal coloured tigers (*Panthera tigris*) which are heterozygous for the gene that confers white coat colour. (b) White tiger (*Panthera tigris*) cubs.

(Continued)

Box 12.2 *(Cont'd)*

Some white tigers are pink-eyed and pure white, while others have ice-blue eyes and black or brown stripes on a white, egg-shell white, or cream background. Tigers of the latter type are well known from the former Indian state of Rewa, an area which is now within Bandhavgarh National Park. A white tiger was captured from this district in 1915 and described in the *Journal of the Bombay Natural History Society*. In 1951 the Maharaja of Rewa began breeding white tigers in captivity and selling them to zoos. During the 1960s specimens were obtained by zoos in Bristol in the UK and Washington DC (Guggisberg, 1975). The last record of a white tiger shot in India is from 1958, near Hazaribagh in Bihar. There appear to have been no reported sightings in the wild since the 1950s.

Table 12.10 The minimum number of individuals required to maintain genetic diversity.

Species	Generation time (years)	Effective population size
Caribbean flamingo	26	37
Indian rhinoceros	18	53
Mauritius pink pigeon	10	95
Arabian oryx	10	95
Siberian tiger	7	136
Bullfrog	7	136
Striped grass mouse	0.75	1275

Adapted from Conway (1986).

process is known as genetic drift. The effective population size is the size of an ideal population that would lose genetic variation by genetic drift at the same rate. In other words, a population of 200 individuals may have an effective population size of, say, 150 because there are too few females in the population. So this population of 200 actually loses genetic variation at the same rate as an ideal population of 150. The effective population size is essentially a measure of the number of individuals that are effectively contributing genes to the next generation.

Effective population size (N_e) may be calculated as

$$N_e = \frac{4(N_f \times N_m)}{(N_f + N_m)}$$

where N_m is the number of breeding males and N_f is the number of breeding females. Strictly speaking this formula only applies to stable, randomly mating populations with non-overlapping generations (Ballou and Foose, 1996).

The importance of the concept of effective population size is best explained by an example. At the end of 1980 there were 558 southern white rhinos (*Ceratotherium simum simum*) in captivity. Of these, only 20 males and 65 females were reproducing. The effective population size was

$$N_e = \frac{4 \times 20 \times 65}{20 + 65} = 61$$

So, although there were 85 viable rhinos in the zoo population, the genetically effective size was 61 (Foose, 1983). The effect of sex ratio on N_e is illustrated in Fig. 12.16. When the sex ratio is 1:1 the effective population size is the actual population size. The further the ratio deviates from this the greater the difference between the actual and the effective population size. When all of the individuals are of the same sex $N_e = 0$.

As a rule of thumb, it is generally considered that an effective population of at least 500 is required for the long-term maintenance of genetic variation. However, it may require 1500–2000 individuals to maintain this (Soulé, 1980).

For a given population size the effective population size will decrease as the sex ratio deviates from 1:1 (Fig. 12.16). In order to achieve an effective population size of 500 individuals, if the size of the breeding male population is small, there is little point in increasing the size of the breeding female population. In a population of 100 males and 400 females, N_e = 320. Adding 10 females to the population would only increase N_e to 322, but adding 10 males would increase N_e to 345.

Effective population size may be substantially lower than actual population size due to:

- variance in the sex ratio
- variance in reproductive success between individuals
- changes in population size with time, which can be influenced by the mating system (Caro and Eadie, 2005).

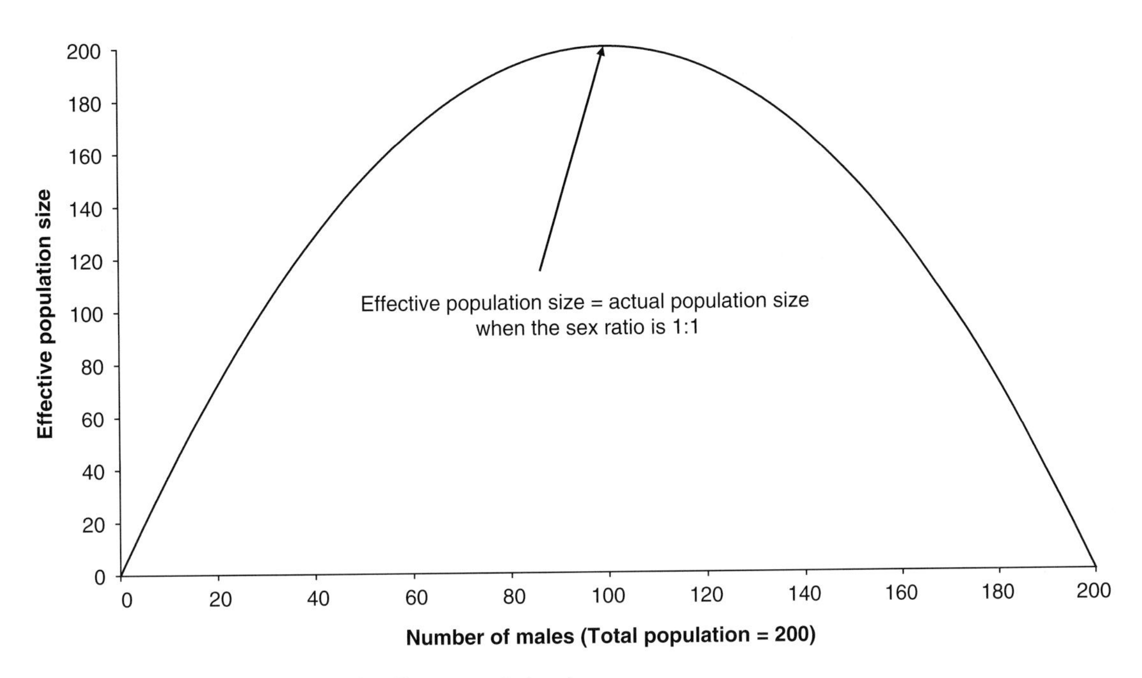

Fig. 12.16 The effect of sex ratio on the effective population size.

As the mating system deviates from monogamy to extremes of polygyny or polyandry, N_e is reduced because fewer males (polygyny) or fewer females (polyandry) contribute genes to the next generation (Fig. 12.17). On the other hand, increased promiscuity leads to an increase in N_e because members of the more abundant sex have more opportunities to breed than they would have in a monogamous system (Parker and Waite, 1997).

Mate selection may have an important effect on N_e. Females rarely choose mates at random. Usually they select on the basis of the presence of a particular trait or traits, such as body symmetry (e.g. symmetrical horns in antelope species). If the body condition of males in a population is poor, fewer males may be acceptable to females as mates and effective population size will fall as a result (Blumstein, 1998).

The degree to which adults in a group do not breed is referred to as reproductive skew. If all adults breed there is no skew; if only one adult of a particular sex breeds and all of the others of that sex do not, then skew is complete.

The figures for effective population size in Table 12.10 do not take into account injuries and other losses of animals from the population so must be considered the bare minimum necessary. A recent study has suggested that conservationists have been over-optimistic about the number of individuals necessary for a species to survive. Traill *et al.* (2010) have concluded, after reviewing the relevant literature, that:

> thousands (not hundreds) of individuals are required for a population to have an acceptable probability of riding-out environmental fluctuation and catastrophic events, and ensuring the continuation of evolutionary processes.

They claim that many of the studies of the impacts of inbreeding depression underestimate its effect on population viability because they are based on juvenile mortality in captive populations and do not consider all of the components of reproduction and survival in the wild (see Section 12.6.5). If they are right, this may have serious implications for the number of species that can be saved from extinction.

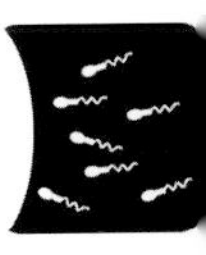

12.6.2.2 Minimum viable population (MVP) size

Population viability analysis (PVA) provides a quantifiable means of predicting the probability that a population will become extinct and it can be used for prioritising conservation needs. PVA allows the

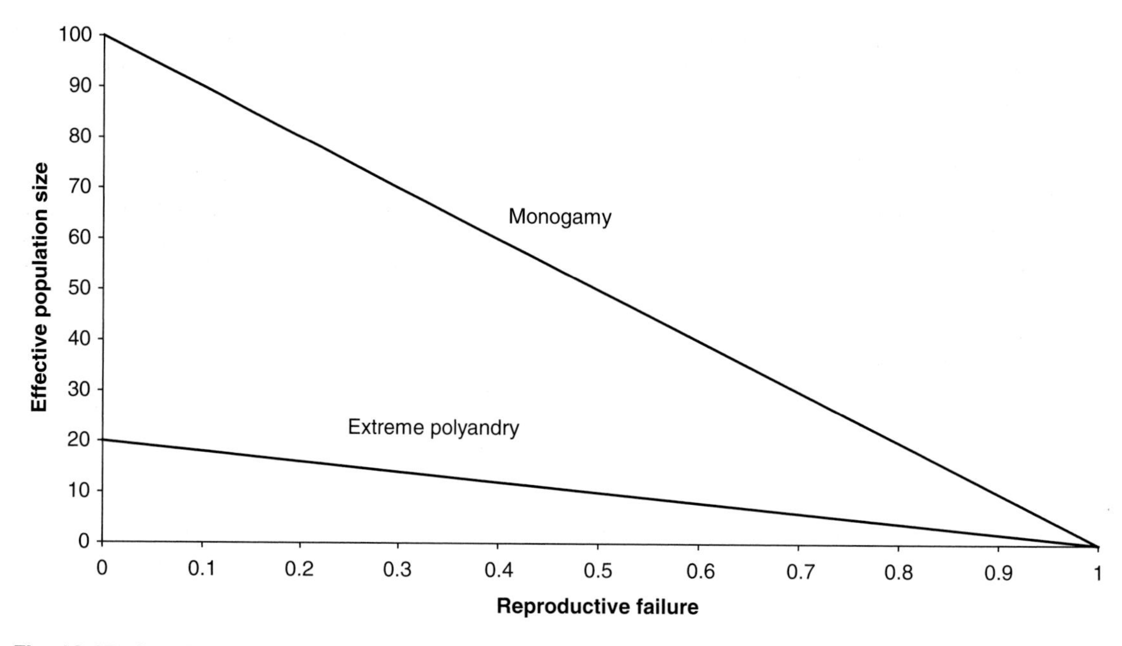

Fig. 12.17 The influence of the mating system on the relationship between the rate of reproductive error and annual effective population size. (Adapted from Parker and Waite, 1997.)

calculation of a minimum viable population (MVP) size for a species. The methodology takes into account both deterministic factors (e.g. habitat loss or over-exploitation) and stochastic (random) factors (e.g. demographic, environmental and genetic factors). This type of analysis may be used to answer questions such as 'What is the risk of extinction of the snow leopard in the next 50 years?'

Reed *et al.* (2003) used PVA to determine MVPs for 102 species of vertebrates (one fish, two amphibians, 18 reptiles, 28 birds and 53 mammals) using VORTEX (see Section 12.6.3). They defined the MVP size as one with a 99% probability of persistence for 40 generations. The models used included age structure, catastrophes, demographic stochasticity, environmental stochasticity, and inbreeding depression.

Reed *et al.* concluded that conservation programmes for wild populations need to be designed to conserve habitat capable of supporting approximately 7000 adult vertebrates – with little variation between major taxa – in order to ensure long-term survival (Table 12.11). This figure is slightly larger than previous estimates of MVP.

Traill *et al.* (2010) believe that the number of individuals required to maintain a small population of animals is generally underestimated. They argue that the targets set by conservation organisations are often too low and that most vulnerable species are not really being managed for long-term viability but for short-term persistence and to accommodate complex political and financial realities. They liken the inability of

Table 12.11 Estimates of MVP for selected vertebrates.

Species	MVP_c (Minimum viable adult population size)*
African elephant (*Loxodonta africana*)	5,474
American alligator (*Alligator mississippiensis*)	3,783
Asian elephant (*Elephas maximus*)	4,737
Asiatic lion (*Panthera leo persica*)	9,405
Black rhinoceros (*Diceros bicornis*)	6,199
Eurasian sparrowhawk (*Accipiter nisus*)	5,244
Gorilla (*Gorilla gorilla beringei*)	11,919
Grey wolf (*Canis lupus*)	6,332
Grivet (*Cercopithecus aethiops*)	19,547
Komodo dragon (*Varanus komodoensis*)	15,283
Song sparrow (*Melospiza melodia*)	9,870

*Corrected to 40 generations worth of data.
Adapted from Reed *et al.* (2003).

conservationists to tackle extinction problems to the inability of governments to tackle climate change. If Traill *et al.* are right then zoos may need to accommodate far larger populations of rare species in the future than their current plans suggest. As zoo accommodation for large species is in short supply, this may mean sacrificing some species to make room for others.

12.6.3 Population analysis software

A number of software packages are available for the demographic and genetic analysis of animal populations. The most important of these are described below.

VORTEX
VORTEX is population viability analysis (PVA) software. It has been used extensively by the Conservation Breeding Specialist Group of the IUCN to identify the threats faced by species and evaluate the likelihood that they will survive.

OUTBREAK
OUTBREAK is modelling software which is used for wildlife disease risk assessment. It is designed to be coupled with VORTEX to provide a more extensive assessment incorporating a consideration of disease, genetic change, demographic stochasticity, environmental variation and management actions.

MateRx
This software is used for prescribing mates in a pedigree population. MateRx is a genetic tool intended as an aid to population management. For each pair (male/female) in the population it calculates an index (mate suitability index (MSI)) that indicates the relative genetic benefit or detriment to the population of breeding from that pair. In calculating the MSI, the program considers the mean kinship values of the pair, the difference in mean kinship values of the male and female, the inbreeding coefficient of the offspring produced, and the amount of unknown ancestry in the pair. In effect the MSI condenses everything known about the genetics of a pair of individuals into a single number. MateRx was developed by staff at the Smithsonian National Zoological Park and Lincoln Park Zoo.

Population Management 2000 (PM2000)
PM2000 is software used for the genetic and demographic analysis and management of pedigrees. It can be used as a stand-alone program but is most easily used in conjunction with the SPARKS studbook management software developed by ISIS (see Section 13.3.3).

GENES
This is software which is used for the genetic analysis and management of pedigrees. GENES may be used as a stand-alone program but is most easily used as an accessory program to the SPARKS software for the management of studbooks.

PARTINBR
This is software which can be used to calculate inbreeding coefficients and partial inbreeding coefficients from pedigrees.

Sebag studbook
This is a Windows based SPARKS-compatible studbook administration program which is used by the European Studbook Foundation and is available in English, German, Dutch and Spanish.

SCBook (SPARKS Compatible studBook)
SCBook is a Windows implementation of SPARKS which implements the functionality required by the European Studbook Foundation.

ZooRisk
This is software which is designed to help managers make scientifically based decisions about the management of captive populations by providing a quantitative assessment of a population's extinction risk due to demographic, genetic and management factors. This assessment is based on the history of the population, the biology of small populations and knowledge of our ability to manage captive populations. ZooRisk was developed by Lincoln Park Zoo and is distributed free as shareware.

12.6.4 Evolutionarily significant units (ESUs)

Which animals should we try to save? An evolutionarily significant unit is a population that is considered distinct for conservation purposes: the minimum unit of conservation management. The term could refer to a species, subspecies, geographical race or population. An ESU should be substantially reproductively isolated from other conspecific populations and it should represent an important component of the evolutionary history of the species. Modern analyses of ESUs rely upon information from molecular genetics, specifically mitochondrial DNA (mtDNA). Mitochondria are cell organelles that are passed from mother to offspring. They contain mtDNA which, because it remains largely unchanged from generation to generation, allows us to trace the evolutionary history of species.

So an ESU is essentially a set of populations that is morphologically or genetically distinct from other similar populations, or a set of populations with a distinct evolutionary history. This causes a dilemma for conservationists. How distinct do populations have to be to qualify as an ESU? If many ESUs can be identified within a single endangered species, can they all be saved from extinction? Some species appear to exist as distinct subspecies in the wild, such as tigers (*Panthera tigris*) and Asian elephants (*Elephas maximus*). However, while captive breeding programmes in zoos separate out tiger subspecies (e.g. Amur tigers (*P. t. altaica*) and Sumatran tigers (*P. t. sumatrae*)) there are too few Asian elephants in zoos to justify attempting to keep the subspecies separate.

From an evolutionary perspective species which are the only representatives of a genus, family or higher taxon are perhaps deserving of greater attention than species which have many close relatives, since the loss of such a species would result in the loss of the higher taxon also. For example, the loss of the aye-aye (*Daubentonia madagascariensis*) would result in the loss of the genus *Daubentonia* and the family Daubentoniidae since this species is the only extant member. By contrast, the loss of the tiger would be the loss of just one of the four extant species of the genus *Panthera*. Obviously, in straightforward numerical terms, both extinctions would be of a single species.

12.6.4.1 Hybridisation between subspecies

In some zoo populations, uncontrolled breeding between subspecies has resulted in the production of generic animals which are of limited conservation value. In some species, for example Asian elephants (*Elephas maximus*), no attempt has been made to maintain separate captive populations of the various subspecies as the total zoo population is too small to make this possible. However, in other species breeding programme coordinators may adopt strategies to maintain the genetic integrity of particular subspecies. Occasionally hybrids are created in zoos unintentionally (Box 12.3).

Case study: The problem of surplus generic chimpanzees in Europe

The chimpanzee (*Pan troglodytes*) exists as three subspecies: the central African form (*P. t. troglodytes*), the western form (*P. t. verus*), and the eastern form (*P. t. schweinfurthi*). Historically, many zoos have kept chimpanzees in mixed subspecies groups, with the result that large numbers of hybrids have been produced.

The Western Chimpanzee EEP Species Committee has produced a strategy for the management of chimpanzees in an attempt to deal with the growing surplus of generic chimpanzees in EAZA zoos caused by interbreeding (Carlsen and de Jongh, 2009). For the western chimpanzee (*Pan troglodytes verus*) the EEP proposed the following:

- A moratorium on breeding including:
  - no breeding with identified subspecies hybrids
  - no breeding between specimens of different subspecies (hybridisation)
  - no breeding with specimens of unidentified subspecies.
- The establishment of additional bachelor groups.
- The keeping of additional males within mixed-sex groups where possible.
- An increase in the EAZA zoo capacity for chimpanzees to accommodate a valuable stock of animals from biomedical laboratories.
- The development of new methods for identifying the subspecific or hybrid status of chimpanzees.

12.6.5 Inbreeding and inbreeding depression

Inbreeding is the process that occurs when closely related animals breed. Inbreeding may be defined as any breeding scheme that results in the sire and the dam of an individual having common ancestors. This is measured by calculating an inbreeding coefficient. This measure was devised by Sewell Wright and is called F (after the statistician and geneticist R. A. Fisher). F is the probability that two alleles in an individual animal are identical by descent (i.e. they have been inherited from the same ancestor).

Consider a situation in which half-siblings mate (Fig. 12.20a). In this example, 'Alex' has mated with 'Mary' to produce 'Jake', and with 'Jane' to produce 'Mimi'. Jake and Mimi mated and produced 'Noah'. Noah has a grandfather who appears on both sides of his pedigree because both of his parents have the same father.

Box 12.3 Hybridisation.

When two species interbreed the resultant offspring are called hybrids (Figs 12.18 and 12.19). In the wild, hybrids are rare because evolution has produced mechanisms to prevent hybridisation:

- Geographical separation (non-overlapping distributions)
- Genetic differences (e.g. incompatible chromosomes)
- Behavioural separation (e.g. different courtship behaviours)
- Ecological separation (e.g. occurring in different habitats).

In the past zoos displayed tigons (male tiger × female lion) and ligers (male lion × female tiger). Some of these were accidental matings but some zoos purchased hybrids for their novelty value. Two tigons arrived at Belle Vue Zoo in Manchester, UK in 1936. They were brought by Hagenbeck from Dresden Zoo. The male died in 1941 and the female in 1949. In 1957 the zoo acquired another tigon from the Paris Zoo at Vincennes. She had originally belonged to the Sultan of Morocco. She died of old age in 1968 (Nicholls, 1992).

Ligers and tigons were long thought to be sterile. However, in 1943 a liger was successfully mated with a lion at Munich-Hellabrunn Zoo and the female cub was raised to adulthood. In Japan, Koshien Zoo produced leopons by mating a male leopard with a lioness. Around the turn of the last century Chicago Zoo produced three cubs from a mating between a male jaguar and a female leopard. They were subsequently sold to a menagerie. The female cubs were eventually mated with a lion and produced several litters. One of these jaguar-leopard-lions went to London Zoo. These hybrids have been described by Guggisberg (1975).

Some zoos keep hybrids between orangutans (*Pongo* spp.) from Borneo and Sumatra which are unsuitable for breeding now that the two forms are considered to be separate species. Some zoos work hard to maintain the genetic integrity of subspecies but for some species there are too few individuals in breeding programmes to make this possible.

Modern zoos sometimes unintentionally create hybrids when similar species are kept together. A false killer whale (*Pseudorca crassidens*) mated with a bottlenose dolphin (*Tursiops truncatus*) producing a 'wholphin' at the Sea Life Park on Oahu, Hawai'i in 1986. On 11 July 1978 a hybrid elephant was born at Chester Zoo in the UK: an African elephant × Asian elephant cross. The baby was a male. He was six weeks premature and 27 kg under weight. Unfortunately, he died of peritonitis after only 10 days: his colon and umbilical area were infected with the bacterium *E. coli*. The hybrid exhibited an interesting combination of African and Asian features. His head, ears and trunk were African, except for the tip which was characteristic of the Asian species, having only one finger. His vertebral column was convex like the Asian, but shaped like the African species above the shoulder. His forelegs possessed five toe nails but his hind legs had only four. This is typical of the Asian elephant but considerable variation is found in African elephants (Rees, 2001c).

We should not forget that hybridisation occurs between closely related species in the wild. In the United States, coyotes hybridise with domestic dogs producing 'coydogs', and with wolves producing 'coywolves'. In Scotland hybrids between the indigenous red deer (*Cervus elephus*) and the introduced sika deer (*C. nippon*) are common, as are hybrids between the wildcat (*Felis silvestris*) and the domestic cat (*F. catus*).

What forms should we conserve? – When is a hybrid worth conserving?

In the United States in 1967 the red wolf (*Canis rufus*) was listed as endangered under the Endangered Species Preservation Act 1966. However, the genetic status of the red wolf has been the subject of considerable controversy, with some authorities claiming that it is a distinct species (e.g. Nowak and Federoff, 1998; Wilson *et al.*, 2000) and others claiming that it is a hybrid between the grey wolf (*C. lupus*) and the coyote (*C. latrans*) (e.g. Wayne and Jenks, 1991). This distinction is important because the US Fish and Wildlife Service has been involved in a multi-million dollar recovery programme for the red wolf (authorised under the Endangered Species Act) but the Act does not cover hybrids.

Although Wayne believes that the red wolf is the result of hybridisation – citing evidence from mitochondrial DNA analysis – nevertheless, he believes that it is worthy of some measure of protection because 'it may be the only living repository of characteristics once held by a valid subspecies of grey

(*Continued*)

Box 12.3 (*Cont'd*).

Fig. 12.18 A zebra × ass hybrid stallion. (Courtesy of Chetham's Library, Manchester.)

Fig. 12.19 A gooder: a goosander (*Mergus merganser*) × eider (*Somateria mollissima*) hybrid unintentionally created at the Wildfowl and Wetlands Trust's facility at Martin Mere when the two species were housed together.

(*Continued*)

Box 12.3 (*Cont'd*)

wolf' (Wayne, 1995). In contrast, he suggests that 'The hybrids resulting from the recent interbreeding of coyotes with grey wolves in the Minnesota region would not deserve protection as a distinct form. They arose as a result of habitat alterations and predator control measures which decimated the wolf population'. These grey wolves are now protected by the ESA and interbreeding with coyotes has ceased. The extent to which hybridisation may affect the protection of the grey wolf under the ESA has been discussed by Adkins Giese (2005).

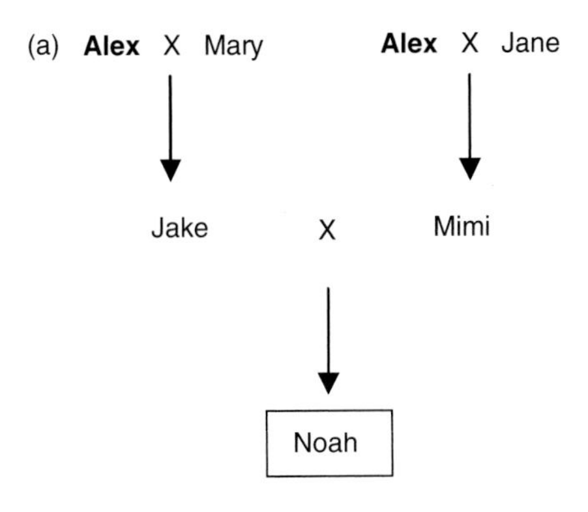

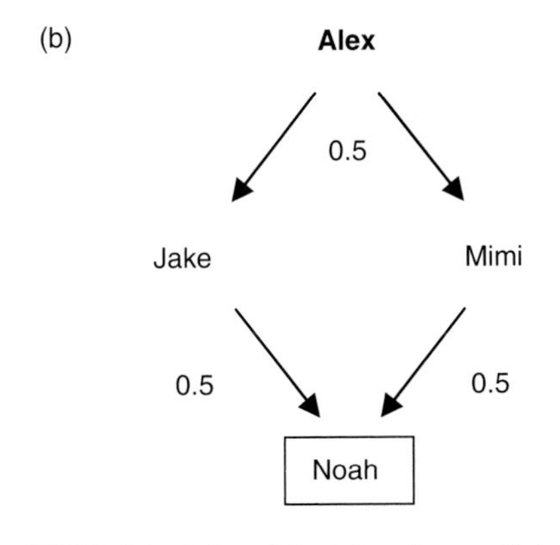

Fig. 12.20 Calculation of the inbreeding coefficient (F) for a mating between half-siblings showing (a) all matings and (b) shared ancestors only. See text for explanation.

If we remove the ancestors that are not shared, the relationships can be simplified (Fig. 12.20b). If Alex possesses two alleles for gene A, A1 and A2, whichever one he passes to Jake has a 50% chance of being passed to Noah. There is also a 50% chance that the same allele will be passed from Alex to Mimi, and if it is, there is a 50% chance of it being passed on to Noah. The probability of all of these events occurring is calculated by multiplying the probabilities together: $0.5 \times 0.5 \times 0.5 = 0.125$, or 12.5%. This is the probability that either A1 or A2 will be homozygous in Noah because of the common grandfather. If this were the only common ancestor, the inbreeding coefficient for Noah would be 0.125.

F may be calculated by counting the links from Noah to the common ancestor and back again on the other side of the pedigree, excluding Noah, i.e. Jake – Alex – Mimi (Fig. 12.20b).

$$F = 0.5^n$$

where n is the number of individuals in the path (excluding the subject of the calculation). In this case $n = 3$, so $F = 0.5 \times 0.5 \times 0.5 = 0.125$.

The higher the value of F, the greater the degree of inbreeding. Studbook keepers use inbreeding coefficients to determine which pairings of animals to recommend to keep inbreeding to a minimum. Of course, this methodology assumes that the ancestry of all of the animals kept in zoos is known and this is not always the case.

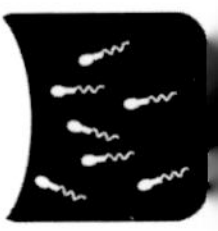

Inbreeding causes an increase in homozygosity in a population. In other words, the proportion of a population which is homozygous dominant or homozygous recessive for a particular gene will increase and the proportion of heterozygotes will decrease.

The effect of inbreeding on homozygosity is best understood by considering a hypothetical example. Imagine a self-fertilising animal that is heterozygous for the allele A and therefore possesses the genotype 'Aa'. Such an animal would be extremely unusual as almost all species undergo at least some cross-fertilisation. However, the example will serve to demonstrate the point. If this animal were to self-fertilise and produce four offspring, one half of them would be heterozygotes (Aa) and therefore identical to the parents, a quarter would be homozygous dominant (AA), and a quarter would be homozygous recessive (aa) (Fig. 12.21a). The homozygous individuals produced will only create offspring with the same genotypes as themselves, so AA individuals will only produce more AA individuals.

For the purpose of this example, assume that each animal produces four offspring in each generation. The result is that the proportion of heterozygotes (Aa) in the population as a whole halves with each generation (Fig. 12.21b). Generation 1 consists of a single heterozygote, so 100% of the population is heterozygous. The second generation consists of four individuals with the genotypes AA, Aa, Aa and aa, so now 50% of the population is heterozygous. The AA individual now produces four AA offspring in generation 3. Likewise, the aa individual produces four aa offspring. The two Aa individuals produce four offspring each with the genotypes AA, Aa, Aa and aa. If we count the genotypes present in generation 3 we find 6 × AAs, 4 × Aa and 6 × aa. So of the 16 individuals, four (25%) are heterozygotes (Aa) and 12 (75%) are homozygotes (i.e. either AA or aa). If we continue with this process, the number of heterozygotes continues to halve with each generation so that by generation 8 the population is 99.2% homozygous (Fig. 12.21c). Inbreeding of this type is producing two homozygous lines and practically eliminating the heterozygotes.

If the recessive allele (a) in this example caused a disease, it would only be exhibited in the double recessive (aa) individuals. So in generation 1 the disease would not be exhibited in the population at all. But, because the founding individual was a carrier of the disease (Aa), by generation 8 almost half the population (49.6%) would have the disease. The pool of deleterious genes in a population – which cause genetic diseases – is known as the 'genetic load'. High juvenile mortality occurs in a number of species which have been inbred (Ralls *et al.*, 1979).

Inbreeding depression is reduced fitness in a population resulting from breeding between close relatives. It results in reduced variation in the population. Inbreeding depression may be the result of:

- an increase in homozygosity, resulting in an increase in the proportion of the population which is homozygous for a deleterious allele; or
- overdominance. Sometimes the heterozygote phenotype lies outside the phenotypic range of the homozygotes, i.e. it does not have the characteristics of either the homozygous dominant or the homozygous recessive. This phenomenon is called overdominance. For example, a newborn which is intermediate in size may have a higher chance of survival than a very large or very small individual. So heterozygotes which influence growth rate show overdominance in this case. A reduction in the proportion of heterozygotes where genes exhibit overdominance and confer a selective advantage may result in inbreeding depression.

It is possible to increase the amount of heterozygosity in a population – a process known as heterosis – by crossing two different inbred lines. This is 'outbreeding' and is effectively the opposite of inbreeding. If an individual that is heterozygous dominant for gene A (AA) is crossed with a double recessive (aa) all of the offspring will be heterozygotes (Aa). This may have a highly beneficial effect on a population by instantly reducing homozygosity in a large number of genes. However, outbreeding may have a negative effect in some cases.

A population may potentially suffer from inbreeding depression (reduced fitness in the offspring of closely related parents) and, because subspecies are not always kept separate in zoos, outbreeding depression (reduced fitness in the offspring of very distantly related parents). Outbreeding may bring together alleles that cause a harmful effect in the offspring.

In a recent study of the effect of inbreeding and outbreeding on juvenile survival in reintroduced Arabian oryx (*Oryx leucoryx*), Marshall and Spalton (2000) warned that outbreeding depression may be more common in vertebrates than previously supposed.

Biography 12.3

Charles Robert Darwin (1809–82)

Darwin was an English biologist who developed the modern theory of evolution. He published his theory of evolution in *On the Origin of Species by Means of Natural Selection* in 1859 after travelling around the world in HMS *Beagle*. His work forms the basis of our modern-day understanding of evolutionary processes and the formation of new species. Darwin was a great supporter of London Zoo.

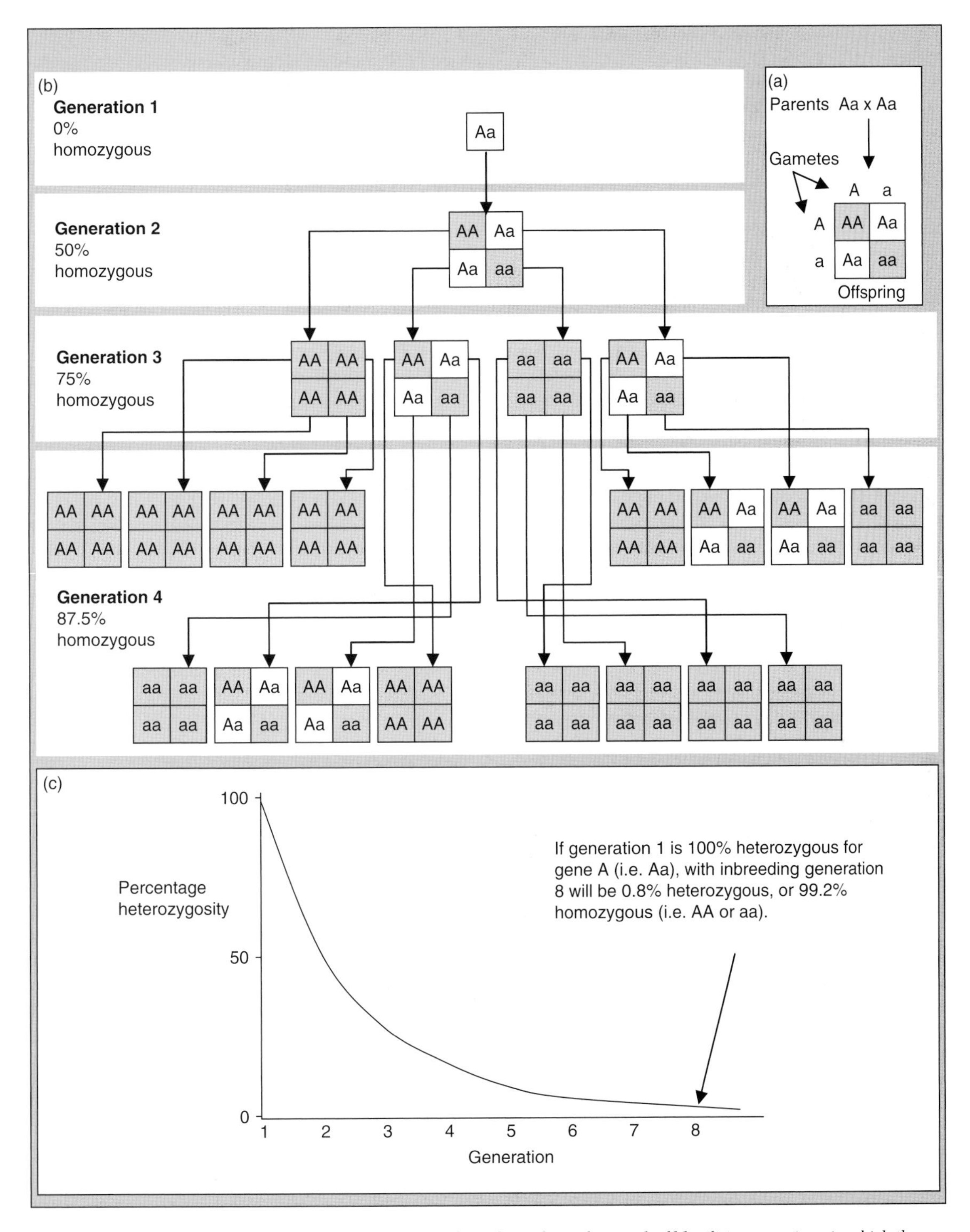

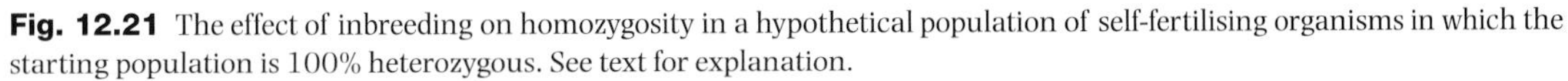
Fig. 12.21 The effect of inbreeding on homozygosity in a hypothetical population of self-fertilising organisms in which the starting population is 100% heterozygous. See text for explanation.

12.6.6 Genetic drift and the founder effect

The allele frequencies in a population may be altered with time by natural selection. However, they may also be altered by chance events (stochasticity) occurring in the environment. This is especially important in small populations and therefore directly relevant to the situation of rare animals in the wild and captive-bred populations in zoos.

When a small, founder population is taken from the wild the genetic composition of future generations derived from this population will be determined to a very large extent by the frequencies of the various alleles possessed by these founders.

The effect of selecting two different founder populations from a source population is illustrated in Fig. 12.22. Founder population 1 contains individuals with a variety of genotypes and, if these individuals mate at random, they are capable of producing offspring with every possible combination of the alleles for genes A and B. However, founder population 2 contains only homozygotes (AAbb) and when these individuals mate at random they can only produce individuals that have exactly the same genotype as themselves. Clearly the individuals selected in any founder population can have a devastating effect on future genetic diversity if insufficient attention is paid to selecting animals carrying a wide range of alleles and the founders have a high level of homozygosity.

This process can lead to the creation of a genetic bottleneck, severely restricting the capacity of the population to evolve in response to future environmental change. There is evidence of bottlenecks substantially increasing the risk of population extinction due to the deleterious effects of inbreeding and increased susceptibility to disease (e.g. Mills and Smouse, 1994; Westemeier *et al.*, 1998). The best known example of this has occurred in wild populations of the cheetah (*Acinonyx jubatus*). Menotti-Raymond and O'Brien (1993) have suggested that, from mitochondrial DNA and other evidence, a bottleneck occurred in this species during the Pleistocene.

The Florida panther is a subspecies of cougar (*Puma concolor*) which has low genetic diversity and is similar in this respect to an inbred strain of laboratory mouse (Roelke *et al.*, 1993). By the mid-1980s only about 40 animals remained in the wild, creating a genetic bottleneck. Individuals suffered from a number of physical and physiological abnormalities including:

- kinked tails
- cowlick (a projecting tuft of hair that grows in a different direction to the rest of the hair)
- sperm defects
- heart defects (atrial septal defects)
- undescended testicles (cryptorchidism)
- vaginal papillomas
- a very high seroprevalence – number of individuals who test positive – to infectious disease.

These characteristics exemplify what happens when genetic diversity is lost in a wild population.

Genetic bottlenecks can affect reintroduction and population restoration efforts. Ramey *et al.* (2000) detected a severe bottleneck in a reintroduced bighorn sheep (*Ovis canadensis*) population in Badlands National Park, South Dakota. They note that the implications of such bottlenecks may not be apparent during the time scale of most management programmes because of the small number of generations that will have elapsed since the initial reintroduction or translocation, and because the effects of inbreeding may be masked by other factors.

Schwartz and Mills (2005) tested the ability of gene flow to alleviate the deleterious effects of inbreeding in the deer mouse (*Peromyscus maniculatus*) and found that the introduction of migrants can reduce inbreeding depression. Five out of eight of their inbred mouse lines went extinct during the inbreeding process.

12.6.7 Population management

Ballou and Foose (1996) have listed a number of guidelines for the management of captive populations (Fig. 12.23). These may be summarised as follows:

1 Attempt to obtain a sufficient number of founder individuals to adequately sample both the heterozygosity and the allelic diversity in the source population – the primary concern is adequate sampling for allelic diversity as this requires more founders than sampling for heterozygosity only. Unfortunately, information on the frequency of alleles in wild populations is often unknown.

2 Expand the population size as quickly as possible to the carrying capacity of the zoos able to support the species – small populations lose genetic diversity faster than large populations.

3 Stabilise the population at the carrying capacity – zoo accommodation is in short supply so fertility and survivorship rates must be managed to stabilise the population at the desired carrying capacity.

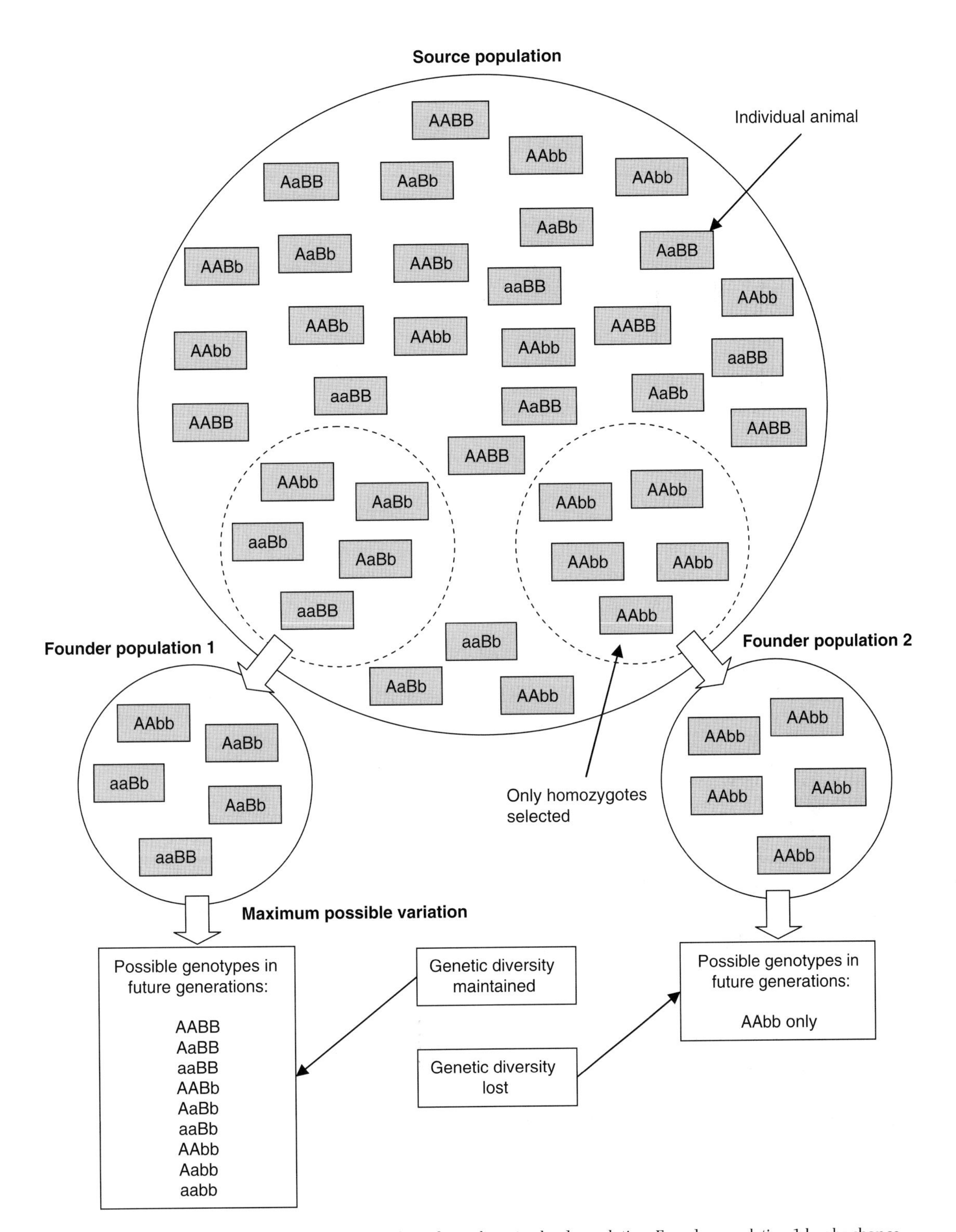

Fig. 12.22 The influence of the founder effect on a hypothetical captive-bred population. Founder population 1 has by chance selected individuals which in future may breed to produce offspring with all of the possible genotypes found in the source population. Founder population 2 by chance contains individuals which can only produce AAbb individuals which are identical to themselves.

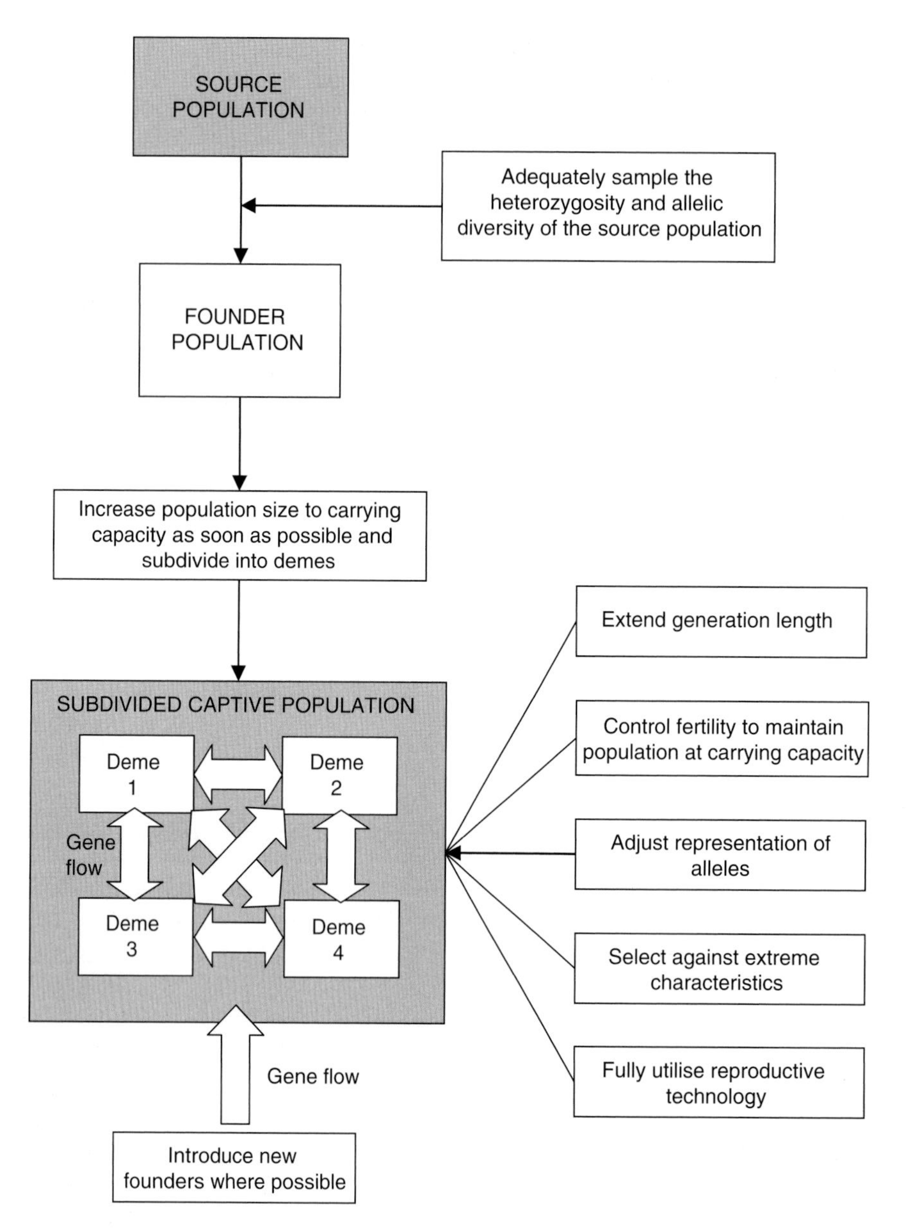

Fig. 12.23 Demographic and genetic management of a hypothetical zoo population. (Based on ideas in Ballou and Foose, 1996.)

4 Extend the generation length as much as possible – genetic diversity is lost with each generation so extending the generation time will reduce the rate of loss. This can be done by shifting the mean age of reproduction to later in life. However, this must be balanced against the possibility of the greater risk of the death of individuals before they have the opportunity to breed and the loss of fertility with age.

5 Adjust the representation of founder lineages to reflect the probable distribution of founder alleles in the living population – determine which animals will breed, with whom and how often in order to compensate for a

skewed founder population whose distribution of alleles does not reflect that found in the wild.
6 Select against individuals with extreme outlying morphological and reproductive characteristics – for example albinism and dwarfism. This will help to reduce genetic load.
7 Consider dividing the population into several subdivisions (demes) among which gene flow is regulated – this would help to protect the population from disease and simplify population management. However, if the subdivisions are too small they risk being affected by genetic drift, reducing genetic diversity.
8 Continually introduce new founders from the wild if possible – this will help to minimise loss of variation due to genetic drift.
9 Utilise available reproductive technology to the fullest possible extent – this technology can extend the reproductive life of individuals (by germ plasm storage), facilitate exchange of germ plasm with wild populations, and increase generation length thereby maintaining genetic diversity.

Meffert *et al.* (2005) tested the effect of alternative captive breeding strategies on inbreeding coefficients, overall fitness and ability to colonise the 'wild' using the housefly (*Musca domestica*) as an experimental model:
a maximum avoidance of inbreeding (MAI) – a strategy to minimise inbreeding and balance founder contributions; versus
b selection against less fit individuals (disregarding relatedness).
The study found that the MAI strategy facilitated some form of purging of inbreeding depression effects, resulting in significantly lower inbreeding coefficients, higher fitness and reduced extinction potential. However, the advantages of the MAI strategy were not apparent during the captive breeding phase of the experiment, suggesting that the long-term benefits of this approach could be underestimated in actual breeding programmes. Meffert *et al.* concluded that the maximum avoidance of breeding approach to the management of captive breeding programmes for threatened species was appropriate, especially for systems with low reproductive potential.

12.7 PROMOTING REPRODUCTION

Although in the past zoos obtained large numbers of animals from the wild, it would be wrong to imagine that no breeding of exotic animals took place until relatively recently. In 1866 the Zoological Society of London presented three of its keepers with the Landseer Bronze Medal and £5, for their success in 'breeding foreign animals in the Society's menagerie' (Chambers, 2007).

Some species reproduce very well in captivity, to the point where zoos need to take active measures to control their populations. Others are difficult to breed either because they have a naturally low fecundity (e.g. giant pandas) or because of difficulties associated with their husbandry in zoos.

The success of particular zoos in captive breeding is often the result of their keeping particularly prolific bulls. In Metro Washington Zoo, a bull Asian elephant (*Elephas maximus*) named 'Thonglaw' sired 15 of the 27 calves born at the zoo. By the age of just 19 years, Chester Zoo's Asian bull 'Chang' had sired eight calves in just seven years, although only five survived (Rees, 2001c). A bull named 'Vance', owned by the Ringling Brothers and Barnum, sired 24 calves (Anon., 2001). There is a long-term risk that the genes of such prolific bulls will be over-represented in the captive population, thus restricting future mating possibilities. In contrast, a bull Asian elephant held at Port Lympne Zoo, Kent has fathered seven calves, none of which has survived (Juniper, 2000).

12.7.1 Sex determination

Sex determination is critical to the efficient operation of breeding programmes. In dimorphic species the sexes exhibit different secondary sexual characteristics such as the presence of large horns in a male antelope or the possession of distinctive plumage colours in male birds. However, many species are monomorphic and it is not possible to determine their sex from their outward appearance. Male and female animals of the same species possess different combinations of sex chromosomes (Table 12.9) so it is possible to sex them by examining their DNA using a technique known as polymerase chain reaction (PCR). This is a molecular biology technique which amplifies small amounts of DNA and uses a polymerase enzyme to assemble new strands of DNA for analysis (Fig. 12.24).

The PCR detects differences in DNA in the sex chromosomes. For example, cock birds possess two Z chromosomes while hen birds possess one Z and one W. Using PCR in the laboratory, a region of DNA that differs in size between the Z and W chromosomes can be amplified. This method yields one PCR product for

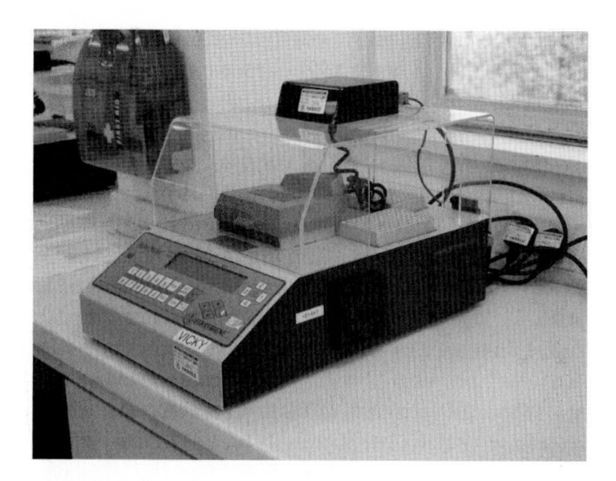

Fig. 12.24 A PCR machine (or DNA amplifier) used to amplify segments of DNA using the polymerase chain reaction.

males and two PCR products for females. For birds, the DNA may be obtained from blood samples or from feathers. Specialist companies can provide a sexing service using DNA analysis for zoos that do not have the facilities to do this themselves.

Individual birds may also be sexed by laparoscopy (Greenwood, 1983). A non-invasive method of sexing anurans has recently been described by Szymanski *et al.* (2006). They used hormone metabolites in the faeces of toads (*Bufo* spp.) to determine their sex. However, they warn that the group of metabolites differentiating sex may not be consistent between species.

behaviour of their animals. The size and quality of enclosures may affect stress levels and reproductive cyclicity in some species.

A study of the effects of different captive housing conditions on adult female tigrinas (*Leopardus tigrinus*) and margays (*L. wiedii*) found that transfer from large enriched enclosures to small barren ones induced stress responses in both species (Moreira *et al.*, 2007). Increases in adrenocortical activity were associated with decreased production of steroid hormones by the ovaries. Enrichment of small barren enclosures with hiding places, tree trunks and plants did not reinitiate normal ovarian steroidogenic activity in margays but it did reduce the stress response in tigrinas.

Simply keeping a male and a female of a species together will not guarantee successful breeding. Individuals of many species like to choose their own mates (see Section 9.2.3). If left in a large group some species will pair up naturally. If the group is to be divided between several enclosures or dispersed to other zoos it is important to identify the pairs and keep them together. With birds this may be possible by using coloured rings for identification.

The presence of dominant individuals may prevent subordinates from reproducing and it may be desirable to subdivide a large group with a dominant male into smaller groups to give other males the opportunity to reproduce. Some species will only reproduce if kept in a large social group due to the Fraser Darling effect.

Biography 12.4

James Watson (1928–) and Francis Crick (1916–2004)

James Watson is an American biologist. Watson and an English molecular biologist, Francis Crick, discovered that DNA consists of a double helix, made up of two parallel chains of alternate sugar and phosphate groups linked by pairs of organic bases, while working at the Cavendish Laboratory at the University of Cambridge. This discovery – published in 1953 – earned Watson and Crick, together with Maurice Wilkins, the Nobel Prize in Physiology or Medicine in 1962. DNA fingerprinting is now an essential tool for conservation biologists.

12.7.2 Creating appropriate physical and social conditions for reproductive success

Reproductive success in zoos is directly related to the extent to which they are able to provide suitable social and physical environments for the normal breeding

12.7.2.1 The 'Fraser Darling effect' – tricking social animals into breeding

In colonial birds the presence of other individuals results in intense social stimulation from group displays which causes the synchronisation of the breeding cycle. This is known as the 'Fraser Darling effect' (Darling, 1938). In captivity, larger flamingo colonies breed more often than smaller ones (Stevens, 1991). Some zoos have used mirrors to simulate the presence of larger flocks. Greater flamingos (*Phoenicopterus ruber*) at Granby Zoological Garden, Quebec, Canada began courtship displays immediately after mirrors were installed in their nesting area (Lanthier, 1995).

Mirrors and 50 cm high glass fibre flamingo statues were used by de Azevedo and Faggioli (2004) in an attempt to stimulate breeding in a flock of 10 Chilean flamingos (*P. chilensis*) at Belo Horizonte Zoo, Brazil. Although, overall, these enrichments were not considered to be very effective, nevertheless, nest building behaviour occurred more often after the enrichments were added to the enclosure and some of the birds began building their own nests rather than using the artificial nests provided.

In some species vocal communication may be important in reproduction. Recordings of calls have been used successfully to stimulate breeding activity in the Puerto Rican crested toad (*Peltophryne lemur*) (Johnson, 1991).

12.7.2.2 Sexual incompetence

Some animals do not reproduce simply because they are incompetent and do not adopt the appropriate position when attempting to mate. Some giant pandas are reproductively incompetent and exhibit poor copulatory positioning.

There is some evidence that juvenile bull Asian elephants (*Elephas maximus*) learn to court and mount females by social facilitation (Rees, 2004a). A young bull at Chester Zoo was recorded mounting a cow calf on days when he observed mounting behaviour in adults but not on days when no adult sexual behaviour was observed (Fig. 12.25).

In Chiang Mai Zoo in Thailand giant pandas (*Ailuropoda melanoleuca*) that have been reluctant to mate have been shown videorecordings of other pandas mating ('panda porn') in an attempt to encourage them to reproduce (Handwerk, 2006).

12.7.3 Assisted reproductive technology (ART)

Many species experience reproductive problems in captivity. Underlying fertility issues may need to be addressed before dealing with the problem of achieving conception and successful birth (Wildt, 2009). These may include issues relating to:

- genetic diversity
- sperm characteristics
- oestrogen surges

Fig. 12.25 Asian elephants (*Elephas maximus*) mating at Chester Zoo. Inset: Juveniles mate shortly after observing mating in adults, suggesting that young elephants may learn some aspects of sexual behaviour via social facilitation.

- the use of fresh versus frozen sperm
- dietary problems
- mood swings and other behavioural problems.

Genetic management and assisted reproductive technology (ART) have been instrumental in addressing reproductive problems in a range of species.

It is important that the amount of genetic diversity in a captive population of a species is similar to that found in a wild population of the same species. This is difficult to achieve because breeding programmes often involve small populations of animals distributed between a large number of zoos spread over a wide geographical area. Problems may arise in small populations due to:

- sexual compatibility
- inbreeding
- loss of heterozygosity
- the expression of deleterious genes.

The key to maintaining genetic diversity is to ensure that animals with the correct (outbred) genotype breed together, thereby retaining heterozygosity (Wildt, 2009).

A number of sophisticated ART techniques are available to assist in the captive breeding of endangered species but those that are most often used are relatively 'low tech'. Often they involve the use of:

- fresh or frozen/thawed sperm
- extensive non-invasive monitoring of urinary or faecal hormones.

Although embryo technologies are available, there is relatively little practical application of these techniques because:

- little is known about the embryology of most rare species
- obtaining a suitable surrogate mother to receive an embryo created *in vitro* is problematic
- interspecific embryo transfer, while possible, is not biologically efficient (Wildt, 2009).

12.7.3.1 Case study: Giant pandas

The giant panda (*Ailuropoda melanoleuca*) is classified as endangered by the IUCN. The species has suffered from the fragmentation and loss of its natural habitat, but is also disadvantaged because of its low fecundity. Giant pandas are monoestrous and come into oestrus once a year for 24–72 hours (Fig. 12.26). This clearly restricts opportunities for breeding in the wild especially where the animals are widely dispersed at low densities. It also makes the establishment of a captive breeding programme extremely challenging.

Fig. 12.26 Giant pandas (*Ailuropoda melanoleuca*) mating. (Courtesy of Dr Glyn Heath.)

A biomedical survey of more than 60 giant pandas at the Wolong National Nature Reserve, the Chengdu Research Base of Giant Panda Breeding and the Beijing Zoo (Wildt *et al.*, 2006) identified the following problems affecting breeding:

- aggression by males
- inadequate dietary fibre resulting from the use of inappropriate artificial foodstuffs
- health problems, many of which could be corrected with minor improvements in preventative health care
- stunted development syndrome in 13% of individuals; these individuals were two-thirds normal size and failed to reproduce.

Once these problems were addressed cub production increased dramatically as a result of natural breeding supported by artificial insemination (AI) using fresh and thawed sperm. The development of artificial insemination for giant pandas has been extremely successful in increasing the numbers of the species in captivity in China and in zoos elsewhere. On 23 July 2009 the first giant panda conceived through AI using frozen sperm was born at the Wolong Giant Panda Research Centre. In 2006 an area of Sichuan Province which is home to pandas was designated as a World Heritage Site.

12.7.3.2 Artificial insemination (AI)

Artificial insemination is the introduction of semen into the female reproductive tract by mechanical means. It has been extensively used in farming for many years and techniques have been developed for a number of exotic species.

There are a number of different ways of collecting sperm. These include electrical stimulation (electro-ejaculation), massage, the use of a 'dummy' animal, or a condom inserted in the vagina of a female and placed on the penis at the last moment before penetration.

Sometimes the sperm used in AI is fresh. However, generally it is more convenient to store it (at −196°C) then thaw it before use. The optimum time of insemination relative to ovulation varies between species. Timing is critical as the female must be about to ovulate. Sometimes the cycle of the female is controlled by injecting prostaglandin hormones which end the previous cycle and bring her into oestrus at a specific time.

Some advantages of AI are:

- It is easier to transport sperm than animals.
- Frozen sperm is easy to maintain.
- Sperm may be removed from anaesthetised wild animals to increase the gene pool, or from zoo animals that die unexpectedly.
- It is theoretically possible to store genes from all species in frozen form.

However, AI also poses some difficulties:

- It is difficult to collect sperm in many species.
- Sperm of some species freezes better than others (e.g. chimpanzee sperm is easier to freeze than that of gorillas or monkeys; giant panda sperm is easier to freeze than red panda (*Ailurus fulgens*) sperm). Sperm needs to be frozen within a suitable medium and then thawed carefully.
- A limited number of males have been used as sperm donors in some species so there is a risk of inbreeding.
- Some species do not mate until just before ovulation, some after. In some species (e.g. squirrels, beavers, raccoons, dogs, weasels, etc.) ovulation is induced by copulation. In the short-tailed opossum (*Monodelphus domestica*) the female does not enter the follicular phase of her cycle until a male is present in the group.
- Reproductive cycles in females need to be carefully monitored.

The development of AI in some species has been difficult and scientists have used model species to perfect their techniques. The European ferret (*Mustela putorius*) is a common species that was used as an animal model to develop a laparoscopic AI technique that has been successfully employed to breed the endangered black-footed ferret (*M. nigripes*).

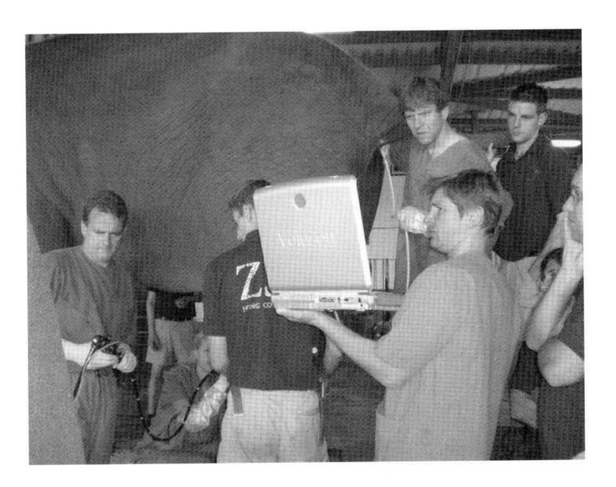

Fig. 12.27 Artificial insemination of an Asian elephant (*Elephas maximus*). (Courtesy of Louise Bell.)

Case study: Artificial insemination in elephants

Studies of the reproductive physiology of elephants, including electroejaculation and semen characteristics (Howard *et al.*, 1989; Mar *et al.*, 1995), have culminated in the successful development of AI techniques for use with these animals.

The first successful elephant pregnancy resulting from AI was reported by Dickerson Park Zoo, Springfield, Missouri, USA (Schmitt, 1998). An Asian cow was inseminated on 25, 26 and 28 January 1998 and successfully gave birth in November 1999. Artificial insemination of elephants is difficult because of the complex anatomy of the cow. However, a breakthrough in this technology was achieved by the use of an endoscope and ultrasound to guide the semen to the correct location (Fig. 12.27).

Ultrasonography of the urogenital tract has been used to assess female and male reproductive function. Hildebrandt *et al.* (2000a) performed ultrasonographic examinations of 280 captive and wild African and captive Asian female elephants. The primary pathological lesions that influenced reproductive rates in these females were uterine tumours, endometrial cysts, and ovarian cysts that resulted in acyclicity. These examinations provide a reference for ultrasound specialists involved in breeding programmes and should assist elephant managers in deciding which animals to include in breeding programmes. An ultrasonographic study of male elephants of both species found that observable reproductive tract pathology in adult males was low (14%), even in older bulls (Hildebrandt *et al.*, 2000b). However, apparent infertility of non-organic cause in these otherwise healthy bulls was high (32%). Semen quality varied markedly in ejaculates from the same bull and in samples taken from different bulls. Hildebrandt *et al.* suggested that this inconsistency raises the question of the reliability of some individuals as participants in breeding programmes, and that the apparent inhibitory effect of suppressive social interactions on male reproductive potential should be investigated.

The African and Asian elephant EEPs recommend ultrasonography of cows in early pregnancy to exclude the possibility of twins, and the examination of every potential breeding bull annually from around the age of eight years to determine the development of the reproductive organs (Dorresteyn and Terkel, 2000).

Artificial insemination has the potential to increase the effective population size of the captive population but it may also have some negative effects (Table 12.12). In the long term, if zoos come to rely upon this technology they may breed generations of sexually inept individuals that have neither witnessed normal adult sexual behaviour nor experienced normal rearing within a herd.

Table 12.12 The positive and negative aspects of artificial insemination in elephants.

Positive features	Negative features
May increase fecundity of zoo population	Not yet widely available
May increase effective population size by allowing mating between animals in geographically separated zoos	May encourage over-dependence on a technological solution to low fecundity
May increase gene pool if semen is imported from non-zoo populations	Reduces experience of normal sexual behaviour in bulls and cows
Allows more accurate prediction of date of birth, thereby allowing zoos to make the necessary preparations	Reduces exposure of calves to normal sexual behaviour during critical periods in their development
Allows better control of timing of births (avoiding winter births in temperate climates)	Requires sophisticated technology and technical expertise
Reduces the need to move animals between zoos for mating	Very expensive
Allows inclusion of sexually incompetent animals in breeding programmes	

On 6 August 2009 the Asian elephant 'Ganesh Vijay' was born to 'Noorjahan' at Twycross Zoo as a result of artificial insemination (Fig. 12.28).

12.7.3.3 *In vitro* fertilisation (IVF) – test-tube babies

In vitro fertilisation is the fertilisation of an egg cell outside the female's body. Once fertilisation has been achieved the resultant embryo is either implanted into a recipient female in order to produce a pregnancy or it is frozen and stored.

In vitro fertilisation is a very useful technique for breeding rare species and it allows some animals that are unable to reproduce naturally to take part in breeding programmes. The use of frozen material has been of particular benefit to breeding programmes because it allows:

- sire selection
- post-mortem sperm and egg collection
- embryo sexing
- interspecies transfer
- embryo splitting and cloning.

The development of IVF for many species has led to the creation of sperm banks. Omaha's Henry Doorly Zoo has the largest gorilla sperm bank in the world. Its reproductive research team helped to produce the world's first test-tube western lowland gorilla, who was born at Cincinnati Zoo in 1995. 'Timu' was born to 'Rosie' after she received IVF treatment, but she was rejected and had to be hand-reared. Timu subsequently gave birth to two offspring but unfortunately failed to bond with either of them (Associated Press, 2005).

12.7.3.4 Embryo transfer (ET)

Embryo transfer involves the insertion of a viable embryo into a recipient female. The embryo may have been taken from a pregnant donor or it may be the result of IVF. If the embryo is inserted into the recipient female at the correct stage of her oestrous cycle it will produce a pregnancy.

This technique allows the repeated harvesting of embryos from the same female donor by curtailing pregnancy within the first few days. Hormone therapy must be used to induce the female to repeatedly produce new embryos and the timing of insemination is also critical so AI techniques are used.

Fig. 12.28 Ganesh Vijay (*Elephas maximus*) was conceived by AI and born to Noorjahan on 6 August 2009 at Twycross Zoo. The semen was donated by a bull ('Emmett') at ZSL Whipsnade Zoo.

Embryo transfer allows the selection of both the female genes and the male genes and is a valuable tool in conservation breeding programmes. It may involve the transfer of an embryo into a female recipient of the same species or a different (surrogate) species. Surrogate species have been successfully used to produce offspring by ET in a number of endangered species. For example, the domestic cat has been used as a surrogate for the African wildcat (*Felis silvestris lybica*) and the eland (*Taurotragus oryx*) as a surrogate for the bongo (*Tragelaphus euryceros*).

12.7.3.5 Cloning

Cloning is the creation of an exact genetic copy of an organism. This occurs naturally whenever identical twins are produced. Cloning has the potential to assist with captive breeding programmes, especially when the total number of individuals of a species that remains is very small. Cloning can be achieved in two different ways:

- artificial embryo twinning, or
- somatic cell nuclear transfer.

Artificial embryo twinning

This is a relatively low-tech type of cloning. The process is essentially the same as that which produces identical twins in nature. It is accomplished in the laboratory by manually separating a very early embryo into individual cells. Each cell is then allowed to divide and develop on its own. Each embryo that results is then implanted in a surrogate mother.

Somatic cell nuclear transfer (SCNT)

The first mammal to be cloned from an adult somatic cell was 'Dolly' the sheep. She was born in 1996 after scientists at the Roslin Institute in Edinburgh created her by a process called somatic cell nuclear transfer (SCNT). In this process the first step is to isolate a somatic cell from an adult sheep. In Dolly's case the cell was from a sheep's udder. The nucleus from this cell is then transferred to an egg cell from which the nucleus has been removed, using microscopic needles. After some chemical manipulation the egg cell will begin to behave as if it has been fertilised. The resulting zygote develops into an embryo and this is then implanted in the uterus of a surrogate mother. The mother later gives birth as normal. A year before Dolly was born two other sheep, 'Megan' and 'Morag', were cloned at the Institute from embryonic (as opposed to adult) cells.

Scientists at the Audubon Nature Institute's Center for Research of Endangered Species (ACRES) in New Orleans created an African wildcat (*Felis silvestris lybica*) by cloning in 2003. The African wildcat was born as a result of the world's first successful interspecies frozen/thawed embryo transfer. The embryo was transferred to a domestic cat (Fig. 12.29). This was the first time a wild carnivore had been cloned. In the same year ACRES produced the world's first caracal (*F. caracal*) from a frozen embryo. It has since produced further cloned African wildcats and Arabian sand cats (*F. margarita*) by cloning.

12.7.3.6 Cryobiology

Cryobiology is the study of the effects of freezing and low temperatures on biological systems, including the cryopreservation of biological specimens. This technology is becoming increasingly important in conservation as scientists endeavour to build collections of preserved DNA and other materials from animals (and plants) in danger of extinction.

The Frozen Zoo

The Frozen Zoo is part of the Center for Conservation Research at the Zoological Society of San Diego. Its mission is 'To help preserve the legacy of life on Earth for future generations by establishing and maintaining genetic resources in support of worldwide efforts in research and conservation'. The Frozen Zoo consists of:

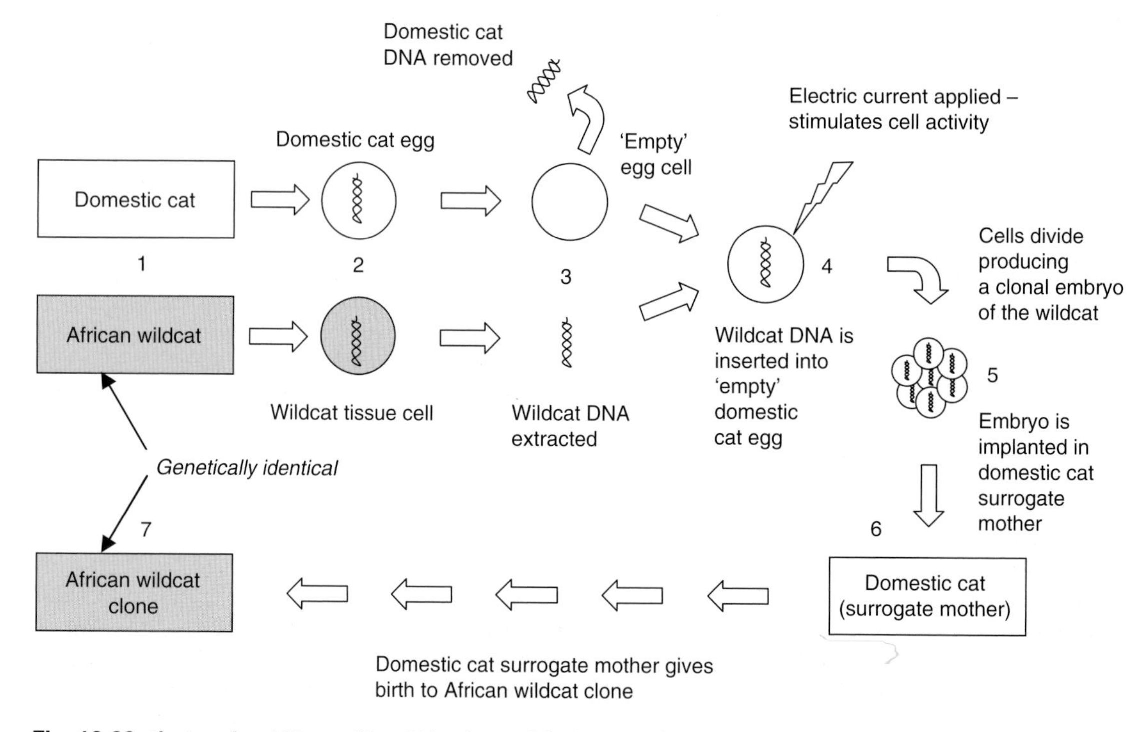

Fig. 12.29 Cloning of an African wildcat (*Felis silvestris lybica*) using a domestic cat (*F. catus*) surrogate mother.

- DNA
- viable cell cultures
- semen
- embryos
- oocytes and ova
- blood and tissue specimens.

A total of more than 800 species and subspecies are represented by over 8400 individual samples. These resources are used in the study of conservation, evolution and human health by the San Diego Zoo's staff and conservation research partners in universities and scientific institutions around the world.

In 2007 scientists at the Frozen Zoo made the first attempts at culturing adult stem cells from the northern white rhino (*Ceratotherium simum cottoni*). Recent additions to the Frozen Zoo have included specimens from the white-bellied tree pangolin (*Manis tricuspis*), Gobi bear (*Ursus arctos gobiensis*), secretary bird (*Sagittarius serpentarius*) and African clawed frog (*Xenopus laevis*). Other projects include work on the recovery of the California condor (*Gymnogyps californianus*). The Frozen Zoo has been involved in whole genome sequencing projects for the African savanna elephant (*Loxodonta africana*), two-toed sloth (*Choloepus* sp.) and western lowland gorilla (*Gorilla gorilla gorilla*). It has also supported the Human Genome Project.

The Frozen Zoo plays an integral role in *Barcoding for Species Conservation*. This project uses DNA barcoding to assign biological samples to their species of origin. This technology is helping to identify specimens involved in the illegal trade in wildlife, including bushmeat products. It will assist in the enforcement of wildlife regulations such as CITES and help in the repatriation of confiscated live specimens.

The Frozen Ark

The Frozen Ark is an international consortium of zoos, museums and university laboratories whose mission is 'to collect, preserve and store DNA and viable cells from animals in danger of extinction'. Samples are collected from zoos, captive breeding programmes and from animals living wild. The aim of the project is to collect DNA for all of the more than 16,000 animals defined as under threat by the IUCN Red Data Lists, and viable cells from as many as possible. These will include somatic cells, eggs, sperm and embryos. This project includes a global database which contains information on the location and content of existing preserved samples.

The Frozen Ark Consortium consists of:

- the University of Nottingham (including the Frozen Ark Office)
- the Natural History Museum, London
- the Zoological Society of London
- the Animal Gene Resource Centre, Monash University, Melbourne, Australia
- the Laboratory for the Conservation of Endangered Species (LaCONES), Hyderabad, India
- the Center for Conservation Research, The Zoological Society of San Diego, California, USA (home of the Frozen Zoo)
- the Ambrose Monell Laboratory, The American Museum of Natural History, New York, USA
- the Wildlife Biological Resource Centre, Endangered Wildlife Trust, The National Zoo, Pretoria, South Africa
- the North of England Zoological Society, Chester, UK
- the Cryobiology Research Group, Luton Institute of Research in the Applied Natural Sciences, University of Bedfordshire, Luton, UK
- the Reproductive Biology Unit, Perth Zoo, Western Australia
- the New Zealand Centre for Conservation Medicine, Auckland Zoological Park, New Zealand.

Some of these facilities have collection and storage facilities, while others hold live specimens from which samples will be taken.

12.7.4 Stimulation of egg production in birds: egg harvesting, double clutching and incubation

Clutch size varies within and between species of birds. Different species lay different numbers of eggs, from just one to 20, and some species produce more than one clutch per year.

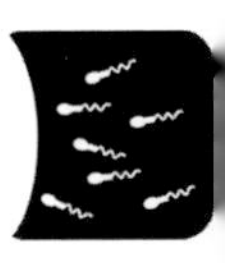

Egg production depends partly on food supply. In the wild, in some species, clutch size will increase when food is abundant.

Birds may be divided into two types on the basis of their egg-laying behaviour: determinate layers and indeterminate layers. In indeterminate layers the number of eggs laid can be changed by adding or removing eggs during or just before the time of laying. In determinate layers this has no effect. Removal of eggs, so as to leave just one or a small number of eggs in the nest, increases the number laid in some species.

In a captive situation it is possible to increase egg production in some species by removing eggs from the

Fig. 12.30 An egg incubator.

nest and incubating them artificially in an incubator (Fig. 12.30). Egg incubation may last from a few days to more than two-and-a-half months, depending on the species (Table 12.13). In some cases chicks may find it difficult to hatch and keepers may need to intervene by helping to break away the egg shell (Fig. 12.31).

In indeterminate species, removing eggs encourages the female to lay replacement eggs when she would have been sitting on her original clutch. This is clearly a useful adaptive mechanism in the wild as it would give a pair a second chance to breed if their first clutch was lost to predators. California condors (*Gymnogyps californianus*) normally raise just one chick every other year. Using double clutching San Diego Zoo is able to raise up to four chicks in a two-year period.

The whooping crane (*Grus americana*) normally lays two eggs but only one chick usually survives. Of 50 eggs which were harvested for captive breeding from wild cranes 82% hatched and 23 (56%) were reared to the age of at least six months (Kepler, 1978). These were

Table 12.13 The incubation times of eggs of selected bird species.

Species	Incubation time (days)
African grey parrot	28
Andean condor	54–58
Bald eagle	35
Chilean teal	26
Chinese pond heron	18–22
Crested caracara	28
Emu	57–62
Greater flamingo	30–32
Great-horned owl	35
Kea	28–29
Kiwi	75–80
Laughing kookaburra	25
Leadbeater's cockatoo	26
Magnificent frigatebird	40
Ostrich	40–42
Red-winged blackbird	11–12
Scarlet ibis	21–23
Scarlet macaw	26
Twin-wattled cassowary	49–56

Source: Blackpool Zoo.

then used to establish a captive population to provide birds for reintroduction. It is also beneficial to remove and artificially incubate the eggs from birds kept in open enclosures in zoos as this will improve the survival rate of eggs and chicks. This is routinely done as part of the captive breeding programmes for ducks and geese kept by the Wildfowl and Wetlands Trust (Fig. 12.32).

12.7.5 Surrogate mothers, adoption and cross-fostering

In some species a young animal that has been rejected by its mother may be adopted by another female of the same species. Captive breeding programmes for some species utilise cross-fostering as a technique to increase the fecundity of a rare species. The young offspring of an individual of a rare species are fostered by adults of another species which is usually closely related. For example, ring-necked parakeets (*Psittacula krameri*) are used to foster echo parakeets (*P. eques*) and Barbary doves (*Streptopelia risoria*) are used for pink pigeons (*Columba mayeri*).

Sometimes quite unrelated species are used. In July 2008 a domestic cat that had just produced four kittens

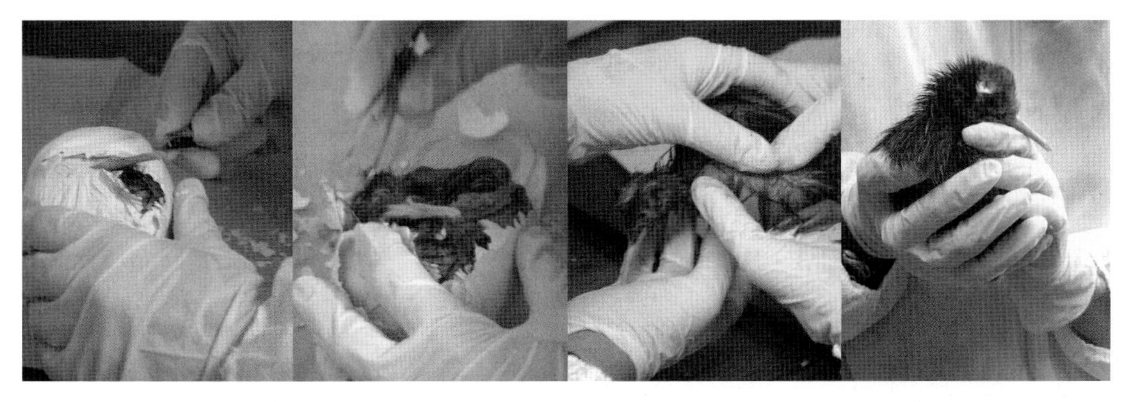

Fig. 12.31 Sometimes chicks need help to hatch. This sequence shows stages in the assisted hatching of a North Island Brown kiwi chick (*Apteryx mantelli*) at Willowbank Wildlife Reserve, Christchurch, New Zealand. (Courtesy of Bethan Shaw.)

was used as a surrogate mother for an abandoned red panda cub (*Ailurus fulgens*) at Artis Zoo in Amsterdam, the Netherlands. A five-year-old female chimpanzee (*Pan troglodytes*) at The Institute of Greatly Endangered and Rare Species (TIGERS) in South Carolina, USA, acted as a surrogate mother to a number of big cats by helping to bottle feed white tiger (*Panthera tigris*) cubs, a leopard (*P. pardus*), four lions (*P. leo*) and a puma (*Puma concolor*).

Cross-fostering is a technique which can be used alongside double clutching in some bird species, whereby offspring are removed from their parents and raised by surrogates, usually similar species. Endangered whooping cranes (*Grus americana*) have been cross-fostered to sandhill cranes (*G. canadensis*) in order to induce the whooping crane parent to produce a second clutch (Kepler, 1978).

Some individuals may have no reproductive value within a social group – because they are too old or barren – yet may serve a useful function by acting as allomothers. A cow elephant at Chester Zoo has acted as an allomother ('auntie') to all of the calves born in the group but has never successfully raised a calf herself.

12.7.6 Hand-rearing

The decision to hand-rear an animal is never an easy one. For many species it is likely to result in problems relating to nutrition, socialisation, and the development of normal behaviour. If normal sexual behaviour does not develop, hand-rearing may make the animal useless for breeding. The young of many mammal species have a long period of dependency on their mother. If they are hand-reared, the chicks of altricial species create the most work for keepers because of their long dependency period. Clearly, the decision to hand-rear an animal may create a great deal of work for keepers. If hand-rearing is to be attempted staff should consider:

- any specialist equipment that might be required
- how the animal will be housed
- the selection and preparation of food
- the techniques that will be used for weaning
- the common medical problems that might occur (Gage, 2002).

12.8 POPULATION CONTROL

When common animals breed well zoos may need to control their populations. Larger populations within a zoo do not necessarily attract more visitors but do cost more to feed and house. Furthermore, these animals may be taking up space required for rarer species.

The control of breeding may be important in maintaining a genetically healthy population. When the genes of a particular animal are over-represented in the captive gene pool it may be necessary to prevent further breeding to reduce the possibility of inbreeding.

Possible solutions to the problem of superfluous animals include:

- contraception
- surgical sterilisation
- keeping sexes separate
- donation to another zoo
- return to the wild
- culling.

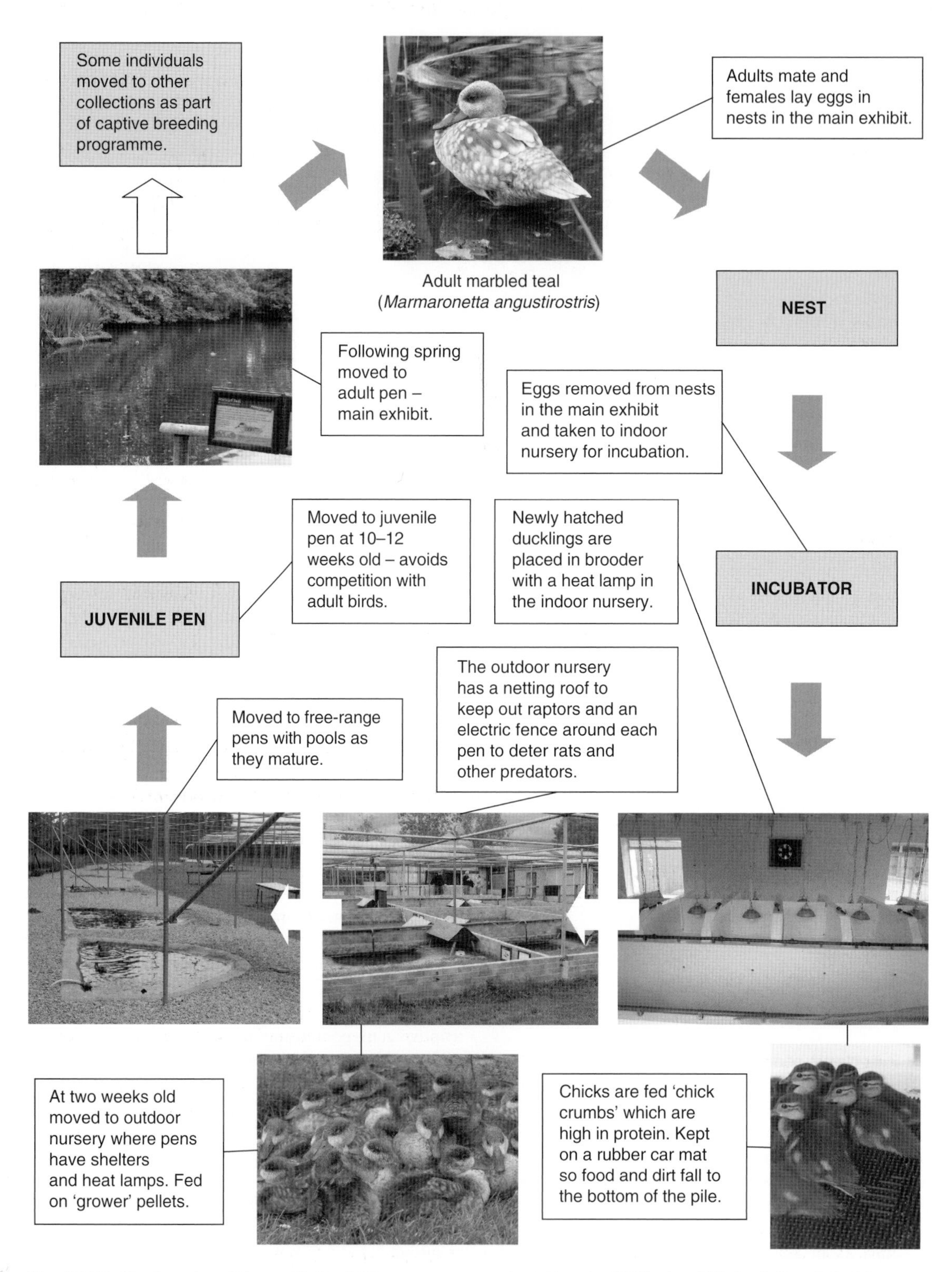

Some individuals moved to other collections as part of captive breeding programme.
Adult marbled teal (*Marmaronetta angustirostris*)
Adults mate and females lay eggs in nests in the main exhibit.
NEST
Following spring moved to adult pen – main exhibit.
Eggs removed from nests in the main exhibit and taken to indoor nursery for incubation.
Moved to juvenile pen at 10–12 weeks old – avoids competition with adult birds.
Newly hatched ducklings are placed in brooder with a heat lamp in the indoor nursery.
INCUBATOR
JUVENILE PEN
The outdoor nursery has a netting roof to keep out raptors and an electric fence around each pen to deter rats and other predators.
Moved to free-range pens with pools as they mature.
At two weeks old moved to outdoor nursery where pens have shelters and heat lamps. Fed on 'grower' pellets.
Chicks are fed 'chick crumbs' which are high in protein. Kept on a rubber car mat so food and dirt fall to the bottom of the pile.

Fig. 12.32 Hand-rearing of the marbled teal (*Marmaronetta angustirostris*) at the Wildfowl and Wetlands Trust's duckling nursery at Martin Mere, UK.

Contraception is a useful technique, especially if it may be desirable to bring an animal back into the breeding population at a later date. Surgical sterilisation is appropriate where there is no longer any possibility that the animal will be required for breeding. This may be:

- because its genes are over-represented in a population
- to prevent breeding with a close relative kept in the same social group (e.g. father–daughter or mother–son matings)
- to prevent interbreeding with a different subspecies where individuals from different subspecies are housed together (e.g. in tigers).

Obviously, keeping the sexes separate will prevent breeding but may be impossible due to shortage of accommodation. It may also be undesirable on welfare grounds if it results in single animals being housed alone.

Some surplus animals may be placed with other zoos and if they are part of coordinated breeding programmes movements between institutions should be part of a management plan. However, when zoos keep common animals which are not subject to the constraints imposed by studbook keepers, surplus individuals may be difficult to place.

Some zoos provide surplus animals to the pet trade. This is a particular problem in relation to primates as there is increasing concern that primates are not suitable to be kept as pets (Soulsbury *et al.*, 2009).

Some zoo-bred animals may be returned to the wild. However, for most species this is impractical. Reintroduction projects should be undertaken for genuine conservation reasons and not merely to reduce the pressure on zoo accommodation (see Section 15.6).

Ultimately, if a zoo cannot find any way of accommodating surplus animals it may be forced to cull them. This last option is distasteful but zoos must focus on the viability of populations.

12.8.1 Contraception

Contraception may be essential in order to control the numbers of some species in a zoo or to prevent certain individuals from breeding because their genes are over-represented in the captive population. This may be achieved in a number of ways, but it is common to use hormone therapy (Seal, 1991). Sometimes it is possible to create single-sex groups, but where this is not a normal social grouping for the species it may cause increased aggression. For males the options are:

- vasectomy – severing of the vasa deferentia to prevent sperm from entering seminal fluid
- castration – removal of the testes
- implants of gonadotrophin-releasing hormone (GnRH) which suppresses sperm production.

In females, contraception may be achieved by:

- tubal ligation – severing and sealing of the Fallopian tubes
- ovarectomy/hysterectomy – removal of the ovaries/uterus
- inhibition of the oestrous cycle using:
  - GnRH implants
  - progestin implants
  - oral or injected progestins.

The AZA Wildlife Contraception Center

The AZA Wildlife Contraception Center is located at St Louis Zoo, Missouri and provides a service to AZA institutions. The Center:

- ensures the safety and effectiveness of contraceptives by organising monitoring programmes
- tests new contraceptive methods
- assists animal managers and vets in the selection and administration of contraceptives.

12.8.2 Euthanasia

Euthanasia is the humane killing of an animal. In a zoo this may happen for one of two general reasons: for health reasons or for population management.

An animal may be euthanised because it is terminally ill or because it is in great discomfort as a result of injury or disease. When an animal is seriously injured, for example as a result of a fall, the mechanics of repair may be relatively simple, but an inherently nervous animal might not be treatable. In some cases the stress that would be caused by attempting treatment after serious injury may not be justifiable and euthanasia may be the best option. This is particularly true of very large, nervous or heavy species such as giraffes or elephants.

Richardson (2000) has discussed a variety of reasons why zoo animals may be euthanised:

- after a serious injury where treatment is impractical
- when an individual is terminally ill and in discomfort or very disabled
- where an individual or group is carrying a particular disease which must be eradicated, especially if it is a threat to human health

• when an individual is very old, possibly much older than would be likely in the wild, and suffering from an age-related health problem
• for population management reasons.

If zoos are to make a significant contribution to the survival of a species it is essential that they actively manage the populations in their care. Euthanasia may be a necessary population management tool in a number of circumstances:

• when an individual of a social species is excluded from the social group and subjected to persistent harassment (and cannot be transferred to another group in the same or a different collection)
• in harem species when males are produced in excessive numbers and surplus to breeding requirements, e.g. in most deer, antelope and equid species
• when young are produced naturally as a form of enrichment for adults but have reached the age when they would naturally disperse from the group and are surplus to breeding requirements
• when young are rejected by the mother and hand-rearing is contraindicated for safety, welfare or other reasons
• when a common species has to be removed from a collection to free up space for a conservation-sensitive species
• to thin out populations which are being kept in overcrowded (stressful) conditions and which cannot be moved to other collections.

Richardson (2000) likens the control of zoo animal populations to the culling of wild deer populations where there is no natural control by predators. Some animals need to be removed to maintain the health and genetic viability of the whole herd, and in order to make the best use of available space.

Zoo staff are often, quite naturally, reluctant to euthanise particular high profile animals. This may lead to an animal being kept alive purely for the sake of staff feelings and public opinion. Richardson (2000) suggests that sometimes zoos delay making hard decisions, and risk prolonging suffering and sacrificing good animal welfare due to sentiment. Veterinarians and other staff involved with euthanasia report experiencing more distress when losing an animal to which they were particularly attached or when healthy animals have to be destroyed than when the animals concerned were unfamiliar to them or sick. The effect of euthanising animals on the staff involved has been discussed by Reeve *et al.* (2004).

Unfortunately, it is not always clear when an animal is suffering because it may not exhibit obvious signs. For example, an elderly animal may be suffering from painful arthritis or worn cartilage in limb joints that will only become apparent after a post-mortem examination. A zoo should have a written protocol that establishes in advance the circumstances in which euthanasia is acceptable and justifiable.

Zoos cannot afford to be sentimental about individual animals that will make no contribution to future gene pools. Millions of animals are culled each day in the wild and for food. Ethically, some people would argue that there is no reason to treat most zoo animals any differently. However, there may be a case for the development of animal sanctuaries for some species, where animals can live out their lives. There are a number of sanctuaries in the USA for ex-circus and ex-zoo elephants, and sanctuaries for other species (e.g. some higher primates) also exist.

BIAZA's Animal Transaction Policy (BIAZA, 2005) discusses the circumstances where euthanasia is appropriate. The American Veterinary Medical Association has produced detailed guidelines on euthanasia which include a section on zoo animals (AVMA, 2007).

12.9 CAPTIVE BREEDING OF INVERTEBRATES

Captive rearing methods for invertebrates have been developed where there has been a commercial interest, for example 'butterfly ranching' and the mariculture of edible molluscs and crustaceans (Nash, 1991). Mariculture is a specialised branch of aquaculture that involves cultivating marine organisms in enclosed sections of the open ocean or in seawater tanks or ponds. The commercial rearing of invertebrates for food and other products has been instrumental in the development of technologies and rearing protocols now used in conservation projects.

Captive breeding of invertebrates has rarely been important in their conservation because habitat loss is the major threat (New, 1995). Unlike the situation in vertebrates, for most invertebrate species individual specimens did not exist in zoos before *ex-situ* breeding programmes were considered necessary.

As with vertebrates, inbreeding should be avoided where possible, but this is extremely difficult with very rare species when efforts must be concentrated on producing large numbers of offspring. Care must be taken not to mix up individuals from widely separated populations when forming breeding groups. Unintentional mixing of genetic stocks which were formerly separated by geographical barriers may mask evolutionary processes such as speciation by hybridisation of distinct biological forms.

Perhaps the best known captive breeding programme for an invertebrate species is that for Partula snails.

12.9.1 Case study: Captive breeding Partula snails

The snail genus *Partula* is a group of small tree snails endemic to the islands of French Polynesia, whose individuals are about 20 mm long. In 1991 an expedition from London Zoo collected the last nine specimens of *Partula hebe* from the island of Raiatea, one of the Leeward Islands in French Polynesia.

Most snails reproduce by producing eggs, but Partula snails give birth to live offspring, and consequently have a relatively low fecundity. This was instrumental in their decline in the wild where they were unable to survive the depredations of a predatory snail (*Euglandina rosea*) introduced in an attempt to control African land snails (*Achatina* spp.).

The Partula snail presents conservationists with a number of practical problems.

1 The genetic integrity of the groups selected for breeding is problematic because the genus exhibits a number of intermediate levels of speciation.

2 High mortality rates can occur due to changes in:
- diet when individuals are transferred between collections
- changes in daily temperature
- the presence or absence of ultraviolet light.

Multiple sites have been used for captive colony formation to spread the risk (New, 1995). The zoos which are involved in the captive breeding of Partula snails include ZSL London, Bristol, *Durrell*, St Louis and Woodland Park, Seattle.

12.10 RESURRECTING LOST SPECIES

Breeding animals that have a novelty value – such as white tigers and white lions – has little conservation value (Rees, 2004c), and good zoos are not generally interested in this. However, the idea that it might be possible to breed back certain lost animals is extremely attractive to some conservationists and some zoos.

12.10.1 Case study: Barbary lion

The lion (*Panthera leo*) was formerly distributed right across Africa, through the Middle East to India. It now exists in fragmented populations in Africa south of the Sahara and in the Gir Forest in India. The Barbary lion (*P. l. leo*) that previously inhabited North Africa no longer exists in the wild but some zoos believe that they hold specimens that contain the remnants of the gene pool of this form. Males of this subspecies had a distinctive thick mane which extended half way along the back and sides, and extended as a fringe along the abdomen (Guggisberg, 1975). Barbary lions were taken to many European zoos in the 19th century.

The Barbary Lion Project was a collaboration between Oxford University and WildLink International. Using DNA samples from Barbary lions in museums across Europe it has been possible to create a DNA fingerprint that identifies the Barbary lion as a distinct subspecies. The project hoped to identify descendants of Barbary lions that currently exist in zoos and to 'breed back' the subspecies by selective breeding. It was hoped that eventually it might be possible to reintroduce these lions back into a national park in the Atlas Mountains of Morocco. For the moment, however, the project has run into financial difficulties.

12.10.2 Case study: The Quagga Project – resurrecting a lost species of equid

The quagga was a distinctive form of zebra which previously occurred in large numbers in the Cape Province of South Africa and the southern part of the Orange Free State, but became extinct in 1883 when the last known specimen died in Amsterdam Zoo. It was originally classified as *Equus quagga* and was distinctive because of the lack of distinct stripes on its hind quarters.

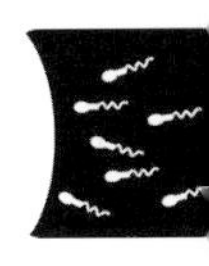

The aim of the Quagga Project is the retrieval of the pelagic characteristics of the quagga by selective breeding from a selected panel of plains zebra (*Equus burchellii*). This is possible because there is conclusive molecular evidence that the quagga and the plains zebra are conspecific. The project started in 1987 and by 2009 it had over 25 third generation progeny and was beginning to produce animals with a degree of reduced striping shown by none of the original founder animals. These striping patterns in individuals approximate to that shown by some museum specimens of the quagga. By the fourth generation the project expected to have achieved its aim of producing illustrations of a phenotype which had disappeared from the extant populations of plains zebra. These animals will form the basis of a herd which can be displayed in the Western Cape (Harley *et al.*, 2009).

Freeman (2009) has questioned the value of projects aimed at the 'resurrection' of lost species using genetic technologies and believes that we should look for alternative solutions to the extinction of species. She suggests that projects to resurrect the quagga and thylacine are expressions of speciesism and that we have not considered sufficiently the fate of the animals produced by such projects. However, others disagree and in May 2010 the First International Conference on the Restoration of Endangered and Extinct Animals was held at the Institute of Genetics and Animal Breeding at the Polish Academy of Sciences in Jastrzebiec near Warsaw.

12.11 FURTHER READING AND RESOURCES

The websites of EAZA, BIAZA, AZA and other zoo organisations are useful sources of information on captive breeding programmes.

The American Association of Zoo Veterinarians has produced *Guidelines for the Euthanasia of Nondomestic Animals* (2006) which is available from their website at www.aazv.org

BIAZA's (2005) *Animal Transaction Policy* is available at www.biaza.org.uk/resources/library/images/ATP%20june%2004.pdf

Much of the software described in this chapter has been developed by Dr Robert Lacy (Department of Conservation Biology, Chicago Zoological Society (Brookfield Zoo)) and his collaborators and is made freely available to zoos, conservation organisations and researchers (http://vortex9.org).

The AZA Wildlife Contraception Center (St Louis Zoo) can provide information on contraception procedures and products: www.stlzoo.org/animals/scienceresearch/contraceptioncenter/

The following volumes of the *International Zoo Yearbook* are relevant to this chapter: volumes 17 and 20, *Breeding Endangered Species in Captivity*; volume 27, *Conservation Science in Zoos*; volume 42, *Amphibian Conservation*.

Asa, C. S. and Porton, I. J. (eds.) (2005). *Wildlife Contraception: Issues, Methods and Applications*. The Johns Hopkins University Press, Baltimore, MD.

Collins, N. M. (ed.) (1990). *The Management and Welfare of Invertebrates in Captivity*. Federation of Zoological Gardens of Great Britain and Ireland, London.

Frankham, R., Ballou, J. D. and Briscoe, D. A. (2010). *Introduction to Conservation Genetics*, 2nd edn. Cambridge University Press, Cambridge, UK.

Gage, L. J. (2002). *Hand-Rearing Wild and Domestic Animals*. Wiley-Blackwell, Oxford.

Gage, L. J. and Duerr, R. S. (eds.) (2007). *Hand-Rearing Birds*. Wiley-Blackwell, Oxford.

Holt, W. V., Pickard, A. R., Rodger, J. C. and Wildt, D. E. (eds.) (2003). *Reproductive Science and Integrated Conservation*. Cambridge University Press, Cambridge, UK.

Nash, C. E. (ed.) (1991). *Production of Aquatic Animals, Crustaceans, Molluscs, Amphibians and Reptiles*. Elsevier, Amsterdam.

Sutherland, W. J., Newton, I. and Green, R. E. (2004). *Bird Ecology and Conservation: A Handbook of Techniques*. Oxford University Press, Oxford.

12.12 EXERCISES

1. How do zoos decide which species to keep?
2. Discuss the arguments for and against keeping common species in zoos as ambassadors.
3. Do zoos need to keep large animals to attract visitors?
4. Describe the methods zoos may use to increase the reproductive potential of a range of animal species.
5. Why is the concept of effective population size important in captive breeding?
6. What are the likely consequences of inbreeding and how do zoos minimise its effects?
7. Should zoos attempt to conserve subspecies?
8. What should zoos do with animals that are no longer able to reproduce?
9. Assess the success of efforts to increase the reproductive performance of a named species (e.g. giant panda, Asian elephant, lowland gorilla).
10. Discuss the importance of the IUCN in conservation.
11. Calculate the effective population size (N_e) of a population of giant pandas which contains 15 males and 26 females.
12. Using the data below for an imaginary population of animals: (a) construct a life table; (b) draw a survivorship curve.

Age (years)	Number
1	250
2	198
3	174
4	150
5	80
6	25
7	4

13. Should scientists be trying to resurrect lost species by selective breeding?
14. To what extent do animals have to learn sexual behaviour?

13 RECORD KEEPING

One thing is certain: conservation is only meaningful in the long term.

Sir Peter Scott

Conservation Status Profile

Asian lion
Panthera leo persica
IUCN status: Endangered D
CITES: Appendix I
Population trend: Unknown

An Introduction to Zoo Biology and Management, First Edition. Paul A. Rees.

13.1 INTRODUCTION

Record keeping is an essential activity in zoos and is a legal requirement for zoos in the European Union under the Zoos Directive. Records kept on individual animals allow zoos to monitor their health and facilitate the operation of effective breeding programmes. The SSSMZP (2004) states that the records should provide the following information:

- Identification and scientific name
- Origin (whether captive-born, identification of parents, previous locations)
- Dates of entry into, and disposal from, the collection and to whom
- Date of birth or hatching (or estimate)
- Sex
- Distinctive markings (tattoos, freeze-brand marks, rings, microchips, etc.)
- Clinical data (including dates of any treatments)
- Behavioural and life history data
- Date of death and results of any post-mortem and laboratory investigations
- Details of any escapes (including damage or injury caused to or by the animal to persons or property, reason for the escape and remedial measures taken)
- Food and diets.

Traditionally, zoos have kept paper records of their animals, often in a card index. Increasingly, these records are being computerised.

An annual stocklist should be kept, recording movements into and out of the collection for each species (Table 13.1). In the past, because zoos bought many of their animals, they would place a value on their holdings. In 1969 London Zoo's most valuable animal was its giant panda, valued at £12,000 (McWhirter and McWhirter, 1969). Putting a financial value on animals is not now normal practice in modern zoos.

13.1.1 Case study: PDAs and penguins at Edinburgh Zoo

Edinburgh Zoo has the largest colony of Gentoo penguins (*Pygoscelis papua*) in the world. In 2009 the colony produced 117 eggs in 50 nests, from which 48 chicks hatched. Staff monitor nests throughout the breeding season and collect data on a personal digital assistant (PDA). This includes information on nest location, sire and dam, and the fate of eggs, with dates (laid, hatched, removed, missing, broken). The reporting functionality of the software used allows keepers to produce automated clutch and egg logs, lists of due dates, a breakdown of egg fates and a record of nest use throughout the breeding season. Use of a PDA to record data has eliminated the duplication of data on different paper record sheets and standardised the recording of the colour coded foot bands used for identification purposes. This database will undoubtedly save keepers time in the future while providing an excellent database of information for future research (Burrill, 2009).

13.2 MARKING AND IDENTIFICATION

A number of computer-based record keeping systems help zoos to create databases of their animals. For these systems to function, each individual needs to possess a unique identification number.

If a zoo is to keep detailed records of animals it is imperative that each individual can be distinguished from the others. Individuals of some species may be identified because they have distinguishing marks or a distinctive appearance and are well known to their keepers (e.g. elephants, chimpanzees and giraffes). These animals are often given names and this sometimes leads to accusations of anthropomorphism,

Table 13.1 An example of stocktaking records for three species in a hypothetical zoo

Common name	Scientific name	Group at 1 January 2010	Arrive	Born	Death within 30 days of birth	Death	Depart	Group at 31 December 2010
Lion	*Panthera leo*	1.2.0	0.1.0	1.3.0	1.0.0	0.0.0	0.1.0	1.5.0
Tiger	*Panthera tigris*	2.4.0	1.0.0	0.0.0	0.0.0	1.0.0	1.0.0	1.4.0
Cheetah	*Acinonyx jubatus*	1.2.0	0.0.0	2.2.0	1.1.0	0.0.0	0.1.0	2.2.0

1.2.0 indicates the presence of 1 male, 2 females and 0 of unknown sex.

or attributing the animals with human characteristics. Jane Goodall was accused of this when she named the wild chimpanzees (*Pan troglodytes*) she studied in Tanzania. This is, of course, nonsense. Names are simply useful labels and whether a chimpanzee is called 'Boris' or #437 is neither here nor there. Suggesting that names cause people to bestow human characteristics on animals is as nonsensical as suggesting that labelling them with numbers assigns them a numerical value, implying that animal #24 is twice as large as #12.

Most animals need to be individually marked, perhaps because they are very numerous or because individuals are difficult to distinguish (e.g. maras (*Dolichotis* spp.), small antelopes, and many bird species). They may also need to be marked with microchips for security reasons so that they can be readily identified if they are stolen.

The type of marking system used depends upon the zoo's requirements. If animals need to be identified without handling them the mark needs to be highly visible. If not, it may be located in an inconspicuous place.

The marking system used for a particular species should:

- be easy to apply with minimal discomfort
- produce a mark that is readily visible or easily accessible
- have no adverse effect on the ecology or behaviour of the animal
- result in no long-term hindrance or irritation from the mark itself or the marking procedure
- be acceptable to zoo visitors.

13.2.1 Types of marks

Marks used to identify animals may be permanent, semi-permanent or temporary, depending upon the purpose of the mark. Wild animals may be marked temporarily if they are part of a field study. In a zoo environment marks generally need to be permanent so that it is possible to identify an animal throughout its life and as it moves from one institution to another.

13.2.1.1 Permanent marks

Branding

Hot branding has been used to imprint identification numbers on the horn and skin of mammals. This is generally unacceptable as it causes burns and the production of scar tissue. Freeze branding (cryo-branding) is less painful and typically involves the use of a branding iron with the required mark cooled to around −96°C by immersion in isopropanol which has been chilled using liquid nitrogen. Reptiles and amphibians may be marked by the electrocauterisation of a number or letter on the skin after applying a local anaesthetic. However, in amphibians the brand marks may not be visible after a few months.

Tattooing

This is used in many mammal species. Tattoos may be applied to the inside of the lip, the ear and the groin. This method has the disadvantage that the animal has to be captured and usually anaesthetised to read the tattoo.

Tattooing may be used to mark amphibians and reptiles. Care must be taken to ensure that the dye contrasts with normal skin pigmentation and that legibility of the mark is not lost due to diffusion or ultraviolet degradation. Paint should not be used on the permeable skin of amphibians. If paints are used they must contain non-toxic pigments, bases and solvents. Although reptile skin is less permeable than that of amphibians some paints may cause problems by distorting shell growth in turtles if applied across shell sutures.

Toe, nail and ear clipping

Toe and nail clipping was formerly widely used to identify wild animals but these methods are now generally considered inappropriate on humane grounds.

Ear clipping is used to identify some mammals, using a code to produce a numerical system based on the number and position of ear notches (Figs 13.1 and 13.2). The system allows the numbers 1 to 99 to be coded by adding the value of the numbers assigned to each notch.

It is important to note that the systems in North America and Europe are different. In North America the animal's left ear is used for the numbers 1, 2, 4 and 7, but in Europe the left ear is used for the numbers 10, 20, 40 and 70.

Shell marking

Turtles may be marked by cutting notches or drilling holes in the marginal scutes of the carapace. Turtles may also be marked with disc-type tags and clamp-on-ear-type tags fixed to the webs between the toes.

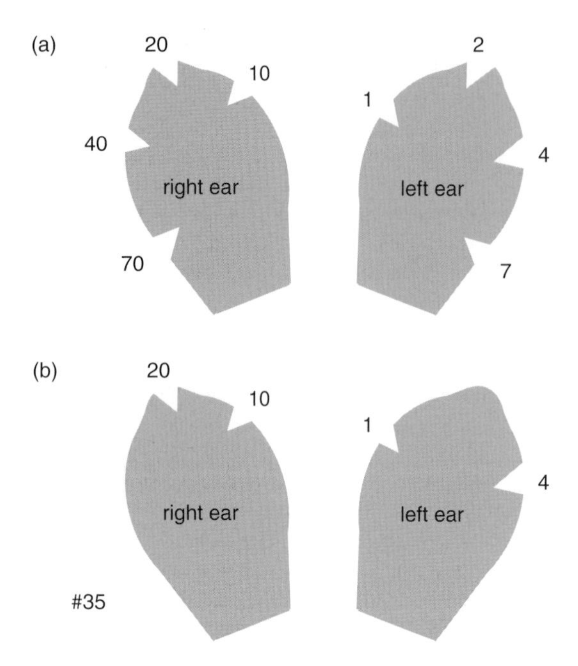

Fig. 13.1 (a) Numbering system using ear notches as used in North America. In Europe the ears are reversed, so the notch for '4' in Europe is the notch for '40' in North America. (b) Notches that code for 35 in North America (i.e. 20 + 10 + 1 + 4). In Europe these notches would code for 53 (i.e. 1 + 2 + 10 + 40).

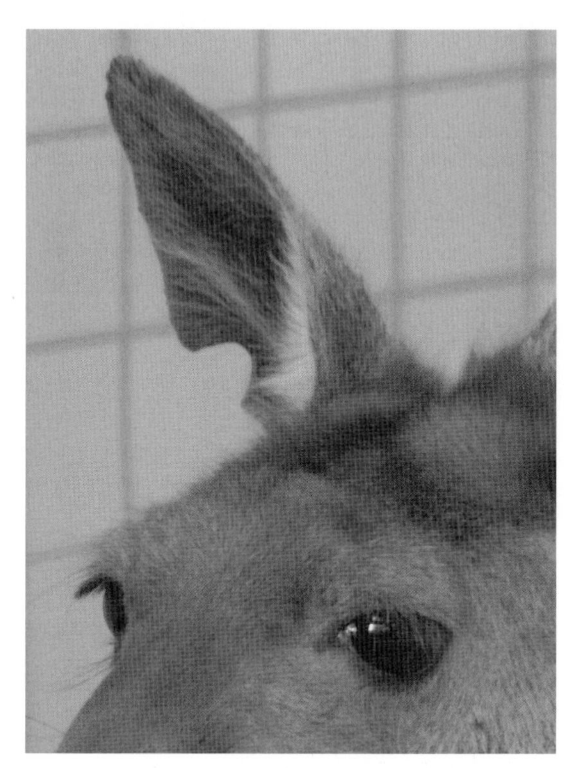

Fig. 13.2 Ear notch identification marks on the ear of a grey kangaroo (*Macropus giganteus*): #7 (in Europe).

Passive integrated transponders (microchips)
Transponders, the size of a grain of rice, can be used to permanently attach an identification mark to an animal that is invisible and can be read using a small, hand-held scanner (Fig. 14.8). The device can be implanted subcutaneously or intraperitoneally (within the body cavity) in a wide variety of species, and complications are rarely reported. Occasionally transponders may migrate if applied subcutaneously, making them difficult to locate in large species.

Transponders are not very useful in species that are difficult to handle, such as porcupines, because it is likely that they will need to be anaesthetised in order to locate the device, thereby making it no easier to identify the animal than if it carried a tattoo or other mark in an inaccessible place. This method is particularly useful for rare species as it allows them to be identified if recovered after being stolen.

Removal of scutes in snakes
Snakes may be permanently marked by removing subcaudal or ventral scutes (scales) based on a standardised numerical code. The scute is removed with surgical scissors or by rapid cauterisation. This appears to have no effect on mortality or locomotion and healing is usually rapid.

13.2.1.2 Semi-permanent marks

Tags
Ear tags are widely used to identify large mammals in zoos and are usually made out of plastic, aluminium or steel (Fig. 13.3). They are available commercially in a wide range of sizes and colours. Care must be taken to select an appropriate tag size for the animal to prevent it from snagging on vegetation, enclosure furniture or fencing. Both ears are sometimes tagged to prevent the loss of identity that would result from the loss of a single tag. Care must be taken to avoid infection when tags are applied. Petersen disc-type tags have been used to mark frogs by placing the tag in the web between the hind toes.

Leg bands
Leg bands made of plastic or metal are widely used to mark birds. Although this is an apparently straightforward method of marking, a number of problems may arise in banded birds:

Fig. 13.3 Plastic identification tag in the ear of an Ankole (*Bos taurus*).

• Some species (e.g. vultures and storks) defaecate on their legs and the resultant build-up of material around the band may cause injury.
• Ice may build up on metal rings on waterfowl in cold climates, causing injury.
• Bands must be the correct size as bands that are too small may cause injury and even loss of a leg.
• Adult birds may react to the presence of a foreign object and eject banded nestlings from the nest.
• Banding may affect the reproductive success of some birds. Swaddle (1996) found that the arrangement of coloured leg bands affected male reproductive success in male zebra finches (*Taeniopygia guttata*). Those that were symmetrically banded left more offspring than those with asymmetrically arranged leg bands. This is a good example of the benefits gained by males of many species when they possess symmetrical physical features.

In large species it may not be necessary to capture an individual in order to read its identification number (Fig. 13.4).

(a)

(b)

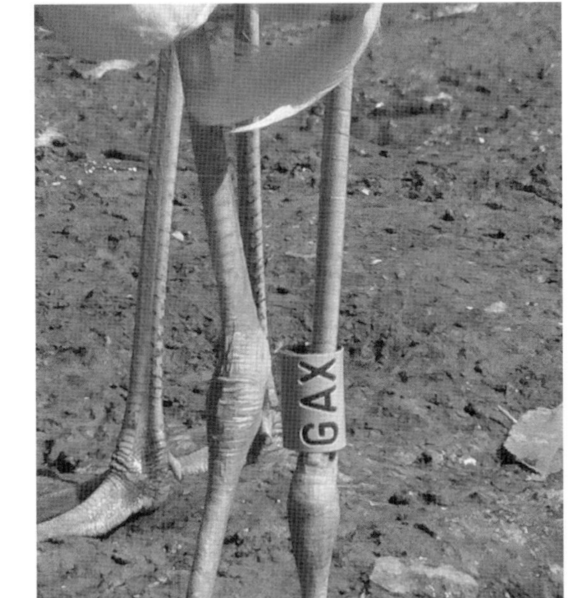

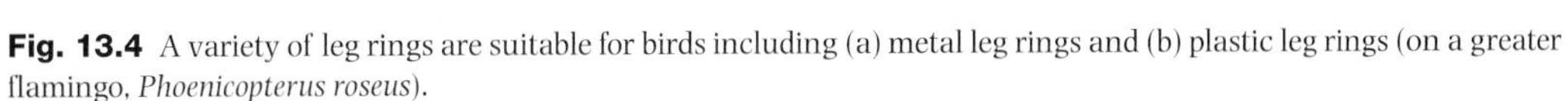

Fig. 13.4 A variety of leg rings are suitable for birds including (a) metal leg rings and (b) plastic leg rings (on a greater flamingo, *Phoenicopterus roseus*).

(a)

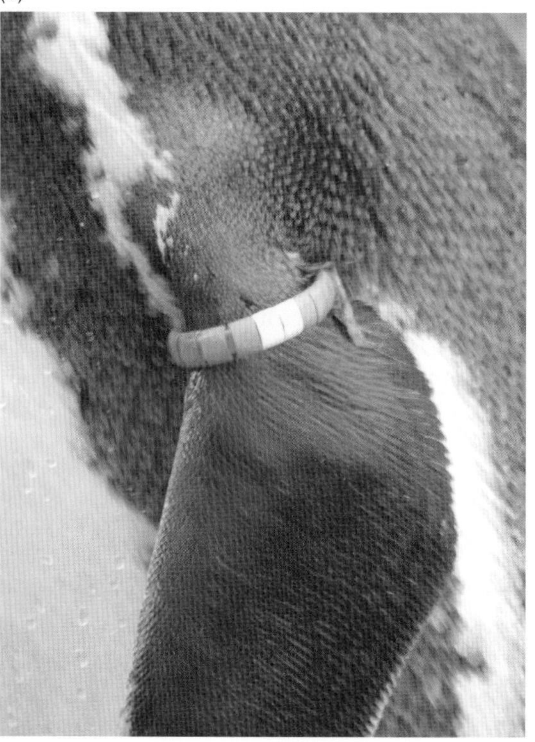

(b)

Fig. 13.5 Marking penguins. (a) Humboldt penguin (*Spheniscus humboldti*) marked with a plastic tag. (b) Silicone rubber tag on an African penguin (*S. demersus*) at Bristol Zoo.

Wing tags

Wing tags are used in penguins. A simple system of tagging each wing with different colours allows a large number of combinations to be produced (Fig. 13.5a).

Bristol Zoo has replaced the metal wing tags previously used to identify its penguins with silicone rubber tags developed by Bristol University. The new tags do not damage the birds' wings and bear identification numbers that are easy to read (Fig. 13.5b).

Unacceptable marking methods

Field ecologists and zoo staff have used methods of marking in the past that are unacceptable now. These have involved, for example, mutilation of small mammals by clipping their toes in such a way as to create a unique combination of missing digits for each animal (Delany, 1974).

13.2.1.3 Do zoos really need to mark animals? – using natural markings

Scientists at Bristol University have developed computer software that can identify individual penguins from the patterns of spots on their plumage. This software is intended for monitoring penguin populations in the wild but there is no reason in principle why it could not be used in a captive situation. Similar software exists which is capable of identifying elephants from the shape of their ears and wildebeest from the patterns of stripes on their flanks. Speed *et al.* (2007) have described the use of photo-identification software to identify individual whale sharks (*Rhincodon typus*).

13.3 RECORD KEEPING SYSTEMS

13.3.1 The International Species Information System (ISIS)

The International Species Information System maintains computer-based information systems about the animals held by around 825 institutions in 76 countries. This information is used to manage breeding programmes and includes data on sex, age, parentage, place of birth, and cause of death. The ISIS central database contains information on over two million animals from almost 15,000 taxa and 10,000 species. Some of this data is openly available via the internet (ISIS Species Holdings) and may be a useful source of up-to-date data for projects undertaken by students and researchers (e.g. Rees, 2009b). Holdings of individual institutions are available, listed by species and subspecies, and indicating the numbers of each sex and recent births (Box 13.1).

Box 13.1 Data held on giant pandas by the ISIS database.

Table 13.2 shows the results of a search made of the ISIS database for holdings of the giant panda on 23 July 2009.

Table 13.2 ISIS records for the giant panda (*Ailuropoda melanoleuca*) retrieved on 23 July 2009.

Institution	Males	Females	Unknowns	Births (last 12 months)[1]
MADRID Z	1	1	0	0
VIENNA	2	1	0	0
{Regional Subtotal}[2]	{3}	{2}	{0}	{0}
ATLANTA	2	2	0	1
MEMPHIS	1	1	0	0
NZP-WASH	2	1	0	0
SANDIEGOZ	1	3	0	0
{Regional Subtotal}[3]	{6}	{7}	{0}	{1}
ABERDE HK	1	2	0	0
{Regional Subtotal}[4]	{1}	{2}	{0}	{0}
Totals	10	11	0	1

[1] Births are included in the adult totals.
[2] Subtotal for Europe.
[3] Subtotal for North America.
[4] Subtotal for South East Asia.
MADRID Z, Zoo Aquarium de Madrid; VIENNA, Zoo Vienna; ATLANTA, Zoo Atlanta; MEMPHIS, Memphis Zoological Garden & Aquarium; NZP-WASH, Smithsonian National Zoological Park; SANDIEGOZ, San Diego Zoo; ABERDE HK, Ocean Park Corporation (Aberdeen, Hong Kong SAR, China).
Source: http://app.isis.org/abstracts/Abs74927.asp. © Copyright ISIS 15 July 2009.

ISIS was founded in 1973 by Drs Ulysses Seal and Dale Makey. Initially 51 zoos in North America and Europe contributed to the database which was hosted by Minnesota Zoo for 15 years. ISIS serves as a centre for the cooperative development of zoological software, thereby keeping the costs to a minimum. It currently distributes a number of different databases which are described briefly below (ARKS4, MedARKS, SPARKS, EGGS, REGASP, ZIMS).

Biography 13.1

Ulysses Seal (1929–2003)

Dr Ulysses Seal was a pioneer in the application of theoretical knowledge in genetics and population biology to practical conservation problems. He trained as a psychologist and then a biochemist. He worked as an endocrinologist at the Veteran's Administration Medical Center in Minneapolis, Minnesota, where he became interested in developing safe techniques for the anaesthesia of wildlife and contraception. In 1973 he founded the International Species Information System (ISIS) and in 1979 he became Chairman of the Captive Breeding Specialist Group (now the Conservation Breeding Specialist Group) of the IUCN. Seal was instrumental in producing the first Species Survival Plans and in saving the black-footed ferret from extinction.

ISIS Specimen Reference DVD
This contains historical and pedigree information on two million specimens of approximately 10,000 species, along with historical data sets ('studbooks') containing all of the data in the ISIS system.

ISIS/World Association of Zoos and Aquariums (WAZA) Studbook and Husbandry Manual CD-ROM
This database contains almost 1100 regional and international studbooks, husbandry manuals and other resources relating to captive animal records and population management. The information is distributed on CD-ROM to save the printing costs of the individual documents.

13.3.2 Animal Record Keeping System (ARKS4)

This software is used for keeping animal records within an individual institution. It is a PC-based application and is multi-lingual. It allows a zoo to produce a number of different reports based on its own records. ARKS4 allows individual institutions to contribute their data to the pooled ISIS database so that it is then available to others through the ISIS website. The ARKS records for part of a population of red-fronted lemurs (*Eulemur fulvus rufus*) kept at Blackpool Zoo are shown in Fig. 13.6.

13.3.3 Other record keeping systems

Medical Animal Record Keeping System (MedARKS)
MedARKS is software that supports the keeping of veterinary medical records and collection management within zoos.

Single Population Analysis and Record Keeping System (SPARKS)
SPARKS is used to manage studbook data sets. The software calculates the genetic relationship between individuals in a population (mean kinship). This helps zoos to decide which animals to use for future matings in order to avoid inbreeding and maximise genetic diversity within the zoo population of a species.

EGGS
This software supports record keeping and egg clutch management for a single institution and augments collection records kept in ARKS and SPARKS.

Regional Animal Species Collection Plan (REGASP)
REGASP is collection planning software available for ISIS under licence from ZAA (formerly ARAZPA). Institutions using REGASP have direct access to plans from other collections, allowing them to make contact to arrange animal movements.

PopLink
PopLink is software developed by Lincoln Park Zoo and is used in the management and analysis of studbook databases (Faust and Bier, 2008). PopLink is shareware and is distributed free of charge by the zoo.

13.3.4 Zoological Information Management System (ZIMS)

ZIMS is a new global database of information on animal health and well-being, and is the first of its kind in the world. It contains pooled information from ISIS member institutions which can be accessed by other members with their permission. This data includes information on veterinary care, animal husbandry and behaviour, and ZIMS will allow zoo professionals access to data from other institutions which was previously only available through personal contacts. The system also has the potential to track new and emerging animal diseases.

ZIMS has been funded by the United States Congress, the US National Science Foundation, the Institute of Museum and Library Services and the world zoological community.

13.4 MEASURING ANIMALS

Zoo staff make measurements of their animals for a number of reasons:

- to monitor their growth and health
- to confirm identification
- as an indicator of sex
- to monitor the effects of the environment on growth (e.g. diet or enclosure size)
- to monitor the effects of genetics (e.g. the breeding regime)
- as baseline information for evolutionary or wildlife biologists (because in many cases measurements from wild individuals are not available).

Measurements need to be standardised so that they may be confidently compared. Lundrigan (1996)

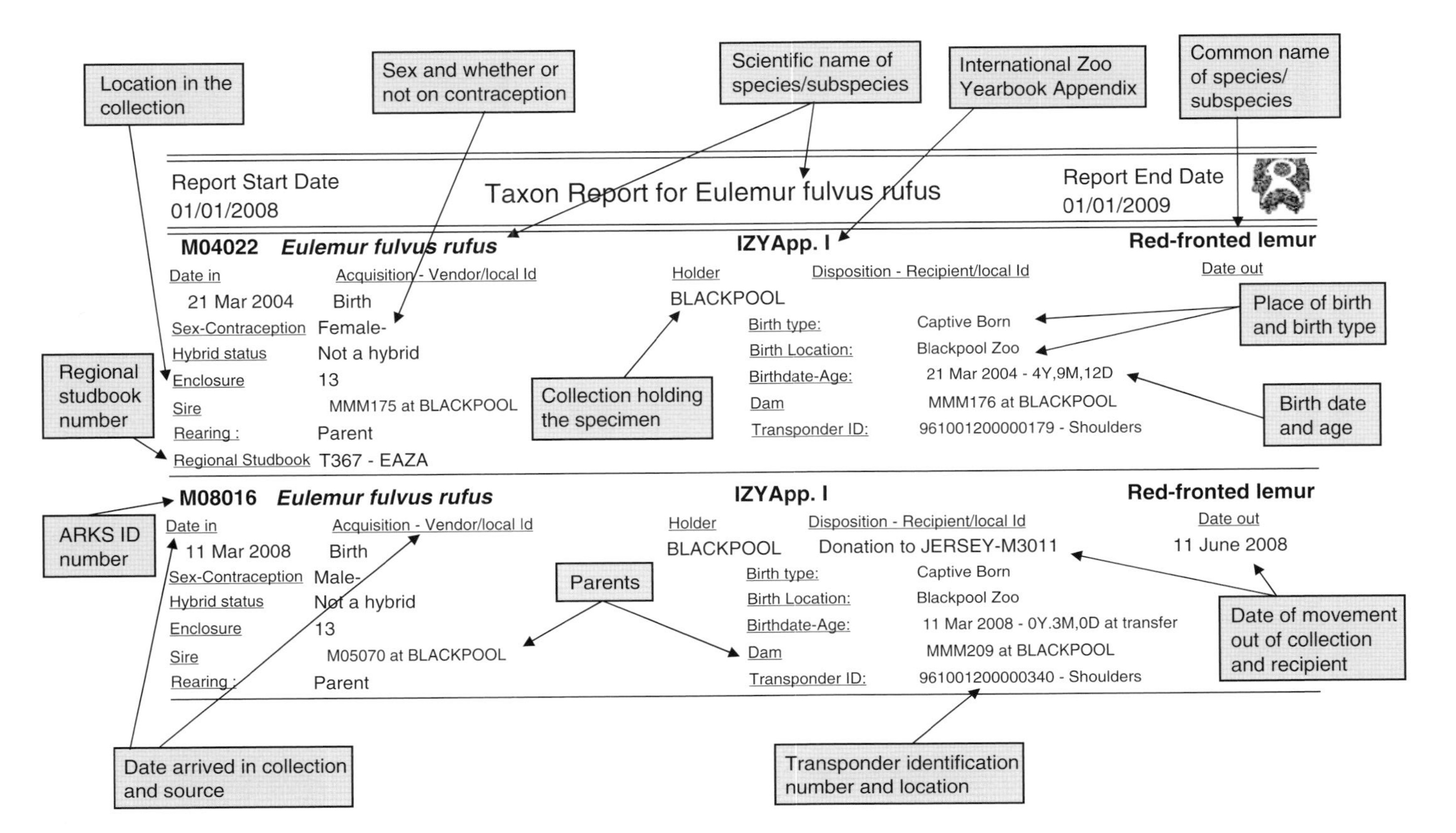

Fig. 13.6 The structure of an ARKS record. (*Source*: Blackpool Zoo.)

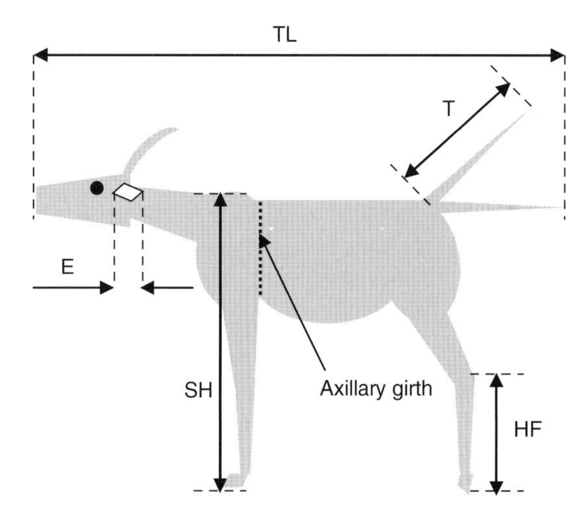

Fig. 13.7 How to measure a large mammal: standard external measurements. See text for explanation.

has described methods used for mammals by North American mammalogists. For small terrestrial mammals, the standard external measurements (in metric units) are:

1 total length (TL)
2 tail length (T)
3 hind foot length (HF)
4 ear length (E)
5 weight (W)

and in addition, for bats only:

6 tragus length (Tr) – the small piece of thick cartilage of the external ear that is immediately in front of the ear canal
7 forearm length (FA).

In Europe, head plus body length (HB) is usually measured instead of total length (which includes the tail).

When large mammals are measured two additional measurements are made in addition to 1–5 above (Fig. 13.7):

8 shoulder height (SH)
9 axillary girth – circumference of the body immediately behind the forelegs.

Measuring pinnipeds (seals, sea lions and walruses) involves five standard measurements:

1 standard length – equivalent to total length in large terrestrial mammals
2 anterior length of front flipper
3 anterior length of hind flipper
4 axillary girth
5 weight.

Measurement of cetaceans is more complex and requires 13 basic measurements. The details of how to make the measurements above and the equipment required may be found in the literature listed in the next section.

For birds the standard measurements are:

1 total length (including tail)
2 tail length
3 wing length
4 tarsus length
5 bill length.

The standard measurement for reptiles and amphibians is total length. Other measurements taken for chelonians (tortoises, turtles and terrapins) include shell height and carapace width.

13.5 FURTHER READING AND RESOURCES

Sources of information on the standardised measurement of animals:

Small terrestrial mammals

Skinner, J. D. and Smithers, R. H. N. (1990). *The Mammals of the Southern African Subregion*. University of Pretoria Press, Pretoria.

Bats

Handley, C. O. (1988). Specimen preparation. In: Kunz, T. H. (ed.) *Ecological and Behavioural Methods for the Study of Bats*, pp. 437–457. Smithsonian Institution Press, Washington DC.

Large terrestrial mammals

Nagorsen, D. W. and Petersen, R. L. (1980). *Mammal Collectors' Manual: A Guide for Collecting, Documenting and Preparing Mammal Specimens for Scientific Research*. Life Sciences Miscellaneous Publications, Royal Ontario Museum, Toronto.

Ungulates

Sachs, R. (1967). Liveweights and body measurements of Serengeti game animals. *East African Wildlife Journal*, 5: 24–27.

Pinnipeds

Committee on Marine Mammals, American Society of Mammalogists (1967). Standard measurements of seals. *Journal of Mammalogy*, 48: 459–462.

Cetaceans

Committee on Marine Mammals, American Society of Mammalogists (1961). Standardized methods for measuring and recording data on the smaller cetaceans. *Journal of Mammalogy*, 42: 471–476.

Websites

The ISIS website can provide a great deal of information on zoo populations and record keeping software, including ARKS, SPARKS, MedARKS and ZIMS (www.isis.org).

Information about PopLink is available from the Alexander Center for Applied Population Biology at the Lincoln Park Zoo: www.lpzoo.com/conservation/Alexander_Center/software/PopLink/index.html

A variety of animals tags are available from the following companies: Ritchey (www.ritchey.co.uk), Ascott (www.ascott.biz) and Roxan iD (www.roxan.co.uk).

13.6 EXERCISES

1. Why do zoos need to keep records about their animals?
2. With reference to a range of taxa, describe the various methods available for marking zoo animals.
3. How do computers help zoo professionals keep records of their animals?
4. Why is important for a zoo to measure the size of its animals?

14 EDUCATION, RESEARCH AND ZOO VISITOR BEHAVIOUR

For the indefinite future more children and adults will continue, as they do now, to visit zoos than attend all major professional sports combined (at least this is so in the United States and Canada)...

Edward O. Wilson

Conservation Status Profile

Bonobo (pygmy chimpanzee)
Pan paniscus
IUCN status:
Endangered A4cd
CITES: Appendix I
Population trend:
Decreasing

An Introduction to Zoo Biology and Management, First Edition. Paul A. Rees.

14.1 INTRODUCTION

There has been considerable recent interest in zoo education – because in many countries zoos are required by law to have an educational function – and in zoo visitor behaviour, because visiting zoos is an important leisure activity for large numbers of people. The two subjects are, of course, inextricably linked because the way in which visitors behave in zoos inevitably has an impact on what they learn from their visit. Also, it is important for designers to take visitor behaviour into account when they create new exhibits and new signage.

14.2 EDUCATION

14.2.1 The history of zoo education

Early zoos had an educational function merely by virtue of the fact that they were the only means by which most people had the opportunity to observe wild animals. Photography, film and video did not exist, so people learned about animals either by reading about them, looking at drawings and paintings, or by visiting the zoo.

The development of zoo education has been summarised by Woollard (1998). He identifies a number of stages, beginning with a pre-19th century 'menagerie mentality' – where visitors simply experienced zoos as living animal exhibits with no interpretation – and evolving through several curiosity-driven and then school curriculum-driven stages, until the concept of ecological immersion arose in the 1990s (Fig. 14.1). By the beginning of the 21st century zoo education had reached a stage that Woollard has called 'holistic empowering', where education is integral to all of the functions and activities of the zoo.

I have added a final stage to this process which I have called the 'ethical-ecoliteracy' stage. At the end of the first decade of the 21st century zoos are adopting a more ethical approach to both the keeping of their animals and the management of their estates. Visitors do not want to be told they are helping to destroy the planet. But, by using electric vehicles, selling Fair Trade products, installing water-saving devices in toilets and placing recycling bins in food service areas they can set a good example and encourage changes in behaviour. By keeping fewer and rarer species zoos are signalling a new focus on conservation-dependent species and applying ethical considerations in their collection planning, often taking difficult decisions about ending the keeping of popular species for whose needs they cannot adequately cater.

14.2.2 How can modern zoo visitors be educated?

Visitors come to zoos to see animals, not to read signs, attend lectures or watch films. The challenge for zoos is to teach people about animals and conservation without them realising it.

The types of learning that may take place within a zoo environment are:

- cognitive (intellectual) learning, e.g. learning facts about the conservation issues that affect the survival of tigers
- affective (emotional) learning, e.g. changing attitudes towards animals or the protection of the environment
- behavioural (changes in behaviour), e.g. increasing the possibility that visitors will donate to conservation organisations, recycle household waste, etc.

Most people visit a zoo to be entertained rather than to be educated. However, zoos have expanded their concerns from entertainment to education and conservation and as a result may set objectives for an exhibit in relation to knowledge acquisition, attitude change and subsequent visitor behaviour (Seidensticker and Doherty, 1996). But is there any evidence that zoos impart knowledge or change the behaviour of visitors in relation to wildlife and the environment?

14.2.3 Are zoos educational?

Modern zoos claim to have an educational role and many include a reference to this in their mission statement (see Section 1.6). In some countries there is a legal requirement for zoos to have an educational role as a condition of their zoo licence. In Member States of the European Union zoos must have an educational function to comply with Article 3 of the Zoos Directive (see Box 5.1).

Zoos claim to have an educational role, but their educational effect is very difficult to measure. They frequently draw attention to the large number of children who visit each year on educational visits but there is usually little attempt to assess how much learning has actually occurred. There is very little

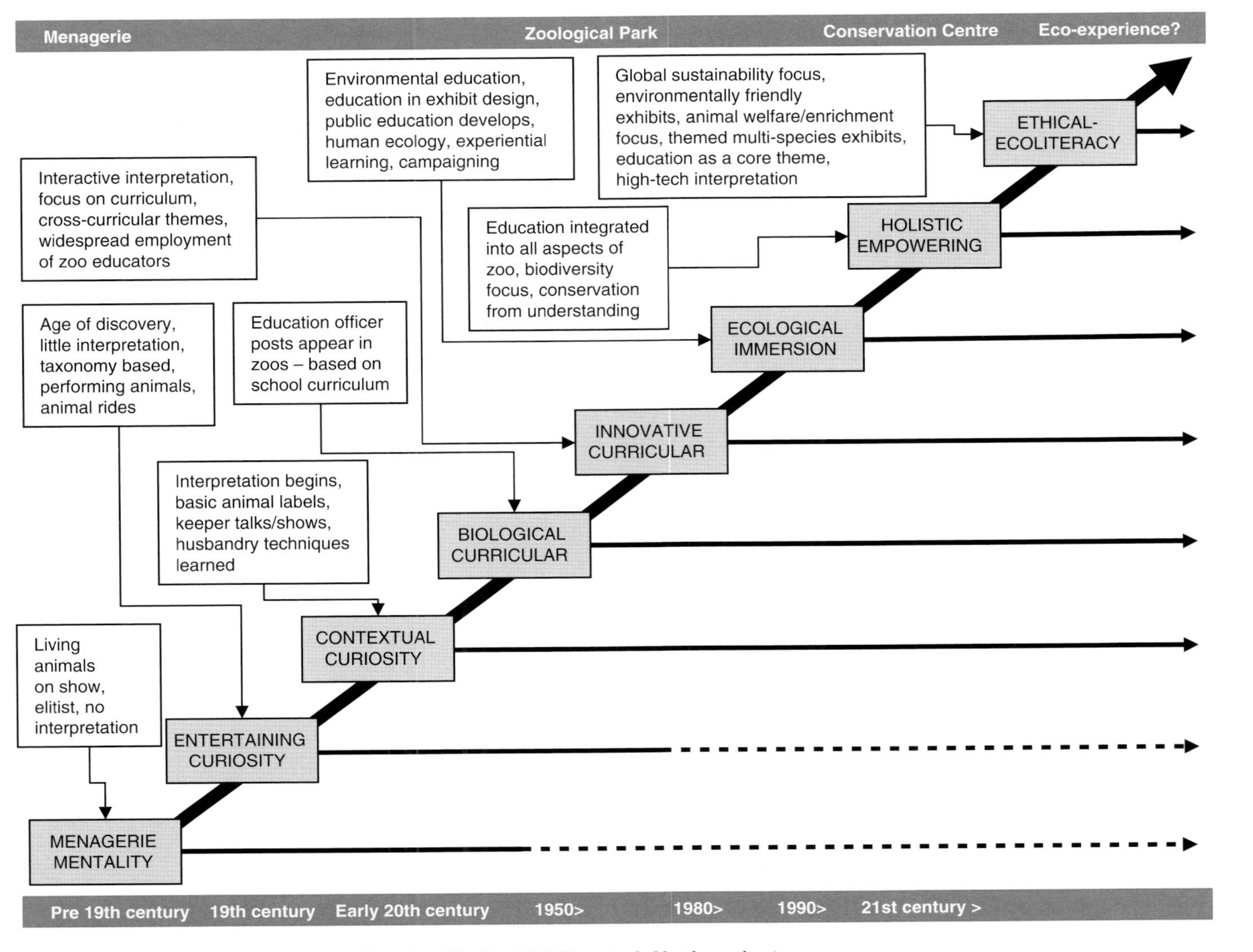
Menagerie
Zoological Park
Conservation Centre
Eco-experience?
Environmental education, education in exhibit design, public education develops, human ecology, experiential learning, campaigning
Global sustainability focus, environmentally friendly exhibits, animal welfare/enrichment focus, themed multi-species exhibits, education as a core theme, high-tech interpretation
ETHICAL-ECOLITERACY
Interactive interpretation, focus on curriculum, cross-curricular themes, widespread employment of zoo educators
Education integrated into all aspects of zoo, biodiversity focus, conservation from understanding
HOLISTIC EMPOWERING
Age of discovery, little interpretation, taxonomy based, performing animals, animal rides
Education officer posts appear in zoos – based on school curriculum
ECOLOGICAL IMMERSION
INNOVATIVE CURRICULAR
Interpretation begins, basic animal labels, keeper talks/shows, husbandry techniques learned
BIOLOGICAL CURRICULAR
CONTEXTUAL CURIOSITY
Living animals on show, elitist, no interpretation
ENTERTAINING CURIOSITY
MENAGERIE MENTALITY
Pre 19th century
19th century
Early 20th century
1950>
1980>
1990>
21st century >

Fig. 14.1 The evolution of zoo education. (Based on Woollard (1998), extended by the author.)

empirical evidence in the scientific literature to either support or refute the contention that zoos have an educational role, and few zoos formally evaluate their conservation education programmes (Dierking *et al.*, 2002). Nevertheless, this has not prevented zoo professionals, philosophers, animal welfare specialists and others from expressing their views. Few balanced discussions exist because authors tend to take one side or the other, depending upon their interests and allegiances.

Broad and Weiler (1998) reviewed the literature on the history of zoo education and identified four key objectives or outcomes, which may be summarised as:

1 an enjoyable, recreational and satisfying educational experience
2 the cognitive learning of facts about animals, zoos or exhibits
3 the development of a concern for wildlife conservation
4 appropriate on-site behaviour and long-term environmentally responsible behaviour.

14.2.3.1 The case 'for' an educational role for zoos

Those who claim an educational role for zoos often have a direct professional interest in promoting zoos. Some appear to have an outdated belief that simple exposure to wild animals in a zoo must result in the visitor learning something. There is little evidence for this; however, research suggests that displaying animals in realistic settings and allowing the public to interact with them may result in some learning.

Kellert and Dunlap (1989) found that zoos that displayed animals in authentic environments and had an educational focus had a positive influence on visitor attitudes towards wildlife while visitors' fear of or indifference to wildlife increased after visits to more traditional zoos.

There is some evidence that interactive experiences in zoos (e.g. birds of prey shows and encounters with snakes) increase learning, awareness, and positive attitudes towards animals (White and Barry, 1984; Morgan and Gramann, 1989; Yerke and Burns, 1991). Sherwood *et al.* (1989) showed that short- and long-term cognitive learning occurred when students handled either live or dead specimens of horseshoe crabs (*Limulus polyphemus*) and sea stars (*Asterias forbesi*), but gains in affective learning (changes in attitude) only occurred when the students handled live animals. Swanagan (2000) has shown that zoo visitors who had an active experience with an elephant show were more likely to support elephant conservation than those who had a passive experience of viewing the animals and reading graphics.

Although the general public is exposed to a great deal of information about wildlife in television documentaries, it has been argued that the behaviour observed in some species in zoos is more natural than that seen in documentaries, especially in modern exhibits (Burgess and Unwin, 1984; Andersen, 2003). Lions in zoos spend a great deal of time resting and sleeping, as they do in the wild, but most documentaries about lions give the impression that they are active for most of the time.

14.2.3.2 The case 'against' an education role for zoos

Studies of zoo visitors' behaviour have shown that they spend very little time at individual zoo exhibits and pay little attention to signage, so the opportunity for learning is extremely limited. Visitors to the Reptile House at the National Zoo in Washington spend just eight minutes looking at the exhibits (Marcellini and Jenssen, 1988), and although most visitors to this zoo read at least some of the exhibit labels, many had erroneous conceptions about specific animals and their habitats (Wolf and Tymitz, 1981).

A number of empirical studies have suggested that a zoo visit has no subsequent positive effect on conservation knowledge or behaviour in relation to the environment. Balmford *et al.* (2007) examined the effects of a single informal visit to one of a number of zoos in the UK, in what they claimed to be the largest study of its kind (1340 respondents at seven facilities). They found very little evidence of any effect on adults' conservation knowledge, concern or ability to do something useful. Broad (1996) found that knowledge about threatened species gained from a zoo visit had not influenced visitors in any way in 80% of cases when they were contacted 7–15 months later. Some zoo visits have even apparently led to a decrease in wildlife knowledge and an increase in 'dominionistic' attitudes to conservation in visitors (Kellert and Dunlap, 1989).

It is often claimed that zoos have become 'educationally redundant' because of the many nature documentaries on television (Margodt, 2000). Broad and Smith (2004) interviewed 125 zoo visitors and found that 42% had obtained their previous knowledge of animals and habitats from the television, 29% had obtained it from documentaries and 21% from books, newspapers and magazines. Only 14% had gained this prior knowledge from previous visits to zoos.

Some zoos exhibit animals in inappropriate social groups (e.g. pairs of coatis) or in inappropriate mixed-species groups (e.g. wallabies and blackbuck) that send a questionable educational message. If a normally solitary animal is kept with others of the same species, or species from different habitats (or even different continents) are housed together, visitors are likely to learn little about normal animal behaviour or ecology. Furthermore, there is considerable evidence that the very presence of visitors influences the behaviour of many species, especially primates, making it impossible to see truly natural behaviour in zoos (e.g. Chamove *et al.*, 1988).

Conclusion

It is too early to assess the educational role of zoos. Relatively little quantitative work has been done and much of what has been done has been superficial with little attention paid to the overall conservation message conveyed by zoos, often focusing on individual exhibits and visitor perceptions (Dierking *et al.*, 2002). Zoos clearly have the potential to provide public education, but if they are providing it they have failed to prove conclusively that it has any long-term effect on visitor behaviour towards wildlife.

Although wildlife documentaries can provide useful information about nature to the public they should not be regarded as an alternative. Broad and Smith (2004) suggest that zoos and wildlife documentaries provide complementary information about animals and conservation.

14.2.3.3 The wrong educational message – creationist zoos

In the EU the law requires zoos to have a conservation function and education is an obligatory element of this. However, the law does not distinguish between scientifically based conservation education and nonsense. Visitors to Noah's Ark Farm Zoo, near Bristol, are exposed to a creationist view of the origin and development of life because this accords with the religious beliefs of the owners. The zoo's website has a page entitled 'Creationist Research' where it states that 'The theme of the zoo is of course Noah and the ark, the biblical story about the great flood...'. The zoo contains a 14 foot scale model of the ark, and its teaching materials include religious education centred around the story of Noah's Ark (Anon., 2010).

14.3 ZOO INTERPRETATION AND SIGNAGE

Zoos have the potential to impart a great deal of information to visitors via signage that forms part of their exhibits. A zoo's signs need to be simple if they are to be effective. If a sign contains too much information it is likely to be ignored by the majority of visitors. Zoo visitors want to spend their time looking at animals, not reading signs.

Studies of the factors which influence visitor reading of exhibit labels have found that:

- Visitors are more likely to read short passages than long passages (Borun and Miller, 1980; Bitgood and Patterson, 1993).
- The perceived costs of reading a long piece of text are more important than interest in the subject matter in determining how much text was read (Bitgood, 2006).
- The probability of reading can be influenced by careful placing of text (Bitgood *et al.*, 1989) and by asking provocative questions (Hirshi and Screven, 1990; Litwak, 1996).

A very basic sign might contain at least the following information:

- Vernacular name or names (e.g. cheetah)
- Scientific name (e.g. *Acinonyx jubatus*)
- Range (e.g. Africa, Middle East, India ...)
- Habitat (e.g. savanna, tropical forest ...).

Fig. 14.2(a) gives an example.

It may also contain additional information about:

- Behaviour
- Reproduction (e.g. courtship and mating, number of young)
- Nutrition (e.g. feeding habits)
- Conservation status (e.g. IUCN Red List category)
- Involvement in zoo breeding programmes (e.g. EEPs, SSPs).

Figs. 14.2(b) and 14.3 give examples.

Some signs are aimed specifically at children and may draw attention to particular features of a species,

(a)

(b)

Fig. 14.2 (a) A group of signs for waterfowl in a multi-species exhibit. Each sign contains the vernacular name of the species, the scientific name, a drawing of its appearance, and a simple map of its breeding and wintering ranges. (b) This sign has been kept simple yet provides information on the vernacular and scientific names (including the subspecies), the appearance of adults and young, feeding and breeding behaviour, predators, hunting, distribution and conservation status.

like the smell of a maned wolf (*Chrysocyon brachyurus*), or the height of a giraffe (*Giraffa camelopardalis*) (Fig. 14.4). Simple signs may be arranged in a series around the edge of an enclosure, perhaps at different viewpoints. Others are more complex. They may be interactive and may even have a scientific content (Fig. 14.5).

Some signs illustrate types of behaviour and can help to interpret behaviours witnessed by visitors (Fig. 14.6). Others help visitors to identify individual animals (Fig. 14.7).

14.3.1 Interactive exhibits

Guidelines for the design of interactive exhibits have been suggested by Bitgood (1991). He describes three types of exhibit:

- Simple hands-on (e.g. touching animal fur, climbing on a statue of an animal)
- Participatory (e.g. assembling an animal skeleton, comparing jumping distance with other animals)
- Interactive (e.g. a label with a flip panel, devices with control buttons which produce sounds, computer tutorials, magnifying glasses or microscopes).

Interactive and participatory devices that may be seen in zoos include:

- hinged information boards with a question on the outside and a hidden answer on the inside
- hinged boards designed as books that tell a simple story about an animal or exhibit
- electronic signs that provide information or link questions with answers when buttons are pressed
- touch boxes – visitors attempt to identify a hidden object by touch
- signs that play sounds when buttons are pressed, e.g. primate vocalisations
- zoo Olympics – devices that measure human abilities such as running speed or jumping distance
- height charts and size comparison boards which allow visitors to compare their size with that of large species
- climbing frames that allow children to mimic the abilities of climbing animals such as monkeys and apes.

At Chester Zoo an exhibition of the work of the veterinary staff includes an exhibit which allows visitors to listen to recordings of the heart beats of a variety of species recorded through a digital stethoscope, including an Asiatic lion (*Panthera leo persica*), a baby Congo buffalo (*Syncerus caffer nanus*), and a buffy-headed capuchin (*Cebus xanthosternos*). It also contains an exhibit which allows visitors to scan microchipped toy chimpanzees and look up their MedARKS records on a computer (Fig. 14.8). The microchips have been inserted in the chimps' wrists and the visitor uses a real scanner to detect the chip. The records of some of the real chimps at the zoo are

Fig. 14.3 A sign describing the role of zoos in the conservation of Przewalski's horse (*Equus ferus przewalskii*).

Fig. 14.4 A very large sign aimed at children, asking them to compare their arm span with the wingspan of an albatross.

then accessed by selecting the ID number from a list on a computer screen.

ZSL Whipsnade Zoo uses a real Land Rover as part of the interpretation for its cheetah exhibit. It contains an

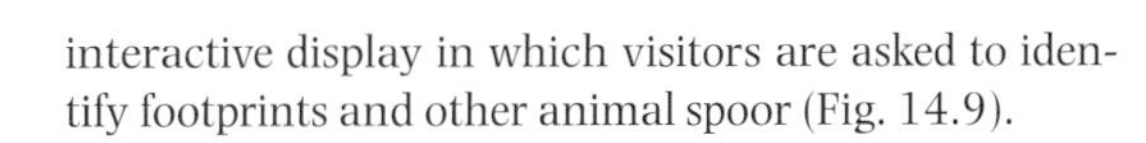

interactive display in which visitors are asked to identify footprints and other animal spoor (Fig. 14.9).

A study which monitored visitors leaving the *African Rock Kopje* exhibit at the San Diego Zoo found that

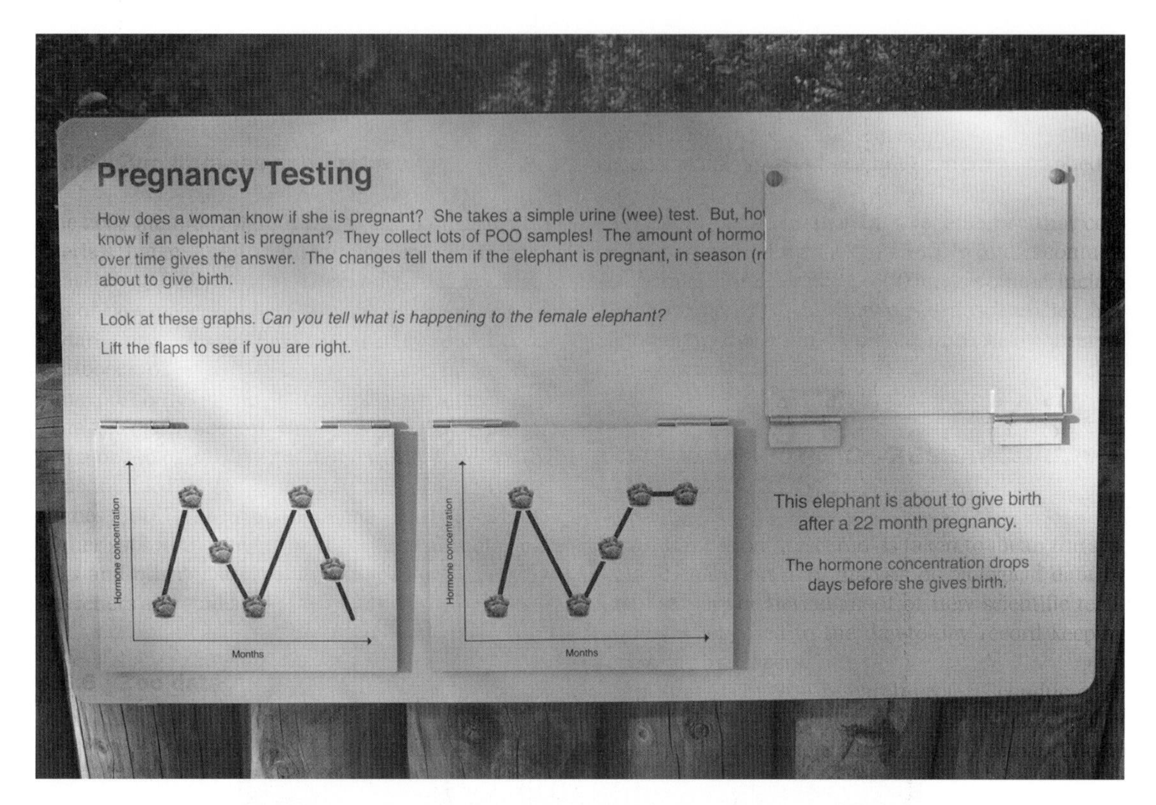

Fig. 14.5 An interactive sign with a scientific content – monitoring pregnancy in elephants.

Fig. 14.6 A sign illustrating courtship behaviour in Asian elephants (*Elephas maximus*) at Chester Zoo. This sign helps visitors to identify sexual behaviour which is part of a courtship sequence. They are unlikely to witness mating itself because it lasts a few seconds. However, they might see various elements of courtship. The information in this sign is based on photographs and text provided by the author.

Fig. 14.7 Signs that help visitors to identify particular animals in a group encourage them to see the animals as distinct individuals with their own personalities and social relationships. Nicky (left) is easily identifiable from his photograph.

Fig. 14.8 An interactive exhibit at Chester Zoo. Visitors can identify individual chimpanzees by reading the ID number in a microchip embedded in their wrists with the hand-held scanner. They can then examine their veterinary records on a computer, just as vets do with real chimpanzees.

they were unable to answer the cognitive questions that were set as objectives for the exhibit – for example describe the kopje as an ecosystem – if they had not read the interpretive signs. Perhaps not surprisingly, the study found that younger visitors were more likely to use the interpretive elements of the exhibit than to read the signs, whereas older visitors were more likely to read the signs (Derwin and Piper, 1988).

Fig. 14.9 A large interactive exhibit that forms part of *Cheetah Rock* at ZSL Whipsnade Zoo.

14.4 THE ROLE OF EDUCATION DEPARTMENTS

The role of a zoo education department will depend upon its size and budget. Some zoos have purpose-built education buildings (Fig. 14.10) and even large lecture theatres where public lectures are held.

Education staff may take part in providing some or all of the following:

- classroom sessions for school children
- lectures for college and university students
- public lectures
- guided tours of the zoo
- teaching and learning materials for teachers
- visits to schools and other organisations
- hands-on animal encounter sessions for visitors
- library and research facilities
- information on the zoo's website
- animal shows
- production of zoo guide books
- design and content of signage and other interpretation associated with zoo exhibits
- management of zoo volunteers working as presenters
- fund-raising for conservation projects
- conservation education overseas
- trade in endangered species exhibitions.

14.4.1 Education staff

Education staff in a zoo may have a number of roles including:

- presenters/explainers
- education officers
- education researchers.

Many zoo educators have been trained as teachers and may have previously worked in a school or with children in some other capacity.

Education staff should be carefully trained if they are to engage with the public. Some presenters oversimplify the information they provide to the public. Ideally they should memorise a prepared script. An explainer who tells crowds of visitors that all of the tiger's problems are caused by the Chinese medicine trade in the presence of people of Chinese origin is not

Fig. 14.10 The Safari School at Knowsley Safari Park.

only providing incorrect information but also risks offending visitors.

Where possible, presenters should use a public address system when they give talks to the public to ensure that they can be heard. Some zoos use portable systems while others have fixed loudspeakers in the ground or mounted on walls or in other suitable locations.

In some zoos, some keepers may have a secondary role as a presenter, especially if they care for animals such as sea lions or birds that are trained to take part in shows.

The International Zoo Educators Association (IZE)

Many zoo educators are members of the International Zoo Educators Association. The IZE is dedicated to expanding the educational impact of zoos and aquariums worldwide. Its mission is to improve zoo education programmes and provide members with access to the latest thinking, techniques and information in conservation education. The IZE publishes the *International Zoo Educators Journal*.

14.4.2 Multimedia and websites

Some zoos provide CD guides to their animals and facilities. At Knowsley Safari Park, visitors who purchase a guide book are provided with a CD commentary to play in their vehicle as they drive through the enclosures. *Durrell* has a cinema in which it shows educational films about the conservation work of the Durrell Wildlife Preservation Trust. Interested visitors can purchase a DVD about the Trust's work.

Interactive multimedia exhibits may include:

- videos – at *The Deep* video images of species from polar regions are projected onto a simulated ice sheet which is cold and wet to the touch
- 3D movies
- simulators (e.g. submarine training simulator, *The Deep*).

The Deep contains a simulated hi-tech undersea research centre which plays recordings of scientists discussing research and environmental problems on large flat screens suspended from the ceiling.

Zoo websites allow zoos to reach out across the world to people who may never be able to visit them. The best sites may contain information about:

- visiting the zoo (e.g. map, location, cost)
- animals kept at the zoos
- the history of the zoo
- the management structure of the zoo
- *in-situ* conservation projects supported by the zoo.

A website may also give access to:

- webcams located within animal enclosures
- the zoo's annual report and other zoo publications
- educational materials.

Zoo websites may be an important source of information for some types of zoo research.

14.4.3 Activities for children and young people

Many hundreds of thousands of children and young people visit zoos every year and it is important for zoos to engage with these groups if they are to influence the way future generations view animals and the natural world. Some individuals visit with their families and friends, while others visit with groups from schools or colleges.

14.4.3.1 Playgrounds and physical activities

Many zoos have a children's playground. The best have an animal theme, such as an ark. Some zoos integrate opportunities for children to engage in physical activities related to animal behaviour into their exhibits: bars to swing on (like a gibbon), a tree trunk to climb (like a gorilla), a running track for measuring speed (to compare with the running speed of a cheetah).

14.4.3.2 Zoo theatres and animal encounters

A zoo theatre provides an opportunity to engage children and their parents in a performance with a conservation theme. The Terrace Theatre at Bristol Zoo is used for a wide range of events from animal encounters to musical performances and has even been used in the evening, when the zoo is closed, to stage a performance of *Romeo and Juliet*.

Many zoos offer opportunities for visitors to come into close contact with a range of animals from lemurs to spiders. Usually this occurs on the zoo's premises, in a special building, marquee or arena. However, sometimes zoos take the animals to the public. Staff from *Durrell* visit hotels on Jersey and provide a 'Bug Show' in the evenings for children.

ZSL London Zoo offers sessions to assist people who suffer from arachnophobia, including hypnosis, cognitive therapy and hands-on sessions with spiders. On summer afternoons visitors to Edinburgh Zoo are treated to a Penguin Parade when keepers allow their penguins to walk along the paths adjacent to their enclosure (Fig. 14.11). In the past, when health and safety regulations were less stringent, some zoos allowed contact with some animals that were potentially very dangerous (Fig. 14.12).

14.4.3.3 Animal rides and shows

Animal rides are rare in zoos now but were once very popular. Some zoos, for example Colombo Zoo in Sri Lanka, still give elephant rides (Fig. 10.29). These were extremely common in zoos in the past and both Asian and African elephants were used for this purpose. However, most zoos would now consider this activity both unacceptable and dangerous in view of the large number of accidents involving elephants that have occurred in zoos, circuses and various elephant tourist attractions in Asia.

The Zoological Society of London's online Print Store contains images of a number of historical photographs of animal rides which would be unacceptable in a modern zoo. These include an Asian elephant giving rides in 1896, four zebras pulling a cart in 1914, two llamas pulling a carriage in 1923 and a camel giving rides in 1924.

Animal shows are still common in zoos but tend to have a much greater educational content than was formerly the case. The best animal shows allow the participants to exhibit their natural behaviour. These include flying shows involving birds of prey or parrots and climbing displays by arboreal species. For some animals, taking part in these shows may be an important enrichment activity. Circus-type acts are generally considered unacceptable, but some zoos still show performing parrots that ride miniature tricycles and other vehicles. Displays using marine mammals such as sea lions are less controversial because often they are used to show off the animals' natural swimming and jumping abilities. Part of their training usually involves the presentation of different parts of the body to the keeper and opening the mouth on demand. Such training may be important in veterinary examinations or treatment.

Busch Gardens Africa in Tampa Bay, Florida has opened a new 1.6 ha (4 acre) attraction, set in the Congo, which is designed to give visitors 'close-up animal interactions'. These include transparent pods which allow visitors to 'enter' a tiger enclosure with their heads at ground level and come face-to-face with tigers. There are other opportunities to have close-up encounters with orangutans, gibbons, fruit bats and crocodiles. This facility contains rides, live entertainment and 'multi-storey family play areas', including climbing nets and rope bridges from which visitors can look down on orangutans. The area includes educational exhibits and is toured by educators and street entertainers, including stiltwalkers dressed as trees, frogs and jungle birds.

Fig. 14.11 The Penguin Parade at Edinburgh Zoo.

Fig. 14.12 In the past zoos allowed visitors close contact with animals which would be considered dangerous today. This photograph is of the author's mother on a visit to London Zoo in the 1930s. (Courtesy of Peter and Enid Rees.)

14.4.3.4 School curricular materials

Many education departments in zoos produce materials and classroom sessions for teachers and students which fulfil some of the requirements of the school or college curriculum. Often these are concerned with animal classification, adaptation to the environment or other topics that are easily demonstrated by reference to species in the zoo's collection. Chester Zoo has devised a route around its grounds that takes students to exhibits of animals that were, or might have been, seen by Charles Darwin on his voyage around the world in HMS *Beagle*.

14.4.3.5 Role of zoos in the training of students in tertiary education

Zoos are an invaluable resource for students studying a wide variety of courses from travel and tourism, business studies and psychology to animal management, veterinary science and zoology. They provide work experience opportunities for college and undergraduate students and research opportunities for students studying for higher degrees (e.g. MSc and PhD) and for academics.

14.4.3.6 Public lectures

Some large zoos provide lectures for their members and members of the public. These lectures may be given by the zoo's own staff, by researchers working on particular taxa or by the staff of conservation NGOs. Well-known zoos are able to attract famous and influential people to contribute to these lectures, thereby attracting greater public interest.

14.4.4 Staff training

Staff training is essential. In the European Union it is one of the means by which a zoo may comply with the Zoos Directive (Art. 3). Training may be provided by a number of means, including:

- in-house, by senior staff and peers
- attendance on specialist programmes provided by colleges and universities
- distance learning programmes provided by educational institutions
- staff exchanges with other zoos
- outside consultants (e.g. vets, animal trainers).

14.4.4.1 Vocational training zoos

In the UK some colleges maintain animal collections that are used for training students following programmes of study in animal management, veterinary nursing and associated subjects. In some cases these collections require a zoo licence because they are open to the public. Reaseheath College in Nantwich (Cheshire) has an animal centre which houses around 125 species of animals and Walford & North Shropshire College near Telford (Shropshire) keeps over 90 species of animals at its Harris Centre, which includes a birds of prey facility. *Durrell* (Jersey Zoo) runs specialist conservation courses in collaboration with the Durrell Institute for Conservation and Ecology at the University of Kent.

Zoo biology is a well-established discipline within the education system in the United States. A very wide variety of courses exists within high schools, community colleges and universities, ranging from zoo technology, animal centre management and captive wildlife management to animal biotechnology and conservation, zoo and aquarium science and zoo and aquarium leadership. Moorpark College is a community college in California which offers an *Exotic Animal Training and Management Program* and styles itself 'America's Training Zoo'. This facility offers hands-on training with animals.

More details of programmes available in the United States are available from the AAZK website (www.aazk.org/zkcareer/training.php).

14.5 VISITOR BEHAVIOUR

If zoos are to provide a meaningful educational experience for their visitors an understanding of visitor behaviour is essential. An appreciation of the implications of research that has been conducted on visitors to zoos and museums can assist in the design of:

- animal enclosures
- zoo layouts
- signage and other interpretation materials.

Studies of zoo visitors have examined a wide range of behaviours including exhibit viewing times (Johnston, 1998), reactions to exhibit design (Bitgood *et al.*, 1988), attitudes to feeding live prey (Ings *et al.*, 1997), reactions to free-ranging animals (Price *et al.*, 1994), the influence of environmental enrichment (Davey *et al.*, 2005) and the effect of visitors on animal welfare (Davey, 2007).

14.5.1 Visitor orientation and circulation

Bitgood (1988) has suggested that orientation and circulation are important areas of the study of visitor behaviour because they influence:

- whether or not people actually visit a zoo
- whether or not they see a particular exhibit
- what they learn
- what they tell their friends and relatives
- whether or not they return.

There are three major elements to visitor orientation and circulation:

- conceptual (or thematic) orientation – an awareness of the themes and the organisation of the subject matter of the zoo
- wayfinding (or topographical or locational) orientation – being able to find places in the zoo
- circulation – the pathways that visitors take while exploring the zoo (Bitgood, 1988).

Knowledge of visitor behaviour should be used when zoos and zoo exhibits are designed. Problems relating to orientation and circulation may be divided into six areas:

- pre-visit or off-site – what information is provided before a visitor arrives?
- arrival at the zoo – can visitors find the entrance?
- finding support facilities (such as toilets, food and shops)
- orientation and circulation while viewing the exhibits in the zoo – how do visitors find their way around?
- exiting the zoo (e.g. finding the car or the bus stop)
- measuring visitor's behaviour in relation to orientation and circulation.

The visitor experience in a zoo is determined as much by the overall design of the facility as it is by the individual exhibits.

Bitgood (2006) has summarised the factors that affect visitor circulation in museums and zoos. He argues that the value of an experience is unconsciously calculated by the ratio between the benefits and costs (the general value principle). In other words, a visitor will only do something if the benefits are perceived as greater than the cost:

- Often visitors to an exhibit do not follow the intended traffic pattern.
- Visitors tend to stay on the main paths.
- They will only approach objects that are attractive or interesting.
- The cost of approach must be perceived as low.
- Visitors tend to turn right at choice points – found in some studies but not others.
- Visitors tend to walk on the right side of paths.
- Visitors tend to walk in a straight line unless something catches their attention and pulls them away.
- Visitors are reluctant to backtrack in an exhibit.
- Visitors tend to move only along one side of a path and are reluctant to move from one side to the other.

Deans *et al.* (1987) found that the most common visitor circulation pattern at the Reid Park Zoo involved turning right and circulating anticlockwise on the periphery of the zoo, but not using paths that connected one part of the outer circle to another. This path was probably perceived as the main path. In a study of North Carolina Zoo, Bitgood (1988) observed that visitors were reluctant to walk down a path to overlook an exhibit that took them away from the main path. Bitgood refers to this phenomenon as 'dominant path security'.

Bitgood *et al.* (1988) compared exhibits of the same species in different zoos. In some of the zoos exhibits were placed on one side of a walkway and in others they were placed on both sides. They found that visitors stopped fewer times in zoo exhibit areas when displays were on both sides of a path. In a study of visitors to the Steinhart Aquarium, Taylor (1986) found that visitors were so unwilling to backtrack that the only visitors who saw the whole aquarium were repeat visitors who were familiar with the layout. Visitors are reluctant to enter dark areas and are more likely to follow lighted pathways in buildings (Loomis, 1987).

Implications of the general value principle for zoo design are:

- Design the zoo so that visitors do not have to walk more than necessary.
- Do not require visitors to backtrack through areas they have already visited.
- Do not include multiple choice points in the design where visitors must choose which path to follow.
- Avoid two-sided designs where exhibits on one side compete with those on the other.

Zoo visitors need to know where they are (orientation) and how to get to where they want to go (wayfinding).

14.5.1.1 Orientation

Orientation is concerned with having an awareness of the themes and organisation of the layout of the zoo. It can be subdivided into:

- pre-visit orientation – prior experiences and expectations, following directions to the zoo
- arrival orientation – parking, finding the entrance, entrance orientation (e.g. maps and information boards)

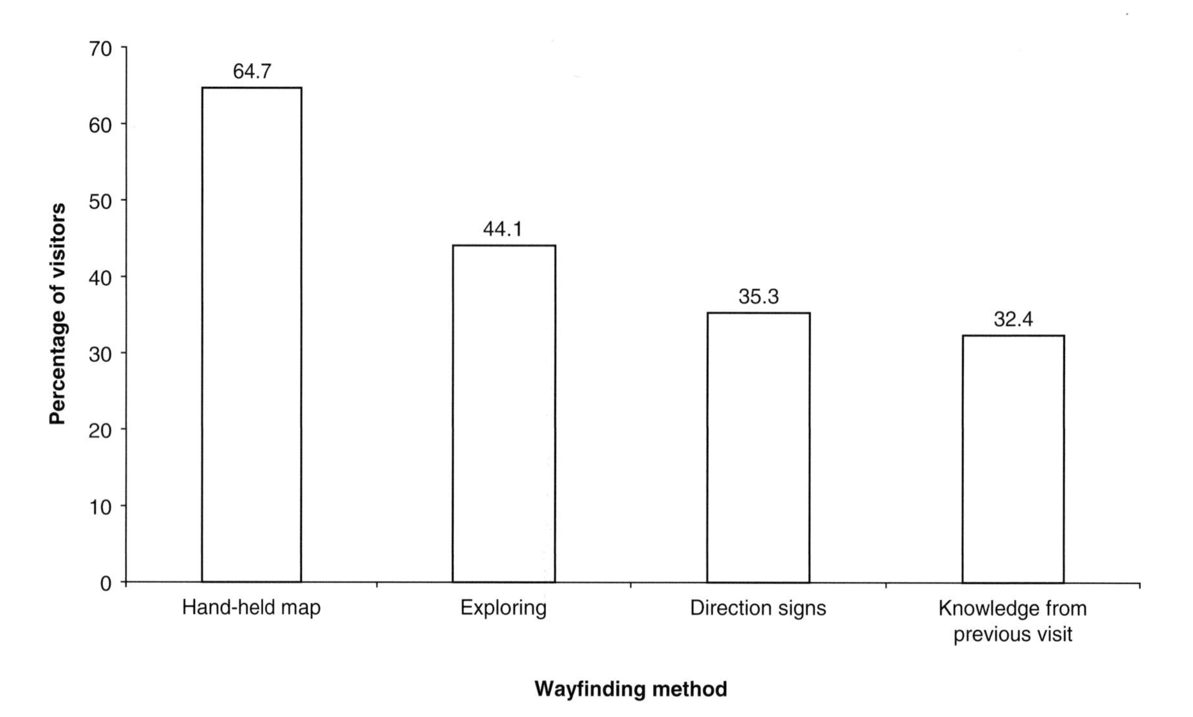

Fig. 14.13 Wayfinding methods used by visitors to the Arizona-Sonora Desert Museum. (Based on information in Shettel-Neuber and O'Reilly, 1981.)

- orientation to support facilities – where are the toilets, shops, food outlets, places to rest?
- orientation and circulation while viewing the exhibits.

Orientation is important in supporting learning in a zoo environment. A study of students who visited the National Zoo, Washington DC found that those who were given pre-visit orientation information – what they would see, when they would eat lunch, etc. – learned more than other groups of students even when they were given more learning-orientated pre-visit preparation (cited in Bitgood, 1988). When visitors to the Arizona-Sonora Desert Museum (a zoo) were asked what kind of information they wanted before their visit most people (55.6%) wanted information on conceptual orientation – information about what they could see – and only 18.5% wanted wayfinding information.

14.5.1.2 Wayfinding

In a study of visitors to the Arizona-Sonora Desert Museum almost 65% of visitors reported using a hand-held map for finding their way (Shettel-Neuber and O'Reilly, 1981) (Fig. 14.13). In a study at the Birmingham Zoo, Alabama 77% of visitors who received hand-held maps were observed using them and those who were given a map viewed a greater percentage of the exhibits (86%) than those who were not (78%) (Bitgood and Richardson, 1986).

Most zoos guide their visitors around their site by providing a map and signage indicating the location of each exhibit. Large zoos may have a complex layout so often a trail of signs is provided from one exhibit to another. Graphics showing types of animals are more useful than signs containing text, especially for children and in zoos that attract a large number of foreign visitors (Fig. 14.14a). An indication of the length of a trail is useful for visitors with small children or mobility difficulties (Fig. 14.14b).

In 1974 the Smithsonian Institution's National Zoological Park in Washington DC adopted a zoo design concept which featured a system of trails marked by pictograms – stylised representations of animal heads – leading to each exhibit. The major exhibits located on each trail were indicated on totems. Each totem

(a)

(b)

Fig. 14.14 Wayfinding signs (a) using words and symbols to indicate the direction to zoo exhibits and (b) indicating the length of each route.

Fig. 14.15 Lines painted on the ground guide visitors around ZSL London Zoo.

indicated the trail length and the amount of time it would take to complete, along with pictograms of the major exhibits along the trail arranged vertically. This system no longer exists but it is considered a classic in zoo graphics (Yew, 1991).

There is evidence that visitors prefer a suggested path (Shettel-Neuber and O'Reilly, 1981). Knowledge of visitors' preferred path can assist a zoo in selecting a suggested route, because people are more likely to follow a route that has been freely chosen (Fig. 14.15). Circulation knowledge can help a zoo to plan exhibits and time interpretive talks to coincide with visitors' circulation patterns.

Some zoos have devised means of rapidly and efficiently moving visitors through popular exhibits. Colombo Zoo operates a one-way system for visitors viewing their big cat enclosures (Fig. 14.16). When the traffic flow in the Reptile House in the Birmingham Zoo was changed from two-way to one-way fewer than 1% of visitors violated the signs (Bitgood *et al.*, 1985).

Aquariums with underwater tunnels are frequently equipped with moving floorways. At *The Deep* visitors must initially ascend to the top of the building and then move downwards via a series of gangways past a number of interactive exhibits before reaching the main tanks. Although a strict one-way system is not operated, nevertheless the general flow of visitors is from the top of the building to the bottom.

Walk-through enclosures such as aviaries, and those containing lemurs or bats, often have a linear structure with a single entrance and a separate exit. In some zoos the timing of educational talks and animal feeding times requires visitors to move rapidly from exhibit to exhibit in a fixed sequence if they are to attend them all.

Drive-through safari parks usually have a single route that visitors must follow so that enclosures are viewed in a fixed sequence. On busy days pressure from following vehicles forces drivers to keep moving. Some parks used land trains to move groups of visitors quickly and efficiently through enclosures. Human behaviour moves most people through traditional zoos and aquariums in a relatively efficient manner simply because they spend so little time looking at each exhibit.

14.5.2 Dwell times

14.5.2.1 How long do visitors spend looking at exhibits?

Studies have shown that visitors spend very little time looking at individual exhibits. In a zoo with a large number of exhibits this is inevitable. North Carolina Zoo keeps over 200 species of animals. In summer the zoo opens for eight hours each day. A visitor who spends eight hours viewing 200 species, and does nothing else, can only spend on average 2.4 minutes looking at each species.

A study of visitors to the Reptile House in the National Zoo, Washington DC showed that they spent less than 30 seconds in front of each exhibit (Mullan and Marvin, 1999). Although they may not be actively encouraged, short dwell times are important to zoos as they avoid congestion at popular exhibits. In San Diego Zoo visitors must queue to see the giant pandas and visitor movements are actively controlled by staff on busy days to avoid long waits to view the animals.

Simple exhibits – old-fashioned cages, enclosures and aquariums mounted in walls – offer the visitor an

Fig. 14.16 Colombo Zoo, Sri Lanka, operates a one-way system around parts of its site. Here visitors are walking along a line of cages containing big cats.

essentially one-dimensional experience. The visitor can walk along the edge of the exhibit and either the animals are visible or they are not. Walk-through exhibits offer at least a two-dimensional experience, and at best a three-dimensional experience as visitors search the habitat for animals horizontally and, in the case of arboreal or flying species, vertically. This undoubtedly increases the dwell time. Indeed, anyone who has been on safari knows that the enjoyment associated with searching for animals and finding their tracks, trails and signs is almost as great as finding the animals themselves.

14.5.2.2 Case study: *Spirit of the Jaguar*, Chester Zoo

In 2001, Chester Zoo opened *Spirit of the Jaguar*, a one-way indoor exhibit sponsored by Jaguar cars, which

Fig. 14.17 The *Spirit of the Jaguar* exhibit at Chester Zoo. Inset: A jaguar (*Panthera onca*) in one of the outside enclosures.

cost $4 million (Fig. 14.17). The exhibit has an indoor visit area of 315.6 m^2.

The exhibit consists of 39 elements:

- two flagship species enclosures (housing jaguars (*Panthera onca*))
- three integral species enclosures (housing poison arrow frogs (*Dendrobates* sp.), butterfly goodeids (*Ameca splendens*) and leafcutter ants (*Atta cephalotes*))
- 18 interactive interpretation elements (including video monitors, soundboards, flip panels, scent signs and signs with tactile models)
- 16 non-interactive interpretation elements (signs consisting only of text and images).

Individual visitors were tracked through the exhibit to measure the effectiveness of the various elements (Francis *et al.*, 2007). The median dwell time in the exhibit was 5 minutes 41 seconds, and ranged from 29 seconds to almost 48 minutes. When at least one jaguar was visible the median dwell time was significantly higher (6 minutes 59 seconds) than when no jaguars were visible (4 minutes 9 seconds). On average visitors stopped at seven of the 39 exhibit elements (18%), and the largest number of exhibit elements stopped at was 21 (54%). The majority of the visitor groups (92.5%) stopped at the flagship species (jaguar enclosures), 36% stopped at the integral species, 17% at the interactive exhibits and just 5.9% stopped at the non-interactive exhibits. Visitors spent 61% of the time engaged with the exhibit elements but were passive – that is, not engaged with any of the exhibit elements – for 39% of the time. The average visitor spent just 1 minute 49 seconds at the jaguar enclosures.

14.5.2.3 Recording visitor circulation and behaviour

Problems encountered in the measurement of orientation and circulation behaviour have been discussed by Bitgood (1988). Some studies have asked visitors to retrace their routes through a zoo at the end of their visit. However, in one study visitors were found to be only 60% accurate in retracing their steps and visitors tended to overestimate the time they spend at an exhibit.

New technology has provided improved methods for circulation studies. Visitor behaviour can now be tracked with GPS equipment. Chester Zoo has used 'Trackstick' – a GPS data logging device – to record visitor movements.

Clearly, the conclusions drawn from the studies quoted here cannot necessarily be generalised to the behaviour of all visitors to all zoos. Each zoo is unique and much more research is necessary before such generalisation will be possible.

14.5.3 Which species do visitors prefer to see?

Moss and Esson (2010) examined the relative interest that visitors showed in 40 species held by Chester Zoo by measuring the proportion of visitors that stopped at exhibits and for how long ('holding time'). They found that visitors were far more interested in mammals than any other group and that visitors found birds least interesting. They also demonstrated that visitors spend more time watching larger species, those that are active and those that were the 'flagship species' in the exhibit. Moss and Esson suggest that species that are brought into collections primarily for their perceived educational value should be selected on the basis of their popularity with visitors.

Ward *et al.* (1998) found that – based on the time they spent viewing zoo exhibits –adults and children preferred larger animals to smaller animals and that children showed a preference for larger groups of mammals. However, other studies have found no difference (e.g. Balmford, 2000). Not surprisingly, live animal exhibits have a higher attracting power than interpretation (Ross and Lukas, 2005).

The evidence for public preferences for predators is equivocal. Some studies have suggested that people dislike and fear predatory and dangerous animals, while others claim that the perceived dangerousness of a species increases the 'attractive power' of a zoo exhibit (Bennett-Levy and Marteau, 1984; Bitgood *et al.*, 1988).

In 2004 a poll of more than 50,000 people in 73 countries found that the tiger was the most popular species (McCarthy, 2004) and the following year the Bengal tiger won a television phone-in competition in the UK in which eight endangered species competed for conservation funding (Endemol, 2005). Nevertheless, even very small animals can be very attractive to visitors. The *BUGS* exhibit at ZSL London Zoo contains insects and other invertebrates and is extremely popular with visitors.

14.6 RESEARCH, RESOURCES AND EXPERTISE

Some zoos are important centres of research and many have excellent libraries and archives which are available to researchers and students.

14.6.1 Zoo research

Zoos are an important source of animals for zoological research and the institutions themselves have become the subject of research in recent years. Some zoos are able to maintain research departments but often, especially in smaller zoos, they may consist of a single person whose job it is to coordinate the work of visiting researchers and students.

A good deal of zoo research is conducted by students and keepers and is never published and as a result some good work is lost to the zoo community. Another consequence of not publishing this work is that zoos and students will often conduct almost identical studies at different times or in different places, unaware that similar work has been done.

Within the European Union zoos may fulfil their obligation to have a conservation function by undertaking research from which conservation benefits accrue (Zoos Directive, Art. 3). This qualification of the type of research required is important because much zoo research has little or nothing to do with conservation (Rees, 2005a). The majority of zoo research is about the effects of captivity on animals, for example nutritional and reproductive problems, abnormal behaviour and behavioural enrichment. While these studies may help to improve animal welfare in zoos, it is probably fair to say that most of the species held in zoos are not in difficulty in the wild because they suffer from nutritional, reproductive or behavioural problems. This does not make most zoo research invalid, it simply means that it does not necessarily fulfil the requirements of the Zoos Directive.

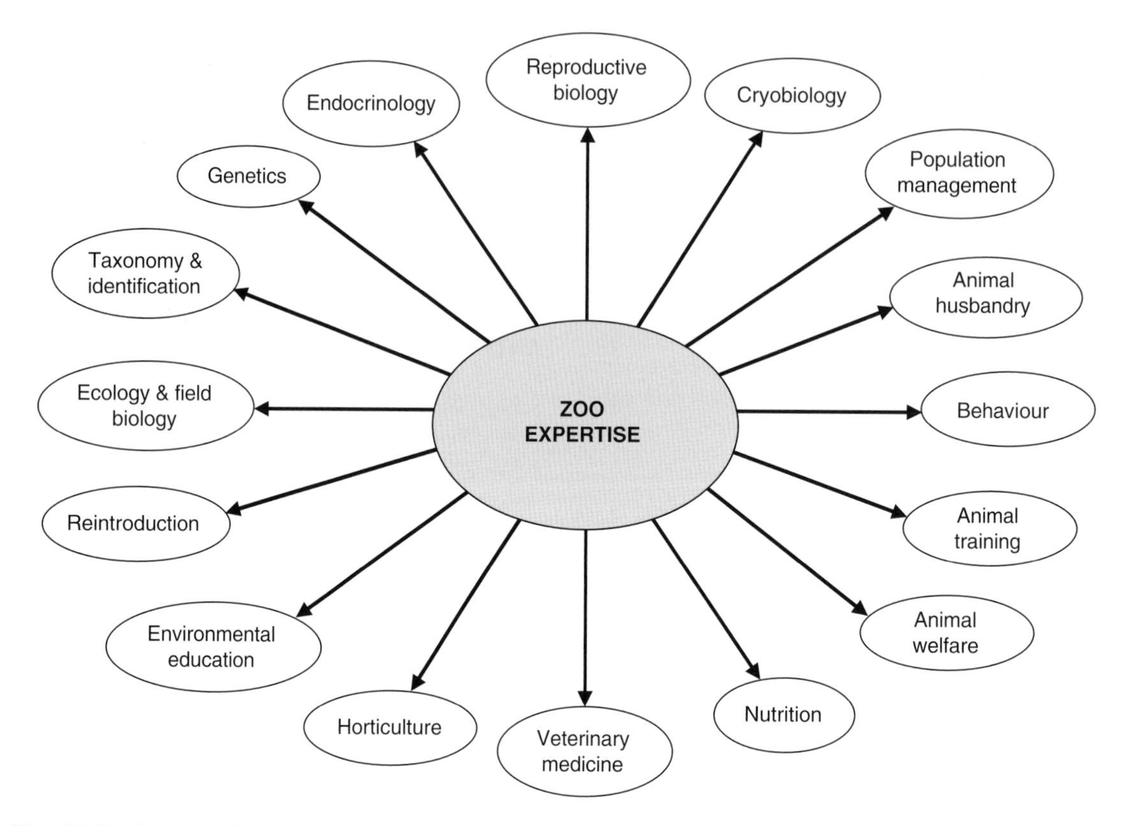

Fig. 14.18 The range of expertise available among the staff of a large modern zoo.

14.6.2 Specialist journals and books

Research concerned with zoo animals and zoos is published in a wide variety of academic journals. The best known of these is *Zoo Biology*, which was first published in 1982. Papers concerned with the welfare of zoo animals may be published by journals whose primary interest is welfare, such as the *Journal of Applied Animal Welfare Science* and *Animal Welfare*, and papers concerned with veterinary aspects of zoo keeping may appear in veterinary journals such as *Veterinary Record* and the *Journal of Zoo and Wildlife Medicine* published by the American Association of Zoo Veterinarians. Other specialised papers are published in journals concerned with animal physiology, nutrition, reproduction and animal behaviour, and in journals concerned with particular taxa, for example primates. The work published in academic journals is peer-reviewed, that is, the papers are reviewed anonymously by experts in the field and the journal's editor before they are accepted for publication. This system of peer-review is intended to ensure that only work of the highest quality is published.

Other journals of a more specialised nature publish articles about zoos and zoo animals, such as *International Zoo News* and *The Shape of Enrichment*. Many of these publications do not operate a system of peer-review and so the quality and scientific validity of the work cannot be assured. Nevertheless, they often contain useful information about specialist topics such as enrichment techniques, animal nutrition, breeding records and enclosure designs. Many of these articles are written by experienced keepers and other zoo professionals. However, it is often difficult to determine the status of the authors. In the absence of a system of peer-review, results of experiments should be treated with caution, especially when articles report studies which may have been poorly designed, used very small sample sizes and have not been subjected to appropriate statistical analysis. There is also a great deal of useful information to be found in documents produced by individual zoos and regional associations such as BIAZA and AZA. These hard-to-find publications are often referred to as the 'grey literature' and would include technical reports from government agencies, working group reports, papers from research groups and other materials that are difficult to locate through the conventional channels such as publishers.

The *International Zoo Yearbook*, published by the Zoological Society of London, contains up-to-date information on most of the world's best zoos, including a summary of the animals held, staff numbers, visitor numbers, and the names of senior staff. Each year the yearbook focuses on a particular aspect of the work of zoos or a particular taxon of animals.

Many books have been written about zoos and zoo animals. Once again, care should be taken when obtaining information from books. Anyone can write a book and the material in many books has not been subjected to peer-review. A book written by an academic who holds a university post is more likely to be authoritative than one written by someone who has an amateur interest in animals or zoos.

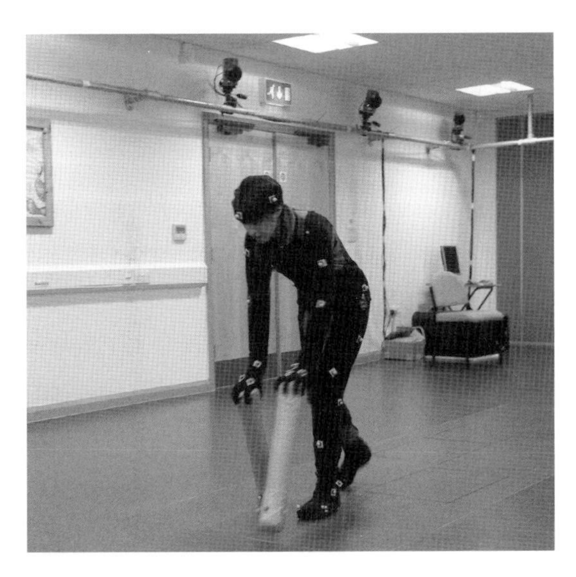

Fig. 14.19 A student who has worked as a chimpanzee keeper and studied their behaviour simulates chimpanzee movements in the Gait Laboratory at the University of Salford. Images of her movements were used to animate a virtual chimpanzee as part of a research project.

14.6.3 Zoo expertise

Large zoos employ staff with a wide range of expertise, and have access to experts in other zoos, regional associations, veterinary practices, research institutes and universities (Fig. 14.18). Zoo professionals may be able to assist with:

- identification of animals and animal products seized by customs officers
- feeding and husbandry advice for other animal collections
- field surveys of animals within their range states
- field studies of animal behaviour and ecology
- establishment of conservation projects
- GIS mapping of conservation areas.

Zoos can be a useful source of information for a very wide range of research projects. Scientists at the University of Salford are creating virtual chimpanzees for educational purposes. In order to animate chimpanzee behaviour a student who had studied chimps in a zoo mimicked various actions, such as walking, grooming, feeding and sleeping, in the University's Gait Laboratory, which is normally used to analyse human movements. The movements of reflective markers on a special suit were detected by cameras mounted high on the walls and their positions were determined by triangulation (Fig. 14.19). The data collected was then used to animate a virtual chimpanzee living in a virtual zoo enclosure.

14.6.4 Major research centres and zoo research facilities

A small number of zoological societies and governments have established major research institutes, for example:

- Institute of Zoology – Zoological Society of London, UK (Box 14.1)
- Smithsonian Conservation Biology Institute (formerly the Conservation and Research Center of the National Zoo) – Smithsonian Institution, USA
- Institute for Conservation Research (formerly the Center for Reproduction of Endangered Species (CRES)) – San Diego Zoo, USA (Box 14.2)

Box 14.1 The Institute of Zoology (IoZ), Zoological Society of London.

Fig. 14.20 The Institute of Zoology, Regent's Park, London.

The Institute of Zoology is located in Regent's Park in London, adjacent to ZSL London Zoo (Fig. 14.20). The staff includes zoologists and wildlife veterinarians and their work is divided into five research themes:

- Biodiversity and Macroecology
- Behavioural and Population Ecology
- Wildlife Epidemiology
- Reproductive Biology
- Genetic Variation, Fitness and Adaptability.

Research projects include:

- The Tsaobis Baboon Project – a long-term study of a desert baboon population in Namibia.
- Understanding the basis of sperm cryopreservation – so that sperm can be maintained as a genetic resource for the future.
- Mechanisms of reproduction in a range of species – including the harbour porpoise, red panda and Mohor gazelle.
- The cause of catastrophic mortality and population declines of *Gyps* spp. vultures in India.
- Investigations into the disease threats to British species, e.g. red squirrels, grey squirrels, garden birds, amphibians, marine mammals and turtles. Other IoZ projects include:
- iBats – a partnership between the IoZ and the Bat Conservation Trust which is collecting data on global changes in bat distributions and abundances.
- The EDGE of Existence Programme aiming to conserve the most Evolutionarily Distinct and Globally Endangered (EDGE) species.
- Bushmeat Research Programme.

Box 14.2 Conservation research at San Diego Zoo.

Institute for Conservation Research

In 2009 San Diego Zoo announced the creation of the Institute for Conservation Research. This represented a significant expansion of the well-established Center for Reproduction of Endangered Species (CRES), which was founded in 1975 and later renamed Conservation and Research for Endangered Species. The institute will provide central organisation and management of the scientific programmes of the zoo. The new institute will include:

- The on-site research efforts of the San Diego Zoo and the zoo's Wild Animal Park
- Laboratory work at the Beckman Center for Conservation Research
- International field programmes, with more than 180 scientists working in 35 countries.

In addition it will include targeted efforts for key endangered species through the work of:

- the Griffin Reptile Conservation Center
- the Botanical Conservation Center
- the Keauhou and Maui Hawaiian Bird Conservation Centers
- management of the California condor programme in Baja, California
- joint operations (in partnership with the USFWS) of the US Bureau of Land Management's Desert Tortoise Conservation Center, Nevada
- the Endangered Species Disease Response Center.

CRES was responsible for a number of conservation successes, including:

- The creation of a sustainable population of California condors
- The development of techniques to improve breeding of cheetahs
- Working to improve cub survival in giant pandas.

Examples of some current conservation research

In the zoo:

- Low-cost testing for the amphibian chytrid fungus (*Batrachochytrium dendrobatidis*) and ranaviruses in zoo collections
- Computer modelling of diseases in captive ruminants
- Reproductive success in the Somali wild ass
- Feline herpesvirus (FHV) in North American captive cheetahs
- Collecting and preserving African elephant semen
- Cardiovascular disease and diet in gorillas.

In the field:

- Vasectomy as a method of birth control in African elephants
- Recovery of Hawai'i's most threatened forest birds
- Use of genetic tools to study the ecology and conservation of drill in Cameroon
- Social dynamics of Sichuan snub-nosed monkeys in China.

Source: www.sandiegozoo.org/conservation/ (accessed 1 July 2010).

- Leibniz Institute for Zoo and Wildlife Research (IZW), Berlin – established by the German Science Council.

In addition to these large organisations, smaller individual research facilities have been established within some zoos.

Many zoo studies involve simple observations of animals made from outside their enclosure. However, some zoos offer scientists the opportunity to undertake complex studies by providing special facilities. The *Living Links to Human Evolution Centre* at Edinburgh Zoo is dedicated to primate research. At this centre researchers are able to study, for example, cognition in chimpanzees using research 'pods' where the animals have been taught to make choices by touching a computer screen. Rewards are received by the animals via a narrow tube near the screen. The chimps can choose to enter the pod and take part in the experiments or ignore them if they so wish. The research pods were

built into the chimpanzee exhibit at Edinburgh as an integral part of the design.

14.6.5 Zoo libraries and archives

Some zoos maintain significant archives of historical materials that may be a useful source of information for research (e.g. Rees, 2001c). These may include collections of newspaper cuttings, annual reports (containing data on species holdings), veterinary records, zoo guide books, photographs, enclosure plans, invoices for animal purchases and all manner of other materials.

The Zoological Society of London (ZSL) owns one of the major zoological libraries in the world (Fig. 14.20). It was founded in 1826 and contains over 200,000 volumes, 5000 journal titles and the Society's archive.

Other zoos also have major collections of zoological books and other materials, and may allow access by researchers and students.

14.6.6 Zoo data

Professionally run zoos will produce an annual report and a statement of their financial accounts, and may have done so for many decades. Such reports are very useful records for the researcher as they may provide information on visitor numbers, numbers and species of animals held, staff numbers, management structure, conservation projects, and so on.

Zoos may keep press cuttings about significant events and these may be a useful historical record, especially for species whose arrival, departure, birth or death are a significant event for the zoo and the local community (e.g. Rees, 2001c).

14.6.7 Zoo photographs and films

Photographs taken in zoos provide us with a record of historical changes in zoo architecture, enclosure design, animal exhibits and the types of animals kept in zoos in the past. They also document various aspects of zoo culture such as animal shows and rides that would now be unacceptable in a modern zoo. From a scientific point of view, zoos have provided photographs and films of animals that would otherwise only be known from drawings and written descriptions because they are now extinct. At the time these images were made telephoto lens technology did not exist so it would have been impossible to photograph such animals in the wild. Some taxonomic texts rely heavily on photographs taken in zoos – and museum specimens – simply because many species have not been photographed in the wild.

The ZSL has a commercial image library that contains drawings dating back to 1560. In total it contains 4000 images online and 12,000 images offline, including artworks, scientific drawings, transparencies, and digital images of animals, architecture, field conservation work and horticulture.

14.7 WHAT TYPE OF RESEARCH IS DONE IN ZOOS?

In this discussion, 'research' is taken to mean the systematic collection and analysis of biological data by scientists, or the development of new scientific techniques, as opposed to the day-to-day record keeping performed by keepers.

An analysis of 904 research projects conducted in British and Irish zoos found a highly skewed distribution across 15 subject categories (Semple, 2002). Behavioural studies represented the largest category (40%), followed by studies of environmental enrichment (18%) and reproduction (8%). Fewer than 5% of projects studied the genetics, ecology or conservation of a species. To date, unfortunately, relatively little zoo research can claim to have conferred a clear conservation benefit on endangered species – as required by the Zoos Directive – although this may be equally true of mainstream academic research on conservation.

Some behavioural work may be of relevance to conservation, for example where it concerns the study of reproductive behaviour (e.g. Laurenson, 1993; Lindburg and Fitch-Snyder, 1994). However, many feeding behaviour studies are concerned with the beneficial effects of enriching the zoo environment with some kind of feeding device (e.g. Jenny and Schmid, 2002). The zoo community is currently preoccupied with environmental enrichment, which, while often conferring some welfare benefit on zoo animals, often has little to do with conservation.

Unless zoos can demonstrate a clear and substantial role in reintroduction programmes much zoo research on reproductive biology is only likely to be of importance in helping zoos to maintain their supply of replacement animals. The recent success in producing elephants by artificial insemination (Schwammer *et al.*, 2001) is beneficial to elephant breeding programmes in zoos, but elephants breed perfectly well in the wild, and the threats to wild populations are not the result of low fecundity. Unless and until zoo animals become an important source for reintroduction projects developments in reproductive technology are irrelevant to conservation efforts, although clearly they may be of great importance for some critically endangered species, for example giant pandas (*Ailuropoda melanoleuca*) (Pérez-Garnelo *et al.*, 2004). Zoo-bred animals are rarely used in reintroduction projects, although notable exceptions include the European bison (*Bison bonasus*), golden lion tamarin (*Leontopithecus rosalia*), Hawaiian goose (*Branta sandvicensis*), Père David's deer (*Elaphurus davidianus*) and the Mauritius kestrel (*Falco punctatus*) (Anon.,1993; Frankham *et al.*, 2002). For many species little more than a protective zoo environment was required in order to encourage reproduction, but in others (e.g. the California condor (*Gymnogyps californianus*)) novel rearing techniques have been developed and there can be no doubt that a conservation benefit accrued as a result.

By its very nature, zoo research is conducted in unnatural conditions and often with very small samples of animals (e.g. Jenny and Schmid (2002) studied the effect of feeding boxes on the behaviour of just two tigers). Relatively few peer-reviewed scientific journals are available to publish this work. There are some notable exceptions, such as *Zoo Biology*, the *Journal of Zoo and Wildlife Medicine* and the *International Zoo Yearbook*, but they inevitably contain accounts of research which is relevant to the keeping of zoo animals rather than conservation *per se*.

14.7.1 Research published in *Zoo Biology*, 1982–2004

An analysis of 353 papers published in *Zoo Biology* between 1982 and 1992 found that 81.3% concerned non-human mammals (Hardy, 1996). Of the 287 papers on mammals, 29.6% were studies of behaviour or behavioural ecology, a further 5.9% involved behavioural/environmental enrichment and 20.2% were studies of reproductive biology. Only 3.8% of papers were concerned with genetics or population biology, and just 2.3% involved wildlife management. The remainder were concerned with nutrition and diet (3.5%), exhibit design and evaluation (1.2%), veterinary medicine (5.6%), captive management (24%) and morphology and development (5.6%). This analysis was undertaken before the Zoos Directive came into force and covers research from a wide geographical area, not just the European Union. However, it gives an indication of the historical focus of zoo research.

An examination of 349 papers published in *Zoo Biology* between 1996 and mid-2004 (Fig. 14.21) suggests a significant change in emphasis in the research carried out by zoos. Reproductive studies now replaced behaviour as the largest category (34%), followed by studies of nutrition, growth and development (19%) and behaviour and enrichment (17%). Studies concerned with ecology, field biology, conservation and reintroduction only accounted for some 2% of the total, but there was an increase in papers on taxonomy, genetics and population biology (10%).

14.7.2 Research in American institutions

In an analysis of 302 research projects carried out on mammals by zoo staff in 40 American zoos (Wiese *et al.*, 1992), behavioural and behavioural ecology studies made up 22.8% of studies undertaken but only 5.3% of studies published in the same period. Studies of reproductive physiology accounted for only 19.5% of studies undertaken but almost 31% of all published studies, while studies of natural history or fieldwork represented 23.1% of all published studies but only 16.6% of studies conducted. Field-based studies are likely to be more relevant to conservation than zoo studies. However, while they had a relatively good publication record, field studies represented a small proportion of the total research conducted by zoo staff in this study.

Examination of more recent data from North America shows a more encouraging picture. Lankard (2001) lists and categorises 957 publications produced in 1999–2000 by the member institutions of the American Zoo and Aquarium Association (AZA). The largest research category by far was ecology/field conservation/reintroduction (27%), although this encompassed a wide range of publications from status reports on individual taxa and recovery plans to papers on how to record the weather and how to use a compass. The next largest category was veterinary medicine/physiology (15%), followed by behaviour/ethology (9%). Studies of reproductive

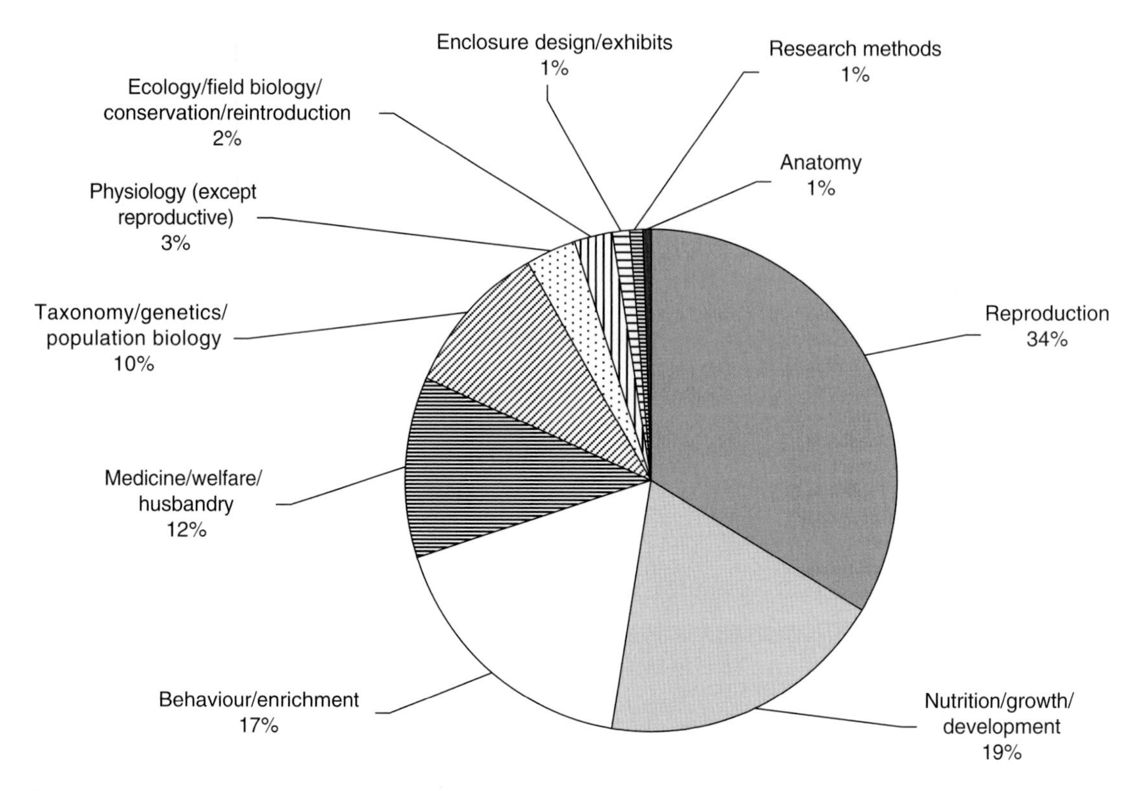

Fig. 14.21 Analysis of the subjects of 349 papers published in *Zoo Biology* between 1996 and 2004. (Based on data originally published in Rees, 2005a.)

physiology/technology amounted to just 7% of the total (compared with 34% of papers in *Zoo Biology* between 1996 and 2004) and nutrition accounted for only 3% (Fig. 14.22).

14.7.3 Taxonomic bias in zoo research

Some taxa are very poorly represented in research programmes. Card *et al.* (1998) conducted a survey of the research activity and conservation programmes of 52 North American zoo reptile and amphibian departments. Of 164 technical papers produced between 1987 and 1997 by the 22 respondent institutions, 79% were conducted by just three institutions and only 16 field studies were reported. Only one institution received funding specifically for research. Card *et al.* concluded that zoo herpetology departments were not realising their potential for formalised research and conservation projects.

14.7.4 Welfare vs. conservation research

The public expects high welfare standards in zoos but does not necessarily expect an obvious conservation role. It is, therefore, perhaps inevitable that zoos will expend a great deal of effort on enrichment projects, thereby creating many opportunities for collecting research data. Much of this behavioural research is of little conservation interest, and is not published. This may be because the data was collected in order to inform management practices within a specific institution. Hardy (1996) has discussed the possible reasons for the poor publication record of zoos, including the possibility that many zoo studies may not produce adequate empirical data for publication. A great deal of zoo research is not published in peer-reviewed scientific journals, and is therefore largely lost to the wider scientific community. One weakness of the Zoos Directive is that it does not require research to be published at all.

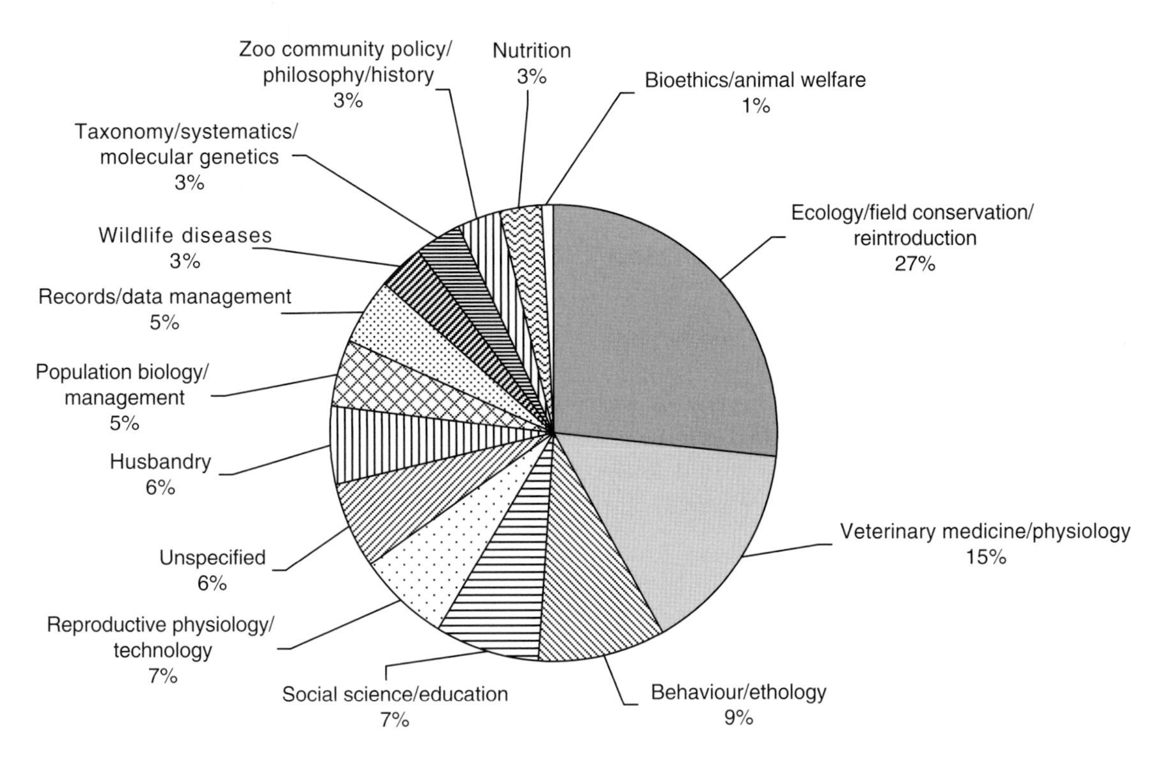

Fig. 14.22 Analysis of the subjects of publications produced by member institutions of the American Zoo and Aquarium Association 1999–2000 (based on publications listed in, and categories used by, Lankard, 2001).

14.7.5 Why is there so little research in zoos?

Very few zoos can afford to build research facilities or employ full-time researchers, and this will continue to limit the quantity and quality of zoo research. There is evidence of an increased emphasis on research in American zoos. A survey of 173 North American zoos and aquariums (Stoinski *et al.*, 1998) showed an increase in the role of research in AZA institutions in the previous decade, and a doubling of the number of researchers per institution since 1986. However, Stoinski *et al.* (1998) found that the most common reasons for American zoos not conducting research were lack of funds, time and qualified personnel. This is likely to be true for most European zoos, so much of the research work is, therefore, likely to continue to be undertaken by keepers and students. Furthermore, university researchers who work in zoos are only likely to conduct work that can be published. Finding conservation-relevant subjects is more difficult than finding behavioural or physiological subjects, and test and control situations are difficult to set up in a zoo environment. Some types of research would raise welfare concerns (e.g. invasive physiological studies) and would require a government licence in many countries.

The systematic scholarly use of European menageries began some 350 years ago and expanded during the Enlightenment (Baratay and Hardouin-Fugier, 2002). Early zoo research was focused on anatomy, physiology and systematics. However, more recently a significant amount of basic research on the biology of zoo animals – which may be of considerable conservation relevance – has been performed by zoological research institutions allied to zoos, for example the Institute of Zoology in London and the Institute for Zoo and Wildlife Research in Berlin. Veterinary studies on zoo animals have led to some developments which have had considerable *ex-situ* conservation benefits, for example in the areas of identification and marking (Rice and Kalk, 1996)

and capture, handling and anaesthesia (Bush, 1996), and zoos are increasingly becoming involved in field research on *in-situ* populations.

14.8 FURTHER READING AND RESOURCES

The *International Zoo Educators Journal* contains many interesting articles about visitor behaviour and the educational work of zoos and is available from the Association's website (www.izea.net).

India's Central Zoo Authority has published a Zoo Education Master Plan which includes examples of many different types of educational methods and projects from zoos around the world. This is available at www.cza.nic.in

Volume 21 of the *International Zoo Yearbook* was entitled *Zoo Display and Information Techniques*.

BIAZA publishes research guidelines which are available from its website (www.biaza.org.uk).

Martin, P. and Bateson, P. (1993). *Measuring Behaviour: An Introductory Guide*, 2nd edn. Cambridge University Press, Cambridge, UK.

14.9 EXERCISES

1. 'Most zoo research has no conservation value.' Discuss.
2. A student wishes to undertake a study of the behaviour of mandrills (*Mandrillus sphinx*) in your zoo. What questions might the zoo's ethics committee ask before giving permission for the project to take place?
3. Discuss the extent to which zoos have a role in educating the public about conservation.
4. Why do zoos study the behaviour of their visitors?
5. Describe the range of zoos' activities that might be considered to be educational.
6. What makes a good zoo interpretation sign?
7. What contribution can the internet make to the education role of zoos?
8. How should schools use zoos?
9. 'Most people learn more about animals from television than from zoos.' Discuss.

15 *IN-SITU* CONSERVATION AND REINTRODUCTIONS

> Is it really so stupid to work for the zebras, lions and men who will walk the earth fifty years from now? And for those in a hundred or two hundred years' time...?
>
> Bernard Grzimek

Conservation Status Profile

Komodo dragon
Varanus komodoensis
IUCN status: Vulnerable B1, 2cde
CITES: Appendix I
Population trend: Unknown

An Introduction to Zoo Biology and Management, First Edition. Paul A. Rees.

15.1 INTRODUCTION

Modern zoos are increasingly emphasising their role in supporting *in-situ* conservation projects: the protection of threatened species in their natural environment. Zoos may contribute to the *in-situ* conservation of a species in a number of ways. These include:

- raising money for specific conservation projects
- providing scientific expertise and personnel
- providing educational expertise and personnel
- providing *ex-situ* captive breeding facilities and releasing animals to the wild.

15.2 RAISING FUNDS FOR CONSERVATION

15.2.1 Competing for the 'animal£'

Many zoos operate as charities – or run specific charities for particular species – and as such compete with other charities for donations. In the UK, most charitable giving goes to medical charities. Animal charities are popular with the public, but much of this giving is directed towards pets and animal welfare organisations rather than conservation (Table 15.1). In 2009 there were 160,515 registered charities competing for funds in England and Wales alone, with a total annual income of almost £52 billion.

Table 15.1 Incomes of selected charities in England and Wales (2009).

	Annual income (£m)			
Charity	Voluntary	Total	Staff	Volunteers
Royal Society for the Protection of Birds	88.49	111.84	1,903	13,500
Marie Curie Cancer Care	83.14	138.77	1,331	5,000
WWF (UK)	36.63	44.17	289	—
Blue Cross	21.21	24.27	430	1,000
Donkey Sanctuary	20.57	22.74	357	—
Born Free Foundation Ltd.	4.66	5.26	39	15
Chester Zoo	3.03	22.40	392	—
Save the Rhino International	0.86	0.87	5	30
Orangutan Foundation	0.50	—	—	—

Source: Charity Commission (2009).

15.2.2 Funds for *in-situ* conservation

Zoos are increasingly emphasising their role in raising funds for *in-situ* conservation projects. Christie (2007) has recently studied zoo-based fund-raising for *in-situ* wildlife conservation and concluded that although it is possible to quantify the contribution of zoos to projects for particular species – for example tigers (*Panthera tigris*) and Amur leopards (*P. pardus orientalis*) – it is impossible to assess their global role due to poor data capture.

Very few studies have considered the role of zoos in generating funds for *in-situ* conservation. A bibliography of 470 publications concerned with 'Zoos and conservation' produced by the North of England Zoological Society (Wilson and Zimmerman, 2005) contained 40 documents with 'captive breeding' in the title (8.51%) and 30 references to 'reintroduction' (6.38%), but just 14 titles (2.98%) contained the term '*in-situ*'. The earliest of these was published relatively recently (Hutchins and Tilson, 1993). Only three documents referred to funding in the title (Bettinger and Quinn, 2000; Hutchins and Ballentine, 2001; Hutchins and Souza, 2001).

An analysis of the funding support for *in-situ* tiger conservation found that zoos supplied or channelled only about 16% of the total NGO budget between 1998 and 2002, and this included grants obtained by zoos from other sources. Over the same period, zoos were found to be particularly significant in supporting Sumatran tiger conservation, supplying or channelling 54% of the funding (Christie, 2007). Some individual zoos and zoo exhibits have the potential to raise substantial amounts of money. By the end of 2003, the Bronx Zoo's *Congo Gorilla Forest* exhibit had raised over US$2 million for the *in-situ* conservation of Congolese wildlife (Baker, 2007). In the UK in 2004 Chester Zoo and the supermarket chain ASDA ran a cause-related marketing promotion, featuring its *Tiger Bread* range, which raised £130,000 for tiger conservation in 12 months (Chester Zoo, 2007).

Zoos may make an important contribution to *in-situ* projects, especially in less developed countries (Balmford *et al.*, 2003), but in financial terms the total contribution is small compared with the cost of running a zoo facility. Furthermore, it is quite clear that many NGOs are quite able to generate funding for wildlife conservation without keeping animals.

In 2005 the Zoological Society of London had a total expenditure of about US$50 million (ZSL, 2006). Of this, 65.1% was spent on the animal collections, 15.0% on science and research, but only about US$2.8 million (5.5%) was spent on conservation programmes. Recently Seneca Park Zoo in the USA spent US$4.4 million on a new elephant facility for just three elephants while Reid Park Zoo, Arizona and North Carolina Zoo have spent US$16.5 million on their elephant facilities (AZA, 2007). In contrast, between 1991 and 2003 AZA centrally administered conservation funds provided just US$1.1 million for *in-situ* projects (Christie, 2007).

Compared with zoos, organisations that do not keep animals provide much larger sums of money for *in-situ* projects. The Global Environment Facility (GEF), which helps developing countries implement the UN Convention on Biological Diversity (Jefferies, 1997), invested US$217.1 million in biodiversity projects in 2007 alone (GEF, 2007). In the UK, the Royal Society for the Protection of Birds spent £81.8 million (US$126 million) on conservation in 2008–09 (RSPB, 2009) and WWF-UK spent £34.8 million (US$53.6 million) in 2007–08 (WWF, 2008).

If zoos did not already exist, no one would now advocate the taking of large rare species into captivity solely to use them to attract funding for conservation. Nevertheless, zoos may well raise money for conservation that would not be generated by other means.

15.2.3 Case study: When disaster strikes – giant pandas and earthquakes

When disaster strikes a species the zoo community is in a strong position to respond quickly because it already has access to financial resources, supporters who will donate more money and international contacts who can help to mobilise a rescue effort.

On 12 May 2008 an earthquake struck Sichuan Province in China. This had a devastating effect upon the Wolong Nature Reserve and the Panda Conservation and Research Centre. Five of the staff were killed and many others were injured. Several pandas (*Ailuropoda melanoleuca*) were injured and had to be evacuated to other centres. Six pandas were taken by lorry to another reserve 120 miles away and another eight were flown to Beijing Zoo (Anon, 2008b).

For those pandas that remained, food quickly became a problem as this is usually collected by the local people. Within a few days of the disaster the Chinese government sent an emergency consignment of bamboo and other feed to Wolong (Lewis, 2008). To make matters worse, the breeding centre suffered an immediate drop in income due to travel restrictions in the area.

The zoo community responded quickly. In the United States, the Smithsonian National Zoological Park worked with the Giant Panda Conservation Fund to raise money for Wolong, and San Diego Zoo and the Association of Zoos and Aquariums launched a Wolong Panda Center Earthquake Relief Fund. The Royal Zoological Society of Scotland quickly donated £10,000 to disaster relief. Pandas International – a non-profit organisation based in Colorado which raises money for panda research and conservation – responded by sending essential supplies to the area for the people and the pandas.

The effect of this disaster serves to highlight the danger of concentrating conservation efforts for a particular species in one location. At the time of the earthquake the Wolong Nature Reserve held 50% of all the giant pandas in captivity.

15.2.4 Case study: A zoo organisation buys a nature reserve – the BIAZA Reserve, Brazil

BIAZA is working with the World Land Trust to contribute to habitat conservation in Brazil. Working with the Brazilian conservation organisation Reserva Ecológica de Guapiaçu (REGUA), BIAZA has purchased land adjacent to an existing REGUA reserve. The reserve is an area of Atlantic Forest and the region is an important biodiversity hotspot. Funding for the reserve assists in the conservation of biodiversity, protecting land, providing employment for the local community and supporting an environmental education programme. This project has allowed zoos that are too small to engage with overseas projects themselves, but are members of BIAZA, to contribute to *in-situ* conservation.

15.2.5 Case study: Chester Zoo and the Assam Haathi Project

Chester Zoo has kept Asian elephants (*Elephas maximus*) since 1941 (Rees, 2001a). In Asia, elephant populations are increasingly threatened as a result of human encroachment and the fragmentation of their habitat. Staff from Chester Zoo are now working with EcoSystems-India and other local NGOs on a community-based project in Assam (Northeast India) aimed at reducing human–elephant conflict using the following approaches:

- Community-based conflict mitigation – This involves assisting the community to develop harmless elephant damage control measures.
- Elephant research and monitoring – Geographical information systems (GIS) techniques are being used to monitor elephant movements and to study their behaviour and nutritional needs.
- Conservation collaboration – A long-term strategy for habitat protection and land-use planning is being developed by bringing together government agencies and conservation groups.

The zoo staff involved with this project have included conservation officers, research assistants, a nutritionist and a specialist elephant keeper. Funding for the project has been received from the Darwin Initiative. This is administered by the Department for Environment, Food and Rural Affairs (DEFRA) as part of the UK's obligation to assist developing countries with biodiversity conservation as required by the UN Convention on Biological Diversity.

The project has produced a handbook for the community entitled *Living with Elephants in Assam*. The mitigation measures that have been developed include the use of trip-wire alarms, watch towers, electric fences, searchlights and chilli (which can be burned to produce an unpleasant smoke which acts as a deterrent). In addition, the project provides conservation education workshops and assistance with government compensation schemes (Zimmerman *et al.*, 2009).

15.3 EDUCATIONAL OUTREACH PROJECTS

An important role for zoo educators is to spread the conservation message to people, particularly children, in developing countries and conservation education programmes in such countries may benefit from collaboration with zoo education departments. Education staff from Chester Zoo have established a classroom in Mkomazi Game Reserve in Tanzania, where they teach local children about their wildlife and conservation issues. Frankfurt Zoo has assisted in the development of a modern visitor centre for visitors to the Serengeti National Park (Rees, 2004d) (Box 15.1).

15.4 PROVISION OF *EX-SITU* CAPTIVE BREEDING AND RESEARCH FACILITIES

Some zoos assist with *in-situ* conservation projects by providing *ex-situ* captive breeding facilities. It should be noted that captive breeding projects are '*ex-situ*' even if they occur within the range state of the species.

Zoos may study and breed animals obtained on loan from one of their range states (e.g. China loans giant pandas to San Diego Zoo). Some small zoos can make a significant contribution to *in-situ* conservation. *Monkey Forest* (Stoke-on-Trent, UK) is a small organisation which keeps a single species – the Barbary macaque (*Macaca sylvanus*) – in a 24 ha (60 acre) broadleaved woodland. The species is indigenous to North Africa and is classified by the IUCN as 'endangered'. There are approximately 10,000 left in the wild. *Monkey Forest* works with three sister organisations in Europe to breed Barbary macaques. Approximately 600 individuals had been reintroduced to the wild by the summer of 2009.

Biography 15.1

Bernard Grzimek (1909–87)

Professor Bernard Grzimek was trained as a vet and was formerly Director of Frankfurt Zoo and President of the Frankfurt Zoological Society. Grzimek was instrumental in the establishment of the Serengeti National Park in Tanzania. He was co-author, with his son Michael, of *Serengeti Shall Not Die* and author of *No Room for Wild Animals*. He was editor-in-chief of the 13-volume *Grzimek's Animal Life Encyclopaedia* which was translated into English in 1975 and became a standard reference work.

Box 15.1 The Serengeti National Park Visitor Centre, Tanzania.

The Frankfurt Zoological Society has a long-standing relationship with the Serengeti in Tanzania. In the 1950s Dr Bernard Grzimek – then Director of Frankfurt Zoo – and his son Michael bought a plane (a Piper Cub), learned to fly, and set off to Africa to count the game animals in the Serengeti National Park for the first time. They made an enormous contribution to the conservation of African wildlife through the publication of their book *Serengeti Shall Not Die* and the release of a film by the same name (Grzimek and Grzimek, 1964).

The Serengeti National Park Visitor Centre (Fig. 15.1) now stands in the Seronera as a tribute to this early work and an outstanding example of what can be achieved when a modern zoo works in collaboration with a national park in a developing country. The Frankfurt Zoological Society has partly funded the centre which is built around a kopje: a small rocky outcrop that is characteristic of the landscape. It contains exhibits that describe the famous wildebeest migration, the work of the Tanzania National Parks Authority (TANAPA) and the place of local tribes in the culture and ecology of the area, with signs in Swahili and English (Rees, 2004d).

To this day the Frankfurt Zoological Society still works in the Serengeti on a number of projects including the Serengeti Wild Dog Conservation Project and the Serengeti Rhino Repatriation Project.

Fig. 15.1 Entrance to the Visitor Centre.

15.4.1 Case study: Amphibian Ark – keeping threatened amphibian species afloat

The world's amphibians are in peril due to overharvesting, habitat loss, climate change, infectious disease (particularly chytrid fungus) and environmental contamination. One third to one half of all amphibian species are threatened with extinction. Hundreds of species face threats that cannot be mitigated in the wild, according to the IUCN's Global Amphibian Assessment (GAA, 2008). In response to these threats the *Amphibian Ark* (AArk) was established as a joint effort between WAZA, the IUCN/SSC Conservation Breeding Specialist Group (CBSG) and IUCN/SSC Amphibian Specialist Group (ASG), and other partners around the world, aimed at ensuring the global survival of amphibians. Since 2006, AArk has been assisting the *ex-situ* conservation community to address the captive components of the Amphibian Conservation Action Plan of the IUCN. This involves taking species at immediate risk of extinction into captivity in order to establish 'captive-survival assurance colonies' (Gascon *et al.*, 2007).

Box 15.2 Aquarium conservation and rescue projects.

A network of 26 aquariums is operated by Sea Life in 13 countries. It supports its own environment programme (SOS Conservation and Rescue) and works with other organisations such as the Shark Trust and the Whale and Dolphin Conservation Society. Its work has included:

- Rehabilitating around 100 common and grey seals every year at Sea Life seal rescue facilities, and returning them to the wild
- Petitioning the European Union to ban shark-finning in EU waters and to protect loggerhead sea turtles
- Lobbying the UK government for a reduction in fishing quotas
- Fund-raising for a sea turtle rescue centre on the island of Zakynthos, Greece
- Campaigning to reduce the bycatch of sea turtles, dolphins and sharks
- Campaigning against whaling.

Many zoos around the world are involved in captive breeding programmes, research projects and reintroduction projects for amphibian species. The focus of AArk's work is to support *ex-situ* conservation work rather than to collect amphibians purely for exhibits in zoos. AArk's primary activities are the rescue of species and capacity building by training individuals and institutions in the range states of endangered amphibians.

The mountain chicken frog (*Leptodactylus fallax*) – so named because it tastes like chicken – is a large frog that is only found in Dominica and Montserrat, in the Caribbean. Between December 2002 and March 2004 its population declined by about 70% due to chytridiomycosis. As part of a rescue effort for this species scientists from ZSL London Zoo travelled to Dominica. They were able to capture seven of the frogs and take them into captivity. A second captive group of 12 specimens of this species, also from Dominica, is held by a private collector in the United States (ZSL, 2010).

15.5 ZOOS, NATIVE SPECIES CONSERVATION AND WILDLIFE RESCUES

Many zoos are involved in the conservation of species which are native to the country in which the zoos are located. They may also become involved in rescue projects when environmental disasters threaten local wildlife. These activities are especially important to small zoos that may not have the resources to engage in *in-situ* conservation projects overseas. However, it should be remembered that native species conservation in the UK may mean involvement in projects which help small species like newts and butterflies but in China it means working with giant pandas. In some cases, zoos can assist by providing rescue facilities for local wildlife, for example stranded or orphaned seals (Box 15.2).

In the UK some zoos are partners in Biodiversity Action Plans for species such as red squirrels (*Sciurus vulgaris*) and sand lizards (*Lacerta agilis*), and BIAZA has a Native Species Working Group. Edinburgh Zoo has been collaborating with the Scottish Wildlife Trust to return beavers (*Castor fiber*) to Scotland. On a smaller scale, the grounds of some zoos may simply act as safe havens for native species. Dudley Zoo has great-crested newts (*Triturus cristatus*) breeding in its grounds and translocates individuals to ponds elsewhere whenever possible.

In the Channel Islands, *Durrell* has constructed several 'bug habitats' – to attract local insect species – from pieces of bamboo, bricks, logs, and other discarded items including computer keyboards (Fig. 15.2). It also feeds the local red squirrels, manages some of its estate to benefit glow-worms and rare orchids, and has built a bird hide to encourage interest in the local avifauna. *Durrell* propagates rare plants – which are the subject of Biodiversity Action Plans – for reintroduction back into the wild by the States of Jersey Environment Division.

Wildflower meadows can easily be created within the grounds of a zoo. Martin Mere (Wildfowl and Wetlands Trust) manages a small area of its grounds as a wildflower meadow, where visitors can see whorled caraway (*Carum verticillatum*), great willowherb (*Epilobium hirsutum*), tufted vetch (*Vicia cracca*), cuckoo flower (*Cardamine pratensis*) and many other species of wild flowers and the insects that they attract. Large zoos can protect large areas for native species. San Diego Zoo's Wild Animal Park includes a 364 ha (900 acre) native species reserve.

Fig. 15.2 A bug habitat, *Durrell.*

Zoos in the United States have been important in the restoration of species that are the subject of Species Recovery Plans, for example the California condor (*Gymnogyps californianus*) and the black-footed ferret (*Mustela nigripes*) (see Section 15.6.5). Chinese zoos are involved in a breeding programme for giant pandas in collaboration with dedicated breeding centres. Beijing Zoo acted as a refuge for eight panda cubs that were left homeless when their home in the Wolong Reserve was destroyed by an earthquake in 2008 (Anon., 2008b) (see Section 15.2.3).

15.5.1 Case study: The AZA's response to the Deepwater Horizon oil spill, 2010

On 20 April 2010 an oil drilling rig called the *Deepwater Horizon* sank in the Gulf of Mexico resulting in a massive oil leak which affected the coast of the United States and its wildlife. AZA-accredited zoos quickly mobilised to help the affected wildlife and habitats, and were identified by key federal agencies as important partners in animal rescue and rehabilitation (AZA, 2010h). Organisations that provided support, facilities and expertise included:

- Audubon Aquarium of the Americas
- Mote Marine Laboratory and Aquarium
- Hubbs-SeaWorld Research Institute
- SeaWorld Parks and Entertainment
- Walt Disney Company
- The Florida Aquarium Center for Conservation.

In addition, 11 AZA-accredited aquariums assisted in their role as part of a national Marine Mammal Stranding Network, and other zoos and aquariums assisted in raising funds to support the rescue efforts.

15.6 RELEASING CAPTIVE-BRED ANIMALS TO THE WILD

Although most modern zoos no longer stress reintroduction of animals back into the wild as one of their major aims, zoos may nevertheless still have an important role to play in the reintroduction of some species.

15.6.1 What is a reintroduction?

The IUCN defines reintroduction as:

> ...an attempt to establish a species in an area which was once part of its historical range, but from which it has been extirpated or become extinct.
>
> IUCN (1995)

Nechay (1996) has drawn attention to the importance of distinguishing between the reintroduction, restocking, reinforcement, translocation and introduction of species. He notes that the term 'reintroduction' is often used in practice to describe the release of animals in order to enhance their existing populations; a process that he prefers to call 'restocking' or 'reinforcement'. Nechay also emphasises that the term 'reintroduction' should not be used where a species is placed in an ecosystem where it has not previously existed, as in these circumstances the term 'introduction' is more precise. He concludes that whether a species is the subject of introduction, reintroduction or reinforcement may have serious implications for its future for a number of reasons,

including the uncontrolled transfer of genes from one area to another and the possible spread of disease.

15.6.2 Guidelines for reintroductions

A number of organisations and individuals have published guidelines for the reintroduction of animals to the wild including the IUCN and the AZA (1992). The latter points out that some of the existing guidelines contradict each other. For example, some suggest that individuals should be trained in survival skills before release while others claim that this is not cost effective.

Criteria for responsible reintroductions have been produced by the Council of Europe and they represent a good summary of the most important considerations (Recommendation Number R (85) 15 (1985)). It recommends that governments of Member States:

1 Carry out reintroduction projects only after conducting research to:

a determine the causes of extinction

b analyse past and present ecological characteristics of the area concerned

c propose remedies for the causes of extinction

d enumerate management measures to be taken before, during and after the reintroduction

e evaluate the chances of success and possible repercussions of reintroduction

f establish which subspecies or ecotypes are the closest to those that are extinct or are best suited to the reintroduction area.

2 Authorise reintroduction only after remedying the causes of extinction and restoring biotopes where necessary.

3 Prohibit reintroduction where adverse effects on the ecosystem are feared.

4 Inform the local population and interested groups of the reintroduction project.

5 Prohibit collecting from a population where this would constitute a threat to it.

6 Limit the duration of reintroduction projects and, if unsuccessful, give up further attempts.

7 Ensure scientific support, supervision and documentation until the reintroduced individuals are integrated into the local biological community.

8 Inform the European Committee for Conservation of Nature and Natural Resources, and if necessary the governments of neighbouring countries, of reintroduction projects and, if possible, coordinate reintroductions among the countries concerned.

15.6.3 Reintroduction successes and failures

Zoo captive breeding programmes have resulted in a number of well-known and successful reintroductions, for example the European bison (*Bison bonasus*) (Box 15.3). However, there have also been a number of spectacular failures. In a review of 45 case studies of carnivore reintroduction projects (in 17 species in five families) Jule *et al.* (2008) concluded that:

- wild-caught carnivores are significantly more likely to survive than captive-born carnivores in reintroductions
- humans were the direct cause of death in over 50% of all fatalities
- reintroduced captive-born carnivores are particularly susceptible to starvation, unsuccessful predator/competitor avoidance and disease.

Two particular areas of behavioural development have direct consequences for restoration projects:

- the acquisition of anti-predator behaviour
- imprinting.

Some projects have trained animals to respond appropriately to predators prior to release. In Australia rufous hare-wallabies (*Lagorchestes hirsutus*) have been trained to avoid a model fox or domestic cat by pairing the appearance of the predator with a conspecific alarm call or loud noise, or with a moving predator puppet and a squirt of water (McLean *et al.*, 1995).

Individuals reared in captivity may imprint on their human caretakers, foster parents of a different species or even their surroundings. This may result in the development of inappropriate breeding behaviour. This phenomenon is well known and elaborate techniques have been developed to deal with it.

In the USA, California condor (*Gymnogyps californianus*) chicks are fed using hand-held puppets that mimic the parent's head to prevent the chicks from forming an attachment to humans, and workers disguised as whooping cranes feed endangered whooping crane (*Grus americana*) chicks that have been cross-fostered to sandhill cranes (*Grus canadensis*).

Caro and Eadie (2005) have noted that most of the progress in mitigating developmental problems relating to the release of captive-bred animals to the wild has been made on a case-by-case basis, and the application

Box 15.3 The return of the European bison (*Bison bonasus*) to the wild.

IUCN Red List
Status: Vulnerable D1
Population trend: Increasing
Source: IUCN (2009).

Fig. 15.3 European bison (*Bison bonasus*).

The European bison or wisent (*Bison bonasus*) is the largest mammal found in Europe (Fig. 15.3). Its historical range extended across western, central and southeastern Europe and it probably occurred in the Asiatic part of the Russian Federation. It was exterminated by hunting and considered extinct in the wild by 1919.

As a result of intensive conservation management, involving captive breeding in zoos and breeding stations, the bison has made a spectacular recovery. The species now occurs in Belarus, Lithuania, Poland, Romania, the Russian Federation, Slovakia and Ukraine. It was also reintroduced into Kyrgyzstan but this population went extinct. Captive populations occur in 30 countries worldwide. In July 2009 ISIS listed 308 individuals in zoos and the European Bison Specialist Group listed 1436 in breeding centres (including those in zoos).

Recent populations consist of two distinct genetic lines: the Lowland line (*B. b. bonasus*) and the Lowland-Caucasian line (*B. b. bonasus* and *B. b. caucasicus*). Pure-bred populations of *B. b. caucasicus* have not survived.

7th century	Bison still existed in NE France
8th century	Started to die out in Gallia
11th century	Extinct in Sweden
11th and 12th centuries	In Poland, bison limited to larger forests and protected as royal game
12th century	Bison reported from Usocin Forest, near Szczecin, Poland
14th century	Still survived in Ardennes and in the Vosges
15th century	Bison still found in Poland in Bialowiez Forest and several other forests
16th century	Became extinct in Hungary

(*Continued*)

Box 15.3 *(Cont'd)*

16th century	Bison transported to Saxony from Prussia and Poland and kept and bred in enclosures. Set free in 1733–46
17th and 18th centuries	Bison still appear to survive in the former Soviet Union
1689	Unsuccessful attempt to release bison from enclosures in Mecklenburg (northern Germany)
1726	Bison numbers in eastern Prussia estimated at 117
1755	Last two bison in eastern Prussia killed by poachers
1762	Last bison in Romania killed in Radnai Mountains, Romania
1793	Survived in enclosures in Kreyern and Liebenwerda, Germany
1919	Last European population – in Bialowiez Forest – becomes extinct (*B. b. bonasus*)
Post First World War	54 individuals (29 males; 25 females) in captivity in zoos and at breeding stations. Of these, 39 came from Bialowiez. They originated from 12 founder animals
1927	Last bison in Caucasus becomes extinct (*B. b. caucasicus*)
1943	Captive population = 160
1946	Captive population = 93
1952	Two bulls released into Bialowiez National Park. A number of females released shortly afterwards
1957	First European bison born in the wild following reintroduction
Early 1990s	Peak of 2000 free-living individuals reached
2006	Global population of free-living animals was 1800

The information above has been compiled from data from the Bison Specialist Group (http://ebac.sggw.pl), Prague Zoo (www.zoopraha.cz/en/about-animals/studbook/) and IUCN (2009) (all accessed 31 July 2009).

of our knowledge of behavioural ontogeny (development) to conservation is largely by trial and error.

15.6.4 Reintroduction techniques

Captive-bred animals should not simply be released back into the wild in an *ad hoc* manner.

When animals are being captive bred and prepared for release a number of precautions should be taken:

- The animals should be genetically similar to those that were previously present in the area where they are to be released.
- Species-typical behaviours should be encouraged by providing appropriate food and enrichment structures, including food recognition and prey recognition and avoidance.
- Human contact should be minimised or, where possible, avoided completely, to reduce the possibility of habituation to or bonding with people.
- Individuals should be released in established social groups where appropriate, or at least have demonstrated the ability to recognise and interact with conspecifics.
- Released animals should be monitored.

Public consultation is important in a reintroduction programme, especially if a release is likely to be controversial because of fears over public safety or the possibility of economic losses. It is important to find suitable founding individuals, precondition them prior to release and monitor their progress thereafter.

15.6.4.1 Public consultation

Reintroductions are likely to be most successful when they have public support. In order to establish public opinion in relation to the reintroduction of the European beaver (*Castor fiber*) to the wild in Scotland, Scottish Natural Heritage undertook an extensive public consultation exercise (Anon., 1998). There were concerns that beavers would do significant damage to fisheries and affect forestry. Almost two-thirds (63%) of the 2141 members of the public consulted were in favour of the reintroduction, with just 12% against (although many of these gave 'lack of interest' as their reason). A further 1944 responses were received from a 'pro-active public' sample consisting of academics, ecologists, zoologists, conservationists, hillwalkers, and others with similar interests, including land managers. Overall, 86% of this sample favoured reintroduction but the survey identified a clear lack of support from those with a fishing or agricultural interest.

15.6.4.2 Suitable founder individuals

A study of oldfield mice (*Peromyscus polionotus subgriseus*) has shown that the more generations a population has been kept in captivity the less likely an individual animal is to hide after seeing a predator (McPhee, 2003). In addition, variance in predator-response increases with generations in captivity. The implications of this study for reintroduction programmes are that survival rates could be lowered, requiring more individuals to be released to reach the target wild population size.

15.6.4.3 Preconditioning

Reintroduction projects commonly involve keeping animals in the release area in holding pens for a period before release to the wild. The black-footed ferret (*Mustela nigripes*) recovery programme in the United States preconditions individuals prior to release. Ferrets are kept in outdoor pens containing naturalistic prairie dog (*Cynomys* sp.) burrows, and exposed to prairie dog prey. This process significantly enhances the survival of released ferrets (Biggins *et al.*, 1998).

Some projects have used model species to develop and test release techniques. On Mauritius, ring-necked parakeets (*Psittacula krameri*) were released to develop techniques for echo parakeets (*P. eques*) and in California release techniques for California condors (*Gymnogyps californianus*) were tested using Andean condors (*Vultur gryphus*) which were later recaptured (Wallace and Temple, 1987).

15.6.4.4 Monitoring

A monitoring scheme should be in place to track individual animals, monitor health and record population data (e.g. survival rates, dispersal patterns, and causes of death). Many deaths of reintroduced animals are the result of behavioural deficiencies. During the reintroduction of golden lion tamarins (*Leontopithecus rosalia*) some individuals were unable to survive because their locomotor skills were deficient; they could not orientate themselves spatially; and they could not recognise natural foods, non-avian predators or other dangerous animals (Kleiman *et al.*, 1990). Failure of early repatriation attempts with captive-reared desert tortoises (*Gopherus agassizii*) has been partly attributed to too great a dependence on humans and the failure of individuals fed on iceberg lettuce to recognise or use wild plants as forage (Cook *et al.*, 1978).

Reintroductions of animals into the wild have met with varying degrees of success. The following case studies provide an insight into the successes and difficulties experienced by projects that have saved three iconic species from extinction: the black-footed ferret, the Arabian oryx and the Hawaiian goose.

15.6.5 Case study: The return of the black-footed ferret

The black-footed ferret (*Mustela nigripes*) is a North American species that feeds on prairie dogs (*Cynomys*). As a result of the dramatic decline in prairie dog habitat the ferret populations fell until in 1985 there were just 18 animals left. These animals were captured and an AZA Species Survival Plan was established for the ferrets in cooperation with the US Fish and Wildlife Service and other wildlife authorities.

The largest breeding facility is the US Fish and Wildlife Service's Black-footed Ferret Conservation Center in Colorado; other breeding groups exist at five AZA zoos. The Smithsonian National Zoological Park's Black-Footed Ferret Reproduction Project studies the biology of the species to enhance reproduction, maintain genetic diversity, and provide individuals for reintroduction to the wild. Scientists at the zoo's Conservation and Research Center have developed artificial insemination techniques for the species, resulting in the birth of over 100 kits.

In total, over 6500 kits have been born in captivity, most of which have been released into the wild.

15.6.6 Case study: Arabian oryx

The Arabian oryx (*Oryx leucoryx*) is an iconic species for the people of the Middle East. It is a desert species that was once widely distributed across the Middle East and the Arabian Peninsula. During the first half of the 20th century the oryx population declined rapidly as a result of hunting which was exacerbated by the increased availability of firearms and motor vehicles. By the 1960s only a handful of oryx remained in Oman. The last one was shot here in 1972.

In 1961 the Fauna Preservation Society (now Fauna and Flora International) launched 'Operation Oryx'. They captured three wild oryx and obtained others from zoos and private collections in the Middle East. Eventually, a captive population was established at Phoenix Zoo in Arizona and Los Angeles Zoo in California. By 1964 there were 13 oryx in captivity in the USA (Tudge, 1991).

The Arabian oryx was first returned to the New Shaumari Reserve in Jordan in 1978. Later oryx were released on the Jiddat al-Harasis plateau in Oman. The first five animals arrived in 1980 and by 1996 there were 450 free-ranging animals. Subsequently, when protection was relaxed – partly because the animals had dispersed over a wide area – organised bands of poachers moved into the area. They captured many animals and smuggled them out of the area to the private zoos of wealthy individuals elsewhere in the Gulf and the Arabian Peninsula. They particularly targeted females, making recovery of the population difficult. Female oryx were sold for £15,000, and animals were often killed during capture or transportation. By 2007 there were just 65 oryx – including just four females – left in Oman. That year the size of the Arabian Oryx Sanctuary was reduced by 90%. This made the area easier to police but also released land for oil prospecting. As a result UNESCO took the unprecedented step of removing the sanctuary from its list of World Heritage Sites.

On 8 April 2010 ISIS listed 47 institutions holding a total of 889 Arabian oryx. The largest number (240) was held by the Wadi Al Safa Wildlife Centre, Dubai, in the United Arab Emirates. Many more individuals are held by institutions that do not contribute to ISIS. Introduced oryx are now present in Saudi Arabia, the United Arab Emirates, Jordan and Israel, some in large fenced areas. An account of the early years of this project is given in Tudge (1991).

15.6.7 Case study: Hawaiian goose

The Hawaiian goose or nēnē (*Branta sandvicensis*) was saved from extinction by the efforts of Sir Peter Scott and the Wildfowl Trust (now the Wildfowl and Wetlands Trust) (Fig. 15.4), Paul Breese (then the Director of Honolulu Zoo), Herbert Shipman and others. By 1952 only 52 individuals survived in the wild, as a result of hunting, habitat loss and predation, particularly by introduced mongooses, rats and feral cats. Birds were taken from the wild to establish a breeding colony at the Wildfowl Trust's facility at Slimbridge in the UK. The birds bred well and now exist in zoos and refuges around the world. However, the reintroduction process has required multiple releases and has not yet managed to establish sustainable populations. Between 1962 and 2003 a total of 598 geese were released but in 2003 the population was estimated to be around 217 and it is not self-sustaining (USFWS, 2004). In 2006, the wild population was estimated to be 1744 and the species is currently categorised as 'vulnerable' by the IUCN (2010).

15.7 IS THERE A LEGAL OBLIGATION TO REINTRODUCE ANIMALS INTO THEIR FORMER HABITATS?

Species reintroduction programmes are an important feature of global conservation efforts, but are governments under any legal obligation to reintroduce species that have been lost? There is evidence within the texts of some international and European laws that they are (Rees, 2001b). However, these obligations are inconsistent between legal instruments, and it is not at all clear exactly what it is they are legislating to recreate

Biography 15.2 **Sir Peter Scott (1909–89)**

Peter Scott was the son of the Antarctic explorer Robert Falcon Scott. He was a renowned wildlife artist and was responsible for the establishment of the Wildfowl Trust (now the Wildfowl and Wetlands Trust). One of its early successes was in the captive breeding and reintroduction of the Hawaiian goose. Scott was a past Chairman of the Survival Service Commission of the IUCN and was largely responsible for establishing the concept of Red Data Books. He was Chairman of the Council of the Fauna Preservation Society (now Fauna and Flora International) and also Chairman of the World Wildlife Fund, which he helped to found.

Fig. 15.4 Hawaiian goose or nēnē (*Branta sandvicensis*).

(see Legal Box 15.1). In particular, definitions of native species are either absent from the law or unclear, especially in a historical context.

Attempts to reintroduce some predators – such as grey wolves (*Canis lupus*) in the USA – have been met with legal challenges, and so it is essential that conservation authorities have a clear mission in their reintroduction activities and that this mission is reflected in their national laws.

15.7.1 Grey wolf reintroductions in the USA

In December 1997 a federal court ruled that a programme to reintroduce grey wolves to Yellowstone National Park and central Idaho was illegal (*Wyoming Farm Bureau Federation v Babbitt, 1997*). An order was made requiring the wolves to be removed, but this order was stayed pending an appeal by conservation organisations.

The court found that assigning the translocated wolves 'experimental and nonessential' status could potentially harm wolves that may naturally migrate from Montana and Canada into the recovery areas, thereby losing their protection under the Endangered Species Act 1973 (ESA). This is because ranchers could legally kill the introduced individuals because of their 'experimental and nonessential status' but would not be able to distinguish them from other wolves.

The case was brought by the Idaho, Montana and Wyoming Farms Bureaus against the Department of the Interior (which oversees the endangered species restoration programmes managed by the US Fish and Wildlife Service (USFWS)). In spite of the existence of a Wolf Compensation Fund set up by the organisation Defenders of Wildlife that compensated ranchers at full market value for verifiable livestock losses, the ranching community was determined to have the wolves removed. Under the ESA wolves cannot be killed legally in the United States. Conferring 'experimental status' on the translocated wolves under s.10(j) of the ESA allowed ranchers legally to kill any wolves found taking livestock on private land. The judge decided, however, that the effect of assigning these wolves 'experimental status' was to make the introductions illegal under the ESA because they were introduced within the range of non-experimental (native) populations. In a second case, brought by the National Audubon Society, attempts have been made to force the USFWS to restore full ESA protection to Idaho's wolves, claiming that when experimental animals and endangered populations overlap, the 'experimental' animals revert to 'endangered'.

Legal Box 15.1
Examples of laws that create a legal obligation to reintroduce animal species into their former habitats.

The **Convention on the Conservation of European Wildlife and Natural Habitats (1979)**, the Berne Convention, was the first wildlife treaty to encourage its Parties to reintroduce native species as a method of conservation.
Under Article 11(2) of the Convention the Contracting Parties undertake:

> *(a) to encourage the reintroduction of native species of wild flora and fauna when this would contribute to the conservation of an endangered species...*

The **Convention on Biological Diversity (1992)** has reaffirmed an international commitment to the recovery of species. The preamble to the treaty states that:

> *the fundamental requirement for the conservation of biological diversity is the in-situ conservation of ecosystems and natural habitats and the maintenance and recovery of viable populations of species in their natural surroundings.*

Article 9(c) creates an obligation to reintroduce threatened species, requiring that:

> *Each Contracting Party shall, as far as possible and as appropriate, and predominantly for the purpose of complementing in-situ measures:*
>
> *(c) Adopt measures for the recovery and rehabilitation of threatened species and for their reintroduction into their natural habitats under appropriate conditions.*

Council Directive 92/43/EEC of 21 May 1992 on the Conservation of Natural Habitats and of Wild Fauna and Flora requires, under Article 22, that Member States shall:

> *(a) study the desirability of re-introducing species in Annex IV, that are native to their territory where this might contribute to their conservation,...*

The American experience with wolf reintroduction programmes shows us the importance of clear legal definitions. The ESA extends the debate on reintroductions by creating a category of animals that are deemed to be 'non-essential experimental populations' within a protected native species. The translocated wolves came originally from Canada. The species was virtually exterminated from the lower 48 states of the USA by an intensive government-funded predator-eradication programme in the early decades of the 20th century. The species has therefore only been absent from the recovery areas for less than 100 years. By any sensible definition the introduced wolves were native species, and their classification as 'experimental' was only instituted to allow ranchers to protect their livestock. If there had been no public objections to the reintroductions these wolves would undoubtedly have been classified as endangered (native) species. Eventually common sense prevailed. In January 2000, after a long legal battle, the 10th Circuit Court of Appeals overturned the original 1997 decision and determined that an overly technical interpretation of the ESA was inappropriate (*Wyoming Farm Bureau Federation v Babbitt 01/13/2000*). The wolves have been allowed to stay and are now an important tourist attraction. From 1995 to the end of 1999 approximately 50,000 visitors observed the Druid Peak pack in Lamar Valley, Yellowstone National Park, making them arguably the most viewed wolf pack in the world (Smith *et al.*, 1999).

Legal challenges have been mounted to a number of other reintroduction projects (Rees, 2001b). Successful reintroductions will be achieved only with public support, and this is more likely where clear objectives have been established after public consultation.

15.7.2 Other examples of legal problems with animal releases

Under Queensland State law the koala (*Phascolarctos cinereus*) is listed as a 'special native animal' (Schedule 4, Nature Conservation (Wildlife Management) Regulation 2006) and its translocation and reintroduction to the wild is strictly controlled. Under s.343.1(b) 'A person must not release an animal into an area of the wild that is not a prescribed natural habitat for the animal', except under licence or other authority (s.343(2)(a)). Koalas must be released in 'prescribed natural (koala) habitat', not more than 5 km from the original capture site (Schedule 2, Nature Conservation (Koala) Conservation Plan 2006 (Subordinate Legislation No. 208, made under the Nature Conservation Act 1992)). The animal hospital at Australia Zoo has recently been accused of ignoring this law on 13 occasions (Larter, 2008), but officials claimed that this had been done to protect koalas from roads and urban developments.

In Spain, the conservation of two evolutionarily significant units of the European wild rabbit (*Oryctolagus cuniculus*) is at risk because there is no legal requirement for the authorities to consider genetic lineages when issuing permits for the capture, transit and release of rabbits (Delibes-Mateos *et al.*, 2008).

It is imperative that government conservation agencies strictly control the release of animals into the wild and the translocation of populations if we are to ensure the genetic integrity of our biological inheritance.

15.8 REINTRODUCTION AND RE-WILDING

There will undoubtedly be more reintroductions in the future and at least some of these will involve zoos or animals descended from individuals kept in zoos. Most reintroduction efforts involve a single species, but some scientists want to recreate whole ecosystems. Recently a group of American scientists (Donlan *et al.*, 2005) has suggested that African mammals should be introduced into wilderness areas in the United States to replace the megafauna that has been lost: a Pleistocene re-wilding. They envisage the creation of novel food chains with cheetah preying on pronghorn and elephants devouring North American native tree species. Introducing species that are ecologically analogous to those that have been lost is scientifically pointless. Would we introduce jaguars into Kenya to replace leopards, or kangaroos into Montana to replace bison? If it were possible to create a fully functional community of African animal species in America it would exist within a habitat containing native American plant species and therefore be of limited ecological value. If the ecosystem was created merely as a temporary sanctuary for rare species it would be no better than a zoo. Donlan *et al.* envisage the need for fences to contain the animals, in which case – if completely enclosed – they would effectively be captive not introduced.

15.9 EVALUATING THE CONTRIBUTION OF THE ZOO AND AQUARIUM COMMUNITY TO *IN-SITU* CONSERVATION

Zoos are increasingly claiming to have a significant role in *in-situ* conservation projects so it is more important than ever that their contribution is measurable. WAZA has recently published an audit of projects supported by the world zoo and aquarium community (Gusset and Dick, 2010). The study examined 113 projects and found that they mainly focused on mammals (50%) and within this taxon mostly on charismatic species of primates (13%) and carnivores (12%). Almost three-quarters of the projects (73%) worked with taxa which are globally threatened with extinction according to IUCN classifications, but amphibians and fishes were significantly under-represented.

The main source of project support from zoos was financial (48%) and most projects would not have been viable without this (59%). Gusset and Dick concluded that zoos and aquariums could make a greater contribution to biodiversity conservation by allocating more resources to *in-situ* conservation projects.

15.10 FURTHER READING

The websites of individual zoos often provide information about their *in-situ* activities as does the site of the Zoological Society of London. The website of the US Fish and Wildlife Service provides information about the reintroduction projects with which it has been associated.

I have discussed in more detail the legal obligations to reintroduce species to the wild in Rees, P. A. (2001). Is there a legal obligation to reintroduce animal species into their former habitats? *Oryx*, 35(3): 216–223.

15.11 EXERCISES

1. To what extent do zoos have a role in the reintroduction of rare species back into the wild?
2. How can a small zoo, with very limited funds, contribute to conservation?
3. Describe the extent to which a zoo can assist with the conservation of species in the country in which it is located.
4. Zoos claim that one of their roles is to maintain 'insurance' populations of rare species. To what extent can this claim be justified?
5. Does the law require that governments reintroduce lost species into their former habitats?
6. Describe a zoo you might visit in 2100.
7. Discuss the measures that should be taken before, during and after attempting to reintroduce an animal species into its former habitat.

APPENDICES

nservation Status
file

lean flamingo
oenicopterus
lensis
N status: Near
reatened
ES: Appendix II
pulation trend:
creasing

APPENDIX 1 ANIMAL CLASSIFICATION

This classification describes the main phyla in the Animal Kingdom and the orders of vertebrates. It is not intended to be comprehensive, but simply to provide a guide to the types of animals that are most likely to be encountered in zoos and aquariums.

KINGDOM ANIMALIA

Phylum Porifera

Sponges are extremely simple aquatic organisms and almost all of them are marine. The body consists of a loose collection of cells arranged around a water-canal system. There are no body organs. Water is taken into the central cavity through tiny pores in special epidermal cells (porocytes) and expelled through a large osculum at the top of the body. Specialised flagellated cells remove particles of food from the current as the water passes over them. Sponges are immobile throughout most of the life cycle and attach themselves to the substratum with a structure called a holdfast. Sponges have the ability to reform the body if they are broken into smaller pieces.

Phylum Cnidaria

Cnidarians are simple aquatic organisms, including coral-forming species and jellyfish. Cnidarians possess body organs. Their tentacles are covered in stinging cells called nematocysts used for catching food. They have a mouth and a blind-ending gastrovascular cavity. Most cnidarians exhibit an alternation of generations in which they alternate in form between a polyp – a fixed (immobile) stage – and a medusa. The polyp has tentacles and is attached to the substratum and the medusa is free-swimming and has a body that looks like the typical jellyfish. Corals are made up of a vast number of tiny polyps.

Phylum Platyhelminthes

The platyhelminths are free-living and parasitic flatworms. They include many economically important parasites such as tapeworms and liver flukes. They are characteristically flat and relatively featureless in form, but some have large eyes and suckers. Cnidarians possess a head where the sensory organs are concentrated, including simple eyes (ocelli), and there is a simple 'brain'. The mouth opens into a blind-ending sac which is often highly branched. There is no real circulatory system but a primitive excretory system is present.

Phylum Nematoda

Nematodes are free-living and parasitic roundworms. They are extremely numerous but rarely seen because most are very small. Some species live in the soil and at the bottom of lakes and streams, while others are parasites and live inside plants or other animals. They are worm-like in appearance and their general form is very similar in all species. Nematodes possess a tough outer cuticle.

Phylum Annelida

Most annelids are free-living segmented worms found in soil and aquatic environments, and include earthworms and ragworms. Some forms are parasitic (leeches). Their bodies are segmented, and constructed of muscular rings separated by thin partitions called septa. Segmentation has allowed the development of advanced systems of coordinated locomotion with some terrestrial species using setae ('hairs') to grip the soil and marine species using a system of paddles (parapodia) attached to each segment. The nervous system is relatively advanced with a simple 'brain' at the head end. The digestive system consists of a tube which runs the length of the body. Annelids have an excretory system made up of individual excretory units (nephridia) in each segment.

Phylum Arthropoda

The arthropods are segmented animals possessing a hard outer skeleton – made of the polysaccharide chitin – with paired, jointed limbs, for example insects, crabs, woodlice and spiders. They are advanced animals, many of which have developed impressive swimming, running or flying abilities. In many groups the exoskeleton is toughened with mineral salts (e.g. calcium carbonate). The muscular system is highly specialised and the legs are hinged, allowing the exoskeleton to move. The exoskeleton provided this group with the necessary structural support to evolve systems for walking on land.

Phylum Mollusca

The molluscs are soft-bodied animals with a muscular 'foot' and often possess a shell, for example snails, slugs, octopuses and squid. The mantle is a sheet of specialised tissue which covers the viscera like a body wall. It secretes a shell, in most groups of molluscs, which protects the body. Within the mantle cavity are the gills (ctenidia) and the floor of the mouth possesses a radula which is used for scraping food. The cephalopods (octopuses, squids and their relatives) possess a well-formed head with a relatively large and advanced brain and eyes, along with arms and tentacles capable of manipulating objects.

Phylum Echinodermata

The echinoderms are marine animals most of which have a radially symmetrical structure, for example starfish, brittlestars, sea cucumbers and sea urchins. They possess a sophisticated water-vascular system which consists of an array of canals and tube feet which terminate in tiny suckers. This system provides both a means of locomotion and a means of catching prey. Echinoderms have an internal skeleton of calcareous plates which form a rigid skeletal box. Spiny or wart-like projections extend outward for protection.

Phylum Chordata

Most of the chordates are vertebrates: mammals, birds, reptiles, amphibians and fishes. All chordates share a number of common features in their embryonic development and at some point in their life cycle they have:

- a dorsal hollow nerve cord
- gill slits used to circulate water during feeding and respiration
- a stiffening hollow rod along the dorsal surface called a notochord.

A vertebrate possesses a series of vertebrae (segmented bones) surrounding the notochord and the nerve cord. The notochord is only present during embryonic development and its protective function is replaced by the vertebrae in the adult animal. Vertebrates possess complex, closed circulatory systems in which blood containing haemoglobin is pumped around the body by a heart. This, and the development of a complex respiratory system, has allowed the evolution of some very large species.

The classification below lists the orders of mammals, birds, reptiles, amphibians and fishes.

ORDERS OF MAMMALS (AFTER NOWAK, 1999)

Monotremata	Echidnas and duck-billed platypus
Didelphimorphia	American opossums
Paucituberculata	'Shrew' opossums
Microbiotheria	Monito del Monte
Dasyuromorphia	Australasian carnivorous mice. Tasmanian devil, marsupial 'mice' and 'cats', numbat, thylacine
Peramelemorphia	Bandicoots
Notoryctemorphia	Marsupial 'mole'
Diprotodontia	Koala, wombats, possums, wallabies, kangaroos
Xenarthra	Sloths, anteaters, armadillos
Insectivora	Insectivores: hedgehogs, gymnures, golden moles, moles, tenrecs, solenodons, shrews, shrew moles, desmans
Scandentia	Tree shrews
Dermoptera	Flying lemurs

(Continued)

(*Cont'd*)

Chiroptera	Bats
Primates	Primates: lorises, pottos, galagos, lemurs, tarsiers, monkeys, apes, humans
Carnivora	Dogs, bears, raccoons, weasels, civets, otters, mongooses, hyenas, cats
Pinnipedia	Seals, sea lions, walrus
Cetacea	Whales, dolphins, porpoises
Sirenia	Dugong, sea cow, manatees
Proboscidea	Elephants
Perissodactyla	Odd-toed ungulates: horses, zebras, asses, tapirs, rhinoceroses
Hydracoidea	Hyraxes
Tubulidentata	Aardvark
Artiodactyla	Even-toed ungulates: pigs, peccaries, hippopotamus, camels, llamas, okapi, giraffe, chevrotains, deer, musk deer, pronghorn, antelopes, cattle, bison, buffalo, goats, sheep
Philodota	Pangolins
Rodentia	Rodents: beavers, squirrels, chipmunks, prairie dogs, mice, rats, gophers, hamsters, lemmings, gerbils, voles, squirrels, porcupines, springhare, mole-rats, cane rats, agoutis, capybaras, chinchillas, coypu, hutias
Lagomorpha	Pikas, rabbits, hares
Macroscelidea	Elephant shrews

ORDERS OF BIRDS (AFTER PETERS *ET AL.*, 1931–87)

Struthioniformes	Ostrich, tinamous, rheas, cassowaries, emus, kiwis
Procellariiformes	Albatrosses, shearwaters, petrels
Sphenisciformes	Penguins
Gaviiformes	Divers
Podicipediformes	Grebes
Pelecaniformes	Pelicans, tropicbirds, cormorants, gannets, boobies, frigatebirds
Ciconiiformes	Herons, storks, hammerhead, ibises, spoonbills, New World vultures
Phoenicopteriformes	Flamingos
Falconiformes	Raptors: osprey, eagles, kites, hawks, Old World vultures, falcons, secretarybird, buzzards, caracaras
Anseriformes	Geese, swans, ducks, screamers
Galliformes	Megapodes, chachalacas, guans, curassows, turkeys, grouse
Opisthocomiformes	Hoatzin
Gruiformes	Cranes, mesites, trumpeters, rails, sunbittern, bustards, kagus, finfoots, seriemas
Charadriiformes	Jacanas, snipes, plovers, dotterels, avocets, coursers, pratincoles, stilts, curlews, sheathbills, gulls, terns, skimmers, skuas, auks
Columbiformes	Sandgrouse, pigeons, doves
Psittaciformes	Cockatoos, parrots
Cuculiformes	Turacos, louries, cuckoos
Strigiformes	Owls

Caprimulgiformes	Oilbird, nightjars, frogmouths, potoos
Apodiformes	Swifts, hummingbirds
Coliiformes	Mousebirds
Trogoniformes	Trogons
Coraciiformes	Kingfishers, bee-eaters, rollers, hoopoe, todies, motmots, hornbills
Piciformes	Woodpeckers, barbets, jacamars, toucans, honeyguides, puffbirds
Passeriformes	Broadbills, pitas, ovenbirds, woodcreepers, tyrant flycatchers, and songbirds, e.g. lyrebirds, pipits, wrens, thrushes, sparrows, titmice, weavers, drongos, crows, bowerbirds, birds of paradise

ORDERS OF REPTILES

Rhynchocephalia	Tuataras
Squamata	Iguanas, lizards, monitor lizards, chameleons, geckos, lacertids, skinks, snakes, adders, vipers
Testudinata	Turtles, terrapins, tortoises
Crocodilia	Alligators, caimans, crocodiles, gharials

ORDERS OF AMPHIBIANS

Caudata	Salamanders, newts, sirens, hellbenders, mudpuppies
Gymnophiona	Caecilians
Anura	Frogs and toads

ORDERS OF FISHES (MODIFIED AFTER NELSON, 1994)

Superclass AGNATHA	Jawless fishes
Class MYXINI	Hagfishes
Order Mixiniformes	Hagfishes
Class CEPHALASPIDOMORPHI	Lampreys and allies
Order Petromyzontiformes	Lampreys
Superclass GNATHOSTOMATA	Jawed fishes
Class CHONDRICHTHYES	Cartilaginous fishes
Subclass ELASMOBRANCHII	Sharks and rays
Order Hexanchiformes	Six- and seven-gill sharks
Order Squaliformes	Dogfish sharks and allies
Order Pristiophoriformes	Saw sharks
Order Heterodontiformes	Horn (bullhead) sharks
Order Orectolobiformes	Carpet sharks and allies
Order Lamniformes	Mackerel sharks and allies
Order Carcharhiniformes	Ground sharks

(Continued)

(*Cont'd*)

Order Squatiniformes	Angel sharks
Order Rhinobatiformes	Shovelnose rays
Order Rajiformes	Skates
Order Pristiformes	Sawfishes
Order Torpediniformes	Electric rays
Order Myliobatiformes	Stingrays and allies
Subclass HOLOCEPHALI	Chimaeras
Order Chimaeriformes	Chimaeras
Class OSTEICHTHYES	Bony fishes
Subclass SARCOPTERYGII	Lungfishes and coelacanth
Order Ceratodontiformes	Australian lungfishes
Order Lepidosireniformes	South American and African lungfishes
Order Coelacanthiformes	Coelacanth
Subclass ACTINOPTERYGII	Ray-finned fishes
Order Polypteriformes	Bichirs
Order Acipenseriformes	Sturgeons and allies
Order Lepisosteiformes	Gars
Order Amiiformes	Bowfin
Order Hiodontiformes	Mooneyes
Order Osteoglossiformes	Bonytongues, elephantfishes and allies
Order Elopiformes	Tarpons and allies
Order Albuliformes	Bonefishes and allies
Order Notacanthiformes	Spiny eels and allies
Ordee Anguilliformes	Eels
Order Clupeiformes	Sardines, herrings, anchovies and allies
Order Gonorynchiformes	Milkfish, beaked salmons and allies
Order Cypriniformes	Carps, minnows and allies
Order Characiformes	Characins and allies
Order Siluriformes	Catfishes and knifefishes
Order Argentiniformes	Herring smelts, barreleyes and allies
Order Salmoniformes	Salmons, smelts and allies
Order Esociformes	Pikes, pickerels and allies
Order Ateleopodiformes	Jellynose fishes
Order Stomiiformes	Dragonfishes, lightfishes and allies
Order Aulopiformes	Lizardfishes and allies
Order Myctophiformes	Lanternfishes and allies
Order Lampridiformes	Oarfishes and allies
Order Polymixiiformes	Beardfishes
Order Percopsiformes	Troutperches and allies
Order Gadiformes	Cods, hakes and allies
Order Ophidiiformes	Cuskeels, pearlfishes and allies

Order Batrachoidiformes	Toadfishes and midshipmen
Order Lophiiformes	Anglerfishes, goosefishes and frogfishes
Order Gobiesociformes	Clingfishes and allies
Order Cyprinodontiformes	Killfishes and allies
Order Beloniformes	Ricefishes, flyingfishes and allies
Order Atheriniformes	Silversides, rainbowfishes and allies
Order Stephanoberyciformes	Pricklefishes, whalefishes and allies
Order Beryciformes	Squirrelfishes and allies
Order Zeiformes	Dories and allies
Order Gasterosteiformes	Sticklebacks, pipefishes and allies
Order Synbranchiformes	Swampeels and allies
Order Scorpaeniformes	Scorpionfishes and allies
Order Perciformes	Perches and allies
Order Pleuronectiformes	Flatfishes
Order Tetraodontiformes	Triggerfishes and allies

Information about animal species can be found in the online *Encyclopedia of Life* (EOL; www.eol.org) (Wilson, 2003). The EOL is an 'online collaborative encyclopedia intended to document all of the 1.8 million living species known to science', compiled from existing databases and from contributions by experts and non-experts throughout the world.

APPENDIX 2 SOME USEFUL WEBSITES

ASSOCIATIONS OF ZOOS, AQUARIUMS AND CURATORS

African Association of Zoos and Aquaria (PAAZAB)	www.paazab.com
Alliance of Marine Mammal Parks and Aquariums (AMMPA)	www.ammpa.org
Asociación de Zoológicos, Criaderos y Acuarios de México AC (AZCARM) – Association of Mexican Zoos and Aquariums	www.azcarm.com.mx
Asociación Espanola de Zoos y Acuarios (AEZA) – Spanish Association of Zoos and Aquariums	www.aeza.es
Asociación Ibérica de Zoos y Acuarios (AIZA) – Iberian Association of Zoos and Aquaria	www.aiza.org.es
Asociación Latinoamericana de Parques Zoológicos y Acuarios (ALPZA) – Latin American Zoo and Aquarium Association	www.alpza.com
Asociación Mesoamericana y del Caribe de Zoológico i Acuarios (AMACZOOA) – Association of Mesoamerican and Caribbean Zoos and Aquariums	www.amaczooa.org
Association Francaise des Parcs Zoologiques (AFdPZ) – Association of French Zoos	www.afdpz.com
Association of Zoos and Aquariums (AZA) – North America	www.aza.org
British and Irish Association of Zoos and Aquariums (BIAZA)	www.biaza.org.uk

(*Continued*)

(Cont'd)

Canadian Association of Zoos and Aquariums (CAZA)	www.caza.ca
Danske Zoologiske Haver & Akvarier – Danish Association of Zoological Gardens and Aquaria (DAZA)	www.daza.dk
Deutsche Tierpark-Gesellschaft (DTG) – German Animal Park Society	www.deutsche-tierparkgesellschaft.de
European Association of Zoos and Aquaria (EAZA)	www.eaza.net
European Union of Aquarium Curators (EUAC)	www.euac.org
Japanese Association of Zoos and Aquariums (JAZA)	www.jazga.or.jp
Nederlandse Vereniging van Deirentuinen (NVD) – Dutch Zoo Federation	www.nvdzoos.nl
Österreichische Zoo Organisation (OZO) – Austrian Zoo Organisation	www.ozo.at
Sociedade de Zoológicos do Brasil (SZB) – Brazilian Zoo Society	www.szb.org.br
South Asian Zoo Association for Regional Cooperation (SAZARC)	www.zooreach.org
South East Asian Zoos Association (SEAZA)	www.seaza.org
Svenska Djurparksföreningen (SDF) – Swedish Association of Zoological Parks and Aquaria (SAZA)	www.svenska-djurparksforeningen.nu
Swiss Association of Scientific Zoos (ZOOSchweiz)	www.zoos.ch/zooschweiz/
Syndicat National des Directeurs de Parcs Zoologiques (SNDPZ) – French zoo directors union	www.sndpz.fr
Union of Czech and Slovak Zoological Gardens (UCSZ)	www.zoo.cz
Unione Italiana dei Giardini Zoologici ed Acquari (UIZA) – Italian Union of Zoos & Aquaria	www.uiza.org
Verband deutscher Zoodirektoren e.V. (VdZ) – German Federation of Zoo Directors	www.zoodirektoren.de
World Association of Zoos and Aquariums (WAZA)	www.waza.org
Zoo and Aquarium Association (formerly Australasian Regional Association of Zoological Parks and Aquaria (ARAZPA))	www.arazpa.org.au
Zoological Association of America (ZAA)	www.zaoa.org

ZOO KEEPER AND TRAINER ORGANISATIONS

American Association of Zoo Keepers (AAZK)	http://aazk.org
Animal Keepers Association of Africa (AKAA)	www.akaafrica.com
Association of British and Irish Wild Animal Keepers (ABWAK)	www.abwak.org
Australasian Society of Zoo Keeping Inc. (ASZK)	www.aszk.org.au
Berufsverband der Zootierpfleger (BdZ) – Union of zookeepers (Germany)	www.zootierpflege.de
Elephant Managers Association (EMA)	www.elephant-managers.com
European Elephant Keeper and Manager Association	www.eekma.org
International Congress of Zookeepers (ICZ)	www.iczoo.org
International Marine Animal Trainers' Association (IMATA)	www.imata.org

International Rhino Keeper Association (IRKA)	www.rhinokeeperassociation.org
International Zoo Educators Association (IZE)	www.izea.net
La Asociación Ibérica de Cuidadores de Animales Salvajes (AICAS) – Iberian Zookeepers Association	www.aicas.org
Stichting de Harpij (The Harpy Foundation) – Organisation for Dutch and Belgian Zoo Professionals	www.deharpij.nl
Zoo Outreach Organisation (ZOO) – India	www.zooreach.org

ZOO VETS, ANIMAL HEALTH, WELFARE AND NUTRITION

American Association of Zoo Veterinarians	www.aazv.org
Avian Rearing Resource	www.avianrearingresource.co.uk
AZA Wildlife Contraception Center	www.stlzoo.org/animals/scienceresearch/contraceptioncenter/
Bear Information Exchange for Rehabilitators, Zoos and Sanctuaries (BIERZS)	www.bearkeepers.net
European Zoo Nutrition Centre	www.eznc.org
Great Ape Health Monitoring Unit (GAHMU)	www.eva.mpg.de/primat/GAHMU/index.htm
International Zoo Veterinary Group	www.izvg.co.uk
United States Department of Agriculture, National Agricultural Library, Animal Welfare Information Center	http://awic.nal.usda.gov
Wildlife Information Network (WIN)	www.wildlifeinformation.org
Wildlife Vets International	www.wildlifevetsinternational.org
World Organisation for Animal Health (Office International des Epizooties, OIE)	www.oie.int/eng/en_index.htm
Zoodent International	www.zoodent.co.uk

ZOO RESEARCH

Audubon Nature Institute's Center for Research of Endangered Species (ACRES)	www.auduboninstitute.org
Behavior Scientific Advisory Group (AZA/Lincoln Park Zoo) – database of ethograms	www.lpzoo.com/ethograms/about.html
Institute for Conservation Research (ZSSD)	www.sandiegozoo.org/conservation/
Institute of Zoology (IoZ) – ZSL	www.zsl.org/science/
Leibniz Institute for Zoo and Wildlife Research	www.izw-berlin.de
Smithsonian Conservation Biology Institute	http://nationalzoo.si.edu/scbi/default.cfm

EQUIPMENT, ENRICHMENT AND TRANSPORTATION

Ascott – animal tagging	www.ascott.biz
Aussie Dog Products – animal enrichment products	www.aussiedog.com.au
Enrichment online, Fort Worth Zoo	www.enrichmentonline.org
Gallagher Powerfence UK Ltd. – electric fencing	www.gallagher.co.uk
Global Wildlife Logistics (Pty) Ltd. – animal transport	www.gwl.co.za:8080/globalwildlife/WebObjects/globalwildlife.woa
Huamei Metal Mesh factory, China	www.ropemesh.com
Intershape Trading Ltd. – disinfectant mats	www.intershape.com
JCS Livestock (James Cargo Services Ltd.) – zoological transport	www.jamescargo.com/livestock_transport/
Mazuri Zoo Foods	www.mazurifoods.com
On Time Wildlife Feeders	www.ontimefeeders.com
Ritchey – animal tagging	www.ritchey.co.uk
Roxan iD – animal tagging	www.roxan.co.uk
Sanctuary Supplies	www.sanctuarysupplies.com
Shape of Enrichment	www.enrichment.org

ZOO DESIGN

Jon Coe Design Pty Ltd.	www.joncoedesign.com
Jones and Jones (architects)	www.jonesandjones.com
Zoolex	www.zoolex.org

CAPTIVE BREEDING AND POPULATION SOFTWARE

Conservation Breeding Specialist Group	www.cbsg.org
European Studbook Foundation	www.studbook.eu
International Species Information System (ISIS)	www.isis.org
MateRx	www.vortex9.org/materx.html
OUTBREAK	www.vortex9.org/outbreak.html
PARTINBR	www.vortex9.org/partinbr.html
PM2000	www.vortex9.org/pm2000.html
PopLink	www.lpzoo.com/conservation/Alexander_Center/software/PopLink/index.html
VORTEX	www.vortex9.org/vortex.html
ZooRisk	www.lpzoo.org/cs_centers_alex_software_zr.php

ANIMAL CLASSIFICATION

ARKive	www.arkive.org
Encyclopedia of Life (EOL)	www.eol.org
Integrated Taxonomic Information System (ITIS)	www.itis.gov
International Commission on Zoological Nomenclature (ICZN)	http://iczn.org
Natural History Museum, London	www.nhm.ac.uk
Royal Botanic Gardens, Kew	www.kew.org
Smithsonian Institution	www.si.edu
Species 2000 and ITIS Catalogue of Life	www.catalogueoflife.org/search.php
Wilson and Reeder's Mammal Species of the World	www.bucknell.edu/MSW3/
World Register of Marine Species (WoRMS)	www.marinespecies.org

WILDLIFE CONSERVATION AND ANIMAL WELFARE ORGANISATIONS

21st Century Tiger	www.21stcenturytiger.org
Advocates for Animals (UK)	www.onekind.org
African Conservation Foundation	www.africanconservation.org
Africat Foundation	www.africat.org
Alaska Wilderness League	www.alaskawild.org
Amboseli Trust for Elephants (Kenya and USA)	www.elephanttrust.org
American Bird Conservancy	www.abcbirds.org
Amphibian Ark	www.amphibianark.org
AMUR	http://amur.org.uk/home.shtml
Amur Leopard and Tiger Alliance (ALTA)	www.amur-leopard.org
Animal Defenders International (UK)	www.ad-international.org/adi_world/
Animal Public e.V. (Germany)	www.animal-public.de
Animal Rights Africa (South Africa)	www.animalrightsafrica.org
Animals Asia Foundation (Hong Kong)	www.animalsasia.org
Animals Australia (Australia)	www.animalsaustralia.org
Ape Alliance	www.4apes.com
Arboricultural Association	www.trees.org.uk
Australian Bush Heritage Fund	www.bushheritage.org
Australian Marine Conservation Society (AMCS)	www.marineconservation.org.au
Australian Wildlife Conservancy	www.australianwildlife.org
Barn Owl Trust	www.barnowltrust.org.uk
Bat Conservation Trust	www.bats.org.uk
BirdLife International	www.birdlife.org
Birdwatch Ireland	www.birdwatchireland.ie

(Continued)

(*Cont'd*)

Bombay Natural History Society	www.bnhs.org
Born Free Foundation	www.bornfree.org.uk
Borneo Orangutan Survival Foundation (BOS)	www.orangutan.org.uk
Botanic Gardens Conservation International	www.bgci.org
British Butterfly Conservation Society	www.butterfly-conservation.org
British Dragonfly Society (BDS)	www.dragonflysoc.org.uk
British Herpetological Society	www.thebhs.org
British Trust for Ornithology (BTO)	www.bto.org
Buglife – The Invertebrate Conservation Trust	www.buglife.org.uk
Campaigns Against the Cruelty To Animals CATCA (Canada)	http://catcahelpanimals.org
Canadian Marine Environment Protection Society (CMEPS) (Canada)	www.cmeps.org
Captive Animals' Protection Society (CAPS)	www.captiveanimals.org
Care for the Wild International (CWI)	www.careforthewild.com
Center for Plant Conservation (CPC)	www.centreforplantconservation.org
Charles Darwin Foundation	www.darwinfoundation.org
Cheetah Conservation Fund (CCF)	www.cheetah.org
Coalition Against Wildlife Trafficking (CAWT)	www.cawtglobal.org
Conchological Society of Great Britain & Ireland	www.conchsoc.org
Conservation International	www.conservation.org
Coral Reef Alliance	www.coral.org
David Shepherd Wildlife Foundation (DSWF)	www.sheldrickwildlifetrust.org
Defenders of Wildlife	www.defenders.org
Dian Fossey Gorilla Fund International	www.gorillafund.org
Durrell Wildlife Conservation Trust	www.durrell.org
Earthwatch Institute	www.earthwatch.org
Elephant Care International	www.elephantcare.org
ElephantVoices (Norway and Kenya)	www.elephantvoices.org
Entomological Foundation	www.entfdn.org
Environmental Investigation Agency (EIA)	www.eia-international.org
Euronatur	www.euronatur.org
Falklands Conservation	www.falklandconservation.com
Fauna and Flora International (FFI)	www.fauna-flora.org
Flora Locale	www.floralocale.org
Friends of Conservation (FOC)	www.foc-uk.com
Friends of the Earth (FoE)	www.foe.co.uk
Friends of the Elephant (Netherlands)	www.elephantfriends.org
Froglife	www.froglife.org
Fundación para la Adopción, Apadrinamiento y Defensa de los Animales FAADA – (Spain)	www.faada.org

Galapagos Conservation Trust	www.savegalapagos.org
Gibraltar Ornithological and Natural History Society	www.gonhs.org
Global Amphibian Assessment (GAA)	www.globalamphibians.org
Global Biodiversity Information Facility (GBIF)	www.gbif.org
Global Biodiversity Outlook 3 (GBO-3)	http://gbo3.cbd.int
Global Environment Facility (GEF)	www.thegef.org/gef/home/
Gorilla Organization	www.gorillas.org
Greenpeace International	www.greenpeace.org/international/
Grevy's Zebra Trust	www.grevyszebratrust.org
Habitat Acquisition Trust	www.hat.bc.ca
Hawk and Owl Trust	www.hawkandowl.org
Herpetological Conservation Trust	www.herpconstrust.org.uk
Humane Society International	www.hsi.org
Humane Society of the United States (HSUS)	www.humanesociety.org
In Defense of Animals (IDA)	www.idausa.org
International Animal Rescue (IAR)	www.internationalanimalrescue.org
International Crane Foundation (ICF)	www.savingcranes.org
International Fund for Animal Welfare (IFAW)	www.ifaw.org/splash.php
International Primate Protection League (IPPL)	www.ippl.org
International Reptile Conservation Foundation (IRCF)	www.ircf.org
Islands Trust Fund	www.islandstrustfund.bc.ca
Lemur Conservation Trust	www.lemurreserve.org
Libera! Asociacion Animalista (Spain)	www.change.org/liberaong/
Mammal Society	www.mammal.org.uk
Marine Conservation Society	www.mcsuk.org
Millennium Seed Bank Project	www.kew.org/msbp/
National Council of SPCAs (NSPCA) (South Africa)	www.nspca.co.za
National Wildflower Centre, UK	www.nwc.org.uk
No Whales in Captivity (Canada)	www.vcn.bc.ca/cmeps/
Norwegian Animal Protection Alliance (NAPA)	www.dyrevern.no/english/
Ocean Care	www.oceancare.org
Ocean Conservancy	www.oceanconservancy.org
Organisation Cetacea	www.orcaweb.org.uk
Otter Trust	www.ottertrust.org.uk
People for the Ethical Treatment of Animals (PETA)	www.peta.org
People's Trust for Endangered Species	www.ptes.org
Performing Animal Welfare Society (PAWS) (USA)	www.pawsweb.org
Plantlife	www.plantlife.org.uk
Pond Conservation	www.pondconservation.org.uk
Pro Natura	www.pronatura.ch
Pro Wildlife	www.prowildlife.de

(Continued)

(*Cont'd*)

ProFauna Indonesia	www.profauna.org
Rainforest Concern	www.rainforestconcern.org
RARE	www.rareconservation.org
Royal Society for the Conservation of Nature (Jordan)	www.rscn.org.jo
Royal Society for the Prevention of Cruelty to Animals (RSPCA)	www.rspca.org.uk
Royal Society for the Protection of Birds (RSPB)	www.rspb.org.uk
Save China's Tigers	www.savechinastigers.org
Save Our Seas Foundation (SOSF)	www.saveourseas.com
Save the Rhino	www.savetherhino.org
Seahorse Trust	www.theseahorsetrust.co.uk
Seal Conservation Society	www.pinnipeds.org
Shark Trust	www.sharktrust.org
Sierra Club	www.sierraclub.org
Stop Animal Exploitation Now! (SAEN) (USA)	www.all-creatures.org/saen/
Sumatran Tiger Trust	www.tigertrust.info
Tiger Research and Conservation Trust (TRACT) (India)	http://tractindia.org
Tigris Foundation	www.tigrisfoundation.nl
UK Wolf Conservation Trust	www.ukwolf.org
Universities Federation for Animal Welfare (UFAW)	www.ufaw.org.uk
Wetlands International	www.wetlands.org
Whale and Dolphin Conservation Society (WDCS)	www.whales.org
Wilderness Society	www.wilderness.org.au
Wildfowl and Wetlands Trust (WWT)	www.wwt.org.uk
Wildlife Alliance	www.wildlifealliance.org
Wildlife Conservation Society	www.wcs.org
Wildlife SOS, India	www.wildlifesos.org
Wildlife Trusts	www.wildlifetrusts.org
Wildscreen	www.wildscreen.org.uk
Wolves and Humans Foundation	www.wolvesandhumans.org
Woodland Trust	www.woodland-trust.org
World Land Trust (WLT)	www.worldlandtrust.org
World Owl Trust	www.owls.org
World Parrot Trust	www.parrots.org
World Pheasant Association	www.pheasant.org.uk
World Society for the Protection of Animals (WSPA)	www.wspa.ca
World Wide Fund for Nature/World Wildlife Fund (WWF)	www.panda.org
Xerces Society for Invertebrate Conservation	www.xerces.org
Yellow-eyed Penguin Trust	www.yellow-eyedpenguin.org.nz

LEGISLATION AND REGULATION

Animal and Plant Health Inspection Service (APHIS)	www.aphis.usda.gov
Center for Wildlife Law (USA)	http://wildlifelaw.unm.edu
Central Zoo Authority (India)	www.cza.nic.in
CITES	www.cites.org
Countryside Council for Wales	www.ccw.gov.uk
Department for Environment, Food and Rural Affairs (DEFRA/defra)	ww2.defra.gov.uk
European Union	http://europa.eu
European Union law	http://eur-lex.europa.eu/en/index.htm
International Air Transport Association (IATA)	www.iata.org
International Union for the Conservation of Nature and Natural Resources (IUCN) – World Conservation Union	www.iucn.org
Metropolitan Police Service, Wildlife Crime Unit, London	www.met.police.uk/wildlife/
Natural England	www.naturalengland.org.uk
Partnership for Action Against Wildlife Crime (PAW), DEFRA	www.defra.gov.uk/paw/
Scottish Natural Heritage	www.snh.gov.uk
TRAFFIC	www.traffic.org
United Nations (UN)	www.un.org/en/
United Nations Educational, Scientific and Cultural Organisation (UNESCO)	www.unesco.org/new/en/unesco/
United Nations Environment Programme (UNEP)	www.unep.org
United States Department of Agriculture (USDA)	www.usda.gov/wps/portal/usda/usdahome/
United States Fish and Wildlife Service (USFWS)	www.fws.gov
USFWS Forensics Laboratory	www.lab.fws.gov
Wildlife Inspectorate	www.defra.gov.uk/animalhealth/cites/wildlifeinspect.htm
World Conservation Monitoring Centre	www.unep-wcmc.org
Zoos Inspectorate	www.defra.gov.uk/animalhealth/CITES/zoo-inspect.htm

ZOO ENTHUSIASTS

Bartlett Society	http://zoohistory.co.uk/html
Independent Zoo Enthusiasts Society (IZES)	www.izes.co.uk
ZooChat	www.zoochat.com

Note: Although the URLs listed here were believed to be accurate at the time of publication organisations often change their URLs from time to time.

ACRONYMS USED IN THE TEXT

AArk	Amphibian Ark
AAZK	American Association of Zoo Keepers
AAZPA	American Association of Zoological Parks and Aquariums (*now* AZA)
ABWAK	Association of British and Irish Wild Animal Keepers
ACRES	Audubon Nature Institute's Centre for Research of Endangered Species
ACTH	Adrenocorticotrophic hormone
ADI	Animal Defenders International
AEWA	African-Eurasian Waterbird Agreement
AI	Artificial insemination
AKAA	Animal Keepers Association of Africa
APHIS	Animal and Plant Health Inspection Service
APP	African Preservation Programme
ARAZPA	Australasian Regional Association of Zoological Parks and Aquaria (*now* ZAA)
ARKS	Animal Record Keeping System
ART	Assisted reproductive therapy
ASMP	Australasian Species Management Programme
ASZK	Australian Society of Zoo Keeping
AWA	Animal Welfare Act (USA)
AWG	Aquarium Working Group (BIAZA)
AZA	(American) Association of Zoos and Aquariums
BAG	Behavior Advisory Group (AZA)
BAP	Biodiversity Action Plan
BBC	British Broadcasting Corporation
BIAZA	British and Irish Association of Zoos and Aquariums
BIERZS	Bear Information Exchange for Rehabilitators, Zoos and Sanctuaries
BSE	Bovine spongiform encephalitis (encephalopathy)
BWG	Bird Working Group (BIAZA)
CAPS	Captive Animals' Protection Society
CBSG	Conservation Breeding Specialist Group
CCTV	Closed-circuit television
CFR	Code of Federal Regulations
CG	Chorionic gonadotrophin
CITES	Convention on International Trade in Endangered Species of Wild Fauna and Flora
COTES	Control of Trade in Endangered Species (Enforcement) Regulations
CRES	Conservation and Research for Endangered Species (*formerly* Center for Reproduction of Endangered Species) (ZSSD)
CZA	Central Zoo Authority (India)
DEFRA (defra)	Department for Environment, Food and Rural Affairs
DNA	Deoxyribonucleic acid
EAZA	European Association of Zoos and Aquaria
EDGE	Evolutionarily Distinct and Globally Threatened
EEKMA	European Elephant Keeper and Manager Association
EEP	European Endangered Species Programme (EAZA)
EMA	Elephant Managers Association
EOL	Encyclopedia of Life
ESA	Endangered Species Act (USA)
ESB	European Studbook (EAZA)
ESF	European Studbook Foundation
ESU	Evolutionarily significant unit
ET	Embryo transfer
EU	European Union
EUAC	European Union of Aquarium Curators
FFI	Fauna and Flora International
FMD	Foot and mouth disease
FONZ	Friends of the National Zoo (USA)
FSH	Follicle-stimulating hormone
GAA	Global Amphibian Assessment
GAE	Gross assimilation efficiency
GAHMU	Great Ape Health Monitoring Unit
GBIF	Global Biodiversity Information Facility

An Introduction to Zoo Biology and Management, First Edition. Paul A. Rees.

GBO-3	Global Biodiversity Outlook 3
GEF	Global Environment Facility
GIS	Geographical information systems
GMT	Greenwich Mean Time
GnRH	Gonadotrophin-releasing hormone
GPS	Global positioning system
HPA	Hypothalamic-pituitary-adrenal
HSUS	Humane Society of the United States
IATA	International Air Transport Association
ICP	Institutional Collection Plan
ICSH	Interstitial cell-stimulating hormone
ICZ	International Congress of Zookeepers
ICZN	International Commission on Zoological Nomenclature
IDA	In Defense of Animals
IoZ	Institute of Zoology (ZSL)
ISIS	International Species Information System
ITIS	Integrated Taxonomic Information System
IUCN	International Union for the Conservation of Nature and Natural Resources
IUDZG	International Union of Directors of Zoological Gardens
IUPN	International Union for the Protection of Nature (*now* IUCN)
IVF	*In vitro* fertilisation
IZE	International Zoo Educators Association
IZES	Independent Zoo Enthusiasts Society
JAZA	Japanese Association of Zoos and Aquariums
JMSC	Joint Management of Species Committee
JMSP	Joint Management of Species Programme
LaCONES	Laboratory for the Conservation of Endangered Species (India)
LCMV	Lymphocytic choriomeningitis virus
LH	Luteinising hormone
MAB	Man and the Biosphere Programme
MAI	Maximum avoidance of inbreeding
MedARKS	Medical Animal Record Keeping System
MOC	Multiple ocular coloboma
MSI	Mate suitability index
mtDNA	Mitochondrial deoxyribonucleic acid
MVP	Minimum viable population
MWG	Mammal Working Group (BIAZA)
NGO	Non-governmental organisation
NSWG	Native Species Working Group (BIAZA)
OIE	Office International des Epizooties (World Organisation for Animal Health)
PAAZAB	African Association of Zoos and Aquaria
PAW	Partnership for Action Against Wildlife Crime (UK)
PCR	Polymerase chain reaction
PDA	Personal digital assistant
PM2000	Population Management 2000
PMP	Population Management Plan (AZA)
PVA	Population viability analysis
PWG	Plant Working Group (BIAZA)
RAWG	Reptile and Amphibian Working Group (BIAZA)
RCP	Regional Collection Plan
REGASP	Regional Animal Species Collection Plan
RSPB	Royal Society for the Protection of Birds
RSPCA	Royal Society for the Prevention of Cruelty to Animals
SAC	Special Area of Conservation
SARS	Severe acute respiratory syndrome
SAZARC	South Asian Zoo Association for Regional Co-operation
SCNT	Somatic cell nuclear transfer
SEAZA	South East Asian Zoo Association
SI	Statutory Instrument
SPA	Special Protection Area
SPARKS	Single Population Analysis and Record Keeping System
SPIDER	Setting goals, Planning, Implementing, Documenting, Evaluating and Readjusting
SR	Statutory Rule
SSC	Species Survival Commission
SSI	Scottish Statutory Instrument
SSP	Species Survival Plan (AZA)
SSSMZP	Secretary of State's Standards of Modern Zoo Practice
TAG	Taxon Advisory Group
TANAPA	Tanzania National Parks Authority
TB	Tuberculosis
TIGERS	The Institute of Greatly Endangered and Rare Species (USA)
TIWG	Terrestrial Invertebrate Working Group (BIAZA)
TRAFFIC	Trade Records Analysis of Flora and Fauna in Commerce
TWG	Taxon Working Group
UFAW	Universities Federation for Animal Welfare
UN	United Nations
UNESCO	United Nations Educational, Scientific and Cultural Organisation
USC	United States Code
USDA	United States Department of Agriculture
USFWS	United States Fish and Wildlife Service
WAZA	World Association of Zoos and Aquariums
WoRMS	World Register of Marine Species
WSI	Welsh Statutory Instrument
WSPA	World Society for the Protection of Animals
WWF	World Wide Fund for Nature/World Wildlife Fund
WWT	Wildfowl and Wetlands Trust
ZAA	Zoo and Aquarium Association (*formerly* ARAZPA)
ZIMS	Zoological Information Management System
ZSL	Zoological Society of London
ZSSD	Zoological Society of San Diego

REFERENCES

AAZK (2010). American Association of Zoo Keepers, Inc. http://aazk.org/about (accessed 28 July 2010).

AAZPA (1986). *The Purposes of Zoos and Aquariums*. American Association of Zoological Parks and Aquariums, Wheeling, WV.

Abee, C. R. (1985). Medical care and management of the squirrel monkey. In: Rosenblum, A. and Coe, C. L. (eds.) *The Handbook of Squirrel Monkey Research*, pp. 447–488. Plenum Publishing, New York.

ABWAK (2010). www.abwak.co.uk/About ABWAK.html (accessed 28 July 2010).

Adkins Giese, C. L. (2005). The Big Bad Wolf hybrid: How molecular genetics research may undermine protection for gray wolves under the Endangered Species Act. *Minnesota Journal of Law, Science & Technology*, 6: 865–872.

AKAA (2010). www.africanconservation.org/explorer/south-africa/553-animal-keepers-association-of-africa-akaa/view-details.html (accessed 28 July 2010).

Alberts, A. C. (1994). Dominance hierarchies in male lizards: Implications for zoo management programs. *Zoo Biology*, 13: 479–490.

Allen, M. E. and Oftedal, O. T. (1996). Essential nutrients in mammalian diets. In: Kleiman, D. G., Allen, M. E., Thompson, K. V. and Lumpkin, S. (eds.) *Wild Mammals in Captivity: Principles and Techniques*, pp. 117–128. University of Chicago Press, Chicago, IL.

ALVA (2009). www.alva.org.uk/visitor_statistics/ (accessed 17 July 2009).

Andersen, L. L. (2003). Zoo education: from formal school programmes to exhibit design and interpretation. *International Zoo Yearbook*, 38: 75–81.

Anderson, U. S., Maple, T. L. and Bloomsmith, M. A. (2004). A close keeper–nonhuman animal distance does not reduce undesirable behavior in contact yard goats and sheep. *Journal of Applied Animal Welfare Science*, 7: 59–69.

Anon. (1906). Bushman shares a cage with Bronx Park apes. *The New York Times*, 9 September.

Anon. (1993). *The World Zoo Conservation Strategy: The Role of the Zoos and Aquaria of the World in Global Conservation*. IUDZG–The World Zoo Organization and the Captive Breeding Specialist Group of IUCN/SSC. www.waza.org/conservation/wczs.php (accessed 21 October 2003).

Anon. (1995a). USA lift wildlife trade embargo on Taiwan. *Traffic Bulletin*, 15: 101.

Anon. (1995b). *Biodiversity: The UK Steering Group Report Meeting the Rio Challenge (Volume I)*. HMSO, London.

Anon. (1998). *Re-introduction of the European beaver to Scotland: Results of a Public Consultation*. Scottish Natural Heritage Research Survey and Monitoring Series No. 121. Scott Porter Research and Marketing.

Anon. (2001). www.elephantcenter.com/pregnancy/ (accessed 26 February 2001).

Anon. (2004). Pandas make zoo attendance double in Thailand. http://english.peopledaily.com.cn/ 200403/08/print20040308_136801.html (accessed 10 August 2004).

Anon. (2005). Row over German zoo's Africa show. BBC News, 8 June. http://news.bbc.co.uk/go/pr/fr/-/1/hi/world/africa/4070816.stm (accessed 6 November 2009).

Anon. (2007). *Standard 154.03.04: Containment Facilities for Zoo Animals*. Ministry of Agriculture and Forestry, Biosecurity New Zealand, Wellington, New Zealand.

Anon. (2008a). Orang utan escapes pen at US zoo. BBC News, 18 May. http://news.bbc.co.uk/2/hi/americas/7407050.stm (accessed 8 June 2008).

Anon. (2008b). Quake zone panda cubs thrive. *The Times*, 23 August, p. 40.

Anon. (2009a). New species of giant rat discovered in crater of volcano in Papua New Guinea. Smithsonian Institution. http://smithsonianscience.org/2009/09/new-species-of-giant-rat-discovered-in-crater-of-volcano-in-papua-new-guinea (accessed 17 March 2010).

Anon. (2009b). Dresden zoo forced to rename primate called 'Obama'. *Telegraph*, 10 July. www.telegraph.co.uk/news/

An Introduction to Zoo Biology and Management, First Edition. Paul A. Rees.

worldnews/northamerica/usa/barackobama/5799568 (accessed 6 November 2009).

Anon. (2009c). *Policy on the Management of Solitary Elephants in New South Wales*. www.dpi.nsw.gov.au/agriculture/livestock/animal-welfare/exhibit/policies/solitary-elephants/ (accessed 1 December 2009).

Anon. (2009d). Rare monkeys stolen. *The Times*, 11 July, p. 31.

Anon. (2009e). Trio of gorillas put in picture about their foreign stud. *The Times*, 28 August, p. 21.

Anon. (2010). www.noahsarkzoofarm.co.uk/pages/research/research.php (accessed 11 February 2010).

Associated Press (2005). Test tube gorilla fails to bond with baby. 10 April. www.cjonline.com/stories/041005/pag_testgorilla.shtml (accessed 13 June 2008).

Attenborough, D. (2003). *Life on Air: Memoirs of a Broadcaster*. BBC, London.

AVMA (2007). *AVMA Guidelines on Euthanasia*. www.avma.org/issues/animal_welfare/euthanasia.pdf (accessed 18 June 2010).

AZA (1992). *Guidelines for Reintroduction of Animals Born or Held in Captivity*. www.aza.org/reintroduction/ (accessed 29 June 2010).

AZA (2007). Top ten AZA elephant success stories. www.aza.org/Newrsroom/PR_Top10ElephStories (accessed 23 April 2007).

AZA (2009). www.aza.org/behavior-advisory-group/ (accessed 15 June 2009).

AZA (2010a). Association of Zoos and Aquariums, Visitor demographics. www.aza.org/visitor-demographics/ (accessed 13 May 2010).

AZA (2010b). *The Accreditation Standards and Related Policies, 2010 edition*. Association of Zoos and Aquariums, Silver Spring, MD.

AZA (2010c). www.aza.org/institutional-collection-plans/ (accessed 8 June 2010).

AZA (2010d). www.aza.org/taxon-advisory-groups/ (accessed 8 June 2010).

AZA (2010e). www.aza.org/ssp-list/ (accessed 4 June 2010).

AZA (2010f). www.aza.org/population-management-plan-programs/ (accessed 4 June 2010).

AZA (2010g). www.aza.org/species-survival-plan-program/ (accessed 4 June 2010).

AZA (2010h). www.aza.org/oilspill/ (accessed 8 June 2010).

Baer, J. F. (1998). A veterinary perspective of potential risk factors in environmental enrichment. In: Shepherdson, D. J., Mellen, J. D. and Hutchins, M. (eds.) *Second Nature: Environmental Enrichment for Captive Animals*, pp. 277–301. Smithsonian Institution Press, Washington DC and London.

Bagemihl, B. (1999). *Biological Exuberance: Animal Homosexuality and Natural Diversity*. Profile Books, London.

Baker, A. (2007). Animal ambassadors: an analysis of the effectiveness and conservation impact of ex-situ breeding efforts. In: Zimmerman, A., Hatchwell, M., Dickie, L. and West, C. (eds.) *Zoos in the 21st Century: Catalysts for Conservation*, pp. 139–154. Cambridge University Press, Cambridge, UK.

Baker, S. J. (1990). Escaped exotic mammals in Britain. *Mammal Review*, 20: 75–96.

Ballou, J. D. and Foose, T. J. (1996). Demographic and genetic management of captive populations. In: Kleiman, D. G., Allen, M. E., Thompson, K. V. and Lumpkin, S. (eds.) *Wild Mammals in Captivity: Principles and Techniques*, pp. 263–283. University of Chicago Press, Chicago, IL.

Balmford, A. (2000). Separating fact from artifact in analyses of zoo visitor preferences. *Conservation Biology*, 14: 1193–1195.

Balmford, A., Gaston, K. J., Blyth, S., James, A. and Kapos, V. (2003). Global variation in terrestrial conservation costs, conservation benefits, and unmet conservation needs. *Proceedings of the National Academy of Sciences of the USA*, 100: 1046–1050.

Balmford, A., Leader-Williams, N., Mace, G. M., Manica, A., Walter, O., West, C. and Zimmerman, A. (2007). Message received? Quantifying the impact of informal conservation education on adults visiting zoos. In: Zimmerman, A., Hatchwell, M., Dickie, L. and West, C. (eds.) *Zoos in the 21st Century: Catalysts for Conservation*, pp. 120–136. Cambridge University Press, Cambridge, UK.

Barataу, E. and Hardouin-Fugier, E. (2002). *Zoo: A History of Zoological Gardens in the West*. Reaktion Books Ltd., London.

Barker, P. (2002). Setting the stage: using murals in the renovation of exhibits. *Communiqué*, August: 42–43.

Barnett, K. C. and Lewis, J. C. M. (2002). Multiple ocular colobomas in the snow leopard (*Uncia uncia*). *Veterinary Ophthalmology*, 5(3): 197–199.

Barnett, S. A. (1964). The concept of stress. In: Carthy, J. D. and Dudington, C. L. (eds.) *Viewpoints in Biology*, 3: 170–218. Butterworth, London.

Bashaw, M. J., Tarou, L. R., Maki, T. S. and Maple, T. L. (2001). A survey assessment of variables related to stereotypy in captive giraffe and okapi. *Applied Animal Behaviour Science*, 73: 235–247.

Bayazit, V. (2009). Evaluation of cortisol and stress in captive animals. *Australian Journal of Basic and Applied Sciences*, 3: 1022–1031.

BBC (2007). *Annual Report 2007*. www.bbc.co.uk/annualreport/2007/pdfs/reviewofyear.pdf (accessed 27 January 2010).

Beck, W. S., Liem, K. F. and Simpson, G. G. (1991). *Life: An Introduction to Biology*, 3rd edn. Harper Collins Publishers, New York.

Beisner, B. A. and Isbell, L. A. (2008). Ground substrate affects activity budgets and hair loss in outdoor captive groups of rhesus macaques (*Macaca mulatta*). *American Journal of Primatology*, 70: 1160–1168.

Benato, L., Eatwell, K. and Stidworthy, M. F. (2010). Necrosis of the pinnae in a grey short-tailed opossum (*Monodelphis domestica*). *Veterinary Record*, 166: 121–122.

Benirschke, K. and Roocroft, A. (1992). Elephant inflicted injuries. *Verhandlungsberichte Erkrankungen Zootiere*, 34: 239–247.

Bennett-Levy, J. and Marteau, T. (1984). Fear of animals: What is prepared? *British Journal of Psychology*, 75: 35–42.

Bentham, J. (1823). *Introduction to the Principles of Morals and Legislation*, 2nd edn. Clarendon Press, Oxford.

Berkson, G., Goodrich, J. and Kraft, I. (1966). Abnormal stereotyped movements of marmosets. *Perceptual and Motor Skills*, 23: 491–498.

Bertram, B. (2004). Misconceptions about zoos. *Biologist*, 51: 199–206.

Bettinger, T. and Quinn, H. (2000). Conservation funds: How do zoos and aquariums decide which project to fund? In: *American Zoo and Aquarium Association Annual Conference Proceedings*, St Louis, MO, pp. 52–54. American Zoo and Aquarium Association, Silver Spring, MD.

BIAZA (2005). *Animal Transaction Policy*. www.biaza.org.uk/resources/library/images/ATP%20june%2004.pdf (accessed 18 June 2010).

Bickert, I. and Meier, J. (2005). Zooselbstverständnis und Kundenerwartungen – Resultate einer Besucherumfrage in Zoo Basel. *Der Zoologische Garten*, 75: 202–208.

Biggins, D., Godbey, J., Hanebury, L., Marinari, P., Matchett, R. and Vargas, A. (1998). Survival of black-footed ferrets. *Journal of Wildlife Management*, 62: 643–653.

Bitgood, S. (1988). Problems in visitor orientation and circulation. In: Bitgood, S., Roper, J. and Benefield, A. (eds.) *Visitor Studies – 1988: Theory, Research, and Practice*, pp. 155–170. Center for Social Design, Jacksonville, AL.

Bitgood, S. (1991). Suggested guidelines for designing interactive exhibits. *Visitor Behavior*, 6: 4–11.

Bitgood, S. (2006). An analysis of visitor circulation movement patterns and the general value principle. *Curator: The Museum Journal*, 49: 463–475.

Bitgood, S. and Patterson, D. (1993). The effects of gallery changes on visitor reading and object viewing. *Environment and Behavior*, 25: 761–781.

Bitgood, S. and Richardson, K. (1986). Wayfinding at the Birmingham Zoo. *Visitor Behavior*, 1: 9.

Bitgood, S., Benefield, A. and Pattersen, D. (1989). The importance of label placement: a neglected factor in exhibit design. *Current Trends in Audience Research*, 4: 49–52.

Bitgood, S., Benefield, A., Patterson, D., Lewis, D. and Landers, A. (1985). Zoo visitors: Can we make them behave? *Annual Proceedings of the 1985 American Association of Zoological Parks and Aquariums*, Columbus, OH.

Bitgood, S., Patterson, D. and Benefield, A. (1988). Exhibit design and visitor behavior: empirical relationships. *Environment and Behavior*, 20: 474–491.

Blaney, E. C. and Wells, D. L. (2004). The influence of a camouflage net barrier on the behaviour, welfare and public perceptions of zoo-housed gorillas. *Animal Welfare*, 13: 111–118.

Bloomsmith, M. A. and Lambeth, S. P. (2000). Videotapes as enrichment for captive chimpanzees (*Pan troglodytes*). *Zoo Biology*, 19: 541–551.

Blumstein, D. T. (1998). Female preference and effective population size. *Animal Conservation*, 1: 173–177.

Boden, E. (ed.) (2007). *Black's Student Veterinary Dictionary*. A & C Black, London.

Boorer, M. (1972). Some aspects of stereotyped patterns of movement exhibited by zoo animals. *International Zoo Yearbook*, 12: 164–166.

Borun, M. and Miller, M. S. (1980). To label or not to label. *Museum News*, 58(4): 64–67.

Bostock, S. St.C. (1993). *Zoos and Animal Rights: The Ethics of Keeping Animals*. Routledge, London and New York.

Boyes, R. (2007). Berlin Zoo culls creator of the cult of Knut. *The Times*, 13 December. www.timesonline.co.uk/tol/news/world/europe/article3042791.ece (accessed 31 July 2009).

Brady, M., Rehling, M., Mueller, J. and Lukas, K. (2010). Giant Pacific octopus behavior and enrichment. *International Zoo News*, 57: 134–145.

Broad, G. (1996). Visitor profile and evaluation of informal education at Jersey Zoo. *Dodo*, 32: 166–192.

Broad, S. and Smith, L. (2004). Who educates the public about conservation issues? Examining the role of zoos and the media. In: Frost, W., Croy, G. and Beeton, S. (eds.) *International Tourism and Media Conference Proceedings*, 24–26 November, Tourism Research Unit, Monash University, Melbourne, pp. 15–23.

Broad, S. and Weiler, B. (1998). Captive animals and interpretation: a tail of two tiger exhibits. *Journal of Tourism Studies*, 9: 14–27.

Brockett, R. C., Stoinski, T. S., Black, J., Markowitz, T. and Maple, T. L. (1999). Nocturnal behaviour in a group of unchained female African elephants. *Zoo Biology*, 18: 101–109.

Brown, T. (2009). *The IZES Guide to British Zoos & Aquariums*. The Independent Zoo Enthusiasts Society, Todmorden, Lancashire.

Burgess, J. and Unwin, D. (1984). Exploring the Living Planet with David Attenborough. *Journal of Geography in Higher Education*, 8: 93–113.

Burrill, L. (2009). *The Application of Technology within Edinburgh Zoo's Penguin Colony*. BIAZA Research Conference, 13–14 July, Blackpool Zoo.

Burt, W. H. (1943). Territoriality and home range concepts as applied to mammals. *Journal of Mammalogy*, 24: 346–352.

Burton, M. (1970). *Dictionary of the World's Mammals*. Sphere Books Limited, London.

Bush, M. (1996). Methods of capture, handling and anesthesia. In: Kleiman, D. G., Allen, M. E., Thompson, K. V. and Lumpkins, S. (eds.) *Wild Mammals in Captivity: Principles and Techniques*, pp. 25–40. University of Chicago Press, Chicago, IL.

Cain, L. P. and Meritt, D. A. (1998). The growing commercialism of zoos and aquariums. *Journal of Policy Analysis and Management*, 17: 298–312.

Calhoun, J. B. (1962). Population density and social pathology. *Scientific American*, 206: 139–148.

Card, W. C., Roberts, D. T. and Odum, R. A. (1998). Does zoo herpetology have a future? *Zoo Biology*, **17**: 453–462.

Carlsen, F. and de Jongh, T. (2009). Getting it right for chimpanzees. *EAZA News*, 67: 22–23.

Carlstead, K. (1996). Effects of captivity on the behaviour of wild mammals. In: Kleiman, D. G., Allen, M. E., Thompson, K. V. and Lumpkins, S. (eds.) *Wild Mammals in Captivity: Principles and Techniques*, pp. 317–333. University of Chicago Press, Chicago, IL.

Carlstead, K., Brown, J. L. and Seidensticker, J. C. (1993). Behavioral and adrenocortical responses to environmental changes in leopard cats (*Felis bengalensis*). *Zoo Biology*, 12: 321–331.

Carlstead, K., Seidensticker, J. C. and Baldwin, R. (1991). Environmental enrichment for zoo bears. *Zoo Biology*, 10: 3–16.

Caro, T. and Eadie, J. (2005). Animal behavior and conservation biology. In: Bolhuis, J. J. and Giraldeau, L. (eds.) *The Behavior of Animals: Mechanisms, Function, and Evolution*, pp. 367–392. Blackwell Publishing Ltd., Oxford.

Chambers, P. (2007). *Jumbo: This Being the True Story of the Greatest Elephant in the World*. André Deutsch, London.

Chamove, A. S., Hosey, G. R. and Schaetzel, P. (1988). Visitors excite primates in zoos. *Zoo Biology*, 7: 358–369.

Charity Commission (2009). www.charity-commission.gov.uk (accessed 1 April 2010).

Cherfas, J. (1984). *Zoo 2000: A Look Beyond the Bars*. BBC, London.

Chester Zoo (2007). www.chesterzoo.org/Support%20Us/Corporate%20Support/Cause-Related%20Marketing.aspx (accessed 1 January 2008).

Chester Zoo Review (2009). *Annual Report of the North of England Zoological Society for the Year Ended 31st December 2009*. North of England Zoological Society, Chester.

Cheung, Y-Y., Chen, T-Y., Yu, P-H. and Chi, C-H. (2010). Observations on the female reproductive cycles of captive Asian yellow pond turtles (*Mauremys mutica*) with radiography and ultrasonography. *Zoo Biology*, 29: 50–58.

Chitty, D. (1960). Population processes in the vole and their relevance to general theory. *Canadian Journal of Zoology*, 38: 99–113.

Christian, J. D. and Radcliffe, H. L. (1952). Shock disease in captive wild animals. *American Journal of Pathology*, 28: 725–737.

Christian, J. J. (1963). The pathology of overpopulation. *Military Medicine*, 128: 571–603.

Christian, J. J. and Davis, E. E. (1966). Adrenal glands in female voles (*Microtus pennsylvanicus*) as related to production and population size. *Journal of Mammalogy*, 47: 1–18.

Christie, S. (2007). Zoo-based fundraising for in situ wildlife conservation. In: Zimmerman, A., Hatchwell, M., Dickie, L. and West, C. (eds.) *Zoos in the 21st Century: Catalysts for Conservation*, pp. 257–274. Cambridge University Press, Cambridge, UK.

Church, J. (ed.) (1995). *Social Trends*. Central Statistical Office, HMSO, London.

Cloete, C., Mogogane, O. and Sebati, M. (2008). A change in perspective: providing enrichment for Hamadryas baboon. *The Shape of Enrichment*, 17: 1–3.

Clubb, R. and Mason, G. J. (2002). *A Review of the Welfare of Zoo Elephants in Europe: A Report Commissioned by the RSPCA*. Animal Behaviour Research Group, Department of Zoology, University of Oxford, Oxford.

Clubb, R. and Mason, G. J. (2007). Natural behavioural biology as a risk factor in carnivore welfare: How analysing species differences could help zoos improve enclosures. *Applied Animal Behaviour Science*, 102: 303–328.

Clubb, R., Rowcliffe, M., Lee, P., Mar, K. U., Moss, C. and Mason, G. J. (2008). Compromised survivorship in zoo elephants. *Science*, 322: 1649.

Cociu, M., Wagner, G., Micu, N. E. and Mihaescu, G. (1974). Adaptational gastro-enteritis in Siberian tiger (*Panthera tigris altaica*) in the Bucharest Zoo. *International Zoo Yearbook*, 14: 171–174.

Coe, J. C., Scott, D. and Lukas, K. E. (2009). Facility design for bachelor gorilla groups. *Zoo Biology*, 28: 144–162.

Colahan, H. and Breder, C. (2003). Primate training at Disney's Animal Kingdom. *Journal of Applied Animal Welfare Science*, 6: 235–246.

Colman, R. J., Anderson, R. M., Johnson, S. C. *et al.* (2009). Caloric restriction delays disease onset and mortality in rhesus monkeys. *Science*, 325: 201–204.

Conway, W. (1986). The practical difficulties and financial implications of endangered species breeding programmes. *International Zoo Yearbook*, 24/25: 210–219.

Cook, J. C., Weber, A. E. and Stewart, G. R. (1978). Survival of captive tortoises released in California. In: *Proceedings of the Desert Tortoise Council Symposium*, pp. 130–135. Desert Tortoise Council, San Diego, CA.

Cooper, M. E. (2003). Zoo legislation. *International Zoo Yearbook*, 38: 81–93.

Croke, V. (1997). *The Modern Ark: The Story of Zoos: Past, Present and Future*. Scribner, New York.

Cusack, O. and Smith, E. (1984). *Pets and the Elderly: The Therapeutic Bond*. The Hayworth Press, New York.

Darling, F. F. (1938). *Bird Flocks and the Breeding Cycle: A Contribution to the Study of Avian Sociality*. Cambridge University Press, Cambridge, UK.

Davey, G. (2005). Is zoo-going a human instinct? Biophilia and zoos. *International Zoo News*, 52: 452–459.

Davey, G. (2007). Visitors' effects on the welfare of animals in the zoo: a review. *Journal of Applied Animal Welfare Science*, 10: 169–183.

Davey, G., Henzi, P. and Higgins, L.T. (2005). The influence of environmental enrichment on Chinese visitor behaviour. *Journal of Applied Animal Welfare Science*, 8: 131–140.

de Azevedo, C. S. and Faggioli, A. B. (2004). Effects of the introduction of mirrors and flamingo statues on the reproductive behaviour of a Chilean flamingo flock. *International Zoo News*, 51: 478–483.

De Grazia, D. (2002). *Animal Rights: A Very Short Introduction*. Oxford University Press, Oxford.

de Waal, F. (1996). *Good Natured: The Origins of Right and Wrong in Humans and Other Animals*. Harvard University Press, Cambridge, MA.

de Wit, J. J. (1995). Mortality of rheas caused by synchamus trachea infection. *Veterinary Quarterly*, 17: 39–40.

Deans, C., Martin, J., Neon, K., Nuesa, B. and O'Reilly, J. (1987). *A Zoo for Who? A Pilot Study in Zoo Design for Children. The Reid Park Zoo.* Center for Social Design, Jacksonville, AL.

D'Eath, R. B. and Keeling, L. D. (2003). Social discrimination and aggression by laying hens in large groups: from peck order to social intolerance. *Applied Animal Behaviour Science*, 84: 197–212.

DEFRA (2003). Zoo Licensing Act 1981 (as amended by The Zoo Licensing Act 1981 (Amendment) (England and Wales) Regulations 2002) ("the 2002 Regulations"). Circular 02/2003, DEFRA, London.

DEFRA (2007). *Reports Received by Defra of Escapes of Non-native Cats in the UK 1975 to Present Day.* www.defra.gov.uk/wildlife-countryside/vertebrates/reports/exotic-cat-escapes.pdf (accessed 1 January 2008).

DEFRA (2010). www.defra.gov.uk/foodfarm/farmanimal/diseases/atoz/notifiable.htm (accessed 4 August 2010).

Delacour, J. T. (1947). Wildlife conservation and zoos. *Parks and Recreation*, 30: 493–494.

Delany, M. J. (1974). *The Ecology of Small Mammals.* Edward Arnold (Publishers) Ltd., London.

Deleu, R., Veenhuizen, R. and Nelissen, M. (2003). Evaluation of the mixed-species exhibit of African elephants and Hamadryas baboons in Safari Beekse Bergen, the Netherlands. *Primate Report*, 65: 5–19.

Delibes-Mateos, M., Ramírez, E., Ferreras, P. and Villafuerte, R. (2008). Translocations as a risk for the conservation of European wild rabbit *Oryctolagus cuniculus* lineages. *Oryx*, 42: 259–264.

Delort, R. (1992). *The Life and Lore of the Elephant.* Thames and Hudson, London.

Dembiec, D. P., Snider, R. J. and Zanella, A. J. (2004). The effects of transport stress on tiger physiology and behavior. *Zoo Biology*, 23: 335–346.

Derwin, C. L. and Piper, J. B. (1988). The African Rock Kopje Exhibit evaluation and interpretive elements. *Environment and Behavior*, 20: 435–451.

Diamond, J. and Bond, A. B. (1991). Social behavior and the ontogeny of foraging in the kea (*Nestor notabilis*). *Ethology*, 88: 128–144.

Dierenfeld, E. S. (1996). Nutritional wisdom: adding the science to the art. *Zoo Biology*, 15: 447–448.

Dierking, L. D., Burknyk, K., Buchner, K. S. and Falk, J. H. (2002). *Visitor Learning in Zoos and Aquariums: A Literature Review.* Institute for Learning Innovation, Annapolis, MD.

Dixon, A. (2005). Meerkat integration at Auckland Zoo: a comparison of two methods. *International Zoo News*, 52: 472–477.

Donlan, J., Greene, H. W., Berger, J. *et al.* (2005). Re-wilding North America. *Nature*, 436: 913–914.

Dorresteyn, T. and Terkel, A. (2000). Captive elephant breeding: what should we do? In: Rietkerk, F., Hiddinga, B., Brouwer, K. and Smits, S. (eds.) *EEP Yearbook 1998/99 including Proceedings of the 16th EAZA Conference*, Basel, 7–12 September 1999, EAZA Executive Office, Amsterdam, pp. 482–483.

Duncan, I. J. H. and Filshie, J. H. (1980). The use of radiotelemetry devices to measure temperature and heart rate in domestic fowl. In: Amlaner, C. J. and Macdonald, D. W. (eds.) *A Handbook of Biotelemetry and Radio Tracking*, pp. 579–588. Pergamon Press, Oxford.

Durrell, G. (1976). *Catch Me a Colobus.* Harper Collins Publishers, London.

DZG (2009). *Annual Review 2008/9.* Dudley Zoological Gardens, Dudley.

EAZA (2010). www.eaza.net/activities/cp/Pages/TAGs.aspx (accessed 1 August 2010).

Elsney, R. M., Joanen, T., McNease, L. and Lance, V. (1990). Growth rate and plasma corticosterone levels in juvenile alligators maintained at different stocking densities. *Journal of Experimental Zoology*, 255: 30–36.

Elzanowski, A. and Sergiel, A. (2006). Stereotypic behavior of a female Asiatic elephant (*Elephas maximus*) in a zoo. *Journal of Applied Animal Welfare Science*, 9: 223–232.

Endemol (2005). *Extinct.* Cheetah Television (Endemol UK). ITV1 and WWF. 9–16 December.

Epstein, R., Lanca, R. P. and Skinner, B. F. (1981). 'Self-awareness' in the pigeon. *Science*, 212: 695–696.

Erwin, J. (1979). Aggression in captive macaques: Interaction of social and spatial factors. In: Erwin, J., Maple, T. and Mitchell, G. (eds.) *Captivity and Behavior: Primates in Breeding Colonies, Laboratories and Zoos*, pp. 139–171. Van Nostrand Reinhold, New York.

Evans, L. T. and Quaranta, J. V. (1951). A study of the social behavior of a captive herd of tortoises. *Zoologica* (New York), 36: 171–181.

Ewer, R. F. (1968). *Ethology of Mammals.* Elek Science, London.

EZNC (2010). *European Zoo Nutrition Centre Feeding Guidelines.* www.eznc.org/PrimoSite/show.do?ctx=7795,21671&anav=21550#21698 (accessed 16 July 2010).

Faust, L. J. and Bier, L. (2008). *PopLink 1.3: User's Manual.* Lincoln Park Zoo, Chicago, IL.

Fell, L. R. and Shutt, D. A. (1986). Adrenal responses of calves to transport stress as measured by salivary cortisol. *Canadian Journal of Animal Science*, 66: 637–641.

Fisken, F. A. (ed.) (2007). *International Zoo Yearbook. Volume 41: Animal Health and Conservation.* Zoological Society of London, London.

Fisken, F. A. (ed.) (2010). Zoos and aquariums of the world. *International Zoo Yearbook*, 44: 251–431.

Fitter, R. (1986). *Wildlife for Man: How and Why We Should Conserve Our Species.* Collins, London.

FOE (2010). www.foe.co.uk/resource/faqs/fund_how_many.html (accessed 27 January 2010).

Foose, T. J. (1983). The relevance of captive populations to the conservation of biodiversity. In: Schonewald-Cox, C. M., Chambers, S. M., MacBryde, B. and Thomas, W. L. (eds.) *Genetics and Conservation: A Reference for Managing Wild Animal and Plant Populations*, pp. 374–401. The Benjamin/Cummings Publishing Company, Inc., Menlo Park, CA.

Fowler, G. S., Wingfield, J. C. and Boersma, P. D. (1995). Hormonal and reproductive effects of low levels of petroleum fouling in Magellanic penguins (*Spheniscus magellanicus*). *Auk*, 112: 382–389.

Fowler, M. E. (1993). Foot care in elephants. In: Fowler, M. E. (ed.) *Zoo and Wild Animal Medicine*, 3rd edn., pp. 448–453. W. B. Saunders Company, Philadelphia, PA.

Francis, D., Esson, M. and Moss, A. (2007). Following visitors and what it tells us: the use of visitor tracking to evaluate 'Spirit of the Jaguar' at Chester Zoo. *International Zoo Educators Journal*, 43: 20–24.

Frankham, R., Ballou, J. D. and Brisoe, D. A. (2002). *Introduction to Conservation Genetics*. Cambridge University Press, Cambridge, UK.

Fraser, D. and Weary, D. M. (2005). Applied animal behaviour and animal welfare. In: Bolhuis, J. J. and Giraldeau, L. (eds.) *The Behavior of Animals: Mechanisms, Function, and Evolution*, pp. 346–366. Blackwell Publishing Ltd., Oxford.

Freeman, C. (2009). Ending extinction: The quagga, the thylacine and the "smart human". In: Gigliotti, C. (ed.) *Leonardo's Choice: Genetic Technologies and Animals*, pp. 235–256. Springer, The Netherlands.

Freyfogle, E. T. and Goble, D. D. (2009). *Wildlife Law: A Primer*. Island Press, Washington DC.

GAA (2008). *Global Amphibian Assessment*. IUCN. www.redlist.org/initiatives/amphibians/ (accessed 18 July 2010).

Gage, L. J. (2002). *Hand-Rearing Wild and Domestic Animals*. Wiley-Blackwell, Oxford.

Gallup Jr, G. G. (1970). Chimpanzees: Self-recognition. *Science*, 167: 86–87.

Gans, C. and Mix, H. (1974). A sequential insect dispenser for behavioral experiments. *BioScience*, 24: 88–89.

Gascon, C., Collins, J. P., Moore, R. D., Church, D. R., McKay, J. E. and Mendelson, J. R. III (eds.) (2007). *Amphibian Conservation Action Plan*, p. 64. IUCN/SSC Amphibian Specialist Group, Gland, Switzerland and Cambridge, UK.

GBO-3 (2010). *Global Biodiversity Outlook 3*. Secretariat of the UN Convention on Biological Diversity, Montreal.

Geertsema, A. A. (1985). Aspects of the ecology of the serval (*Leptailurus serval*) in the Ngorongoro Crater, Tanzania. *Netherlands Journal of Zoology*, 35: 527–610.

GEF (2007). *Fertile Ground: Seeding National Actions for the Global Environment*. Global Environment Facility Report 2007. GEF, Washington DC.

Goodall, J. (1971). *In the Shadow of Man*. William Collins & Sons Co. Ltd., Glasgow.

Goodall, J. (1986). *The Chimpanzees of Gombe: Patterns of Behavior*. Harvard University Press, Cambridge, MA.

Gordon, L. (2010). Man jailed for rhino horn smuggling bid in Manchester. www.bbc.co.uk/news/uk-11474166 (accessed 30 October 2010).

Gozalo, A. and Montoya, E. (1991). Mortality causes of the moustached tamarin (*Saguinus oedipus*) in captivity. *Journal of Medical Primatology*, 21: 35–38.

Grandin, T. (1997). Assessment of stress during handling and transport. *Journal of Animal Science*, 75: 249–257.

Greenwood, A. G. (1983). Avian sex determination by laparoscopy. *Veterinary Record*, 112: 105.

Gregor, M. (ed.) (1998). *Kant: Groundwork of the Metaphysics of Morals*. Cambridge University Press, Cambridge, UK.

Gruber, T. M., Friend, T. H., Gardner, J. M., Packard, J. M., Beaver, B. and Bushong, D. (2000). Variation in stereotypic behavior related to restraint in circus elephants. *Zoo Biology*, 19: 209–221.

Grützner, F., Rens, W., Tsend-Ayush, E. *et al.* (2004). In the platypus a meiotic chain of ten sex chromosomes shares genes with the bird Z and mammal X chromosomes. *Nature*, 432: 913–917.

Grzimek, B. and Grzimek, M. (1964). *Serengeti Shall Not Die*. William Collins & Sons Co. Ltd., Glasgow.

Gupta, B. K. (2008). *Barrier Designs for Zoos*. Central Zoo Authority, Ministry of Environment & Forests, India (available from www.cza.nic.in).

Guggisberg, C. A. W. (1975). *Wild Cats of the World*. David & Charles, London.

Gusset, M. and Dick, G. (2010). 'Building a Future for Wildlife'? Evaluating the contribution of the world zoo and aquarium community to *in-situ* conservation. *International Zoo Yearbook*, 44: 183–191.

Haas, G. (1958). 24-Stunden-Periodik von Grosskatzen im Zoologischen Garten. *Säugetierk. Mitt.* 6: 113–117.

Hambler, C. (2004). *Conservation*. Cambridge University Press, Cambridge, UK.

Hancocks, D. (1996). The design and use of moats and barriers. In: Kleiman, D. G., Allen, M. E., Thompson, K. V. and Lumpkin, S. (eds.) *Wild Mammals in Captivity: Principles and Techniques*, pp. 191–203. University of Chicago Press, Chicago, IL.

Handwerk, B. (2006). Panda "porn" to boost mating efforts at Thai zoo. http://news.nationalgeographic.com/news/2006/11/061113-panda-mate.html?source=rss (accessed 18 June 2010).

Hardy, D. F. (1996). Current research activities in zoos. In: Kleiman, D. G., Allen, M. E., Thompson, K. V. and Lumpkin, S. (eds.) *Wild Mammals in Captivity: Principles and Techniques*, pp. 531–536. University of Chicago Press, Chicago, IL.

Harley, E. H., Knight, M. H., Lardner, C., Wooding, B. and Gregor, M. (2009). The Quagga Project: progress over 20 years of selective breeding. *South African Journal of Wildlife Research*, 39: 155–163.

Harris, C. L. (1992). *Concepts in Zoology*. Harper Collins Publishers Inc., New York.

Harris, M., Sherwin, C. and Harris, S. (2008). *The Welfare, Housing and Husbandry of Elephants in UK Zoos*. Final Report, 10 November. University of Bristol, Bristol.

Hayes, M. P., Jenning, M. R. and Mellin, J. D. (1998). Beyond mammals: Environmental enrichment for amphibians and reptiles. In: Shepherdson, D. J., Mellin, J. D. and Hutchins, M. (eds.) *Second Nature: Environmental Enrichment for Captive Animals*, pp. 205–235. Smithsonian Institution Press, Washington DC and London.

Hediger, H. (1950). *Wild Animals in Captivity*. Butterworth Scientific Publications Ltd., London.

Hemsworth, P. H. and Coleman, G. J. (1998). *Human–Livestock Interactions: The Stockperson and the Productivity and Welfare of Intensively Farmed Animals*. CAB International, London.

Hemsworth, P. H., Price, E. O. and Borgwardt, R. (1996). Behavioural responses of domestic pigs and cattle to humans and novel stimuli. *Applied Animal Behaviour Science*, 50: 43–56.

Heywood, V. H. (ed.) (1995). *Global Biodiversity Assessment*. United Nations Environment Programme. Cambridge University Press, Cambridge, UK.

Hildebrandt, T. B., Göritz, F., Pratt, N. C. *et al.* (2000a). Ultrasonography of the urogenital tract in elephants (*Loxodonta africana* and *Elephas maximus*): an important tool for assessing female reproductive function. *Zoo Biology*, 19: 321–332.

Hildebrandt, T. B., Hermes, R., Pratt, N. C. *et al.* (2000b). Ultrasonography of the urogenital tract in elephants (*Loxodonta africana* and *Elephas maximus*): an important tool for assessing male reproductive function. *Zoo Biology*, 19: 333–345.

Hinshaw, K. C., Amand, W. B. and Tinkelman, C. L. (1996). Preventative medicine. In: Kleiman, D. G., Allen, M. E., Thompson, K. V. and Lumpkin, S. (eds.) *Wild Mammals in Captivity: Principles and Techniques*, pp. 16–24. University of Chicago Press, Chicago, IL.

Hirshi, K. and Screven, C. (1990). Effects of questions on visitor reading behavior. *ILVS Review: A Journal of Visitor Behavior*, 1: 50–61.

HMSO (1965). *Report of the Technical Committee to Enquire into the Welfare of Animals kept under Intensive Livestock Husbandry Systems*. HMSO, London.

Hoage, R. J. and Deiss, W. A. (eds.) (1996). *New Worlds, New Animals: From Menagerie to Zoological Park in the Nineteenth Century*. The Johns Hopkins University Press, Baltimore, MD and London.

Hoage, R. J., Roskell, A. and Mansour, J. (1996). Menageries and zoos to 1900. In: Hoage, R. J. and Deiss, W. A. (eds.) *New Worlds, New Animals: From Menagerie to Zoological Park in the Nineteenth Century*, pp. 8–18. The Johns Hopkins University Press, Baltimore, MD and London.

Hodges, J. K. (1996). Determining and manipulating female reproductive parameters. In: Kleiman, D. G., Allen, M. E., Thompson, K. V. and Lumpkin, S. (eds.) *Wild Mammals in Captivity: Principles and Techniques*, pp. 418–428. University of Chicago Press, Chicago, IL.

Hollén, L. I. and Manser, M. B. (2007). Persistence of alarm-call behaviour in the absence of predators: a comparison between wild and captive-born meerkats (*Suricata suricatta*). *Ethology*, 113: 1038–1047.

Howard, J. G., Bush, M., de Voss, V. and Wildt, D. E. (1989). Electroejaculation, semen characteristics and serum testosterone concentration of free ranging African elephants (*Loxodonta africana*). *Journal of Reproduction and Fertility*, 72: 187–195.

Hrdy, S. B. (1977). *The Langurs of Abu: Female and Male Strategies of Reproduction*. Harvard University Press, Cambridge, MA.

Hutchins, M. (2006). Variation in nature: its implications for zoo elephant management. *Zoo Biology*, 25: 161–171.

Hutchins, M. and Ballentine, J. (2001). Fueling the conservation engine: fund-raising and public relations. In: Conway, W. G., Hutchins, M., Souza, M., Kapetanakos, Y. and Paul, E. (eds.) *The AZA Field Conservation Resource Guide*, pp. 268–271. Wildlife Conservation Society and Zoo, Atlanta, GA.

Hutchins, M. and Keele, M. (2006). Elephant importation from range countries: ethical considerations for accredited zoos. *Zoo Biology*, 25: 219–233.

Hutchins, M. and Souza, M. (2001). AZA's conservation endowment fund: Zoos and aquariums supporting conservation action. In: Conway, W. G., Hutchins, M., Souza, M., Kapetanakos, Y. and Paul, E. (eds.) *The AZA Field Conservation Resource Guide*, pp. 291–302. Wildlife Conservation Society and Zoo, Atlanta, GA.

Hutchins, M. and Tilson, R. (1993). What are modern, professionally-managed zoological parks and aquariums doing for in situ conservation? *Annual Conference of the Society for Conservation Biology*, Arizona State University, Tempe, AZ.

Hyson, J. (2000). Jungles of Eden: the design of American zoos. In: Conan, M. (ed.) *Environmentalism in Landscape Architecture*, pp. 23–44. Harvard University Press, Washington DC.

IATA (2006). *IATA Live Animal Regulations*, 33rd edn. International Air Transport Association.

ILAR (1998). *The Psychological Well-Being of Nonhuman Primates*. Committee on Well-Being of Nonhuman Primates, Institute for Laboratory Animal Research, Commission on Life Sciences, National Research Council, National Academy Press, Washington DC.

Inglis, I. R. and Fergusson, N. J. K. (1986). Starlings search for food rather than eat freely-available, identical food. *Animal Behaviour*, 34: 614–617.

Ings, R., Waran, N. K. and Young, R. J. (1997). Attitude of zoo visitors to the idea of feeding live prey to zoo animals. *Zoo Biology*, 16: 343–347.

Inskipp, T. and Wells, S. (1979). *International Trade in Wildlife*. Earthscan, London.

Ipsos MORI (2010). *The Political State of Play: The Current Political Scene going into General Election 2010*. April 2010. Ipsos Mori, London.

Ironmonger, J. (1992). *The Good Zoo Guide*. Harper Collins Publishers, London.

Itoh, K., Ide, K., Kojima, Y. and Terada, M. (2010). Hibernation exhibit for Japanese black bear *Ursus thibetanus japonicus* at Ueno Zoological Gardens. *International Zoo Yearbook*, 44: 55–64.

IUCN (1980). *World Conservation Strategy*. International Union for the Conservation of Nature and Natural Resources, Gland, Switzerland.

IUCN (1995). *IUCN/SSC Guidelines for Re-introductions.* SSC Re-introduction Specialist Group. www.iucnsscrsg.org/policy_guidelines.php (accessed 21 October 2010).

IUCN (2001). *IUCN Red List Categories and Criteria, Version 3.1.* IUCN Species Survival Commission, Gland, Switzerland and Cambridge, UK.

IUCN (2002). *IUCN Technical Guidelines on the Management of Ex-situ Populations for Conservation.* IUCN, Gland, Switzerland.

IUCN (2009). *IUCN Red List of Threatened Species, Version 2009.1.* www.iucnredlist.org (accessed 31 July 2009).

IUCN (2010). *Branta sandvicensis.* In: *IUCN Red List of Threatened Species, Version 2010.1.* www.iucnredlist.org (accessed 27 April 2010).

IUDZG (1993). *The World Zoo Conservation Strategy: The Role of Zoos and Aquaria of the World in Global Conservation.* International Union of Directors of Zoological Gardens. Chicago Zoological Society, Brookfield, IL.

Jackson, D. W. (1996). Horticultural philosophies in zoo exhibit design. In: Kleiman, D. G., Allen, M. E., Thompson, K. V. and Lumpkin, S. (eds.) *Wild Mammals in Captivity: Principles and Techniques,* pp. 175–179. University of Chicago Press, Chicago, IL.

Jalil, J. (2004). Gorilla escape confounds US zoo. BBC News. http://news.bbc.co.uk/1/hi/world/americas/3551427.stm (accessed 4 August 2010).

Jamieson, D. (1986). Against zoos. In: Singer, P. (ed.) *In Defense of Animals,* pp. 108–177. Harper and Row, New York.

Janofsky, M. (1996). Philadelphia Zoo fire is tied to heating cable. *The New York Times,* 14 March. www.nytimes.com/1996/03/14/us/philadelphia-zoo-fire-is-tied-to-heating-cable.html?pagewanted=1 (accessed 28 April 2010).

Jarvis, C. and Morris, D. (eds.) (1960). *International Zoo Yearbook. Volume 2: Elephants, Hippopotamuses and Rhinoceroses in Captivity.* Zoological Society of London, London.

Jefferies, M. J. (1997). *Biodiversity and Conservation.* Routledge, London and New York.

Jenny, S. and Schmid, H. (2002). Effect of feeding boxes on the behavior of stereotyping Amur tigers (*Panthera tigris altaica*) in the Zurich Zoo, Zurich, Switzerland. *Zoo Biology,* **21**: 573–584.

Jensen, R. G. (ed.) (1995). *Handbook of Milk Composition.* Academic Press, San Diego, CA.

Jensvold, M. L. B. (2008). Chimpanzee (*Pan troglodytes*) responses to caregiver use of chimpanzee behaviors. *Zoo Biology,* 27: 345–359.

Johnson, B. R. (1991). Conservation of threatened amphibians: The integration of captive breeding and field research. In: Staub, R. E. (ed.) *Proceedings of the Conference on Captive Propagation and Husbandry of Reptiles and Amphibians, Special Publication* 6, pp. 33–38. Northern California Herpetological Society, Davis, CA.

Johnston, R. J. (1998). Exogenous factors and visitor behavior: a regression analysis of exhibit viewing time. *Environment and Behavior,* 30: 322–347.

Jolly, L. (2003). *Giraffe Husbandry Manual.* Australian Regional Association of Zoological Parks and Aquaria, Mosman, NSW.

Jones, B. and McGreevy, P. (2007). How much space does an elephant need? The impact of confinement on animal welfare. *Journal of Veterinary Behavior,* 2: 185–187.

Jones, G. R. (1982). Design principles for the presentation of animals and nature. In: *AAZPA Annual Conference Proceedings,* pp. 184–192. American Association of Zoological Parks and Aquariums, Wheeling, WV.

Jones, M. L. (1962). Mammals in captivity – primate longevity. *Laboratory Primate Newsletter,* 1: 3–13.

Jule, K. R., Leaver, L. A. and Lea, S. E. G. (2008). The effects of captive experience on reintroduction survival in carnivores: A review and analysis. *Biological Conservation,* 141: 355–363.

Juniper, P. (2000). The management of Asian elephants (*Elephas maximus*) at Port Lympne Wild Animal Park. *Ratel,* 27: 168–175.

Kamberg, M. (1989). Pet prescription. *Current Health,* 15: 10–12.

Karstad, L. and Sileo, L. (1971). Causes of death in captive wild waterfowl in the Kortright Waterfowl Park, 1967–1970. *Journal of Wildlife Diseases,* 7: 236–241.

Kashiwayanagi, M. (2003). Chemical communication via high molecular weight pheromones in mammals. *Journal of Biological Macromolecules,* 3: 83–88.

Kellert, S. R. (1984). Urban American perceptions of animals and the natural environment. *Urban Ecology,* 8: 209–228.

Kellert, S. R. and Dunlap, J. (1989). *Informal Learning at the Zoo: A Study of Attitude and Knowledge Impacts.* Zoological Society of Philadelphia, Philadelphia, PA.

Kepler, C. B. (1978). Captive propagation of whooping cranes: a behavioural approach. In: Semple, S. A. (ed.) *Endangered Birds: Management Techniques for Preserving Threatened Species,* pp. 231–241. University of Wisconsin Press, Madison, WI.

Kertesz, P. (1993). *A Colour Atlas of Veterinary Dentistry and Oral Surgery.* Wolfe Publishing, Aylesbury.

King, T. J. (1980). *Ecology.* Thomas Nelson & Sons Ltd., Walton-on-Thames, Surrey.

Kingdon, J. (1997). *The Kingdon Field Guide to African Mammals.* A & C Black Publishers Ltd., London.

Kisling, V. N. (1996). The origin and development of American zoological parks to 1899. In: Hoage, R. J. and Deiss, W. A. (eds.) *New Worlds, New Animals: From Menagerie to Zoological Park in the Nineteenth Century,* pp. 109–125. The Johns Hopkins University Press, Baltimore, MD and London.

Kitchener, A. C. (2004). The problem of old bears in zoos. *International Zoo News,* 51: 282–293.

Kleiman, D., Beck, B., Baker, A., Ballou, J., Dietz, L. and Dietz, J. (1990). The conservation program for the golden lion tamarin, *Leontopithecus rosalia. Endangered Species UPDATE* 8(1).

Kohler, I. V., Preston, S. H. and Lackey, L. B. (2006). Comparative mortality levels among selected species of captive animals. *Demographic Research,* 15: 413–434.

Kreger, M. D., Hutchins, M. and Fascione, N. (1998). Context, ethics and environmental enrichment in zoos and aquariums. In: Shepherdson, D. J., Mellen, J. D. and Hutchins, M. (1998). *Second Nature: Environmental Enrichment for Captive Animals*, pp. 59–82. Smithsonian Institution Press, Washington DC and London.

Kurt, F. (1995). Asian elephants (*Elephas maximus*) in captivity and the role of captive propagation for maintenance of the species. In: Spooner, N. G. and Whitear, J. A. (eds.) *Proceedings of the Eighth UK Elephant Workshop*, 21 September 1994, pp. 69–96. North of England Zoological Society, Chester Zoo, Chester.

Lankard, J. R. (ed.) (2001). *AZA Annual Report on Conservation and Science 1999–2000. Volume III: Member Institution Publications*. American Zoo and Aquarium Association, Silver Spring, MD.

Lanthier, C. (1995). Successful breeding in a small colony of greater flamingos. *Avicultural Magazine*, 101: 52–57.

Larter, P. (2008). Irwin hospital flouted law by releasing its rescued koalas too far from home. *The Times*, 15 March, p. 58.

Latham, N. R. and Mason, G. J. (2008). Maternal deprivation and the development of stereotypic behaviour. *Applied Animal Behaviour Science*, 110: 84–108.

Laurenson, M. K. (1993). Early maternal behavior of wild cheetahs: implications for captive husbandry. *Zoo Biology*, 12: 31–43.

Law, C. (ed.) (2010). *Jaguar Species Survival Plan: Guidelines for Captive Management of Jaguars*. Association of Zoos and Aquariums. www.jaguarssp.org/Animal%20Mgmt/JAGUAR%20HUSBANDRY%20MANUAL.pdf (accessed 25 May 2010; document undated).

Leakey, R. and Lewin, R. (1996). *The Sixth Extinction: Patterns of Life and the Future of Humankind*. Anchor Books, New York.

Lees, C. and Barlow, S. (2008). Collection planning in Australia. *EAZA News*, 64: 27.

Lehnhardt, J. (1991). Elephant handling: A problem of risk management and resource allocation. In: *AAZPA Regional Conference Proceedings*, pp. 569–575. American Association of Zoological Parks and Aquariums, Wheeling, WV.

Leong, K. M., Terrell, S. P. and Savage, A. (2004). Causes of mortality in captive cotton-top tamarins (*Saguinus oedipus*). *Zoo Biology*, 23: 127–137.

Levine, S. (1983). Coping: an overview. In: Ursin, H. and Murison, R. (eds.) *Biological and Psychological Basis of Psychosomatic Disease*, pp. 15–26. Pergamon Press, Oxford.

Lewis, L. (2008). Pandas moved out amid fears over flood. *The Times*, 24 May, p. 47.

Leyhausen, P. (1979). *Cat Behavior: The Predatory and Social Behavior of Domestic and Wild Cats*. Garland Press, New York.

Li, C., Jiang, Z., Tang, S. and Zeng, Y. (2007). Influence of enclosure size and animal density on fecal cortisol concentration and aggression in Père David's deer stags. *General and Comparative Endocrinology*, 151(2): 202–209.

Lindburg, D. G. and Fitch-Snyder, H. (1994). Use of behavior to evaluate reproductive problems in captive mammals. *Zoo Biology*, 13: 433–445.

Litchfield, P. (2005). Leaders and matriarchs – a new look at elephant social hierarchies. *International Zoo News*, 52: 338–339.

Litwak, J. (1996). Visitors learn more from labels that ask questions. *Current Trends in Audience Research*, 10: 40–50.

Loisel, G. (1912). *Histoire des ménageries de l'antiquité à nos jours* (3 vols.). Octave Doin et Fils and Henri Laurens, Paris.

Loomis, R. (1987). *Museum Visitor Evaluation: New Tool for Management*. American Association for State and Local History, Nashville, TN.

Lundrigan, B. (1996). Standard methods for measuring mammals. In: Kleiman, D. G., Allen, M. E., Thompson, K. V. and Lumpkin, S. (eds.) *Wild Mammals in Captivity: Principles and Techniques*, pp. 566–570. University of Chicago Press, Chicago, IL.

Luo, S. J., Kim, J. H., Johnson, W. E. *et al.* (2004). Phylogeography and genetic ancestry of tigers (*Panthera tigris*). *PLoS Biology*, 2: 2275–2293.

Macdonald, D. (1984). *The Encyclopedia of Mammals*. Andromeda Oxford Ltd., Abingdon, UK.

Macdonald, D. W. (1980). *Rabies and Wildlife: A Biologist's Perspective*. Oxford University Press, Oxford.

Macdonald, D. W. (1983). The ecology of carnivore social behaviour. *Nature*, 301: 379–384.

MacQuarrie, B. and Belkin, D. (2003). Franklin Park gorilla escapes, attack 2. *The Boston Globe*. www.boston.com/news/local/massachusetts/articles/2003/09/29/franklin_park_gorilla_escapes_attacks_2 (accessed 4 August 2010).

Mallapur, A., Qureshi, Q. and Chellam, R. (2002). Enclosure design and space utilization by Indian leopards (*Panthera leo*) in four zoos in Southern India. *Journal of Applied Animal Welfare Science*, 5: 111–124.

Manning, A. (1972). *An Introduction to Animal Behaviour*, 2nd edn. Edward Arnold (Publishers) Ltd., London.

Mar, K. U., Maung, M., Thein, M., Khaing, A. T., Tun, W. and Nyunt, T. (1995). Electroejaculation and semen characteristics in Myanmar timber elephants. In: Daniel, J. C. and Datye, H. (eds.) *A Week with Elephants: Proceedings of the International Seminar on the Conservation of Asian Elephants*, pp. 473–482. Bombay Natural History Society. Oxford University Press, Oxford.

Marcellini, D. L. and Jenssen, T. A. (1988). Visitor behavior in the National Zoo's reptile house. *Zoo Biology*, 7: 329–338.

Margodt, K. (2000). *The Welfare Ark: Suggestion for a Renewed Policy in Zoos*. VUB University Press, Brussels.

Markowitz, H. (1982). *Behavioral Enrichment at the Zoo*. Van Nostrand Reinhold, New York.

Marmie, W., Kuhn, S. and Chizar, D. (1990). Behavior of captive-raised rattlesnakes (*Crotalus enyo*) as a function of rearing conditions. *Zoo Biology*, 9: 241–246.

Marshall, T. C. and Spalton, J. A. (2000). Simultaneous inbreeding and outbreeding depression in reintroduced Arabian oryx. *Animal Conservation*, 3: 241–248.

Marten, K. and Psarakos, S. (1995). Using self-view television to distinguish between self-examination and social behaviour in the bottlenose dolphin (*Tursiops truncatus*). *Consciousness and Cognition*, 4: 205–224.

Martin, P. and Bateson, P. (1993). *Measuring Behaviour: An Introductory Guide*, 2nd edn. Cambridge University Press, Cambridge, UK.

Martin, P. S. (1971). Prehistoric overkill. In: Detwyler, T. R. (ed.) *Man's Impact on Environment*, pp. 612–624. McGraw-Hill Book Company, New York.

Mason, G. J. (1991). Stereotypies: a critical review. *Animal Behaviour*, 41: 1015–1037.

Mason, J. W. (1971). A reevaluation of the concept of "non-specificity" in stress theory. *Journal of Psychiatric Research*, 8: 323–333.

Masters, B. (1989). *The Passion of John Aspinall*. Hodder and Stoughton, London.

Masters, N. J., Stidworthy, M. F., Lewis, J. C. M., Greenwood, A. G. and Gopal, R. (2007). First confirmed outbreak of callitrichid Hepatitis/LCMV at a zoo in the UK: Preliminary results. *Proceedings of BVZS Autumn Meeting 2007*, p. 50.

McCarthy, M. (2004). Endangered tiger earns its stripes as the world's most popular beast. *The Independent*, 6 December. http://findarticles.com/p/articles/mi_qn4158/is_20041206/ai_n12814678 (accessed 3 March 2007).

McGregor, P. K. (2005). Communication. In: Bolhuis, J. J. and Giraldeau, L. (eds.) *The Behavior of Animals: Mechanisms, Function, and Evolution*, pp. 226–250. Blackwell Publishing Ltd., Oxford.

McKay, G. M. (1973). *Behavior and Ecology of the Asiatic Elephant in Southeastern Ceylon*. Smithsonian Institution Press, Washington DC.

McKillop, I. G. (No date). *Electric Fence Reference Manual*. Research and Development Surveillance Report 607. DEFRA, London.

McKnight, C. M. and Gutzke, W. H. N. (1993). Effects of the embryonic environment and of hatchling housing conditions on growth of young snapping turtles (*Chelydra serpentina*). *Copeia*, 1993: 474–482.

McLaughlin, R. (1970). *Aspects of the biology of cheetahs Acinonyx jubatus (Schreber) in Nairobi National Park*. MSc thesis, University of Nairobi, Nairobi, Kenya.

McLean, I. G., Lundie-Jenkins, G. and Jarman, P. J. (1995). Teaching an endangered mammal to recognize predators. *Biological Conservation*, 75: 51–62.

McPhee, M. E. (2003). Generations in captivity increases behavioral variance: considerations for captive breeding and reintroduction programs. *Biological Conservation*, 115: 71–77.

McWhirter, N. and McWhirter, R. (1969). *The Guinness Book of Records*. Guinness Superlatives Limited, London.

Meffert, L. M., Mukana, N., Hicks, S. K. and Day, S. B. (2005). Testing alternative captive breeding strategies with the subsequent release into the wild. *Zoo Biology*, 24: 375–392.

Mellen, J. D. (1997). *Minimum Husbandry Guidelines for Mammals: Small Felids*. Association of Zoos and Aquariums, Silver Spring, MD.

Meller, C. L., Croney, C. C. and Shepherdson, D. (2007). Effects of rubberized flooring on Asian elephant behavior in captivity. *Zoo Biology*, 26: 51–61.

Mench, J. A. and Kreger, M. D. (1996). Ethical and welfare issues associated with keeping wild mammals in captivity. In: Kleiman, D. G., Allen, M. E., Thompson, K. V. and Lumpkin, S. (eds.) *Wild Mammals in Captivity: Principles and Techniques*, pp. 5–15. University of Chicago Press, Chicago, IL.

Mendel, G. (1866). Versuche über Pflanzenhybriden. *Verhandlungen des naturforschenden Vereines in Brünn, Bd. IV für das Jahr 1865*, Abhandlungen, 3–47.

Menotti-Raymond, M. and O'Brien, S. J. (1993). Dating the genetic bottleneck of the African cheetah. *Proceedings of the National Academy of Sciences of the USA*, 90: 3172–3176.

Mikota, S. K., Larsen, R. S. and Montali, R. J. (2000). Tuberculosis in elephants in North America. *Zoo Biology*, 19: 393–403.

Miller, G. T. (1994). *Living in the Environment: Principles, Connections, and Solutions*. Wadsworth Publishing Company, Belmont, CA.

Mills, L. S. and Smouse, P. E. (1994). Demographic consequences of inbreeding in remnant populations. *American Naturalist*, 144: 412–431.

Montali, R. J., Richman, L. K. and Hildebrandt, T. B. (1998). Highly fatal disease of Asian elephants in North America and Europe is attributed to a newly recognised endotheliotropic herpesvirus. *Elephant Journal*, 1: 3.

Moreira, N., Brown, J. L., Moraes, W., Swanson, W. F. and Monteiro-Filho, E. L. A. (2007). Effect of housing and environmental enrichment on adrenocortical activity, behavior and reproductive cyclicity in the female tigrina (*Leopardus tigrinus*) and margay (*Leopardus wiedii*). *Zoo Biology*, 26: 441–460.

Morgan, J. M. and Gramann, J. H. (1989). Predicting effectiveness of wildlife education programs: a study of students' attitudes and knowledge towards snakes. *Wildlife Society Bulletin*, 17: 501–509.

Morris, D. (1960). Automatic seal feeding apparatus at London Zoo. *International Zoo Yearbook*, 2: 70.

Morris, D. (1990). *The Animal Contract*. Virgin Books, London.

Moss, A. and Esson, E. (2010). Visitor interest in zoo animals and the implications for collection planning and zoo education programmes. *Zoo Biology*, 28: 1–17.

Moss, C. (1989). *Elephant Memories: Thirteen Years in the Life of an Elephant Family*. William Collins & Sons Co. Ltd., Glasgow.

Mullin, B. and Marvin, G. (1999). *Zoo Culture*. University of Illinois Press, Urbana and Chicago, IL.

Myers, N., Mittermeier, R. A., Mittermeier, C. G., da Fonseca, G. A. B. and Kent, J. (2000). Biodiversity hotspots for conservation priorities. *Nature*, 403: 853–858.

Nadler, R. D. (1980). Reproductive physiology and behaviour of gorillas. *Journal of Reproduction and Fertility, Supplement*, 28: 79–89.

Nash, C. E. (ed.) (1991). *Production of Aquatic Animals, Crustaceans, Mollusks, Amphibians and Reptiles*. Elsevier, Amsterdam.

Nechay, G. (1996). Editorial. *Naturopa*, 82: 3.

Nelson, J. S. (1994). *Fishes of the World*. Wiley-Interscience, New York.

Neuringer, A. J. (1969). Animals respond to food in the presence of free food. *Science*, 166: 399–401.

New, T. R. (1995). *Introduction to Invertebrate Conservation Biology*. Open University Press, Oxford.

Nicholls, R. (1992). *The Belle Vue Story*. Neil Richardson (Publisher), Radcliffe, Manchester.

Niewold, F. J. (1976). Aspecten van het sociale leven van de vos. *Overdruk van Natura*, 73: 1–8.

Nissani, M. and Hoefler-Nissani, D. (2007). Absence of mirror self-referential behavior in two Asian elephants. *Journal of Veterinary Science*, 1(1). www.scientificjournals.org/journals2007/articles/1043.htm (accessed 10 March 2010).

Nowak, R. M. (1991). *Walker's Mammals of the World*, 5th edn. The Johns Hopkins University Press, Baltimore, MD and London.

Nowak, R. M. (1999). *Walker's Mammals of the World*, 6th edn. The Johns Hopkins University Press, Baltimore, MD and London.

Nowak, R. M. and Federoff, N. E. (1998). Validity of the red wolf: Response to Roy *et al. Conservation Biology*, 12: 722–725.

Nyahongo, J. W. (2007). Flight initiation distances of five herbivores to approaches by vehicles in the Serengeti National Park, Tanzania. *African Journal of Ecology*, 46: 227–229.

Nyhuis, A. W. and Wassner, J. (2008). *America's Best Zoos: A Travel Guide for Fans and Families*. The Intrepid Traveler, Branford, CT.

Ödberg, F. O. (1978). Abnormal behavior: stereotypies. In: *Proceedings of the First World Congress of Ethology Applied to Zootechnics*, Madrid, Industrias Grafices Espana, pp. 475–480.

Olney, P. J. S. (ed.) (2005). *Building a Future for Wildlife: The World Zoo and Aquarium Conservation Strategy*. World Association of Zoos and Aquariums. WAZA Executive Office, Bern, Switzerland.

Olney, P. J. S., Ellis, P. and Fiskin, F. A. (1994). Census of rare animals in captivity. *International Zoo Yearbook*, 33: 408–453.

Paine, R. T. (1969). A note on trophic complexity and community structure. *American Naturalist*, 103: 91–93.

Parker, P. G. and Waite, T. A. (1997). Mating systems, effective population size and conservation of natural populations. In: Clemmons, J. R. and Buchholz, R. (eds.) *Behavioural Approaches to Conservation in the Wild*, pp. 243–261. Cambridge University Press, Cambridge, UK.

Parr, L. A. and Waller, B. M. (2006). Understanding chimpanzee facial expression: insights into the evolution of communication. *Social Cognitive and Affective Neuroscience*, 1: 221–228.

Patrick, P. G., Tunnicliffe, S. D., Matthews, C. E. and Ayers, D. F. (2007). Mission statements of AZA-accredited zoos: do they say what we think they say? *International Zoo News*, 54: 90–98.

Pelletier, F., Hogg, J. T. and Festa-Bianchet, M. (2004). Effect of chemical immobilization on social status of bighorn rams. *Animal Behaviour*, 67: 1163–1165.

Penning, M., Reid, G. McG., Koldewey, H. *et al.* (eds.) (2009). *Turning the Tide: A Global Aquarium Strategy for Conservation and Sustainability*. WAZA, Bern, Switzerland.

Pérez-Garnelo, S. S., Garde, J., Pintado, B., Borque, C., Talavera, C., Delclaux, M., López, M. and Martinez, J. (2004). Characteristics and in vitro fertilizing ability of giant panda (*Ailuropoda melanoleuca*) frozen-thawed epididymal spermatozoa obtained 4 hours post-mortem: A case report. *Zoo Biology*, 23: 279–285.

Peters, J. L., Mayr, E., Greenway, J. C. *et al.* (1931–87). *Checklist of Birds of the World. Volumes 1–16*. Museum of Comparative Zoology, Cambridge, MA.

Plotnik, J. M., de Waal, F. B. M. and Reiss, D. (2006). Self-recognition in an Asian elephant. *Proceedings of the National Academy of Sciences of the USA*, 103: 17053–17057.

Pluháček, J. and Bartoš, L. (2005). Further evidence for male infanticide and feticide in captive plains zebra, *Equus burchelli*. *Folia Zoologica*, 54: 258–262.

Polakowski, K. J. (1987). *Zoo Design: The Reality of Wild Illusions*. University of Michigan, School of Natural Resources, Ann Arbor, MI.

Povinelli, D. J. (1989). Failure to find self-recognition in Asian elephants (*Elephas maximus*) in contrast to their use of mirror cues to discover hidden food. *Journal of Comparative Psychology*, 103: 122–131.

Povinelli, D. J. and Preuss, T. M. (1995). Theory of mind: evolutionary history of a cognitive specialization. *Trends in Neurosciences*, 18: 418–424.

Powell, D. M., Carlstead, K., Tarou, L. R., Brown, J. L. and Monfort, S. L. (2006). Effects of construction noise on behavior and cortisol levels in a pair of captive giant pandas (*Ailuropoda melanoleuca*). *Zoo Biology*, 25: 391–408.

Price, E. C., Ashmore, L. A. and McGivern, A. (1994). Reactions of zoo visitors to free-ranging monkeys. *Zoo Biology*, 13: 355–373.

Prior, H., Schwarz, A. and Güntürkün, O. (2008). Mirror-induced behavior in the magpie (*Pica pica*): evidence of self-recognition. *Public Library of Science Biology*, 6(8): e202. doi:10.1371/journal.pbio.0060202.

Pullar, T. (1996). *Kea (Nestor notabilis) Captive Management Plan and Husbandry Manual*. Threatened Species Occasional Publication No. 9. Department of Conservation, Threatened Species Unit, Wellington, New Zealand.

Purse, B. V., Mellor, P. S., Rogers, D. J., Samuel, A. R., Mertens, P. P. C. and Baylis, M. (2005). Climate change and the recent emergence of bluetongue in Europe. *Nature Reviews Microbiology*, 3: 171–181.

Ralls, K., Brugger, K. and Ballou, J. (1979). Inbreeding and juvenile mortality in small populations of ungulates. *Science*, 206: 1101–1103.

Ramey II, R. B., Luikart, G. and Singer, F. J. (2000). Genetic bottlenecks resulting from restoration efforts: the case of bighorn sheep in Badlands National Park. *Restoration Ecology*, 8: 85–90.

Ramsey, M. A. and Stirling, I. (1988). Reproductive biology and ecology of female polar bears (*Ursus maritimus*). *Journal of Zoology*, 214: 601–634.

Randerson, J. (2003). Wide-roaming carnivores suffer most in zoos. *New Scientist*, 1 October. www.newscientist.com/article/dn4221-wideroaming-carnivores-suffer-most-in-zoos.html (accessed 27 May 2008).

Ratto, M. H., Huanca, W., Singh, J. and Adams, J. P. (2005). Local versus systemic effect of ovulation-inducing factor of the seminal plasma of alpacas. *Reproductive Biology and Endocrinology*, 3: 29.

Raup, D. M. (1986). Biological extinction in early history. *Science*, 231: 1528–1533.

Raup, D. M. and Boyajian, G. E. (1988). Patterns of generic extinction in the fossil record. *Palaeobiology*, 14: 109–125.

Rayers, R. (2009). Exotic species. In: Williams, J. (ed.) *The Complete Textbook of Animal Health and Welfare*, pp. 301–325. Saunders Elsevier, London.

Reed, D. H., O'Grady, J. J., Brook, B. W., Ballou, J. D. and Frankham, R. (2003). Estimates of minimum viable population sizes for vertebrates and factors influencing those estimates. *Biological Conservation*, 113: 23–34.

Rees, P. A. (1977). *Some aspects of the feeding ecology of the African elephant (Loxodonta africana africana, Blumenbach 1797) in captivity*. Unpublished BSc dissertation, University of Liverpool.

Rees, P. A. (1982a). Gross assimilation efficiency and food passage time in the African elephant. *African Journal of Ecology*, 20: 193–198.

Rees, P. A. (1982b). *The ecology and management of feral cat colonies*. Unpublished PhD thesis, University of Bradford.

Rees, P. A. (2000a). The introduction of a captive herd of Asian elephants (*Elephas maximus*) to a novel area. *Ratel*, 27: 120–126.

Rees, P. A. (2000b). Are elephant enrichment studies missing the point? *International Zoo News*, 47: 369–371.

Rees, P. A. (2001a). Captive breeding of Asian elephants (*Elephas maximus*): the importance of producing socially competent animals. In: Hossetti, B. B. and Venkateshwarlu, M. (eds.) *Trends in Wildlife Biodiversity, Conservation and Management*, 1: 76–91. Daya Publishing House, Delhi, India.

Rees, P. A. (2001b). Is there a legal obligation to reintroduce animal species into their former habitats? *Oryx*, 35(3): 216–223.

Rees, P. A. (2001c). The history of the National Elephant Centre, Chester Zoo. *International Zoo News*, 48: 170–183.

Rees, P. A. (2002a). *Urban Environments and Wildlife Law: A Manual for Sustainable Development*. Blackwell Science, Oxford.

Rees, P. A. (2002b). Asian elephants (*Elephas maximus*) dust bath in response to an increase in environmental temperature. *Journal of Thermal Biology*, 27: 353–358.

Rees, P. A. (2004a). Some preliminary evidence of the social facilitation of mounting behaviour in the Asian elephant (*Elephas maximus*). *Journal of Applied Animal Welfare Science*, 7: 49–58.

Rees, P. A. (2004b). Low environmental temperature causes an increase in stereotypic behaviour in captive Asian elephants (*Elephas maximus*). *Journal of Thermal Biology*, 29: 37–43.

Rees, P. A. (2004c). Are white lions ambassadors or conservation white elephants? *International Zoo News*, 51: 484–489.

Rees, P. A. (2004d). The Serengeti Visitor Centre – a model of cooperation between a zoological society and a national park. *International Zoo News*, 51: 94–97.

Rees, P. A. (2005a). Will the EC Zoos Directive increase the conservation value of zoo research? *Oryx*, 39: 128–131.

Rees, P. A. (2005b). The EC Zoos Directive: a lost opportunity to implement the Convention on Biological Diversity. *Journal of International Wildlife Law and Policy*, 8: 51–62.

Rees, P. A. (2005c). Towards a research-based identity for zoos – a reply to Wehnelt and Wilkinson and Thomas. *Oryx*, 39: 135–136.

Rees, P. A. (2009a). Activity budgets and the relationship between feeding and stereotypic behaviors in Asian elephants (*Elephas maximus*) in a zoo. *Zoo Biology*, 28: 79–97.

Rees, P. A. (2009b). The sizes of elephant groups in zoos: implications for elephant welfare. *Journal of Applied Animal Welfare Science*, 12: 44–60.

Reeve, C. L., Spitzmüller, C., Rogelberg, S. G., Walker, A., Schultz, L. and Clark, O. (2004). Employee reactions and adjustments to euthanasia-related work: identifying turning-point events through retrospective narratives. *Journal of Applied Animal Welfare Science*, 7: 1–25.

Regan, J. (2004). *Manifesto for Zoos*. John Regan Associates Ltd., Manchester.

Regan, T. (1988). *The Case for Animal Rights*. Routledge, London.

Reinhardt, V. and Roberts, A. (1997). Effective feeding enrichment for non-human primates: A brief review. *Animal Welfare*, 6: 265–272.

Reuters (2008). A tiger devoured a mentally ill man who entered the animal's zoo cage in northeast China, local media reported on Friday. http://uk.reuters.com/article/idUKPEK9713720080404 (accessed 10 November 2010).

Rice, C. G. and Kalk, P. (1996). Identification and marking techniques. In: Kleiman, D. G., Allen, M. E., Thompson, K. V. and Lumpkins, S. (eds.) *Wild Mammals in Captivity: Principles and Techniques*, pp. 56–66. University of Chicago Press, Chicago, IL.

Rich, C. N. and Talent, L. G. (2008). The effects of prey species on food conversion efficiency and growth of an insectivorous lizard. *Zoo Biology*, 27: 181–187.

Richardson, D. M. (2000). Euthanasia: a nettle we need to grasp. *Ratel*, 27: 80–88.

Rickman, L. K., Montali, R. J. and Hayward, G. S. (2000). Review of a newly recognised disease of elephants caused by endotheliotropic herpesviruses. *Zoo Biology*, 19: 383–392.

Rickman, L. K., Montali, R. J., Garber, R. L. *et al.* (1999). Novel endotheliotropic herpesviruses fatal for Asian and African elephants. *Science*, 283: 1171.

Roelke, M. E., Martenson, J. and O'Brien, S. J. (1993). The consequences of demographic reduction and genetic depletion in the endangered Florida panther. *Current Biology*, 3: 340–350.

Rogers, L. J. (1997). *Minds of Their Own: Thinking and Awareness in Animals*. Allen & Unwin, St Leonards, NSW, Australia.

Roocroft, A. (2009). A clear definition of protected contact elephant handling. *International Zoo News*, 55: 210–217.

Ross, S. and Lukas, K. (2005). Zoo visitor behavior at an African ape exhibit. *Visitor Studies Today*, 8(3): 1–10.

RSPB (2009). *RSPB Trustees' Report and Accounts 2009*. Royal Society for the Protection of Birds, The Lodge, Sandy, Bedfordshire.

RSPB (2010). www.rspb.org.uk/about/ (accessed 27 January 2010).

RSPCA (2008). *Campaigns Home Page – Exotic Animals*. www.rspca.org.uk/servlet/Satellite?pagename=RSPCA/Page/RSPCAContentTemplate&cid=1123153965004&articleId=997353167090 (accessed 23 June 2008).

Rushen, J. (1984). Stereotyped behaviour, adjunctive drinking and the feeding period of tethered sows. *Animal Behaviour*, 32: 1059–1067.

Rushen, J. (1993). The "coping" hypothesis of stereotypic behaviour. *Animal Behaviour*, 45: 613–615.

Rushen, J., de Passillé, A. M. and Munksgaard, L. (1999a). Fear of people by cows and effects on milk yield, behavior and heart rate at milking. *Journal of Dairy Science*, 82: 720–727.

Rushen, J., Taylor, A. A. and de Passille, A. M. (1999b). Domestic animals' fear of humans and its effect on their welfare. *Applied Animal Behaviour Science*, 65: 285–303.

Russow, L. M. (1994). Why do species matter? In: Westphal, D. and Westphal, F. (eds.) *Planet in Peril: Essays in Environmental Ethics*, pp. 149–170. Holt, Reinhart and Winston, Orlando, FL.

Ryder, R. (1975). *Victims of Science: Use of Animals in Research*. Davis-Poynter Limited, London.

Sannen, A., Van Elsacker, L. and Eens, M. (2004). Effect of spatial crowding on aggressive behavior in a bonobo colony. *Zoo Biology*, 23: 383–395.

Savage-Rumbaugh, S. and Lewin, R. (1994). *Kanzi: The Ape at the Brink of the Human Mind*. Doubleday, London.

Schaller, G. B. (1972). *The Serengeti Lion: A Study of Predator–Prey Relations*. University of Chicago Press, Chicago, IL.

Schmid, J. (1995). Behavioural effects of keeping circus elephants in paddocks. In: Spooner, N. G. and Whitear, J. A. (eds.) *Proceedings of the Eighth UK Elephant Workshop*, , 21 September 1994, pp. 19–27. North of England Zoological Society, Chester Zoo, Chester.

Schmid, J. and Zeeb, K. (1994). The introduction of paddocks in circus elephant husbandry. *Dtsch. Tierarztl. Wochenschr.*, 101: 50–52.

Schmid, J., Heistermann, M., Glansloßer, U. and Hodges, J. K. (2001). Introduction of foreign female Asian elephants (*Elephas maximus*) into an existing group: behavioural reactions and changes in cortisol levels. *Animal Welfare*, 10: 357–372.

Schmitt, D. L. (1998). Report of a successful artificial insemination in an Asian elephant. In: *Proceedings of the Third International Elephant Research Symposium*, Springfield, MO, USA, p. 7.

Schomberg, G. (1957). *British Zoos: A Study of Animals in Captivity*. Allan Wingate, London.

Schomberg, G. (1970). *The Penguin Guide to British Zoos*. Penguin Books Ltd., Harmondsworth, Middlesex.

Schwaibold, U. and Pillay, N. (2001). Stereotypic behaviour is genetically transmitted in the African striped mouse *Rhabdomys pumilio*. *Applied Animal Behaviour Science*, 74: 273–280.

Schwammer, H. M., Hildebrandt, T. and Göritz, F. (2001). First successful artificial insemination in an African elephant in Europe. *International Zoo News*, 48: 424–429.

Schwartz, M. K. and Mills, L. S. (2005). Gene flow after inbreeding leads to higher survival in deer mice. *Biological Conservation*, 123: 413–420.

Seal, U. S. (1991). Fertility control as a tool for regulating captive and free-ranging wildlife. *Journal of Zoo and Wildlife Medicine*, 22: 1–5.

Seidensticker, J. and Doherty, J. G. (1996). Integrating animal behavior and exhibit design. In: Kleiman, D. G., Allen, M. E., Thompson, K. V. and Lumpkins, S. (eds.) *Wild Mammals in Captivity: Principles and Techniques*, pp. 180–190. University of Chicago Press, Chicago, IL.

Selye, H. (1936). A syndrome produced by diverse nocuous agents. *Nature*, 138: 32.

Semple, S. (2002). Analysis of research projects conducted in Federation collections to 2000. *Federation Research Newsletter*, 3 (January): 3. Federation of Zoological Gardens of Great Britain & Ireland.

Sharp, J. C. M. and McDonald, S. (1967). Effects of rabies vaccine in man. *British Medical Journal*, 3: 20–21.

Shepherdson, D. J., Carlstead, K., Mellen, J. D. and Seidensticker, J. (1993). The influence of food presentation on the behavior of small cats in confined environments. *Zoo Biology*, 12: 203–216.

Shepherdson, D. J. (1998). Introduction. Tracing the Path of Environmental Enrichment in Zoos. In: Shepherdson, D. J., Mellen, J. D. and Hutchins, M. (eds.). *Second Nature: Environmental Enrichment for Captive Animals, pp. 1–12.* Smithsonian Institution Press, Washington DC and London.

Sherwood, K. P., Rallis, S. F. and Stone, J. (1989). Effects of live vs. preserved specimens on student learning. *Zoo Biology*, 8: 99–104.

Shettel-Neuber, J. and O'Reilly, J. (1981). *Now Where? A Study of Visitor Orientation and Circulation at the Arizona-Sonora Desert Museum*. Technical Report No. 87-25. Psychology Institute, Jacksonville State University, Jacksonville, AL.

Shyne, A. (2006). Meta-analytic review of the effects of enrichment on stereotypic behavior in zoo mammals. *Zoo Biology*, 25: 317–337.

Sierra Club (2010). www.sierraclub.org/welcome/ (accessed 27 January 2010).

Simon, M. P. (1984). The influence of conspecifics on egg and larval mortality in amphibians. In: Hausfater, G. and Hrdy, S. B. (eds.) *Infanticide: Comparative and Evolutionary Perspectives*, pp. 65–86. Aldine, New York.

Singer, P. (1995). *Animal Liberation*, 2nd edn. Pimlico, London.

Smith, B. and Hutchins, M. (2000). The value of captive breeding programmes to field conservation: elephants as an example. *Pachyderm*, 28: 101–109.

Smith, D. W., Murphy, K. M. and Guernsey, D. S. (1999). *Yellowstone Wolf Project Annual Report 1999*. National Park Service. Yellowstone Center for Resources, Yellowstone National Park, WY.

Soulé, M. E. (1980). Thresholds for survival: maintaining fitness and evolutionary potential. In: Soulé, M. E. and Wilcox, B. A. (eds.) *Conservation Biology*, pp. 151–170. Sinauer Associates Inc., MA.

Soulsbury, C. D., Iossa, G., Kennell, S. and Harris, S. (2009). The welfare and suitability of primates kept as pets. *Journal of Applied Animal Welfare Science*, 12: 1–20.

Speed, C. W., Meekan, M. G. and Bradshaw, C. J. A. (2007). Spot the match – wildlife photo-identification using information theory. *Frontiers in Zoology*, 4: 2, doi:10.1186/1742-9994-4-2.

Spoon, T. R., Millam, J. R. and Owings, D. H. (2006). The importance of mate behavioural compatibility in parenting and reproductive success by cockatiels, *Nymphicus hollandicus*. *Animal Behaviour*, 71: 315–326.

Spotte, S. and Clark, C. (2004). A knowledge-based survey of adult aquarium visitors. *Human Dimensions of Wildlife*, 9: 143–151.

SSSMZP (2004). *Secretary of State's Standards of Modern Zoo Practice*. DEFRA, London.

Stanley, S. M. (1984). Marine mass extinctions: a dominant role for temperature. In: Nitecki, M. H. (ed.) *Extinctions*, pp. 67–117. University of Chicago Press, Chicago, IL.

Stanley Price, M. R. and Fa, J. E. (2007). Reintroductions from zoos: a conservation guiding light or a shooting star? In: Zimmerman, A., Hatchwell, M., Dickie, L. and West, C. (eds.) *Zoos in the 21st Century: Catalysts for Conservation*, pp. 155–177. Cambridge University Press, Cambridge, UK.

Stevens, E. F. (1991). Flamingo breeding: the role of group displays. *Zoo Biology*, 13: 501–507.

Stevenson, M. F. (1983). The captive environment: Its effects on exploratory and related behavioural responses in wild animals. In: Archer, J. and Birke, L. (eds.) *Exploration in Animals and Man*, pp. 176–179. Van Nostrand Reinhold, Berkshire.

Stoinski, T. S., Daniel, E. and Maple, T. L. (2000). A preliminary study of the behavioral effects of feeding enrichment on African elephants. *Zoo Biology*, 19: 485–493.

Stoinski, T. S., Lukas, K. E. and Maple, T. L. (1998). A survey of research in North American zoos and aquariums. *Zoo Biology*, **17**: 167–180.

Suzuki, M., Hirako, K., Saito, S., Suzuki, C., Kashiwabara, T. and Koie, H. (2008). Usage of high-performance mattresses for transport of Indo-Pacific bottlenose dolphin. *Zoo Biology*, 27: 331–340.

Suzuki, S., Matsui, T., Kawahara, H. and Gotoh, S. (2009). Development of a noncontact and long-term respiration monitoring system using microwave radar for hibernating black bear. *Zoo Biology*, 28: 259–270.

Swaddle, J. P. (1996). Reproductive success and symmetry in zebra finches. *Animal Behaviour*, 51: 203–210.

Swanagan, J. S. (2000). Factors influencing zoo visitors' conservation attitudes and behavior. *Journal of Environmental Education*, 31: 26–31.

Szdzuy, K., Dehnhard, M., Strauss, G., Eulenburger, K. and Hofer, H. (2006). Behavioural and endocrinological parameters of female African and Asian elephants *Loxodonta africana* and *Elephas maximus* in the peripartal period. *International Zoo Yearbook*, 40: 41–50.

Szymanski, D. C., Gist, D. H. and Roth, T. L. (2006). Anuran gender identification by fecal steroid analysis. *Zoo Biology*, 25: 35–46.

Taylor, P. (2005). *Beyond Conservation: A Wildland Strategy*. Earthscan Publications Ltd., London.

Taylor, S. M. (1986). *Understanding processes of informal education: A naturalistic study of visitors to a public aquarium*. Unpublished PhD dissertation, University of California at Berkeley.

Taylor, V. J. and Poole, T. B. (1998). Captive breeding and infant mortality in Asian elephants: A comparison between twenty western zoos and three eastern elephant centers. *Zoo Biology*, 17: 311–322.

Tedmanson, S. (2008). Feeding time at the zoo: intruder aged 7 gives crocodile a feast of rare reptiles. *The Times*, 4 October, p. 13.

Temeles, E. J. (1987). The relative importance of prey availability and intruder pressure in feeding territory size regulation by harriers, *Circus cyaneus*. *Oecologia*, 74: 286–297.

Tennessen, T. (1989). Coping with confinement – features of the environment that influence animals' ability to adapt. *Applied Animal Behaviour Science*, 22: 139–149.

Thomas, W. D. and Maruska, E. J. (1996). Mixed-species exhibits with mammals. In: Kleiman, D. G., Allen, M. E., Thompson, K. V. and Lumpkin, S. (eds.) *Wild Mammals in Captivity: Principles and Techniques*, pp. 204–211. University of Chicago Press, Chicago, IL.

Tinbergen, N. (1963). On aims and methods in ethology. *Zeitschrift für Tierpsychologie*, 20: 410–433.

Tobler, I. (1992). Behavioral sleep in the Asian elephant in captivity. *Sleep*, 15(1): 1–12.

Tover, T. C., Moore, D. and Dierenfeld, E. (2005). Preferences among four species of local browse offered to colobus monkeys (*Guereza kikuyuensis*) at the Central Park Zoo. *Zoo Biology*, 24: 267–274.

Traill, L. W., Brook, B. W., Frankham, R. R. and Bradshaw, C. J. A. (2010). Pragmatic population viability targets in a rapidly changing world. *Biological Conservation*, 143: 28–34.

Tudge, C. (1991). *Last Animals at the Zoo: How Mass Extinction Can Be Stopped*. Hutchinson Radius, London.

Turley, S. K. (2001). Children and the demand for recreational experiences: the case for zoos. *Leisure Studies*, 20: 1–18.

USDL (1997). *CF AR 6/97. US Department of Labor, Bureau of Labor Statistics, Census of Occupational Injuries 1995*. www.bls.gov./iif/oshwc/cfar0020.txt (accessed 1 August 2010).

USFWS (2004). *Draft Revised Recovery Plan for the Ne-Ne or Hawaiian Goose (Branta sandvicensis)* (first revision, July 2004). Region 1, US Fish & Wildlife Service, Portland, OR.

USFWS (2008). *Annual Report FY 2007, US Fish & Wildlife Service, Office of Law Enforcement*. www.fws.gov/le/pdffiles/FinalAnnualReportFY2007.pdf (accessed 3 June 2010).

Varner, G. E. and Monroe, M. C. (1991). Ethical perspectives on captive breeding: is it for the birds? *Endangered Species Update*, 8: 27–29.

Vester, B. M., Burke, S. L., Dikeman, C. L., Simmons, L. G. and Swanson, K. S. (2008). Nutrient digestibility and fecal characteristics are different among captive exotic felids fed a beef-based raw diet. *Zoo Biology*, 27: 126–136.

Videan, E. N., Fritz, J. and Murphy, J. (2007). Development of guidelines for assessing obesity in captive chimpanzees (*Pan troglodytes*). *Zoo Biology*, 26: 93–104.

Wallace, M. P. and Temple, S. A. (1987). Releasing captive-reared Andean condors to the wild. *Journal of Wildlife Management*, 51: 541–550.

Ward, M. P., Ramer, J. C., Proudfoot, J., Garner, M. M., Juan-Sallès, C. and Wu, C. C. (2003). Outbreak of salmonellosis in a zoological collection of Lorikeets and Lories (*Trichoglossus, Lorius*, and *Eos* spp.). *Avian Diseases*, 47: 493–498.

Ward, P. I., Mosberger, N., Kistler, C. and Fischer, O. (1998). The relationship between popularity and body size in zoo animals. *Conservation Biology*, 12(6): 1408–1411.

Warin, R. and Warin, A. (1985). *Portrait of a Zoo: Bristol Zoological Gardens 1835–1985*. Redcliffe Press Ltd., Bristol.

Watson-Smyth, K. (2000). Elephant crushes keeper in fifth Aspinall zoo death. *The Independent*, 8 February. www.independent.co.uk/news/uk/this-britain/elephant-crushes-keeper-in-fifth-aspinall-zoo-death-726641.html (accessed 31 July 2009).

Watts, C. H. S. (1968). The foods eaten by wood mice (*Apodemus sylvaticus*) and bank voles (*Clethrionomys glareolus*) in Wytham Woods, Berkshire. *Journal of Animal Ecology*, 37: 25–41.

Watts, P. C., Buley, K. R., Sanderson, S., Boardman, W., Ciofi, C. and Gibson, R. (2006). Parthenogenesis in komodo dragons. *Nature*, 444: 1021–1022.

Wayne, R. K. (1995). Red wolves: to conserve or not to conserve. *Canid News*, 3: 7–12. IUCN/SSC Specialist Group. www.canids.org/PUBLICAT/CNDNEWS3/2conserv.htm (accessed 21 April 2008).

Wayne, R. K. and Jenks, S. M. (1991). Mitochondrial DNA analysis implying extensive hybridization of the endangered red wolf *Canis rufus*. *Nature*, 351: 565–568.

WAZA (2009). www.waza.org/en/site/conservation/breeding/conservation/breeding/conservation-breeding-programs/ (accessed 3 December 2009).

Wehnelt, S. and Wilkinson, R. (2005). Research, conservation and zoos: the EC Zoos Directive – a response to Rees. *Oryx*, 39: 132–133.

Weisz, I., Wuestenhagen, A. and Schwammer, H. (2000). Research on nocturnal behaviour of African elephants at Schönbrunn Zoo. *International Zoo News*, 47(4): 228–233.

Weldon, C., du Preez, L. H., Hyatt, A. D., Muller, R. and Speare, R. (2004). Origin of the amphibian chytrid fungus. *Emerging Infectious Diseases*. www.cdc.gov/ncidod/EID/vol10no12/03-0804.htm (accessed 18 May 2010).

Weller, R. E. (1994). Infectious and noninfectious diseases of owl monkeys. In: Baer, J. F., Weller, R. E. and Kokoma, I. (eds.) *Aotus: The Owl Monkey*, pp. 178–215. Academic Press, San Diego, CA.

Wells, D. L. and Egli, J. M. (2004). The influence of olfactory enrichment on the behaviour of black-footed cats, *Felis nigripes*. *Applied Animal Behaviour Science*, 85: 107–119.

Wells, D. L. and Irwin, R. M. (2009). The effect of feeding enrichment on the moloch gibbon (*Hylobates moloch*). *Journal of Applied Animal Welfare Science*, 12: 21–29.

Wells, D. L., Coleman, D. and Challis, M. G. (2006). A note on the effect of auditory stimulation on the behaviour and welfare of zoo-housed gorillas. *Applied Animal Behaviour Science*, 100: 327–332.

Wemmer, C., Krishnamurthy, V., Shrestha, S., Hayek, L-A., Thant, M. and Nanjappa, K. A. (2006). Assessment of body condition in Asian elephants (*Elephas maximus*). *Zoo Biology*, 25: 187–200.

Westemeier, R. L., Brawn, J. D., Simpson, S. A. *et al.* (1998). Tracking the long-term decline and recovery of an isolated population. *Science*, 282: 1695–1698.

Wheatley, S. (1995). *Freedom from Extinction: Conservation and Development Under International Law*. Presented at a Conference on 'Nature Conservation Areas and Development', University of Central Lancashire, Preston, May 1995.

White, J. and Barry, S. (1984). *Science Education for Families in Informal Learning Settings: An Evaluation of the Herp Lab Project*. National Zoological Park, Washington DC.

Wiedenmayer, C. (1998). Food hiding and enrichment in captive Asian elephants. *Applied Animal Behaviour Science*, 56: 77–82.

Wielebnowski, N. (1996). Reassessing the relationship between juvenile mortality and genetic monomorphism in captive cheetahs. *Zoo Biology*, 15: 353–369.

Wienker, W. R. (1986). Giraffe squeeze cage procedures. *Zoo Biology*, 5: 371–377.

Wiese, R. J. (2000). Asian elephants are not self-sustaining in North America. *Zoo Biology*, 19: 299–309.

Wiese, R. J., Hutchins, M., Willis, K. and Becker, S. (eds.) (1992). *AAZPA Annual Report on Conservation and Science*. American Association of Zoological Parks and Aquariums, Bethesda, MD.

Wildlife Trusts (2010). www.wildlifetrusts.org (accessed 27 January 2010).

Wildt, D. (2009). Rescuing endangered animals with assisted reproductive technology. *Sexuality, Reproduction and Menopause*, 7: 21–25.

Wildt, D. E., Zhang, A., Zhang, H., Janssen, D. L. and Ellis, S. (2006). *Giant Pandas: Biology, Veterinary Medicine and Management*. Cambridge University Press, Cambridge, UK.

Williams, L. E., Abee, C. R., Barnes, S. R. and Ricker, R. B. (1988). Cage design and configuration for an arboreal species of primate. *Laboratory Animal Science*, 38: 289–291.

Williams, T. (2006). Introduction of silverback gorillas (*Gorilla gorilla gorilla*): A new success in bachelor group formation. *Association of Zoos and Aquariums (AZA) Regional Conference Proceedings*. www.aza.org/AZAPublications/2006Proceedings/Documents/2006RegMtg2.pdf (accessed 12 April 2008).

Wilson, D. E. and Reeder, D. M. (eds.) (2005). *Mammal Species of the World: A Taxonomic and Geographic Reference*, 3rd edn. The Johns Hopkins University Press, Baltimore, MD and London.

Wilson, E. O. (1992). *The Diversity of Life*. The Penguin Press, London.

Wilson, E. O. (1993). Biophilia and the conservation ethic. In: Kellert, S. R. and Wilson, E. O. (eds.) *The Biophilia Hypothesis*, pp. 31–40. Island Press, Washington DC.

Wilson, E. O. (2003). The encyclopedia of life. *Trends in Ecology and Evolution*, 18: 77–80.

Wilson, M. L., Bashaw, M. J., Fountain, K., Kieschnick, S. and Maple, T. L. (2006). Nocturnal behavior in a group of female African elephants. *Zoo Biology*, 25: 173–186.

Wilson, M. L., Bloomsmith, M. A. and Maple, T. L. (2004). Stereotypic swaying and serum cortisol concentrations in three captive African elephants (*Loxodonta africana*). *Animal Welfare*, 13: 39–43.

Wilson, P. J., Grewal, S., Lawford, I. D. *et al.* (2000). DNA profiles of the eastern Canadian wolf and the red wolf provide evidence for a common evolutionary history independent of the gray wolf. *Canadian Journal of Zoology*, 78: 2156–2166.

Wilson, S. and Zimmerman, A. (2005). *Zoos and Conservation Bibliography 2005*. North of England Zoological Society. www.chesterzoo.org/conservation.asp (accessed 1 June 2006).

Wolf, R. L. and Tymitz, B. L. (1981). Studying visitor perceptions of zoo environments: a naturalistic view. *International Zoo Yearbook*, 21: 49–53.

Woollard, S. P. (1998). The development of zoo education. *International Zoo News*, 45: 422–426.

WoRMS (2008). www.marinespecies.org/news.php?p=show&id=349 (accessed 3 June 2010).

WWF (2008). *WWF-UK Annual Review 2008*. WWF-UK, Panda House, Godalming, Surrey.

WWF (2010). www.wwf.org.uk/what_we_do/about_us/faqs/ (accessed 27 January 2010).

Wyatt, J. R. and Eltringham, S. K. (1974). The daily activity of the elephant in Rwenzori National Park, Uganda. *East African Wildlife Journal*, 12: 273–289.

Yerke, R. and Burns, A. (1991). Measuring the impact of animal shows on visitor attitudes. In: *AAZPA Annual Conference Proceedings*, pp. 532–539. American Association of Zoological Parks and Aquariums, Wheeling, WV.

Yew, W. (1991). *Noah's Art: Zoo, Aquarium, Aviary and Wildlife Park Graphics*. Quon Editions, Singapore.

Zimmermann, A., Davies, T. E., Hazarika, N., Wilson, S., Chakrabarty, J., Hazarika, B. and Das, D. (2009). Community-based conflict management in Assam. *Gajah*, 30: 34–40.

ZSL (2006). *Living Conservation 2005–2006: Building for the Future. Financial Statements Year Ended 31 December 2005*. Zoological Society of London, Regent's Park, London.

ZSL (2008). *The Zoological Society of London.: Trustees' Report and Financial Statements, 31 December 2008*. http://static.zsl.org/files/zsl-trustees-and-financial-statements-311208-779.pdf (accessed 4 May 2010).

ZSL (2010). *Mountain Chicken Frog*. www.zsl.org/conservation/animals/amphibians/mountain-chicken,830,AR.html (accessed 18 May 2010).

ZSSD (2008). *Zoological Society of San Diego. Financial Statements, Fiscal Years 2008 and 2007. With Report of Independent Auditors*. Ernst & Young LLP, San Diego, CA. www.sandiegozoo.org/pressbox/annualreport/ZSSD_AFS_2008_final.pdf (accessed 4 March 2010).

INDEX